Bifidobacteria

Genomics and Molecular Aspects

Edited by

Baltasar Mayo

Departamento de Microbiología y Bioquímica
Instituto de Productos Lácteos de Asturias (CSIC)
Carretera de Infiesto, s/n
33300-Villaviciosa
Asturias
Spain

and

Douwe van Sinderen

Department of Microbiology & Alimentary Pharmabiotic Centre
University College Cork
Western Road
Cork
Ireland

Caister Academic Press

Caister Academic Press
Norfolk, UK

www.caister.com

British Library Cataloguing-in-Publication Data
A catalogue record for this book is available from the British Library

ISBN: 978-1-904455-68-4

Cover image adapted from Figure 8.5a

Printed and bound in Great Britain

Contents

Contributors

Pablo Álvarez Martín
Departamento de Microbiología y Bioquímica
Instituto de Productos Lácteos de Asturias
CSIC
Carretera de Infiesto, s/*n*
33300-Villaviciosa
Asturias
Spain

Alberto Amaretti
Department of Chemistry
University of Modena and Reggio-Emilia
Modena
Italy

Katie L. Blackett
Microbiology and Gut Biology Group
The University of Dundee
Ninewells Hospital Medical School
Dundee DD1 9SY
UK

K.L.Blackett@dundee.ac.uk

Francesca Bottacini
Department of Genetics, Biology of Microorganisms, Anthropology and Evolution
University of Parma
Italy

francesca@bottacini@gmail.com

Stephen J. Davids
Nutraceuticals and Functional Foods Institute (INAF)
Université Laval
Quebec City, PQ
G1K 7P4
Canada

Susana Delgado
Departamento de Microbiología y Bioquímica
Instituto de Productos Lácteos de Asturias (CSIC)
Carretera de Infiesto, s/*n*
33300-Villaviciosa
Asturias
Spain

Konrad J. Domig
Department of Food Science and Technology
Division of Food Microbiology and Hygiene
BOKU-University of Natural Resources and Applied Life Sciences
Gregor Mendel Straβe 33
A-1180 Vienna
Austria

Ismail Fliss
Département des Aliments et de Nutrition
Nutraceuticals and Functional Foods Institute (INAF)
Université Laval
Quebec City, PQ
G1K 7P4
Canada

ismail.fliss@fsaa.ulaval.ca

Elena Foroni
Department of Genetics, Biology of Microorganisms, Anthropology and Evolution
University of Parma
Italy

Vanessa Giubellini
Department of Genetics, Biology of Microorganisms, Anthropology and Evolution
University of Parma
Italy

Kamila Goderska
Department of Fermentation and Biosynthesis
Faculty of Food Science and Nutrition
Poznań University of Life Sciences
Poznań
Poland

kamilag@au.poznan.pl

Simone Guglielmetti
Dipartimento di Scienze e Tecnologie Alimentari e Microbiologiche
Sezione di Microbiologia Industriale
Università degli studi di Milano
Via Celoria 2
20133-Milan
Italy

Ehab Kheadr
Nutraceuticals and Functional Foods Institute (INAF)
Université Laval,
Quebec City, PQ
G1K 7P4
Canada

Department of Dairy Science and Technology
Faculty of Agriculture
University of Alexandria
Alexandria
Egypt

Sampo Lahtinen
Danisco, Health & Nutrition
Sokeritehtaantie 20
FIN-02460 Kantvik
Finland

George T. Macfarlane
Microbiology and Gut Biology Group
The University of Dundee
Ninewells Hospital Medical School
Dundee DD1 9SY
UK

g.t.macfarlane@dundee.ac.k

Sandra Macfarlane
Microbiology and Gut Biology Group
The University of Dundee
Ninewells Hospital Medical School
Dundee DD1 9SY
UK

s.macfarlane@dundee.ac.uk

Abelardo Margolles
Instituto de Productos Lácteos de Asturias (IPLA-CSIC)
33300 Villaviciosa
Asturias
Spain

amargolles@ipla.csic.es

Diego Matteuzzi
Department of Pharmaceutical Sciences
University of Bologna
Italy

diego.matteuzzi@unibo.it

Baltasar Mayo
Departamento de Microbiología y Bioquímica
Instituto de Productos Lácteos de Asturias (CSIC)
Carretera de Infiesto, s/*n*
33300-Villaviciosa
Asturias
Spain

baltasar.mayo@ipla.csic.es

Sigrid Mayrhofer
Department of Food Science and Technology
Division of Food Microbiology and Hygiene
BOKU-University of Natural Resources and Applied Life Sciences
Gregor Mendel Straβe 33
A-1180 Vienna
Austria

David A. Mills
Robert Mondavi Institute for Wine and Food Science
University of California
Davis, CA
USA

damills@ucdavis.edu

Eiji Miyauchi
Graduate School of Biosphere Science
Hiroshima University
1-4-4 Kagamiyama
Higashi-Hiroshima
Hiroshima 739–8528,
Japan

emiyauchi@hiroshima-u.ac.jp

Arthur C. Ouwehand
Danisco, Health & Nutrition
Sokeritehtaantie 20
FIN-02460 Kantvik
Finland

Neil P. J. Price
National Center for Agricultural Utilization Research
Agricultural Research Service
US Department of Agriculture
1815 N. University St.
Peoria, IL
USA

neil.price@ars.usda.gov

Maddalena Rossi
Department of Chemistry
University of Modena and Reggio-Emilia
via Campi 183
41100, Modena
Italy

maddalena.rossi@unimore.it

Lorena Ruiz
Instituto de Productos Lácteos de Asturias (IPLA-CSIC)
33300 Villaviciosa
Asturias
Spain

Borja Sánchez
Instituto de Productos Lácteos de Asturias (IPLA-CSIC)
33300 Villaviciosa
Asturias
Spain

Maria H. Saarela
VTT Technical Research Centre of Finland
PO Box 1000
02044 VTT
Finland

maria.saarela@vtt.fi

David A. Sela
Microbiology Graduate Group
University of California
Davis, CA
USA

Current contact information:
Microbiology and Immunology
Stanford University
Palo Alto, CA
USA

sela@stanford.edu

Douwe van Sinderen
Department of Microbiology & Alimentary Pharmabiotic Centre
University College Cork
Western Road
Cork
Ireland

d.vansinderen@ucc.ie

Catherine Stanton
Teagasc
Moorepark Food Research Centre
Moorepark
Fermoy
Co. Cork
Ireland
Alimentary Pharmabiotic Centre
Cork
Ireland

Soichi Tanabe
Graduate School of Biosphere Science
Hiroshima University
1-4-4 Kagamiyama
Higashi-Hiroshima
Hiroshima 739-8528
Japan
stanabe@hiroshima-u.ac.jp

Gerald W. Tannock
Department of Microbiology and Immunology
University of Otago
PO Box 56
Dunedin
New Zealand

gerald.tannock@stonebow.otago.ac.nz

Francesca Turroni
Department of Genetics, Biology of Microorganisms, Anthropology and Evolution
University of Parma
Italy

francesca.turroni@gmail.com

Marco Ventura
Department of Genetics, Biology of Microorganisms, Anthropology and Evolution
University of Parma
Italy

marco.ventura@unipr.it

Jin-zhong Xiao
Food Science and Technology Institute
Morinaga Milk Industry Co. Ltd
5-1-83 Higashihara
Zama 228-8583
Japan

j_xiao@morinagamilk.co.jp

Aldert L. Zomer
Alimentary Pharmabiotic Centre and Department of Microbiology
National University of Ireland
Cork
Ireland

Preface

Because of their health-associated benefits, species of the genus *Bifidobacterium* are undoubtedly among the bacteria with more increasing industrial potential. Bifidobacteria are majority gut populations which are widely used as probiotic organisms in a vast array of formulations for alleviation and treatment of many intestinal disorders. Scientific knowledge on the physiology and genetics of this group of microorganisms has enormously expanded during the last two decades. Moreover, recent genomic and proteomic techniques have gathered an impressive amount of data that could not be envisaged a few years ago. These reasons prompted us to considerer collecting recent (molecular-oriented) data of bifidobacteria into this book, which is intended to fulfil the growing interest and recent research advances in basic and applied aspects on this microbial group. Authors and contributors to the chapters were selected among well-recognised experts working in the cutting edge of research in bifidobacteria. The reported knowledge is thought to engage the attention of industrial, academic and medical professionals, and to be useful for the many graduate and PhD students interested in this exciting and important microbial genus.

Baltasar Mayo and Douwe van Sinderen

Foreword

Among the myriad of bacterial species that inhabit the human gut and the gut of many animals, bifidobacteria are almost certainly the microbial group that has the greatest effect on the health of the host. In most people, bifidobacteria are present in high numbers (10^8–10^9 cells/g of intestinal content) throughout their lives, although each individual permanently harbours only several species or specific biotypes. Apart from lactobacilli, bifidobacteria is the only intestinal microbial group that is generally recognized to possess such positive characteristics through its capacity to produce short chain fatty acids, vitamins, bacteriocins and antibiotic-like substances, and of exerting immunomodulating and immunostimulating activities. Moreover, the bifidobacterial species so far identified lack enzyme urease, azo- and nitro-reductase, beta-glucuronidase and alpha-dehydrolase that exert enzymatic and metabolic effects that are perceived to be toxic to the host. Consequently, bifidobacteria can be considered as valuable probiotics and today they are not only used in the food industry to relieve and treat many intestinal disorders, but they are increasingly attracting the scientific interest of clinicians and researchers. Therefore, interest in bifidobacteria has burgeoned in the last two decades particularly because of their potential use in the food industry, in the medical field as well as in a purely scientific context.

Bifidobacteria were first discovered in the faeces of infants by H. Tissier in 1900, who named them *Bacillus bifidus communis* and up to the early 1960s they were classified, from a taxonomic point of view, among lactic acid bacteria as a unique anaerobic species designated *Lactobacillus bifidus*. It was only in 1965 that Scardovi at the University of Bologna, Italy, and de Vries, Gerbrandy and Stouthamer in the Netherlands demonstrated that this group was taxonomically distinct from lactobacilli because they possessed the so-called 'bifid shunt' or 'fructose-6-phosphate shunt'. Subsequently, using a DNA–DNA hybridization technique, Scardovi and co-workers identified 17 of the 32 currently known species of bifidobacteria. In this context, first G. Reuter (Germany) in 1963, T. Mitsuoka (Japan) in 1969 and the Chinese researcher W. Jian and his collaborators in 2001 have made relevant contributions to the taxonomy of this bacterial group.

More recently, highly sophisticated molecular techniques have led to a more accurate and precise identification of bifidobacterial species, allowing the direct examination of bifidobacteria in intestinal content, faeces and bioptic specimen. In addition, genome sequencing is a powerful means allowing the identification of the genetic determinants involved in the health-promoting effects of probiotic bacteria and characterization of the molecular pathways underlying the production of biomolecules in industrial processes.

Genome sequencing of bifidobacterial strains also constitutes an essential step to generate primary information for downstream comparative and functional purposes, such as comparative genomics, transcriptomics and/or proteomics, which in turn can be employed to address many fundamental and applied questions that need to be answered before the probiotic concept can be fully accepted.

Since the seminal volume by Jeremija Lj. Rašić and Joseph A. Kurmann *Bifidobacteria and their Role* (Birkhäuser Verlag, Basel) that appeared in 1983, there have been no other books on bifidobacteria, and, in view of the advances made in this area, the time has surely come to update this topic. It was in this scenario that the editors of this volume gathered together the contributions of a multidisciplinary group of internationally recognized experts, all of whom are active in this field, to give an up-to-date overview of this topic.

The contents range from molecular identification to genetic manipulation, to the genomics of bifidobacteria, their metabolism and applicative aspects as well as to their interaction with their host. The bibliography that follows each chapter is intended to guide the reader to original data and to relevant publications pertaining to the particular area of interest. We believe that this volume will be of interest to microbiologists, gastroenterologists, paediatricians, scientists, nutritionists, technologists and students.

It is hoped this book will prove of value to all people interested in improving their knowledge on this important intestinal bacterial group that impinges so much on human health.

Professor Diego Matteuzzi

Analysis of Bifidobacterial Populations in Bowel Ecology Studies

1

Gerald W. Tannock

Abstract

Members of the genus *Bifidobacterium* are fermentative bacteria that utilize a variety of carbohydrates, especially oligosaccharides, resulting in the formation of acetic and lactic acids. Hexoses are utilized by an unusual metabolic pathway known as the 'bifid shunt'. The ecology of bifidobacteria can be determined in the laboratory using culture-dependent and culture-independent methods. Nine bifidobacterial species that can be considered to be autochthonous to the bowel have been detected in human faeces. Bifidobacteria achieve large population sizes in the bowels of humans during the first few months after birth. Bifidobacterial populations are probably enriched in the bowel by some of a huge variety of oligosaccharides present in breast milk. The variety of oligosaccharides in milk may partially dictate the kinds of bifidobacteria that establish in the bowel of a given child. Naïve immune cells harvested from cord blood respond differentially to bifidobacterial species. Thus, the occurrence of immunological diseases such as childhood allergies may be influenced by the kinds of bifidobacteria to which a child is exposed in early life. Therefore, methods by which particular, immunologically potent bifidobacteria can be enriched in the infant bowel warrant intensive research.

The bowel biome and bifidobacteria

A microbial community is a constituent of the colonic biome of humans and other mammals (Dethlefsen *et al.*, 2007). The community is largely bacterial in content and is fuelled principally from dietary residues (digestion-resistant carbohydrates), intestinal secretions such as mucus, and the remnants of enterocytes shed from bowel epithelia (Cummings and Macfarlane, 1991; Roberton and Corfield, 1999). The bacterial community (microbiota) of the large bowel is phylogenetically diverse, homeostatic, metabolically active, and interactive with host tissues (Hooper *et al.*, 2001; Eckburg *et al.*, 2005). The majority of the bacteria in the community are represented by four broad taxonomic groups (phyla): *Firmicutes* (Gram-positive), *Bacteroidetes* (Gram-negative), *Proteobacteria* (Gram-negative), and *Actinobacteria* (Gram-positive) (Dethlefsen *et al.*, 2007).

The genus *Bifidobacterium* belongs to the phylum *Actinobacteria*, class *Actinobacteria*, sub-class *Actinobacteridae*, order *Bifidobacteriales*, family *Bifidobacteriaceae* (Dellaglio and Felis, 2005). The bifidobacteria are Gram-positive polymorphic, branched rods that occur singly, in chains, or in clumps when cultured *in vitro*. They are non-spore-forming, non-motile, and non-filamentous. The bifidobacteria are anaerobic chemoorganotrophs with fermentative metabolism and which produce acids but not gas from a variety of carbohydrates. Bifidobacteria catabolize hexoses using an unusual pathway termed the 'fructose-6-phosphate shunt', 'bifid pathway', or 'bifid shunt' (de Vries *et al.*, 1967; Sgorbati *et al.*, 1995). The key enzyme of this pathway is fructose-6-phosphate phosphoketolase which splits the hexose phosphate to erythrose-4-phosphate and acetyl phosphate. Pentose phosphates are formed from tetrose and hexose phosphates by the action of transaldolase and transketolase by 2–3 cleavage, overall resulting in end products acetic and lactic acids in the theoretical ratio of 3:2 (Fig. 1.1). Although often included in the 'Lactic Acid Bacteria', bifidobacteria do not strictly belong to this group because they produce major amounts of acetic acid as well as lactic acid.

Genomic sequence analysis has shown that bifidobacteria are well endowed genetically to produce hydrolytic enzymes for the degradation of oligosaccharides and glycoproteins (Schell *et al.*, 2002; Ventura *et al.*, 2007, 2008). There are also many genes associated with signal transduction by which the bacteria can detect and react to changes in the availability of nutrients. Responses to qualitative changes in substrate availability, and well regulated metabolic pathways that conserve cellular energy, help to explain the ecological fitness of these bacteria for life in the bowel biome where the nature of digestion-resistant carbohydrates delivered each day to the bacteria will vary according to what kinds of food have been consumed by the human host.

Bifidobacteria have been cultured from human faeces, vagina, dental caries; sewage; bovine rumen; and honeybee intestine (Dellaglio and Felis, 2005). There are approximately 30 species of bifidobacteria of which nine appear to be autochthonous (long term association with a particular host species, form stable populations of characteristic size in a particular region of the gut, demonstrated ecological function) to the human bowel (Table 1.1). *Bifidobacterium animalis* subspecies *lactis* is probably of dietary origin but has been detected in the faeces of about 15% of children when sampled at 12 and 24 months of age (Wickens *et al.*, 2008). Bifidobacteria can be divided phylogenetically into five clusters and several distinct branches (Fig. 1.2). Officially, *B. longum* and *B. infantis* have been merged to form a single species (*B. longum*) containing three biotypes – *infantis*, *longum* and *suis* (Dellaglio and Felis, 2005). It has been proposed recently that the three biotypes be reclassified as subspecies (Mattarelli *et al.*, 2008). Because of different biological effects associated with *longum* and *infantis*, the proposed reclassification is likely to be preferred. The human–bifidobacterial relationship has been best studied and will therefore be the focus of this chapter. Information concerning bifidobacteria associated with non-human hosts can be obtained from Sgorbati *et al.*, 1995; Dellaglio and Felis, 2005; and Ventura *et al.*, 2008.

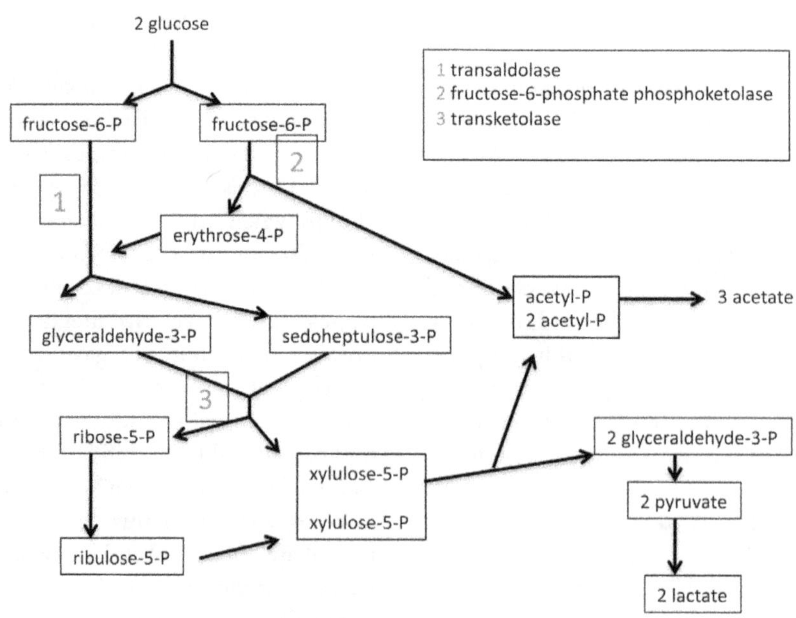

Figure 1.1 Simplified diagram of hexose metabolism in bifidobacteria. After Sgorbati *et al.* (1995). A colour version of this figure is located on the cover of this book.

Table 1.1 Bifidobacterial species commonly detected in human faeces

B. adolescentis
B. bifidum
B. angulatum
B. breve
B. catenulatum
B. gallicum
B. infantis (*B. longum* subspecies *infantis*)
B. longum (*B. longum* subspecies *longum*)
B. pseudocatenulatum
B. animalis subspecies *lactis* (possibly of dietary origin)

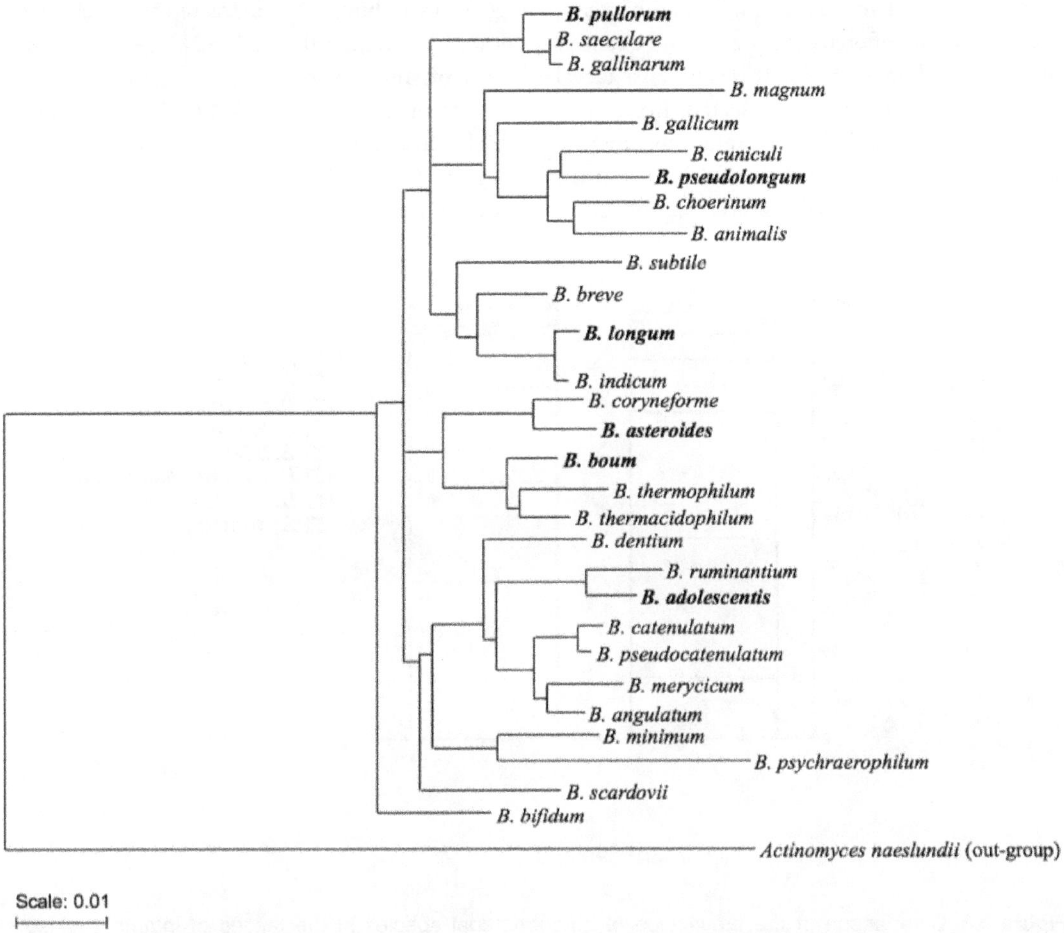

Figure 1.2 Phylogenetic tree showing the relationships of the 16S rDNA sequences of type strains of *Bifidobacterium* species. Sequences were obtained from the Ribosomal Database Project (http://rdp.cme.msu.edu/) and the tree was constructed by an on-line tree-builder resource that uses a weighted version of Neighbour Joining algorithm and the Jukes-Cantor corrected distance model. The 16S rDNA sequence of the type strain of *Actinomyces naeslundii* was selected as an out-group. In bold, species used to designate bifidobacterial clusters. Dendrogram courtesy of Baltasar Mayo.

Analysis of bifidobacterial populations

Bifidobacteria are easily cultivated in the laboratory provided good quality anaerobic procedures are followed. A variety of culture media have been devised for the selective cultivation of bifidobacteria but Beeren's medium and Rogosa SL medium are noteworthy selective, solid media (Beerens, 1990; Requena *et al.*, 2002). Agar plate cultures generally require three days incubation at 37°C to develop colonies of adequate size. Genomic/phenotypic identifications made decades previously on isolates from faeces collected from adults and infants are in accord with more modern analyses of bifidobacterial populations (Fig. 1.3) (Biavati *et al.*, 1984, 1986). Nowadays, culture-independent methods based on bacterial nucleic acids are often used to detect and quantify bifidobacterial populations. This is because faecal and other samples can be collected, frozen and dispatched to an analytical laboratory that may be far distant from the location in which the human subjects reside. Bacterial culture, on the other hand, requires that samples be processed without freezing within a few hours of collection.

Five nucleic acid-based methods have current utility in the analysis of the bacterial communities of the gut with respect to bifidobacterial studies: analysis of 16S rRNA gene sequences, fluorescence *in situ* hybridization (FISH), PCR coupled with polyacrylamide gel electrophoresis, species-specific PCR, and genotyping of cultured strains. It should be noted that sequences from both dead and living bacterial cells are detected by methods based on the extraction of bulk DNA from human faeces.

16S rRNA gene phylogeny

In older studies, PCR amplification of 16S rRNA genes from bulk DNA extracted from faeces followed by cloning the 16S rRNA gene sequences in a plasmid vector in an *Escherichia coli* host, prior to sequencing, was utilized. More recently, high throughput bead/emulsion-based sequencing of PCR-amplified DNA or random sequencing of DNA fragments derived from bacterial

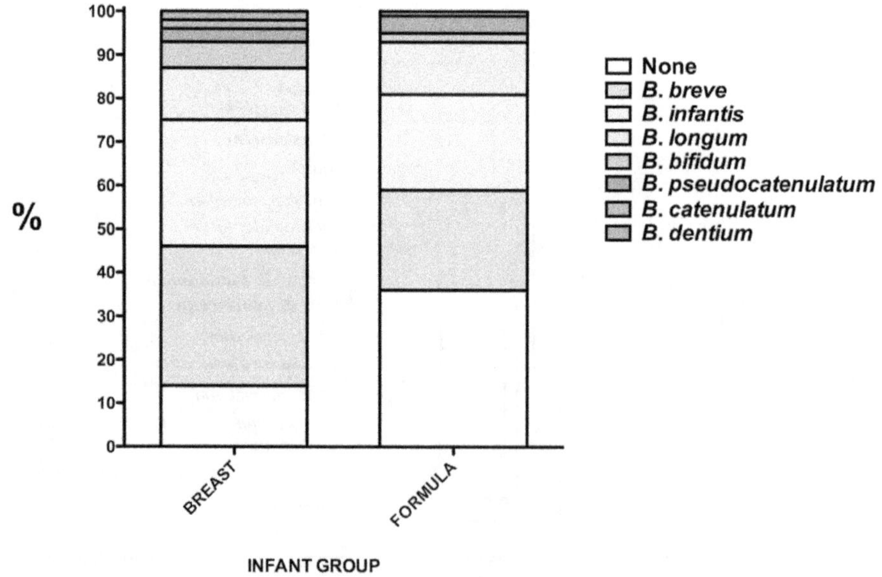

Figure 1.3 Comparison of the prevalence of bifidobacterial species in the faeces of infants suckled at the breast or fed infant formula. The number of faecal samples that contained the given bifidobacterial species is shown as a percentage of all samples. Nine hundred and seven isolates of bifidobacteria from 128 breast milk-fed and 70 formula-fed infants were identified by DNA:DNA hybridization and carbohydrate fermentation tests. Note the greater prevalence of *B. bifidum* in breast milk-fed infants. Note also the greater proportion of samples collected from formula-fed infants that did not contain bifidobacteria (none). Age of infants was not recorded in the original article. After Biavati *et al.* (1984). A colour version of this figure is located on the cover of this book.

communities in faeces (metagenome) has been used. These approaches provide catalogues of the constituent bacterial types (usually broad phylogenetic groups) of the community when analysed in relation to a databank of 16S rRNA gene sequences (such as the Ribosomal Database Project pyrosequencing pipeline tools).

Eckburg *et al.* (2005), for example, examined 13,355 16S rRNA gene sequences from multiple colonic mucosal sites and faeces of three, healthy adult subjects. *Actinobacteria* sequences were relatively rare (0.2% of total). *Actinomyces odontolyticus, Actinomyces graevenitzii, Bifidobacterium adolescentis, Bifidobacterium bifidum, Collinsella aerofaciens, Corynebacterium sundsvallense,* and *Rothia mucilaginosa* were detected.

Careful choice of PCR primers is of course crucial to detection of bifidobacterial sequences in bulk faecal DNA. It is noteworthy that PCR primer 8F commonly has three mismatches with bifidobacterial 16S rRNA genes. Detection of bifidobacteria using this primer is inadequate (Palmer *et al.*, 2007).

FISH

DNA (oligonucleotide) probes target specific rRNA sequences (16S or 23S) within ribosomes to which they hybridize. The probes are 5′ labelled with a fluorescent dye which permits both detection and quantification of specific bacterial populations. Bacterial cells within which hybridization with a probe has occurred fluoresce and hence can be detected and counted by epifluorescence microscopy (preferably automated) or fluorescence-activated flow cytometry. Continual reassessment of the specificity and coverage of FISH probes is essential in order to update and confirm their continuing specificity and hence reliability. This is because new 16S rRNA gene sequences are constantly added to databases. Epifluoresence microscopic detection is laborious and time consuming, and manual microscopic enumeration requires careful attention by the operator. A lower detection limit of about 10^6 bacteria per gram of faeces can be achieved (Welling *et al.*, 1997). An automated method of counting fluorescent bacterial cells has been developed by coupling fluorescence microscopy to a computerized system of image analysis (Jansen *et al.*, 1999). Using this automated counting device, the lower detection threshold has been estimated to be 10^7 bacteria per gram of faeces. Therefore only the more numerous members of the bacterial community can be detected. Nevertheless, identification of individual bacterial cells, as well as morphological and topographical information are valuable characteristics of fluorescence microscopy. Combined with flow cytometry, FISH provides a high throughput quantitative and qualitative method of analysis. Flow cytometry combines quantitative and multiparametrics analysis (size, internal granularity, fluorescence signal). A lower threshold of detection of 0.4% relative to the total number of bacteria determined with the universal bacterial probe EUB338 has been demonstrated (Christophe Lay, PhD thesis, 2004).

Langendijk *et al.* (1995) pioneered the use of FISH for the enumeration of bifidobacteria in faeces. They tested three genus-specific probes and compared FISH/video microscopy quantification of bifidobacterial populations to colony counts of bifidobacteria on a selective agar medium. There was good correlation between FISH and colony counts which indicated that most bifidobacterial cells in faeces were cultivable. As other examples, Gore *et al.* (2008) used FISH/flow cytometry to determine the proportion of bifidobacteria constituting the faecal community of infants with eczema and in matched, healthy controls. The proportions of bifidobacteria in the faeces of the two groups did not differ. Mah *et al.* (2007) also used FISH/flow cytometry to determine the proportion of bifidobacteria constituting the faecal community of formula-fed infants with familial risk of allergies. Bifidobacteria constituted about 50% of the total bacterial community of faeces when the babies were aged 3 months.

PCR/polyacrylamide gel electrophoresis (denaturing gradient gel electrophoresis, temporal temperature gel electrophoresis (DGGE, TTGE))

DNA or RNA is extracted directly from faecal samples. Hypervariable gene sequences (most often 16S rRNA) are amplified using PCR primers that anneal with conserved sequences that span the selected hypervariable regions. One of the PCR primers has a GC-rich 5′ end (GC clamp) to prevent complete denaturation of the DNA fragments during gradient gel electrophoresis. Using 16S rRNA gene sequences as example, amplified fragments from different types of bacteria and present in the PCR product are separated using polyacrylamide gel electrophoresis. In DGGE, the double-stranded 16S fragments migrate through a polyacrylamide gel containing a gradient of urea and formamide until they are partially denatured by the chemical conditions. The fragments do not completely denature because of the GC clamp, and migration is radically slowed when partial denaturation occurs. Because of the variation in the 16S sequences of different bacterial species, chemical stability is also different; therefore different 16S 'species' can be differentiated by this electrophoretic method. Similarly, in TTGE, the 16S sequences can be separated by gradually increasing the temperature of the polyacrylamide gel during electrophoresis. Separation is achieved on the basis of differing temperature stability of the 16S fragments. These methods generate a profile of the numerically predominant members of the bacterial community (Muyzer and Smalla, 1998). Individual fragments of DNA can be cut from DGGE/TTGE gels, further amplified and cloned, then sequenced. The sequence can be compared to those in gene databanks in order to obtain identification of the bacterium from which the 16S sequence originated. Depending on the length of the sequence, identification to at least bacterial phylum can be made (Zoetendal *et al.*, 1998). PCR primers have been derived for the specific detection of bifidobacteria in relation to DGGE/TTGE (Satokari *et al.*, 2001; Requena *et al.*, 2002).

Murray *et al.* (2005), for example, used PCR primers based on the species-specific region of the transaldolase gene to produce amplicons that were then profiled by DGGE. Identification of the bifidobacterial species was carried out with reference to identification ladders run in the gels, or by elution of DNA fragments from the gels, cloning and sequencing. The results of the study provided information about the composition of the bifidobacterial population in the faeces of four-year-old children with atopic wheeze (Table 1.2), either sensitized (positive skin prick test to common allergens) or non-sensitized. There was no difference in the prevalence of bifidobacterial species in the two groups of children but the common detection of *B. adolescentis* once infancy has past is emphasized by this data.

Species-specific PCR

PCR primers targeting nucleic acid sequences, usually 16S rRNA gene sequences, which are unique to particular bacterial species are used with DNA extracted from faeces. The PCR amplicons of sizes characteristic of each species are detected in agarose gels for qualitative scoring (present/absent). Mah *et al.* (2007), for example, used PCR primers specific to *B. longum*, *B. breve*, *B. bifidum*, *B. adolescentis*, and *B. infantis* to detect these species in infant faeces. The results showed the increased prevalence of *B. longum* and *B. breve* in faeces as the children grew older (Table 1.3).

Table 1.2 Bifidobacterial species detected in the faeces of healthy and atopic wheezy children, 4 years of age[1]

Bifidobacterium species	Sensitized[2], wheezy children	Non-sensitized, non-wheezy children
B. adolescentis	85[3]	91
B. bifidum	12	15
B. longum	39	36
B. catenulatum	3	3
B. pseudocatenulatum	6	6

[1]After Murray *et al.* (2005).

[2]Sensitized, positive skin test to common allergens.

[3]Per cent of children.

Table 1.3 Bifidobacterial species detected in faeces of formula-fed infants at risk of allergies using species-specific PCR primers[1]

	Bifidobacterial species detected in the faeces:				
Age of infant	*B. longum*	*B. breve*	*B. bifidum*	*B. adolescentis*	*B. infantis*
3 days	41[2]	24	30	6	6
1 month	53	29	41	12	6
3 months	59	47	47	24	6
1 year	93	65	35	18	12

[1]After Mah *et al.* (2007).

[2]Per cent of infants.

Real time quantitative PCR an also be used to quantify bifidobacterial populations using specific PCR primers and probes. Requena *et al.* (2002) showed that real time quantitative PCR measurements of bifidobacterial populations in faeces from adult humans correlated well with results obtained by culture when either a 16S rRNA gene sequence (Pearson linear correlation $r=0.818$) or a transaldolase gene sequence ($r=0.896$) was targeted.

Genotyping of isolates

Differentiation between bacterial strains requires the use of genetic fingerprinting (genotyping) methods. These methods usually involve the generation of DNA fragment polymorphic profiles in agarose gels from DNA extracted from pure cultures of bacteria. In general, fingerprinting methods have relied on generating DNA fingerprints by restriction endonuclease digestion (such as in the case of pulsed field gel electrophoresis of digested DNA; the 'gold standard'), or by PCR of random or specific sequences followed by agarose gel electrophoresis. McCartney *et al.* (1996) used ribotyping (restriction digests probed with 16S rRNA gene sequence) to determine the strain composition of bifidobacterial populations in faeces. Samples of faeces were collected from two adults each month for 12 months. One subject harboured a relatively simple and stable collection of bifidobacterial strains (five strains detected during the 12 month period). The other subject harboured 32 strains, some of which appeared, disappeared, and sometimes reappeared during the course of the study. The collections of strains detected in each subject were unique to the individual in that a strain common to both subjects was not detected. A later study by Kimura *et al.* (1997) examined the strain composition of bifidobacterial populations in faecal samples collected from 10 adults. Approximately half of the subjects had a simple collection of strains in their faeces whereas the bifidobacterial population was more complex in the remainder. Unique collections of bifidobacterial strains that persisted throughout the study period were detected in each subject. These observations were among the first reports revealing the uniqueness of bacterial communities harboured by individual humans, and the constancy of composition of bacterial populations in many subjects.

Sources of primer/probe details

General information on primers for PCR detection, quantification, and typing of bifidobacterial isolates is summarized in Table 1.4. Species-specific primers have been derived largely due to the efforts of Matsuki and colleagues at the Yakult Central Institute for Microbiological Research. Details of these primers can be obtained from Matsuki *et al.* (1999, 2003).

Enrichment of bifidobacteria in the baby bowel

Although bifidobacteria can be detected in the faeces of adults, they form a relatively small proportion of the total bacterial community, usually less than 5% when quantified by FISH (Lay *et al.*, 2005). Bifidobacterial numbers may be lower in the faeces of elderly humans in some circumstances (Mitsuoka and Hayakawa, 1973). The

Table 1.4 Sources of PCR primers and probes for bifidobacterial investigations

Application	Reference
Genus-specific PCR primers	Langendijk *et al.* (1995)
	Kaufmann *et al.* (1997)
	Turroni *et al.* (2009)
PCR/DGGE group-specific primers	Satakari *et al.* (2001)
	Requena *et al.* (2002)
PCR/DGGE nested PCR	Temmerman *et al.* (2003)
Species-specific PCR primers	Matusuki *et al.* (2001)
	Matsuki *et al.* (2003)
Strain-specific PCR	Kok *et al.* (1996)
Random amplification of polymorphic DNA (RAPD) PCR	Sakata *et al.* (2002)
	Fanedl *et al.* (1998)
	Vincent *et al.* (1998)
	Delgado *et al.* (2006)
Amplified ribosomal DNA restriction analysis (ARDRA) PCR	Roy and Sirios (2000)
Repetitive DNA element PCR fingerprinting (rep-PCR)	Masco *et al.* (2003)
PCR-enterobacterial repetitive intergenic consensus sequences (ERIC) PCR	Ventura *et al.* (2007)
Real time quantitative PCR	Requena *et al.* (2002)
	Matsuki *et al.* (2004)

results of most studies have indicated, however, that bifidobacteria are numerous in the faeces during the first year of life and are among the pioneers of the bacterial succession that occurs in the large bowels of babies. Facultative anaerobes and *Bacteroides–Prevotella* are numerically predominant in very early life but bifidobacteria then rapidly increase in numbers so that, by the age of three months, they are the numerically dominant members of the bacterial community. Bifidobacteria are particularly numerous in the faeces of infants fed at the breast and form up to about 90% of the faecal microbiota (Harmsen *et al.*, 2000; Favier *et al.*, 2002; Mariat *et al.*, 2009). Formula-fed infants also harbour large numbers of bifidobacteria, comprising on average about 50% of the faecal microbiota (Fig. 1.4).

The bifidobacterial inoculum for the infant bowel has historically been assumed to be the faeces of the mother in particular, and other family members and close contacts (Tannock *et al.*, 1990; Mikami *et al.*, 2009). PCR detection of bifidobacterial DNA sequences in breast milk has been reported to be relatively common (Gueimonde *et al.*, 2007; Gronlund *et al.*, 2007), and culture of bifidobacteria from breast milk has been achieved for a small number of samples (Martin *et al.*, 2009). Contamination of the milk during collection is a considerable risk but bifidobacterial DNA has apparently also been detected within peripheral blood monocytes of lactating women and in placental tissues (Perez *et al.*, 2007; Satokari *et al.*, 2008). Therefore at least the remnants of bifidobacterial cells are transmitted from

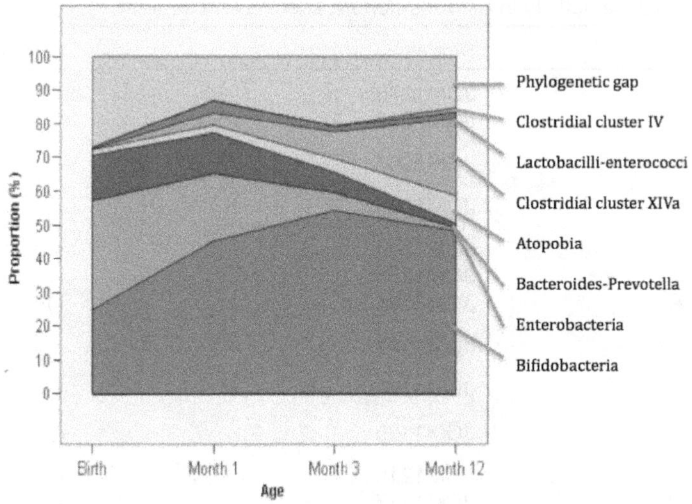

Figure 1.4 The proportions of bacterial phylogenetic groups in the stool community of formula-fed infants (n=17, as determined by FISH and flow cytometry, during the first year of life (sampled at three days of age (birth), one month, three months and 12 months of age. Note the preponderance of bifidobacteria, enterobacteria and *Bacteroides–Prevotella* during the first month of life, and of bifidobacteria thereafter. The phylogenetic gap refers to the proportion of bacteria not detected by phylotype-specific probes but detected by the universal bacterial probe (Eub338). From Mah *et al.* (2007), published with permission. A colour version of this figure is located on the cover of this book.

mother to infant by way of breast milk. Whatever the origin of the bacteria, the physicochemical milieu of the large bowel provides a 'fertile soil' for the proliferation of bifidobacteria that are therefore enriched in the bowel biome.

Lactose, galactose and sucrose are commonly fermented by bifidobacteria but the principal focus on bifidobacterial 'fertilizers' has concerned human milk oligosaccharides (HMO). Human colostrum and milk contain high concentrations (20–23 g/l and 10–13 g/l) of a diversity of soluble oligosaccharides formed from a small 'core' of monosaccharides (Coppa *et al.*, 2004). About 200 unique oligosaccharides have been detected in breast milk. The oligosaccharide diversity is limited in samples collected from some mothers (23 HMO) whereas it is expansive in others (130 HMO) (German *et al.*, 2008). Curiously, some of these HMO are recalcitrant to digestion by the infant and thus pass to the large bowel where they are available for bacterial nutrition (Coppa *et al.*, 2004). HMO constitute the 'bifidus factor' reported long ago in relation to bifidobacterial nutrition (Gyorgy, 1953). HMO characteristically have a lactose moiety at the reducing end with structural diversity arising from assorted configurations of glucose, galactose, *N*-acetyl-glucosamine, fucose, and sialic acid residues. *B. infantis* (*B. longum* subspecies *infantis*) ATCC 15687 has a genomic gene cluster encoding import and processing functions, including four glycosyl hydrolases (sialidase, fucosidase, *N*-acetyl-beta-hexosaminidase, beta-galactosidase) (Sela *et al.*, 2008). These kinds of genetic specialization probably indicate why at least some bifidobacteria can catabolize HMO efficiently. Some bifidobacteria utilize isomeric forms of lacto-*N*-tetraose (Gal(β1–3) GlcNac(β1–3)Gal(β1–4)Glc) which is the core structure of HMO. This structure is catabolized via a lacto-*N*-biose I intermediate (Gal(β1–3] GlcNac) using a pathway with lacto-*N*-biosidase activity (Wada *et al.*, 2008; Kiyohara *et al.*, 2009). The operation of this pathway appears to be species- and even strain-specific (Table 1.5). Lacto-*N*-biosidase activity may be more common among *B. bifidum* strains, perhaps offering an explanation of the particular association of this species with breast-fed infants (Gore *et al.*, 2008; Wada *et al.*, 2008; Kiyohara *et al.*, 2009). Taken together, source of inoculum, qualitative differences in HMO in breast milk between lactating

Table 1.5 Lacto-*N*-biosidase activity in *Bifidobacterium* species and strains[1]

Species	Strain	Activity
B. adolescentis	JCM1275	–[2]
	JCM7046	–
B. angulatum	JCM7096	–
B. animalis	JCM10602	–
B. bifidum	JCM1254	+
	JCM1255	+
	JCM7004	+
B. breve	JCM1192	––
B. catenulatum	JCM1194	–
B. dentium	JCM1195	–
B. longum	JCM1210	-
	JCM1217	+
	JCM1222	+
	JCM7054	+
B. pseudocatenulatum	JCM1200	–
B. pseudolongum	JCM1205	–
B. scardovii	JCM12489	–

[1]After Wada *et al.* (2008).

[2]=, lacto-*N*-biosiase activity detected; –, activity not detected.

mothers, species- and strain-specific catabolic abilities among bifidobacteria probably explain the enrichment of different bifidobacterial species in the bowel of different infants.

Bifidobacteria, babies, and allergies

Many affluent countries have recorded an increase in the prevalence of allergic diseases in recent decades, including eczema, asthma, and hay fever (Hopkin, 1995). Colonization of the infant bowel by specific species of bacteria may be important in the initial regulation of the developing immune system (Murray and Woodcock, 2002). Alterations to neonatal exposure to particular bacteria may have occurred because of the effects of hospital deliveries, caesarean sections, special care baby unit admissions, smaller family size, widespread use of antibiotics, good hygiene, and differences in maternal diet in modern times. Changes in the composition of the bowel microbiota might result in exposure, or lack of exposure, of infants to specific bacteria. This would alter the exposure of children to important antigens at a critical time in the maturation of the immune system (Brandtzaeg, 2002). Members of the genus *Bifidobacterium* have been considered to be important bacteria for investigation in this respect because they form a major portion of the bowel bacterial community in early life (Harmsen *et al.*, 2000; Mah *et al.*, 2007).

Bifidobacteria present in the faeces of children aged 25–35 days in Ghana (a country with a low prevalence of allergies), New Zealand and the United Kingdom (high-prevalence countries) have been reported to differ (Young *et al.*, 2004). The majority of faecal samples of children from Ghana contained *Bifidobacterium infantis*, whereas those of the children from the other countries did not. In contrast, *Bifidobacterium adolescentis*, *Bifidobacterium bifidum*, *Bifidobacterium longum*, and *Bifidobacterium pseudocatenulatum* were detected in the faeces of the New Zealand and British children. Additionally, *B. pseudocatenulatum* has been detected more commonly in children with eczema (27%) than in matched controls (4%) (Gore *et al.*, 2008).

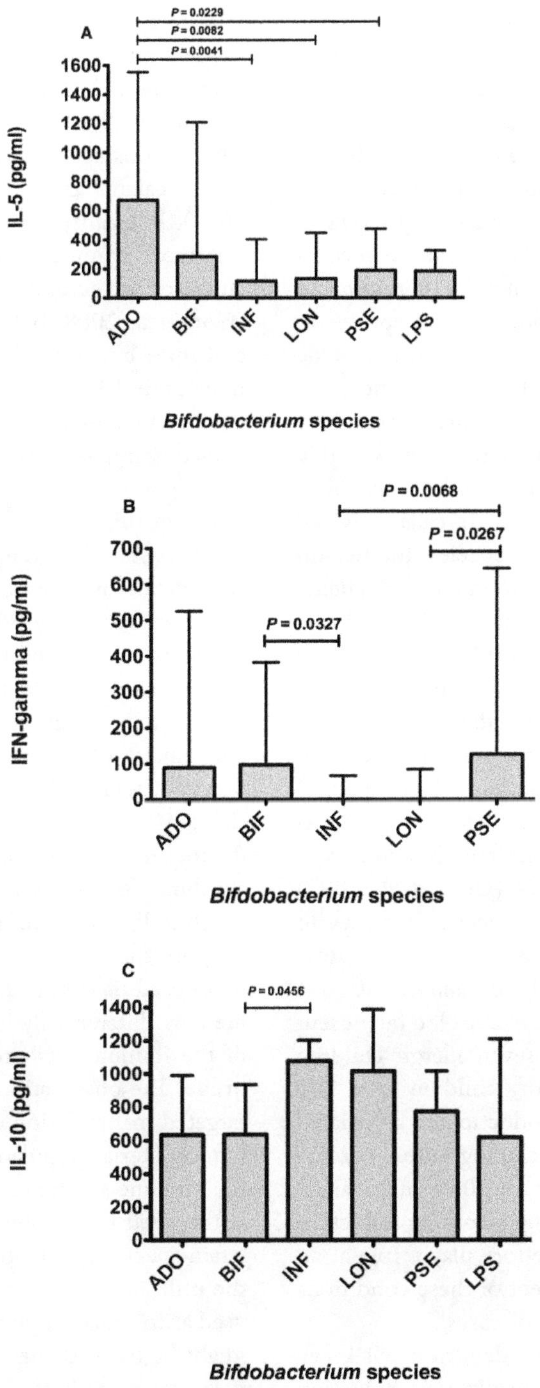

Figure 1.5 Cytokine production. Medians and inter-quartile ranges are shown. ADO, *B. adolescentis*; BIF, *B. bifidum*; LON, *B. longum*; PSE, *B. pseudocatenulatum*; LPS, lipopolysaccharide (positive control). (A) *B. adolescentis*-stimulated preparations produced more interleukin 5 than did those stimulated by other bifidobacteria. $N=14$ infants per group. (B) *B. infantis*- and *B. longum*-stimulated preparations produced little interferon-gamma compared to the other bifidobacterial species. $N=12$ infants per group. Median response to LPS (positive control) was 2695 pg/ml, inter-quartile range 5586). (C) There was a trend for *B. infantis*- and *B. longum*-stimulated preparations to produce more interleukin 10 than was the case for other species. $N=14$ infants per group. Statistically significant *P*-values (Mann–Whitney test) are shown (Baird, Taylor and Tannock unpublished data).

Dendritic cells (antigen presenting cells) derived from umbilical cord blood react differently to heat-killed preparations of type cultures of different bifidobacterial species (Young *et al.*, 2004). CD83 expression has been reported to be increased and interleukin 10 production was induced by exposure of the dendritic cells to heat-killed cells of *B. bifidum*, *B. longum*, and *B. pseudocatenulatum*, but not by *B. infantis*. Therefore, the presence of particular bifidobacterial species in the infant bowel might differentially stimulate the immune system and direct its development.

Recent unpublished work (Baird *et al.*, 2009) has revealed differential cytokine responses of T cells harvested from cord blood and stimulated by exposure to autologous dendritic cells activated by different bifidobacterial species Interleukin 5 was produced in greater amounts by *B. adolescentis*-dendritic cell-T cell combinations relative to other preparations (Fig. 1.4A). In a previous study, *B. adolescentis* was not detected in the faeces of infants 25 to 35 days old that had been born in Ghana, but was detected in randomly selected New Zealand and UK children (8% of infants) (Young *et al.*, 2004). Infants aged 3–6 months are much more likely to harbour *B. adolescentis* in their faeces (about 25%) (Gore *et al.*, 2008) and colonization with this species increases to about 90% of children aged 4 years (Murray *et al.*, 2005). Interestingly, *B. adolescentis* was reported by Ouwehand *et al* (2001) to be present in the faeces of six out of seven allergic children but absent from six healthy children aged 2–7 months. Interleukin 5 production is associated with allergic responses including atopic eczema and asthma (Rosenberg *et al.*, 2007; Incorvaia *et al.*, 2008). *B. adolescentis*, by favouring induction of this cytokine, may therefore play a role at an early age in the development of these conditions in children at risk of atopic diseases.

B. infantis and *B. longum*-dendritic cell-T cell combinations resulted in negligible production of interferon-gamma relative to *B. bifidum* and *B. pseudocatenulatum* (Fig. 1.4B). This observation was not the result of T cell anergy because *B. infantis* and *B. longum*-dendritic cell-T cell combinations produced relatively large amounts of interleukin 10 (Fig. 1.4C). The differential effects relating to interferon-gamma and interleukin 10 production by *B. infantis* and *B. pseudocatenulatum* are of interest because *B. infantis* was detected in the faeces of 72% of Ghana babies, whereas this species was absent in the faeces of New Zealand and UK children (Young *et al.*, 2004). *B. pseudocatenulatum*, on the other hand, was more commonly detected in the faeces of children with eczema compared to control infants (Gore *et al.*, 2008) but was absent from the faeces of babies born in Ghana (Young *et al.*, 2004). Interleukin 10 is associated with the initiation and maintenance of anti-inflammatory responses, so *B. infantis* might contribute to the control of allergic responses through induction of this cytokine (Romagnani, 2006).

There is a need to test multiple strains of each species because the observed phenomena could be strain specific. Additionally, the stimulatory activities of bifidobacterial cells killed by heat or irradiation should be compared because the method of preparation of the cells might affect the antigenicity of the bacteria. The best approach, however, will be to fractionate the bifidobacterial cells, especially cell surface components, to determine the molecular basis of stimulation of the immune response.

Overall, these immunological observations support the view that the composition of the bifidobacterial population of the bowel in early life may differentially influence the development of the immune system during the first months of life. The observations also provide a basis for targeted manipulation of the composition of the bifidobacterial population and the reactivity of the immune system to allergens. Coupled to this is the need to investigate the species- and even strain-specificity of bifidobacterial species in the utilization of oligosaccharides that might be used as formula supplements. These supplements might be used to enrich for specific bifidobacterial populations in the infant bowel. The nature of the epitopes that drive the differential immunological responses that have been reported should also be identified. Purification of these bacterial factors might provide an alternative approach to the modification of the bowel community since they might be administered to infants as pharmacological substances.

Conclusion

Members of the genus *Bifidobacterium* are numerically important members of the bowel biome of infants during early life. These bacteria may, in the future, be shown to interact with the bowel mucosa and to have long lasting influences on the physiology, especially immunology, of the human host.

References

Beerens, H. (1990). An elective and selective isolation medium for *Bifidobacterium* spp. Lett. Appl. Microbiol. *11*, 155–157.

Biavati, B., Castagnoli, P., Crociani, F., and Trovatelli, L.D. (1984). Species of the *Bifidobacterium* in the feces of infants. Microbiologica *7*, 341–345.

Biavati, B., Castagnoli, P., and Trovatelli, L.D. (1986). Species of the genus *Bifidobacterium* in the feces of human adults. Microbiologica *9*, 39–45.

Brandtzaeg, P. (2002). Current understanding of gastrointestinal immunoregulation and its relation to food allergy. Ann. N. Y. Acad. Sci. *964*, 13–45.

Coppa, G.V., Pierani, P., Zampini, L., Bruni, S., Carloni, I., and Gabrielli, O. (2001). Characterization of oligosaccharides in milk and feces of breast-fed infants by high performance anion-exchange chromatography. In Bioactive Components of Human Milk, Newburg, D. ed. (Kluwer Academic/Plenum Publishers, New York), pp. 307–314.

Cummings, J.H., and Macfarlane, G.T. (1991). The control and consequences of bacterial fermentation in the human colon. J. Appl. Bacteriol. *70*, 443–459.

Delgado, S., Suarez, A., and Mayo, B. (2006). Bifidobacterial diversity determined by culturing and by 16S rDNA sequence analysis in feces and mucosa from ten healthy Spanish adults. Dig. Dis. Sci. *51*, 1878–1885.

Dellaglio, F., and Felis, G.E. (2005). Taxonomy of lactobacilli and bifidobacteria. In Probiotics and Prebiotics: Scientific Aspects. Tannock, G.W. , ed. (Caister Academic Press, Wymondham, UK), pp. 25–49.

Dethlefsen, L., McFall-Ngai, M., and Relman, D.A. (2007). An ecological and evolutionary perspective on human-microbe mutualism and disease. Nature *449*, 811–818.

Eckburg, P.B., Bik, E.M., Bernstein, C.N., Purdom, E., Dethlefsen, L., Sargent, M., Gill, S.R., Nelson, K.E., and Relman, D.A. Diversity of the human intestinal microbial flora. Science *308*, 1635–1638.

Fandl, L., Nekrep, F.V., and Avgustin, G. (1998). Random amplified polymorphic DNA analysis and demonstration of genetic variability among bifidobacteria isolated from rats fed with raw kidney beans. Can. J. Microbiol. *44*, 1094–1101.

Favier, C.F., Vaughan, E.E., de Vos, W.M., and Akkermans, A.D. (2002). Molecular monitoring of succession of bacterial communities in human neonates. Appl. Environ. Microbiol. *68*, 219–226.

German, J.B., Freeman, S.L., Lebrilla, C.B., and Mills, D.A. (2008). Human milk oligosaccharides: evolution, structures and bioselectivity as substrates for intestinal bacteria. Nestle Nutr. Workshop Ser. Pediatr. Program *62*, 205–218.

Gore C., Munro, K., Lay, C., Bibiloni, R., Morris, J., Woodcock, A., Custovic, A., and Tannock, G.W. (2008). *Bifidobacterium pseudocatenulatum* is associated with atopic eczema: a nested case-control study investigating the fecal microbiota of infants. J. Allergy Clin. Immunol. *121*, 135–140.

Gronlund, M-M., Gueimonde, M., Laitinen, K., Kociubinski, G., Gronroos, T., Salminen, S., and Isolauri, E. (2007). Maternal breast-milk and intestinal bifidobacteria guide the compositional development of the *Bifidobacterium* microbiota in infants at risk of allergic disease. Clin. Exp. Allergy *37*, 1764–1772.

Gueimonde, M., Laitinen, K., Salminen, S., and Isolauri, E. (2007). Breast milk: a source of bifidobacteria for infant gut development and maturation? Neonatology *92*, 64–66.

Gyorgy, P. (1953). A hitherto unrecognized biochemical difference between human milk and cow's milk. Pediatr. *11*, 98–103.

Harmsen, H.J., Wildeboer-Veloo, A.C., Raangs, G.C., Wagendorp, A.A., Klijn, N., Bindels, J.G., and Welling, G.W. (2000). Analysis of intestinal flora development in breast-fed and formula-fed infants by using molecular identification and detection methods. J. Pediatr. Gastroenterol. Nutr. *30*, 61–67.

Hooper, L.V., Wong, M.H., Thelin, A., Hansson, L., Falk, P.G., and Gordon, J.I. (2001). Molecular analysis of commensal host microbial relationships in the intestine. Science *291*, 881–884.

Hopkin J.M. (1997). Mechanisms of enhanced prevalence of asthma and atopy in developed countries. Curr. Opin. Immunol. *9*, 788–792.

Incorvaia, C., Frati, F., Verna, N., D'Alo, S., Motolese, A., and Pucci, S. (2008). Allergy and the skin. Clin. Exp. Immunol. *153*, 27–29.

Jansen, G.J., Wildeboer-Veloo, A.C., Tonk, R.H., Franks, A.H., and Welling, G.W. (1999). Development and validation of an automated, microscopy-based method for enumeration of groups of intestinal bacteria. J. Microbiol. Methods 37, 215–221.

Kaufmann, P., Pfefferkorn, A., Teuber, M., Meile, L. Identification and quantification of *Bifidobacterium* species isolated from food with genus-specific 16S rRNA-targeted probes by colony hybridization and PCR. Appl. Environ. Microbiol. *63*, 1268–1273.

Kimura, K., McCartney, A., McConnell, M.A., and Tannock, G.W. (1997). Analysis of fecal populations of bifidobacteria and lactobacilli and investigation of the immunological responses of their human hosts to the predominant strains. Appl. Environ. Microbiol. *63*, 3394–3398.

Kiyohara, M., Tachizawa, A., Nishimoto, M., Kitaoka, M., Ashida, H., and Yamamoto, K. (2009). Prebiotic effect of Lacto-*N*-biose I on bifidobacterial growth. Biosci. Biotechnol. Biochem. 73, 1175–1179.

Kok, R.G., de Waal, A., Achut, F., Welling, G.W., Weenk, G., and Hellingwerf, K.J. (1996). Specific detection and analysis of a probiotic *Bifidobacterium* strain in infant feces. Appl. Environ. Microbiol. 62, 3668–3672.

Langendijk, P.S., Schut, F., Jansen, G.J., Raangs, G.C., Kamphuis, G.R., Wilkinson, M.H., and Welling, G.W. (1995). Quantitative fluorescence *in situ* hybridization of *Bifidobacterium* spp. with genus-specific 16S rRNA-targeted probes and its application in fecal samples. Appl. Environ. Microbiol. *61*, 3069–3075.

Lay, C., Rigottier-Gois, L., Holmstrom, K., Rajilic, M., Vaughan, E.E., de Vos, W.M., Collins, M.D., Thiel, R., Namsolleck, P., Blaut, M., and Dore, J. (2005). Colonic microbiota signatures across five northern European countries. Appl. Environ. Microbiol. *71*, 4153–4155.

Mah, K.A.W., Chin, V.I.L., Wong, W.S., Lay, C., Tannock, G.W., Shek, L.P., Aw, M.M., Chua, K.Y., Wong, H.B., Panchalingham, A., and Lee, B.W. (2007). Effect of a milk formula containing probiotics on the fecal microbiota of Asian infants at risk of atopic diseases. Pediatr. Res. *62*,674–679.

Mariat, D., Firmessse, O., Levenez, F., Guimaraes, V.D., Sokol, H., Dore, J., Corthier, G., and Furet, J.-P. (2009). The *Firmicutes/Bacteroidetes* ratio of the human microbiota changes with age. BMC Microbiol. *9*, 123. Doi:10.1186/1471-2180-9-123.

Martin, R., Jimenez, E., Heilig, H., Fernandez, L., Marin, M.L., Zoetendal, E.G. and Rodriguez, J.M. (2009). Isolation of bifidobacteria from breast milk and assessment of the bifidobacterial population by PCR-denaturing gradient gel electrophoresis and quantitative real-time PCR. Appl. Environ. Microbiol. *75*, 965–969.

Masco, L., Huys, G., Gevers, D., Verbrugghen, L., and Swings, J. (2003). Identification of *Bifidobacterium* species using rep-PCR fingerprinting. System. Appl. Microbiol. *26*, 557–563.

Matsuki, T., Watanabe, K., Fujimoto, J., Kado, Y., Takada, T., Matsumoto, K., and Tanaka, R. (2004). Quantitative PCR with 16S rRNA-gene-targeted species-specific primers for analysis of human intestinal bifidobacteria. Appl. Environ. Microbiol. *70*, 167–173.

Matsuki, T., Watanabe, K., and Tanaka, R. (2003). Genus- and species-specific PCR primers for the detection and identification of bifidobacteria. Curr. Issues Intest. Microbiol. *4*, 61–69.

Matsuki, T., Watanabe, K., Tanaka, R., Fukuda, M., and Oyaizu, H. (1999). Distribution of bifidobacterial species in human intestinal microflora examined with 16S rRNA-gene-targeted species-specific primers. Appl. Environ. Microbiol. *65*, 4506–4512.

Mattarelli, P., Bonaparte, C., Pot, B., and Biavati, B. (2008). Proposal to reclassify the three biotypes of *Bifidobacterium longum* as three subspecies: *Bifidobacterium longum* subsp. *longum* subsp. nov., *Bifidobacterium longum* subsp. *infantis* comb. Nov. and *Bifidobacterium longum* subsp. *suis* comb. nov. Int. J. System. Evol. Microbiol. *58*, 767–772.

McCartney, A.L., Wang, W., and Tannock, G.W. (1996). Molecular analysis of the composition of the bifidobacterial and lactobacillus microflora of humans. Appl. Environ. Microbiol. *62*, 4608–4613.

Mikami, K., Takahashi, H., Kimura, M., Isozaki, M., Izuchi, K., Shibata, R., Sudo, N., Matsumoto, H., and Koga, Y. (2009). Influence of maternal bifidobacteria on the establishment of bifidobacteria colonizing the gut of infants. Pediatr. Res. *65*, 669–674.

Mitsuoka, T., and Hayakawa, K. (1973). The fecal flora in man. Composition, I. of the fecal flora of various age groups. Zentralbl. Bakteriol. Orig. A *223*, 333–342 (in German).

Murray, C.S., Tannock, G.W., Simon, M.A., Harmsen, H.J.M., Welling, G.W., Custovic, A., and Woodcock, A. (2005). Fecal microbiota in sensitized wheezy and non-wheezy children: a nested case-control study. Clin. Exp. Allergy *35*, 741–745.

Murray, C.S., and Woodcock, A. (2002). Gut microflora and atopic disease. In GW Tannock, ed. Probiotics and Prebiotics: Where Are We Going? (Caister Academic Press, Wymondham, UK), pp. 239–261.

Muyzer, G., and Smalla, K. (1998). Application of denaturing gradient gel electrophoresis (DGGE) and temperature gradient gel electrophoresis (TGGE) in microbial ecology. Antonie van Leewenhoek *73*, 127–141.

Ouwehand, A.C., Isolauri, E., He, F., Hashimoto, H., Benno, Y., and Salminen, S. (2001). Differences in *Bifidobacterium* flora composition in allergic and healthy infants. J Allergy Clin. Immunol. *108*, 144–145.

Palmer, C., Bik, E.M., DiGiulio, D.B., Relman, D.A., and Brown, P.O. (2007). Development of the human intestinal microbiota. PLoS Biology *5*, e177.doi:10.1371/journal.bbio.0050177.

Penders, J., Vink, C., Driessen, C., London, N., Thijs, C., and Stobberingh, E.E. (2005). Quantification of *Bifidobacterium* spp., *Escherichia coli* and *Clostridium difficile* in faecal samples of breast-fed and formula-fed infants by real-time PCR. FEMS Microbiol. Lett. *243*, 141–147.

Perez, P.F., Dore, J., Leclerc, M., Levenez, F., Benyacoub, J., Serrant, P., Segura-Roggero, I., Schiffrin, E.J., and Donnet-Hughes, A. (2007). Bacterial imprinting of the neonatal immune system: lessons from maternal cells? Pediatrics *119*, e724–32.

Roberton, A.M., and Corfield, A.P. (1999). Mucin degradation and its significance in inflammatory conditions of the gastrointestinal tract. In Tannock, G.W. ed. Medical Importance of the Normal Microflora (Kluwer Academic Publishers, Dordrecht, The Netherlands), pp. 222–261.

Romagnani, S. (2006). Regulation of the T cell response. Clin. Exp. Allergy *36*, 1357–1366.

Rosenberg, H.F., Phipps, S., and Foster, P.S. (2007). Eosinophil trafficking in allergy and asthma. J. Allergy Clin. Immunol. *119*, 1303–1310.

Roy, D., and Sirois, S. (2000). Molecular differentiation of *Bifidobacterium* species with amplified ribosomal DNA restriction analysis and alignment of short regions of the *ldh* gene. FEMS Microbiol. Lett. *191*, 17–24.

Sakata, S., Kitahara, M., Sakamoto, M., Hayashi, H., Fukuyama, M., and Benno, Y. (2002). Unification of *Bifidobacterium infantis* and *Bifidobacterium suis* as *Bifidobacterium longum*. Int. J. System. Evol. Microbiol. *52*, 1945–1951.

Satokari, R., Gronroos, T., Laitinen, K., Salminen, S., and Isolauri, E. (2008). *Bifidobacterium* and *Lactobacillus* DNA in the human placenta. Lett. Appl. Microbiol. *48*, 8–12.

Satokari, R.M., Vaughan, E.E., Akkermans, A.D., Saarela, M., and de Vos, W.M. (2001). Bifidobacterial diversity in human feces detected by genus-specific PCR and denaturing gradient gel electrophoresis. Appl. Environ. Microbiol. *67*, 504–513.

Schell, M.A., Karamirantzou, M., Snel, B., Vilanova, D., Berger, B., Pessi, G., Zwahlen, M.C., Desiere, F., Bork, P., Delby, M., and Pridmore, R.D. (2002). The genome sequence of *Bifidobacterium longum* reflects its adaptation to the human gastrointestinal tract. PNAS *99*, 14422–14427.

Sela, D.A., Chapman, J., Adeuya, A., Kim, J.H., Chen, F., Whitehead, T.R., Lapidus, A., Rokhsar, D.S., Lebrilla, C.B., German, J.B., Price, N.P., Richardson, P.M., and Mills, D.A. (2008). The genome sequence of *Bifidobacterium longum* subsp. *infantis* reveals adaptations for milk utilization within the infant microbiome. PNAS *105*, 18964–18969.

Sgorbati, B., Biavati, B., and Palenzona, D. (1995). The genus *Bifidobacterium*. In The Genera of Lactic Acid Bacteria, volume 2. Wood, B.J.B. and Holzapfel, W.H. , ed. (Blackie Academic and Professional, London), pp. 279–306.

Tannock, G.W., Fuller, R., Smith, S.L., and Hall, M.A. (1990). Plasmid profiling of members of the family *Enterobacteriaceae*, lactobacilli, and bifidobacteria to study the transmission of bacteria from mother to infant. J. Clin. Microbiol. *28*, 1225–1228.

Temmerman, R., Masco, L., Vanhoutte, T., Huys, G., and Swings, J. (2003). Development and validation of a nested-PCR-denaturing gradient gel electrophoresis method for taxonomic characterization of bifidobacterial communities. Appl. Environ. Microbiol. *69*, 6380–6385.

Turroni, F., Foronic, E., Pizzetti, P., Giubellini, V., Ribbera, A., Merusi, P., Cagnasso, P., Bizzarri, B., de'Angelis, G.L., Shanahan, F., van Sinderen, D., and Ventura, M. (2009). Exploring the diversity of the bifidobacterial population in the human intestinal tract. Appl. Environ. Microbiol. *75*, 1534–1545.

Ventura, M., Canchaya, C., Fitzgerald, G.F., Gupta, R.S., van Sinderen, D. (2007). Genomics as a means to understand bacterial phylogeny and ecological adaptation: the case of bifidobacteria. Antonie ven Leeuwenhoek *91*, 351–372.

Ventura, M., Meylan, V., and Zink, R. (2003). Identification and tracing of *Bifidobacterium* species by use of enterobacterial repetitive intergenic consensus sequences. Appl. Environ. Microbiol. *69*, 4296–4301.

Ventura, M., Turroni, F., Ribbera, A., Foroni, E., and van Sinderen, D. (2008). Bifidobacteria: the model human gut commensal. In Therapeutic Microbiology: Probiotics and Related Strategies, Versalovic, J. and Wilson, M. , ed. (ASM Press, Washington, DC), pp. 35–50.

Vincent, D., Roy, D., Mondou, F., and Dery, C. (1998). Characterization of bifidobacteria by random DNA amplification. Int. J. Food Microbiol. *43*, 185–193.

de Vries, W., Gerbrandy, S.J., and Stouthamer, A.H. (1967). Carbohydrate metabolism in *Bifidobacterium bifidum*. Biochim. Biophys. Acta *136*, 415–425.

Wada, J., Ando, T., Kiyohara, M., Ashida, H., Kitaoka, M., Yamaguchi, M., Kumagai, H., Katayama, T., and Yamamoto, K. (2008). *Bifidobacterium bifidum* lacto-*N*-biosidase, a critical enzyme for the degradation of human milk oligosaccharides with a type 1 structure. Appl. Environ. Microbiol. *74*, 3996–4004.

Welling, G.W., Elfferich, P., Raangs, G.C., Wildeboer-Veloo, A.C., Jansen, G.J., and Degener, J.E. (1967). 16S ribosomal RNA-targeted oligonucleotide probes for monitoring of intestinal tract bacteria. Scand. J. Gastroenterol. 222, 17–19.

Young, S.L., Simon, M.A., Baird, M.A., Tannock, G.W., Bibiloni, R., Spencely, K., Lane, J.M., Fitzharris, P., Crane, J., Town, I., Addo-Yobo, E., Murray, C.S., and Woodcock, A. (2004). Bifidobacterial species differentially affect expression of cell surface markers and cytokines of dendritic cells harvested from cord blood. Clin. Diag. Lab. Immunol. *11*, 686–690.

Genomics of Bifidobacteria: Where We Stand Now and Where We Are Going

2

Marco Ventura, Francesca Turroni, Francesca Bottacini, Elena Foroni and Douwe van Sinderen

Abstract

During recent years microbiological research has been fundamentally changed by the ever-increasing number of publicly available bacterial whole-genome sequences. This sequence information has largely affected our understanding of the metabolic capabilities, genetics and phylogeny of bacteria. Bifidobacteria constitute one of the key microbial groups of the human intestinal microbiota, owing to their perceived positive contribution to maintain a balanced gut homeostasis. In recent years bifidobacteria have drawn much scientific attention because of their use as live bacteria in numerous food preparations with several health-related claims. For this reason these bifidobacteria represent a developing area of scientific interest with respect to genomics, molecular biology, genetics and physiology. Recent genome sequencing of different bifidobacterial species has provided the complete genetic make-up of these bacteria. In this chapter we will discuss how genomic data has provided insights into bifidobacterial evolution, while also revealing genetic functions that explain their presence in the particular ecological environment of the gastrointestinal tract.

Genomics and bacteria: the concept

Having access to a large number of fully sequenced bacterial genomes presents a very exciting opportunity to reconstruct events of genome evolution. The main forces that drive bacterial genome evolution are horizontal gene transfer, gene duplication, gene loss and chromosomal rearrangements. Additionally, point mutations are considered to play a crucial role in shaping bacterial genomes (Philippe and Douady, 2003). Despite efforts to predict the relative contribution of each of these processes, no model can yet reliably explain and trace the evolutionary development of bacteria based on their current genome structure.

A mechanism of genome evolution which ensures functional divergence in bacterial genomes and which may be responsible for fast genetic adaptation of bacteria to new environments is represented by horizontal gene transfer (HGT). The introduction of new genes (also called alien genes) by HGT allows for rapid niche-specific adaptation, which is expected to lead to bacterial diversification and speciation (Cohan, 2001). The currently accepted view of bacterial genome evolution is based on the interplay between vertically generated and horizontally acquired variation (Woese, 2000). Taking this concept to its extreme, it has been suggested that two bacterial groups are more similar than a third one, not because they share a more recent ancestor but because they exchange genes more frequently (Gogarten *et al.*, 2002). However, not all genes appear to be subject to the same frequency of HGT. In fact, according to what was originally described (Philippe and Douady, 2003), three different gene categories can be identified based on the frequency by which they are subjected to HGT: (i) a so-called hard core of genes, which are recalcitrant to HGT; (ii) a soft core, which includes genes that are not commonly susceptible to HGT; and (iii) the shell genes, which are relatively frequently subjected to HGT. The latter gene category encompasses genes found on plasmids, bacteriophages, genomic islands and transposable elements, such as insertion sequences (IS), transposons (Tn) and integrons. HGT events are held responsible for the introduction of a variety of functions in bacterial genomes, and have consequently improved the competitiveness of microorganisms in their natural ecological niches. For example, genes encoding enzymes involved in the metabolism of particular carbohydrates (e.g. sugar transporters and glycosyl hydrolases) can represent a large proportion of strain-specific and HGT-acquired genes. In this context, it has been suggested that particular genes involved in sugar metabolism and the production of exopolysaccharides in bifidobacteria have been acquired by HGT as part of an adaptation to a specific ecological niche (e.g. the human oral cavity by *Bifidobacterium dentium* Bd1 or the human intestine by *Bifidobacterium longum* subsp. *longum* NCC2705) (Schell *et al.*, 2002; Ventura *et al.*, 2007a, 2009b). Also other chromosomal regions, such as genes that encode restriction-modification (R/M) systems, appear to have been acquired by bifidobacteria through HGT (Ventura *et al.*, 2007a). The identification of potentially HGT-acquired DNA regions is challenging and is generally performed on the basis of divergence in the G+C content (GC skew), and/or dinucleotide differences and/or codon usage. Estimations of HGT frequency in bacteria point out that it is low (below 10% of the total gene complement) (Kunin and Ouzounis, 2003; Snel *et al.*, 2002). In the case of bifidobacteria the level of HGT seems to vary considerably between species as is illustrated by the genomes of *B. longum* subsp. *longum* NCC2705 and *B. dentium* Bd1, where it is estimated that less than 5% and 1% of their respective chromosome sequences have been acquired by recent and therefore recognizable HGT (Ventura *et al.*, 2007a, 2009b).

Another important force that drives bacterial genome evolution is represented by bacterial gene duplication. Gene duplication is considered a central mechanistic precursor of gene innovation that has promoted adaptation to new environments and exploitation of new ecological niches. In the pre-genomics era it was hypothesized that bacterial genomes had originated from a much smaller chromosome ancestor by numerous gene duplication events (Kunisawa, 1995). However, an analysis based on the currently available bacterial genome data does not appear to support this theory and instead shows that gene duplication is a phenomenon that contributes only marginally to genome evolution (Gevers *et al.*, 2004).

Besides the events that influence variation in gene content, genome reshuffling is a force that affects genome organization. Gene synteny between two bacteria is retained much less stringently than their average protein sequence similarity (Wolf *et al.*, 2001). This indicates that the relative gene position on the chromosome is not essential for gene function. Furthermore, it has been found that the degree of genome reorganization constantly increases as a function of time from the point of divergence, and thus gene order conservation can be used as a phylogenetic measure to investigate bacterial relationships (Korbel *et al.*, 2002). Interestingly, genomes containing an elevated number of repeat sequences display a greater level of genome reorganization through presumed homologous recombination events that must have accelerated loss of gene order (Rocha, 2003). Furthermore, chromosome development is influenced by large genomic reshuffling events, especially large inversions, which lead to the phenomenon of X-shaped patterns in the alignments of whole genomes. The most likely mechanism that underlies the generation of such X-shaped alignments includes large chromosomal inversions that overturn the genomic sequence symmetrically around the origin of replication or replication terminus. Notably, inversions that become fixed in natural populations tend to display such a symmetrical arrangement, thus reducing the disruption of the chromosomal structure relative to replication. Non-symmetrical large inversions are not often detected, indicating that they reduce the fitness of the bacterium in which they have occurred. A feature that could lead to the reduced impact of such chromosomal rearrangements is the physical distance of the site of inversion from the origin of replication, as this determines the relative copy number of the gene in each cell of fast growing cultures of bacteria (Tillier and Collins, 2000). Genes such as rRNA operons or insertion sequences (IS), which are often present in multiple identical copies at different positions on the chromosome, can easily cause genome inversions through homologous recombination.

Genomics and bifidobacteria

The first complete genome of a bifidobacterial strain, *B. longum* subsp. *longum* NCC2705, was made public in 2002 (Schell *et al.*, 2002). From that time onwards the era of genomics has flourished for bifidobacteria generating complete sequences of another eight bifidobacterial genomes (Table 2.1 and Fig. 2.1) (Lee *et al.*, 2008; Sela *et al.*, 2008). Furthermore, the partly assembled genomes of 17 other bifidobacterial strains or species (Table 2.1) have become available in the NCBI database. These genomes range in size from 2.0 to 2.8 Mb, and possess a large number of tRNA molecules (from 54 to 58), while displaying common architectural features of a typical bacterial chromosome, i.e. the co-orientation between gene transcription and DNA replication (McLean *et al.*, 1998) and the asymmetric bias in the nucleotide composition of the leading DNA strand (Frank and Lobry, 1999), which is generally enriched in guanines and depleted in cytosines (Fig. 2.2). Some of these genomes are derived from different strains within the same species, i.e. *Bifidobacterium breve*, *B. longum*, *Bifidobacterium animalis* subsp. *lactis* and *Bifidobacterium adolescentis*. Comparative studies on these genomes would be very helpful in recognizing strain-specific DNA sequences that have recently been evolved or acquired, and may consequently be responsible for adaptations of these strains to their specific niche/lifestyle. Comparative studies using all available bifidobacterial genomic information is thus expected to provide valuable information regarding genetic characteristics that are shared by bifidobacterial species and that distinguish them from other bacteria.

The activities of genome sequencing and *in silico* genome analysis of a probiotic strain is also collectively referred to as probiogenomics (Ventura *et al.*, 2009a), and represent a crucial step in the provision of primary information for downstream comparative and functional purposes, such as comparative genomics, transcriptomics and/or proteomics, which in turn can be utilized to address many fundamental and applied questions that need to be answered before the probiotic concept can be fully accepted.

Biosynthetic capabilities

Genomics data are extremely useful in order to reconstruct the metabolic capabilities of the organism under investigation. Furthermore, the analysis of genomic information is important to facilitate the development of new or improved methods for growth and isolation. In the case of bifidobacteria it is well known that these microorganisms can be rather difficult to cultivate in defined media, apparently owing to their highly exigent nutrient demands.

On the basis of available bifidobacterial genome information, it is deduced that bifidobacteria have the genetic capability to synthesize at least 19 amino acids from ammonium and major biosynthetic precursors (e.g. pyruvate, fumarate, oxaloacetate and oxoglutarate) supplied by the bifid shunt (Scardovi and Trovatelli, 1965) and by the partially present tricarboxylic acid cycle (TCA), which lacks fumarase, oxoglutarate dehydrogenase and malate dehydrogenase (Schell *et al.*, 2002). Whether and how bifidobacteria can synthesize cysteine is not fully clear, as genes for the sulphate/sulphite assimilation pathway appear to be absent in *B. longum* subsp. *longum* NCC2705 (Schell *et al.*, 2002). However, enteric bifidobacteria may be able to synthesize cysteine using the homologues of cysteine synthase/cysthathione β synthase, O-acetylhomoserine aminocarboxypropyltransferase and cystathionine γ synthase starting from a reduced sulphur source (e.g. hydrogen sulphide), which may be supplied by sulphur-reducing colonic microbiota, suggesting a synergistic relationship. With respect to the metabolic capabilities of bifidobacteria to synthesize pyrimidines and purine nucleotides, analysis of various bifidobacterial genomes has indicated that such microorganisms possess all the homologues needed for the *de novo* production from glutamine. The genomes of *B. longum* subsp. *longum* NCC2705, *B. longum* subsp. *longum* DJO10A and *B. longum* subsp. *infantis* ATCC15697 are predicted to encode all necessary enzymes for the synthesis of folic acid, nicotinate and thiamin (Lee *et al.*, 2008; Schell *et al.*, 2002; Sela *et al.*, 2008). In contrast, pathways needed for the *de novo* biosynthesis of biotin, cobalamin, pantothenate, lipoate and pyridoxine appear to be partially or completely absent (Lee *et al.*, 2008; Schell *et al.*, 2002; Sela *et al.*, 2008).

Table 2.1 Bifidobacterial genome sequencing projects

Species	Status	Length (bp)	ORFs	% GC	tRNA	rRNA	Accession number	Institution
B. adolescentis ATCC 15703	*Complete*	2089645	1630	59	54	5	NC_008618.1	Gifu University, Life Science Research Centre, Japan
B. animalis subsp. *lactis* AD011	*Complete*	1933695	1528	60	52	2	NC_011835.1	Korea Research Institute of Bioscience and Biotechnology
B. bifidum 317B	*Complete*	2214650	1848	63	52	3	*Unpublished data*	University of Parma, Italy
B. breve 246B	*Complete*	2397148	2028	59	53	2	*Unpublished data*	University of Parma, Italy
B. breve UCC2003	*Complete*	–	–	–	–	–	*Unpublished data**	University College of Cork, Ireland
B. dentium Bd1	*Complete*	2636368	2143	59	55	4	CP001750	University of Parma, Italy; University College of Cork, Ireland
B. longum subsp. *infantis* ATCC 15697	*Complete*	2832748	2498	60	79	4	NC_011593.1	University of California Davis, USA
B. longum subsp. *longum* DJO 10A	*Complete*	2375792	1990	60	58	4	NC_010816.1	University of Minnesota
B. longum subsp. *longum* NCC 2705	*Complete*	2256640	1727	60	57	4	NC_004307.2	University of Minnesota
B. adolescentis L2–32	*Unfinished*	2385710	2428	59	–	–	NZ_AAXD00000000	Washington University (WashU)
B. angulatum DSM 20098	*Unfinished*	2007108	1846	59	–	–	NZ_ABYS00000000	Washington University (WashU)
B. animalis subsp. *lactis* Bb12	*Unfinished*	–	–	61	–	–	*Unpublished data*	Integrated Genomics Inc.
B. animalis subsp. *lactis* Bl–04	*Unfinished*	–	–	–	–	–	*Unpublished data*	Danisco USA Inc.
B. animalis subsp. *lactis* DSM 10140	*Unfinished*	–	–	–	–	–	*Unpublished data*	Danisco USA Inc.
B. animalis subsp. *lactis* HN019	*Unfinished*	1915892	1578	61	–	–	NZ_ABOT00000000	Fonterra Research Centre

B. animalis subsp. *lactis* V9	*Unfinished*	–	–	–	–	–	*Unpublished data*	Inner Mongolia Agricultural University, China
B. asteroides ATCC 25910	Unfinished	2152141	–	60	–	–	*Unpublished data*	University of Parma, Italy; University College of Cork, Ireland
B. bifidum BGN4	Unfinished	–	–	–	–	–	*Unpublished data*	Korea Research Institute of Bioscience and Biotechnology
B. bifidum DSM 20456	Unfinished	–	–	–	–	–	*Unpublished data*	Washington University (WashU)
B. bifidum NCIMB 41171	Unfinished	2186140	1833	63	–	–	NZ_ABQP00000000	Broad Institute
B. breve DSM 20213	Unfinished	2297799	2251	59	–	–	NZ_ACCG00000000	Washington University (WashU)
B. catenulatum DSM 16992	Unfinished	2058429	1950	56	–	–	NZ_ABXY00000000	Washington University (WashU)
B. dentium ATCC 27678	Unfinished	2642081	2430	59	–	–	NZ_ABIX00000000	Washington University (WashU)
B. dentium JCVIHMP022	Unfinished	–	–	–	–	–	*Unpublished data*	J. Craig Venter Institute
B. dentium JCVIHMP023	Unfinished	–	–	–	–	–	*Unpublished data*	J. Craig Venter Institute
B. gallicum DSM 20093	Unfinished	2016380	1983	58	–	–	NZ_ABXB00000000	Washington University (WashU)
B. longum subsp. *infantis* ATCC 55813	Unfinished	–	2109	60	–	–	NZ_ACHI00000000	Baylor College of Medicine
B. longum subsp. *infantis* CCUG 52486	Unfinished	–	2120	60	–	–	NZ_ABQQ00000000	Broad Institute
B. longum subsp. *longum* strain BORI	Unfinished	–	–	–	–	–	*Unpublished data*	Korea Research Insttitute of Bioscience and Biotechnology
B. pseudocatenulatum DSM 20438	Unfinished	–	2151	56	–	–	NZ_ABXX00000000	Washington University (WashU)

[*]Leahy *et al.* (2005)

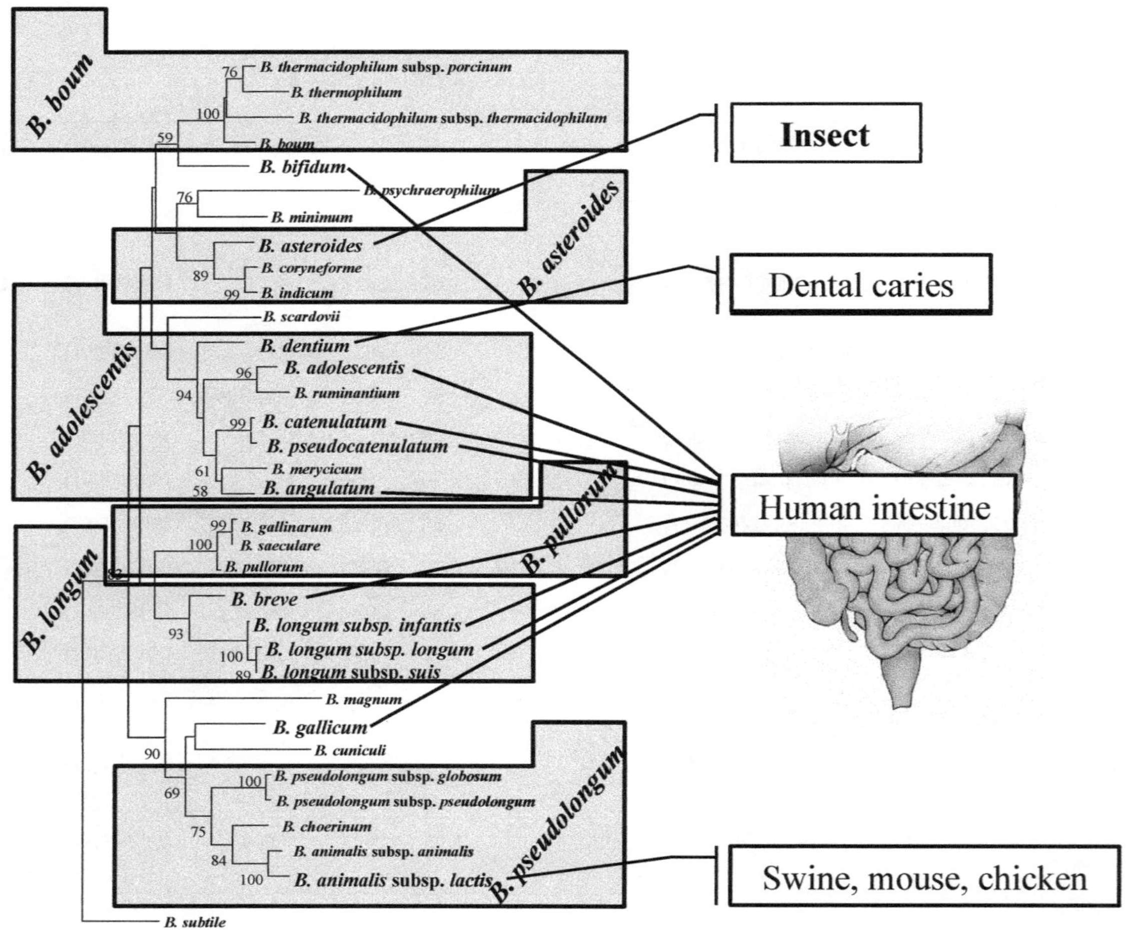

Figure 2.1 Schematic representation of the bifidobacterial species that have been targeted by genome sequencing efforts. The phylogenetic position as well as the ecological origin of each species is indicated. The various observed clusters indicate the phylogenetic groups, some of which have been previously allocated specific names (Ventura *et al.*, 2007a) as indicated on the left-hand margin of the figure.

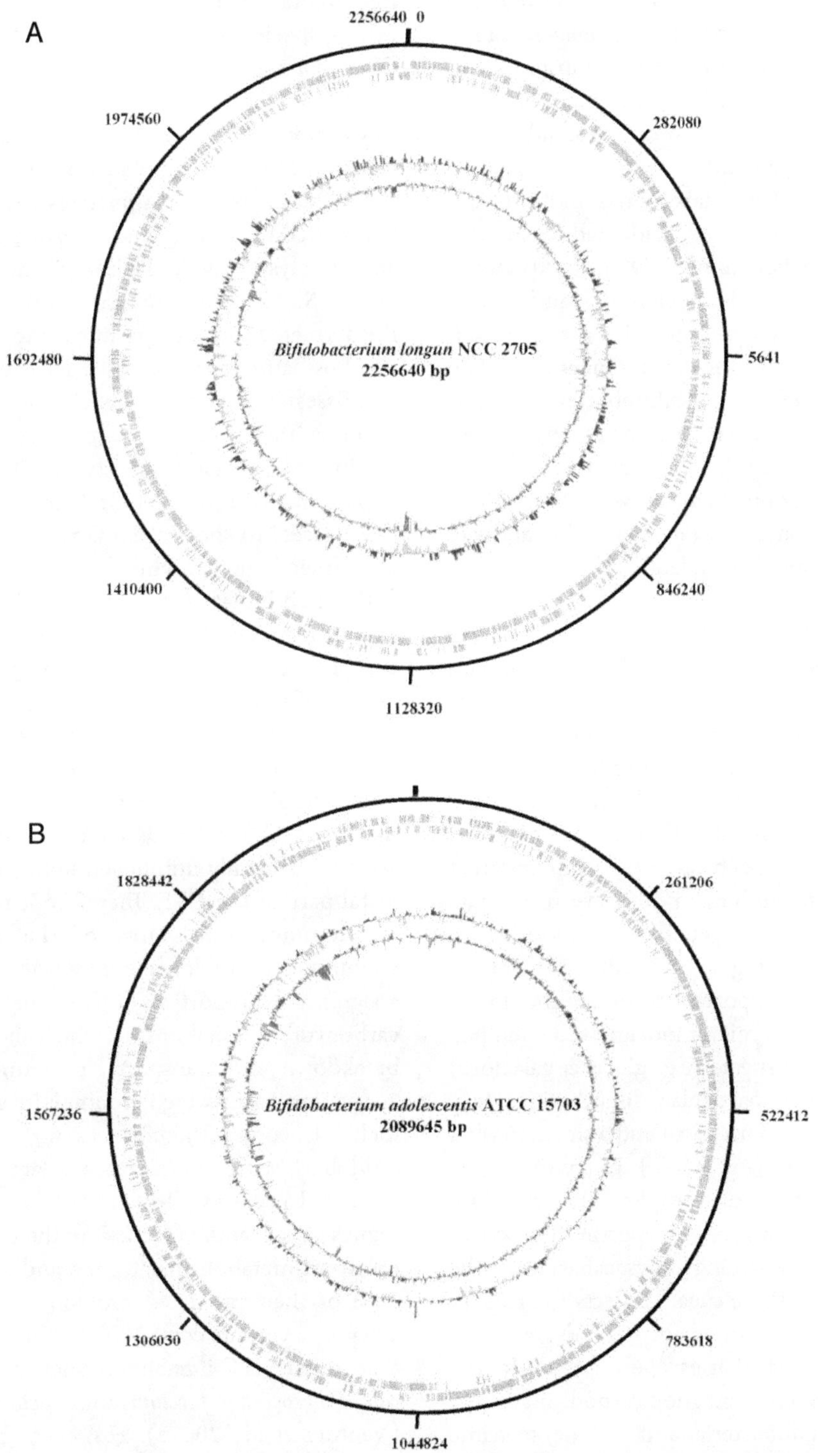

Figure 2.2 Atlas of the intestinal bifidobacteria *B. longum* subsp. *longum* NCC2705 strain (a) and *B. adolescentis* ATCC15703 (b). From inside to outside: ring 1, GC content; ring 2, G+C skew; ring 3 and 4, ORFs distribution.

Furthermore, the biosynthetic pathway required for riboflavin seems absent in *B. longum* subsp. *longum*, but present in *B. longum* subsp. *infantis* (Lee *et al.*, 2008; Schell *et al.*, 2002; Sela *et al.*, 2008), indicating that this property is subspecies or even strain dependent.

These biosynthetic capabilities indicate that *Bifidobacterium* species have adapted to an environment where they cannot rely on an extraneous source of amino acids, nucleotides and certain vitamins. Preliminary comparative genome studies complemented by the development of defined synthetic media for other bifidobacteria, such as *B. breve* UCC2003, have confirmed many of these predicted prototrophic and auxotrophic characteristics (O'Connell-Motherway M., Curley P., Forde A., Zomer A., Fitzgerald G.F. and van Sinderen D., unpublished data).

Carbohydrate utilization

Genome sequencing provides the gold standard approach for the examination of the full genetic blue print of an organism, and provides predictive evidence concerning the genetic basis encoding adaptive functions specific to the environment in which the organism naturally lives. For example, food ingredients such as carbohydrates that are available in the human gut have more than likely influenced the genomic composition of the microbiota living in this compartment. In this context, it is important to realize that mammalian hosts of the microbiota are well equipped to adsorb simple sugars (e.g. glucose, galactose) and to hydrolyse particular disaccharides (e.g. sucrose, lactose and maltose) and a small number of polysaccharides (e.g. starch). In contrast, they cannot degrade more complex, dietary fibre-derived polysaccharides (e.g. pectin, and xylan- and arabinose-containing polysaccharides). The indigestibility of these sugars reflects the paucity of host enzymes required for their degradation. Many of the sugars that escape digestion by the host's enzymes represent good carbon and energy sources for bifidobacteria, and include, to name but a few, fructooligosaccharides (FOS), galacto-oligosaccharides (GOS), gluco-oligosaccharides, xylo-oligosaccharides and raffinose (Guarner and Malagelada, 2003). Notably, these dietary compounds would have been lost if the distal intestine of mammals had not been colonized by a diverse microbiota, members of which include, among others, species of the *Bacteroides*, *Bifidobacterium*, *Clostridium* and *Eubacterium* genera (Eckburg *et al.*, 2005), who possess the metabolic capacity to degrade such oligo- and polysaccharides. The energy needs of these bacteria are met by the fermentation of these substrates that have been left undigested in the small intestine (Cummings and Englyst, 1987; Delzenne and Williams, 2002; Sinha and Kumria, 2001). Moreover, the fermentation end products can be used by the host and in some cases have been shown to possess additional beneficial properties (e.g. immune modulation, anti-atherosclerotic, anti-carcinogenic and anti-obesity activities) (Belury, 2002). In this mutualistic relationship, the host gains access to short-chain fatty acids for energy and other benefits, while bacteria are provided with a rich buffet of carbohydrates in a relatively constant environment. The aptitude to degrade complex carbohydrates, with respect to the degree of polymerization and type of glycosidic bonds, contributes to the competitiveness of a given bacterium in the gut (Turroni *et al.*, 2008). Thus, simple sugars are metabolized rapidly in the upper GIT, whereas complex carbohydrates remain essentially untouched until the reach the distal part of the GIT. Therefore it makes sense for common inhabitants of the mammalian colon, such as bifidobacterial species, to dedicate a significant proportion of its coding capacity to carbohydrate metabolism (which includes both breakdown and transport), reflecting a critical adaptation to this highly competitive ecological niche. It comes therefore as no surprise that bifidobacterial genomes have a large arsenal of predicted proteins allocated to clustered orthologues genes family (COGs) in the carbohydrate transport-metabolism category, and in fact about 10% of their predicted proteome fall into this category (Ventura *et al.*, 2007b). This is 30% more than other GIT inhabitants such as *Escherichia coli*, *Enterococcus faecium* and *Lactobacillus* spp. (Ventura *et al.*, 2007a). However, the deduced fermentative capacities of bifidobacteria are less expanded as compared to those predicted from the genome of *Bacteroides thetaiotaomicron* which is also a commensal of the GIT (Xu *et al.*, 2003).

The enzymatic arsenal encoded by a given micro organism that is predicted to be used for the degradation and utilization of carbohydrates as deduced from its genome content, is also known as the glycobiome and normally includes a variety of glycosyl hydrolases, glycosyl transferases and glycosyl lyases (Coutinho and Henrissat, 1999). The analysis of the glycobiome of bifidobacterial genomes according to the sequence-based classification of carbohydrate-active enzymes (CAZy) (Coutinho and Henrissat, 1999), revealed an extensive but variable set of glycosyl hydrolases and glycosyl transferases among bifidobacteria. Notably, the largest bifidobacterial glycobiome found so far is represented by the oral dweller *Bifidobacterium dentium* Bd1 (Ventura *et al.*, 2009b), whereas the enteric bifidobacterial genomes (e.g. *B. longum* subsp. *longum* NCC2705, *B. longum* subsp. *longum* DJO10A, *B. longum* subsp. *infantis* ATCC15697 and *B. adolescentis* ATCC15703) possess a smaller glycobiome. This suggests that these bacteria exhibit a differential behaviour relating to sugar utilization according to the different ecological niche occupied (e.g. human mouth vs. human colon). Several of these glycosyl hydrolases are assumed to exert their activity outside the cytoplasm as secreted enzymes.

Altogether, about 5% of the total bifidobacterial gene content is devoted to sugar internalization, mostly through ATP-binding cassette (ABC) transporters, permeases and proton symporters, rather than through phosphoenolpyruvate phosphotransferase systems (PTS) (Maze *et al.*, 2007). Moreover, analyses of bifidobacterial genomes allowed the identification of genes or operons specifically involved in the breakdown of complex polysaccharides, such as starch, amylopectin, glycogen and pullulan (O'Connell Motherway *et al.*, 2008; Ryan *et al.*, 2006). Also, arabinofuranosyl-containing oligosaccharides obtained from plant cell wall polysaccharides, such as arabinan, arabinogalactan and arabinoxylan, can be fermented by bifidobacteria (van Laere *et al.*, 2000). However, fermentation studies of bifidobacteria with arabinoxylan and arabinogalactan showed that the rate of degradation of these polymers is rather low compared to the oligosaccharides that can be derived from this polymer (van den Broek *et al.*, 2008). It has been argued that, despite the high number of glycosidases, other bacteria are needed for the hydrolysis of polymeric arabinoxylan and arabinogalactan into oligosaccharides (Degnan and Macfarlane, 1995; Hopkins *et al.*, 2003). Bifidobacteria are generally able to utilize isomeric lacto-N-tetraose (LNT, Gal(β1–3)GlcNac(β1–3)Gal(β1–4)Glc), which constitutes the core of the human milk oligosaccharides (HMOs). Such sugars are metabolized through the activity of a lacto-N-biose (LNB) phosphorylase and other enzymes, whose genes appear to be clustered (Nishimoto and Kitaoka, 2007). Notably, the operon responsible for LNB metabolism is conserved in the genomes of the *B. longum* group while being absent in the genome of *B. adolescentis* ATCC15703 (Ward *et al.*, 2006). Additional gene clusters predicted to metabolize HMO and derivates (e.g. sialic acid) appear to be present in the genome of the infant bifidobacterial isolate *B. longum* subsp. *infantis* ATCC15697. In this latter organism, a 43 kb gene cluster was identified which encompasses a large set of genes predicted to be involved in the metabolism of HMOs, for example genes encoding fucosidase, sialidase, β-galactosidase, β-hexosaminidase activities, which in turn are associated with genes that are predicted to encode ABC transporters specific for such HMOs.

Prebiotics and bifidobacterial genomes

A prebiotic is a food ingredient which cannot be (directly) metabolized by the host and which beneficially affects the host by promoting selective growth or metabolic activity of health-promoting commensals such as bifidobacteria (Roberfroid, 2001). Various oligosaccharides have been claimed to act as a prebiotic, such as FOS (fructooligosaccharides) and GOS (galacto oligosaccharides). FOS constitutes the most widely used commercial prebiotic, being formed of short to medium size chains of fructose moieties (degree of polymerization, 4–60) connected by β(2–1) linkages, which are in turn attached to a terminal glucose unit by a β(2–1) bond. Owing to these β(2–1) linkages FOS are resistant to mammalian enzymes and are thus able to reach the distal compartments of the GIT intact (Gibson, 1999). Inulin, a long-chain FOS, has drawn a significant amount of scientific attention owing to the fact that it is a naturally occurring polysaccharide found in many plants (e.g. chicory), representing a common component of the Mediterranean diet. Inspection of bifidobacterial genomes for the presence of genes that may be involved in prebiotic metabolism revealed a rich arsenal of such genes. For example, various bifidobacterial β-fructofuranosidases involved in FOS metabolism have been characterized, which show different substrate kinetics and biochemical properties (Ehrmann *et al.*, 2003; Imamura *et al.*, 1994; Janer *et al.*, 2004; Warchol *et al.*, 2002). Additionally, in the genome of *B. breve* UCC2003, a *fos* operon was identified, encompassing among others a gene specifying a β-fructofuranosidase, which was shown to partially degrade FOS, although this strain cannot grown on FOS (Ryan *et al.*, 2005).

Human milk is rich in soluble oligosaccharides that have been proposed to function as prebiotics (Ward *et al.*, 2006). Genome investigation of *B. longum* subsp. *infantis* ATCC15697 has identified several chromosomal loci, which are apparently responsible for the genetic adaptation to allow human milk utilization (Sela *et al.*, 2008) (see above).

Apart from HMOs, human milk peptides represent another class of diet-derived compounds, which have been associated with a prebiotic effect (Liepke *et al.*, 2002). These include small peptides, which are derived from the breakdown of milk protein with the gastric protease pepsin, and which contain a pair of cysteine residues forming a disulfide bond and two small hydrophobic domains located C-terminally to the two cysteines (Liepke *et al.*, 2002). The presence of such peptides was shown to be more effective on a molar basis in the stimulation of the growth of bifidobacteria as compared to certain HMO, and it has been speculated that the bifidogenic activity of breast milk may be predominantly due to the presence of specific peptides as opposed to the availability of HMOs (LoCascio *et al.*, 2007; Ward *et al.*, 2007).

Genomic data and host interactions

It is well know that extracytoplasmatic proteins and extracellular structures play crucial roles in the molecular interactions between a bacterium and its environment. In bifidobacteria these could mediate key functions affecting the host, like adhesion, nutrient availability, immune system modulation or pathogen inhibition/exclusion. Intriguing cell surface-encoding proteins are fimbriae-like structures, which were identified owing to the presence of their encoding genes on the genome of intestinal (Schell *et al.*, 2002) as well as oral bifidobacteria (Ventura *et al.*, 2009b). It should be noted that such structures have not yet been truly demonstrated in bifidobacteria, however, they may be of great significance owing to the fact that they are cell-surface filaments which in other microorganisms have been shown to mediate microbial adhesion to and colonization of epithelial cells of the GIT mucosa, or other host cell surfaces (Kubiet *et al.*, 2000; Li *et al.*, 1999). Furthermore, the genomes belonging to certain bifidobacterial species (e.g. *B. breve, B. longum* subsp. *longum, B. longum* subsp. *infantis*) have been shown to contain a gene, whose predicted product exhibits similarity to serine-like protease inhibitor (serpin) proteins found in other bacteria and eukaryotes. Members of the serpin family regulate various signalling pathways in eukaryotes and some are known for their ability to suppress inflammatory responses by inhibiting elastase activity (Potempa *et al.*, 1994). Recent findings have demonstrated that the bifidobacterial serpin-like protein may indeed perform such an immunomodulatory role as shown in a murine colitis model where its presence reduces intestinal inflammation (Ivanov *et al.*, 2006).

Another possibility for interactions between bifidobacteria and its host is provided by the bifidobacterial ability to produce cell surface-decorating sugar structures, known as exopolysaccharides (EPS). The role of EPS produced by non-pathogenic colonic bacteria is not clear, but it is possible that these extracellular structures are important for bacteria to establish themselves within the host (Ruas-Madiedo *et al.*, 2006). Interestingly, the human intestinal microorganism *Bact. thetaiotamicron* has developed a genetic mechanism allowing it to modify its surface make-up by synthesizing at least eight distinct EPS, which may provide the microorganism with opportunities to escape recognition by the host's immune system (Krinos *et al.*, 2001). In bifidobacteria, only a few reports have described the presence of such extracellular structures (Habu *et al.*, 1987; Ruas-Madiedo *et al.*, 2007, 2009; Salazar *et al.*, 2008).

Genome analysis of *B. longum* subsp. *longum* NCC2705 revealed the presence of two regions related to EPS biosynthesis. Interestingly, both regions are flanked by IS elements and show a strong divergence in G+C content relative to the remainder of the genome, indicating that these regions have been acquired by HGT. Similarly, analyses targeting the genomes of oral bifidobacteria such as *B. dentium* Bd1 revealed the presence of DNA regions that encode enzymes required for the synthesis of strain-specific EPS. However, analysis of the so far completed bifidobacterial genomes indicates that not all bifidobacterial genomes contain such genetic clusters. Such findings could be the consequence of a genome reduction experienced by strains after continued cultivation under extraneous ecological culture conditions (e.g. in the laboratory or in batch industrial fermentors), as has been demonstrated for *B. longum* subsp. *longum* DJO10A (Lee *et al.*, 2008).

Mobile genetic elements in bifidobacteria

The contribution of mobile elements to the strain specificity of genomic sequences is widely variable between bifidobacterial genomes. In fact, in *B. longum* subsp. *infantis* ATCC15697 (Sela *et al.*, 2008), *B. longum* subsp. *longum* NCC2705 (Schell *et al.*, 2002) and *B. longum* subsp. *longum* DJO10A (Lee *et al.*, 2008) the impact of mobile genetic elements such as insertion sequences (IS), (episomal) plasmids, and prophage-like elements have shaped their genomes to a significant extent. In contrast, genome evolution in the *B. dentium* species appears to have only moderately been affected by mobile elements (Ventura *et al.*, 2009b). The proliferation of mobile genetic elements in the *B. longum* group may be attributable in part to the presence of *rex*AB homologues, which encode enzymes that participate in double-strand break repair, thereby promoting chromosomal rearrangements and integration of heterologous sequences through recombination (Sela *et al.*, 2008). The presence of a large number of repetitive sequences such as IS elements is expected to prompt chromosomal reshuffling and increase the chance of repositioning functionally related genes into clusters and operons.

Among mobile genetic elements in bifidobacterial genomes, the presence of prophage like-elements is in particular noticeable (Ventura *et al.*, 2005). Recently, a bioinformatic screening for these mobile parasitic elements among all the so far available bifidobacterial genomes sequences allowed the identification of nineteen prophage-like elements displaying high sequence identity to genes of double stranded DNA bacteriophages spanning a broad phylogenetic range of host bacteria (*Actinobacteria* and *Firmicutes*) (Ventura *et al.*, unpublished data). Surprisingly, genome-based analyses revealed that a small number of these prophage-like elements identified in the genome of *B. dentium* Bd1 and *B. longum* subsp. *infantis* ATCC15697 displayed a close phylogenetic relationship with phages infecting low GC-content, Gram-positive bacteria (e.g. lactococcal and staphylococcal phages), perhaps because these bacteria and their phages have shared a common ecological niche during evolution, thereby allowing DNA exchange. This may therefore point to DNA transfer events between low and high G+C-content Gram-positive bacteria. Notably, although most of the prophage-like elements identified in bifidobacterial genomes appear to represent defective prophages, a small number of these prophage-like elements, i.e. Bdent-2 from *B. dentium* Bd1, Binf-4 from *B. longum* subsp. *infantis* ATCC15697, Blj-1 from *B. longum* subsp. *longum* DJO10A and Ban-1 from *B. animalis* subsp. *lactis* ADO11, were shown to be inducible (Ventura *et al.*, 2005, unpublished data) and present opportunities for further studies on the biology of these so-called bifidophages.

Concluding remarks

Many issues regarding the sequences of complete bifidobacterial genomes remain unresolved at present. A first important issue is that only a small number of genomes from very closely related bifidobacterial species have so far been sequenced. Our current set of bifidobacterial genomes therefore does not represent the true extent of bifidobacterial diversity, and, so far, such sequences have been largely biased towards organisms of biotechnological importance (Fig. 2.1). It is probable that with the accumulation of more genome data, not merely representing bifidobacterial strains with industrial interest, or isolated only from the human gut, we may shed light on the general and common metabolic capabilities of the *Bifidobacterium* genus and how these could have been affected by the different ecological niches from which the microorganism originally were isolated (e.g. the human intestine vs. the animal (be they herbivore, carnivore or omnivore] intestine, or the oral cavity vs. the lower compartment of the GIT). The sequencing of a larger number of bifidobacterial genomes will also assist in understanding the evolutionary development of this group of microorganisms. The application of genomics-based methodologies to trace phylogeny (phylogenomics) will provide a more robust image that will explain the phylogenetic relationships between the different bifidobacterial species, while it may provide answers to the question as to how the genus *Bifidobacterium* has developed as an independent taxonomic unit within the eubacteria.

Commensals such as bifidobacteria must coexist with their host and must evade or survive the diversity of responses that the host has developed to eliminate unwanted parasitic and pathogenic bacteria. Understanding the molecular mechanisms underlying the fascinating ability of the human immune system to differentiate between beneficial and harmful bacteria is a significant scientific challenge for the future.

Despite the significantly increased insights into microbial composition, bifidobacterial activity and bifidobacteria–host interactions in the GIT, these new findings also underline our limited understanding of the processes ongoing in this highly complex environmental niche. Recently, genes induced upon exposure to acid (Azcarate-Peril *et al.*, 2004) and bile salts (Bron *et al.*, 2004) have been identified in lactic acid bacteria that might confer important roles in GIT survival and/or colonization. Follow-up *in vivo* studies aimed at evaluating the survival of mutants of acid- and bile-tolerance genes, relative to wild-type strains, together with the corresponding host-cell responses, would provide a useful adjunct to understand the importance of acidic and/or bile tolerance on probiotic functionality in these bacteria. Another useful means to explore the genetic determinants involved in bacterial–host interaction is represented by whole-genome transcriptional profiling with DNA microarrays or by a whole transcriptome sequencing approach. Clinically relevant, global transcriptional profiles in response to ingestion of commensal strains are emerging for several bacterial groups (e.g. *Lactobacillus*; Di Caro *et al.*, 2005). Similar studies for bifidobacteria will be highly useful not only to understand how stable gut residence is achieved but also to assign functions to the many unknown proteins encoded by bifidobacterial genomes.

References

Azcarate-Peril, M.A., Altermann, E., Hoover-Fitzula, R.L., Cano, R.J., and Klaenhammer, T.R. (2004). Identification and inactivation of genetic loci involved with *Lactobacillus acidophilus* acid tolerance. Appl. Environ. Microbiol. *70*, 5315–5322.

Belury, M.A. (2002). Dietary conjugated linoleic acid in health: physiological effects and mechanisms of action. Annu. Rev. Nutr. *22*, 505–531.

Bron, P.A., Marco, M., Hoffer, S.M., van Mullekom, E., de Vos, W.M., and Kleerebezem, M. (2004). Genetic characterization of the bile salt response in *Lactobacillus plantarum* and analysis of responsive promoters *in vitro* and *in situ* in the gastrointestinal tract. J. Bacteriol. *186*, 7829–7835.

Cohan, F.M. (2001). Bacterial species and speciation. Syst. Biol. *50*, 513–524.

Coutinho, P.M., and Henrissat, B. (1999). Life with no sugars? J. Mol. Microbiol. Biotechnol. *1*, 307–308.

Cummings, J.H., and Englyst, H.N. (1987). Fermentation in the human large intestine and the available substrates. Am. J. Clin. Nutr. *45*, 1243–1255.

Degnan, B.A., and Macfarlane, G.T. (1995). Arabinogalactan utilization in continuous cultures of *Bifidobacterium longum*: effect of co-culture with Bacteroides thetaiotaomicron. Anaerobe *1*, 103–112.

Delzenne, N.M., and Williams, C.M. (2002). Prebiotics and lipid metabolism. Curr. Opin. Lipidol. *13*, 61–67.

Di Caro, S., Tao, H., Grillo, A., Elia, C., Gasbarrini, G., Sepulveda, A.R., and Gasbarrini, A. (2005). Effects of *Lactobacillus* GG on genes expression pattern in small bowel mucosa. Dig. Liver Dis. *37*, 320–329.

Eckburg, P.B., Bik, E.M., Bernstein, C.N., Purdom, E., Dethlefsen, L., Sargent, M., Gill, S.R., Nelson, K.E., and Relman, D.A. (2005). Diversity of the human intestinal microbial flora. Science *308*, 1635–1638.

Ehrmann, M.A., Korakli, M., and Vogel, R.F. (2003). Identification of the gene for beta-fructofuranosidase of *Bifidobacterium lactis* DSM10140^{T} and characterization of the enzyme expressed in *Escherichia coli*. Curr. Microbiol. *46*, 391–397.

Frank, A.C., and Lobry, J.R. (1999). Asymmetric substitution patterns: a review of possible underlying mutational or selective mechanisms. Gene *238*, 65–77.

Gevers, D., Vandepoele, K., Simillon, C., and van de Peer, Y. (2004). Gene duplication and biased functional retention of paralogs in bacterial genomes. Trends Microbiol. *12*, 148–154.

Gibson, G.R. (1999). Dietary modulation of the human gut microflora using the prebiotics oligofructose and inulin. J. Nutr. *129*, 1438S–1441S.

Gogarten, J.P., Doolittle, W.F., and Lawrence, J.G. (2002). Prokaryotic evolution in light of gene transfer. Mol. Biol. Evol. *19*, 2226–2238.

Guarner, F., and Malagelada, J.R. (2003). Gut flora in health and disease. Lancet *361*, 512–519.

Habu, Y., Nagaoka, M., Yokokura, T., and Azuma, I. (1987). Structural studies of cell wall polysaccharides from *Bifidobacterium breve* YIT 4010 and related *Bifidobacterium* species. J. Biochem. *102*, 1423–1432.

Hopkins, M.J., Englyst, H.N., Macfarlane, S., Furrie, E., Macfarlane, G.T., and McBain, A.J. (2003). Degradation of cross-linked and non-cross-linked arabinoxylans by the intestinal microbiota in children. Appl. Environ. Microbiol. *69*, 6354–6360.

Imamura, L., Hisamitsu, K., and Kobashi, K. (1994). Purification and characterization of beta-fructofuranosidase from *Bifidobacterium infantis*. Biol. Pharm. Bull. *17*, 596–602.

Ivanov, D., Emonet, C., Foata, F., Affolter, M., Delley, M., Fisseha, M., Blum-Sperisen, S., Kochhar, S., and Arigoni, F. (2006). A serpin from the gut bacterium *Bifidobacterium longum* inhibits eukaryotic elastase-like serine proteases. J. Biol. Chem. *281*, 17246–17252.

Janer, C., Rohr, L.M., Peláez, C., Laloi, M., Cleusix, V., Requena, T., and Meile, L. (2004). Hydrolysis of oligofructoses by the recombinant beta-fructofuranosidase from *Bifidobacterium lactis*. Syst. Appl. Microbiol. *27*, 279–285.

Korbel, J.O., Snel, B., Huynen, M.A., and Bork, P. (2002). SHOT: a web server for the construction of genome phylogenies. Trends Genet. *18*, 158–162.

Krinos, C.M., Coyne, M.J., Weinacht, K.G., Tzianabos, A.O., Kasper, D.L., and Comstock, L.E. (2001). Extensive surface diversity of a commensal microorganism by multiple DNA inversions. Nature *414*, 555–558.

Kubiet, M., Ramphal, R., Weber, A., and Smith, A. (2000). Pilus-mediated adherence of *Haemophilus influenzae* to human respiratory mucins. Infect. Immun. *68*, 3362–3367.

Kunin, V., and Ouzounis, C.A. (2003). The balance of driving forces during genome evolution in prokaryotes. Genome Res. *13*, 1589–1594.

Kunisawa, T. (1995). Identification and chromosomal distribution of DNA sequence segments conserved since divergence of *Escherichia coli* and *Bacillus subtilis*. J. Mol. Evol. *40*, 585–593.

Leahy, S.C., Higgins, D.G., Fitzgerald, G.F., and van Sinderen, D. (2005). Getting better with bifidobacteria. J. Appl. Microbiol. *98*, 1303–1315.

Lee, J.H., Karamychev, V.N., Kozyavkin, S.A., Mills, D., Pavlov, A.R., Pavlova, N.V., Polouchine, N.N., Richardson, P.M., Shakhova, V.V., Slesarev, A.I., *et al.* (2008). Comparative genomic analysis of the gut bacterium *Bifidobacterium longum* reveals loci susceptible to deletion during pure culture growth. BMC Genomics *9*, 247.

Li, T., Johansson, I., Hay, D.I., and Stromberg, N. (1999). Strains of *Actinomyces naeslundii* and *Actinomyces viscosus* exhibit structurally variant fimbrial subunit proteins and bind to different peptide motifs in salivary proteins. Infect. Immun. *67*, 2053–2059.

Liepke, C., Adermann, K., Raida, M., Magert, H.J., Forssmann, W.G., and Zucht, H.D. (2002). Human milk provides peptides highly stimulating the growth of bifidobacteria. Eur. J. Biochem. *269*, 712–718.

LoCascio, R.G., Ninonuevo, M.R., Freeman, S.L., Sela, D.A., Grimm, R., Lebrilla, C.B., Mills, D.A., and German, J.B. (2007). Glycoprofiling of bifidobacterial consumption of human milk oligosaccharides demonstrates strain specific, preferential consumption of small chain glycans secreted in early human lactation. J. Agric. Food Chem. *55*, 8914–8919.

Maze, A., O'Connell-Motherway, M., Fitzgerald, G.F., Deutscher, J., and van Sinderen, D. (2007). Identification and characterization of a fructose phosphotransferase system in *Bifidobacterium breve* UCC2003. Appl. Environ. Microbiol. *73*, 545–553.

McLean, M.J., Wolfe, K.H., and Devine, K.M. (1998). Base composition skews, replication orientation, and gene orientation in 12 prokaryote genomes. J. Mol. Evol. *47*, 691–696.

Nishimoto, M., and Kitaoka, M. (2007). Identification of N-acetylhexosamine 1-kinase in the complete lacto-N-biose I/galacto-N-biose metabolic pathway in *Bifidobacterium longum*. Appl. Environ. Microbiol. *73*, 6444–6449.

O'Connell Motherway, M., Fitzgerald, G.F., Neirynck, S., Ryan, S., Steidler, L., and van Sinderen, D. (2008). Characterization of ApuB, an extracellular type II amylopullulanase from *Bifidobacterium breve* UCC2003. Appl. Environ. Microbiol. *74*, 6271–6279.

Philippe, H., and Douady, C.J. (2003). Horizontal gene transfer and phylogenetics. Curr. Opin. Microbiol. *6*, 498–505.

Potempa, J., Korzus, E., and Travis, J. (1994). The serpin superfamily of proteinase inhibitors: structure, function, and regulation. J. Biol. Chem. *269*, 15957–15960.

Roberfroid, M.B. (2001). Prebiotics: preferential substrates for specific germs? Am. J. Clin. Nutr. *73*, 406S-409S.

Rocha, E.P. (2003). DNA repeats lead to the accelerated loss of gene order in bacteria. Trends Genet. *19*, 600–603.

Ruas-Madiedo, P., Gueimonde, M., Arigoni, F., de los Reyes-Gavilán, C.G., and Margolles, A. (2009). Bile affects the synthesis of exopolysaccharides by *Bifidobacterium animalis*. Appl. Environ. Microbiol. *75*, 1204–1207.

Ruas-Madiedo, P., Gueimonde, M., de los Reyes-Gavilán, C.G., and Salminen, S. (2006). Short communication: effect of exopolysaccharide isolated from 'viili' on the adhesion of probiotics and pathogens to intestinal mucus. J. Dairy Sci. *89*, 2355–2358.

Ruas-Madiedo, P., Moreno, J.A., Salazar, N., Delgado, S., Mayo, B., Margolles, A., and de Los Reyes-Gavilán, C.G. (2007). Screening of exopolysaccharide-producing *Lactobacillus* and *Bifidobacterium* strains isolated from the human intestinal microbiota. Appl. Environ. Microbiol. *73*, 4385–4388.

Ryan, S.M., Fitzgerald, G.F., and van Sinderen, D. (2005). Transcriptional regulation and characterization of a novel beta-fructofuranosidase-encoding gene from *Bifidobacterium breve* UCC2003. Appl. Environ. Microbiol. *71*, 3475–3482.

Ryan, S.M., Fitzgerald, G.F., and van Sinderen, D. (2006). Screening for and identification of starch-, amylopectin-, and pullulan-degrading activities in bifidobacterial strains. Appl. Environ. Microbiol. *72*, 5289–5296.

Salazar, N., Gueimonde, M., Hernandez-Barranco, A.M., Ruas-Madiedo, P., and de los Reyes-Gavilán, C.G. (2008). Exopolysaccharides produced by intestinal *Bifidobacterium* strains act as fermentable substrates for human intestinal bacteria. Appl. Environ. Microbiol. *74*, 4737–4745.

Schell, M.A., Karmirantzou, M., Snel, B., Vilanova, D., Berger, B., Pessi, G., Zwahlen, M.C., Desiere, F., Bork, P., Delley, M., *et al.* (2002). The genome sequence of *Bifidobacterium longum* reflects its adaptation to the human gastrointestinal tract. Proc. Natl. Acad. Sci. U.S.A *99*, 14422–14427.

Sela, D.A., Chapman, J., Adeuya, A., Kim, J.H., Chen, F., Whitehead, T.R., Lapidus, A., Rokhsar, D.S., Lebrilla, C.B., German, J.B., *et al.* (2008). The genome sequence of *Bifidobacterium longum* subsp. *infantis* reveals adaptations for milk utilization within the infant microbiome. Proc. Natl. Acad. Sci. U.S.A *105*, 18964–18969.

Sinha, V.R., and Kumria, R. (2001). Polysaccharides in colon-specific drug delivery. Int. J. Pharm. *224*, 19–38.

Snel, B., Bork, P., and Huynen, M.A. (2002). Genomes in flux: the evolution of archaeal and proteobacterial gene content. Genome Res. *12*, 17–25.

Tillier, E.R., and Collins, R.A. (2000). Genome rearrangement by replication-directed translocation. Nat. Genet. *26*, 195–197.

Turroni, F., Ribbera, A., Foroni, E., van Sinderen, D., and Ventura, M. (2008). Human gut microbiota and bifidobacteria: from composition to functionality. Antonie van Leeuwenhoek *94*, 35–50.

van den Broek, L.A., Hinz, S.W., Beldman, G., Vincken, J.P., and Voragen, A.G. (2008). *Bifidobacterium* carbohydrases-their role in breakdown and synthesis of (potential) prebiotics. Mol. Nutr. Food Res. *52*, 146–163.

van Laere, K.M., Hartemink, R., Bosveld, M., Schols, H.A., and Voragen, A.G. (2000). Fermentation of plant cell wall derived polysaccharides and their corresponding oligosaccharides by intestinal bacteria. J. Agric. Food Chem. *48*, 1644–1652.

Ventura, M., Canchaya, C., Bernini, V., Del Casale, A., Dellaglio, F., Neviani, E., Fitzgerald, G.F., and van Sinderen, D. (2005). Genetic characterization of the *Bifidobacterium breve* UCC 2003 *hrcA* locus. Appl. Environ. Microbiol. *71*, 8998–9007.

Ventura, M., Canchaya, C., Fitzgerald, G.F., Gupta, R.S., and van Sinderen, D. (2007a). Genomics as a means to understand bacterial phylogeny and ecological adaptation: the case of bifidobacteria. Antonie van Leeuwenhoek *91*, 351–372.

Ventura, M., Canchaya, C., Tauch, A., Chandra, G., Fitzgerald, G.F., Chater, K.F., and van Sinderen, D. (2007b). Genomics of *Actinobacteria*: tracing the evolutionary history of an ancient phylum. Microbiol. Mol. Biol. Rev. *71*, 495–548.

Ventura, M., O'Flaherty, S., Claesson, M.J., Turroni, F., Klaenhammer, T.R., van Sinderen, D., and O'Toole, P.W. (2009a). Genome-scale analyses of health-promoting bacteria: probiogenomics. Nat. Rev. Microbiol. 7, 61–71.

Ventura, M., Turroni, F., Zomer, A., Foroni, E., Giubellini, V., Canchaya, C., Bottachini, F., Claesson, M.J., Zhang, Z., Mantzourani, M., *et al.* (2009b). The *Bifidobacterium dentium* Bd1 genome sequence reflects its genetic adaptation to the human oral cavity. Plos. Genetics *12*, el000785.

Warchol, M., Perrin, S., Grill, J.P., and Schneider, F. (2002). Characterization of a purified beta-fructofuranosidase from *Bifidobacterium infantis* ATCC 15697. Lett. Appl. Microbiol. 35, 462–467.

Ward, R.E., Ninonuevo, M., Mills, D.A., Lebrilla, C.B., and German, J.B. (2006). *In vitro* fermentation of breast milk oligosaccharides by *Bifidobacterium infantis* and *Lactobacillus gasseri*. Appl. Environ. Microbiol. *72*, 4497–4499.

Ward, R.E., Ninonuevo, M., Mills, D.A., Lebrilla, C.B., and German, J.B. (2007). *In vitro* fermentability of human milk oligosaccharides by several strains of bifidobacteria. Mol. Nutr. Food Res. *51*, 1398–1405.

Woese, C.R. (2000). Interpreting the universal phylogenetic tree. Proc. Natl. Acad. Sci. U.S.A *97*, 8392–8396.

Wolf, Y.I., Rogozin, I.B., Kondrashov, A.S., and Koonin, E.V. (2001). Genome alignment, evolution of prokaryotic genome organization, and prediction of gene function using genomic context. Genome Res. *11*, 356–372.

Xu, J., Bjursell, M.K., Himrod, J., Deng, S., Carmichael, L.K., Chiang, H.C., Hooper, L.V., and Gordon, J.I. (2003). A genomic view of the human-*Bacteroides thetaiotaomicron* symbiosis. Science *299*, 2074–2076.

Bifidobacterial Ecology and Comparative Genomics: Perspectives and Prospects

3

Marco Ventura, Francesca Turroni, Francesca Bottacini, Vanessa Giubellini and Douwe van Sinderen

Abstract
The gastrointestinal tract (GIT) of humans and other mammals are colonized by a vast number of bacterial cells, including microorganisms such as bifidobacteria, which have been attributed an important role in maintaining a healthy GIT. However, despite this generally accepted beneficial role of bifidobacteria, the underlying molecular mechanisms by which these commensal bacteria exert this health-promoting or probiotic activity is far from understood. Recent genome sequencing has provided detailed insights into the genetic make-up of some members of the genus *Bifidobacterium*, although it represents only the first step in moving towards a better understanding of the biology and metabolic capabilities of these bacteria. Furthermore, comparative analyses of these genomes may increase our understanding of the physiological and metabolic behaviour of these microorganisms, in particular with regards to the way in which bifidobacteria have adapted to the intestinal environment.

History and general features

Bifidobacteria were first isolated by Tissier in 1899 from faeces of a breast-fed infant and originally termed *Bacillus bifidus* (Tissier, 1900). Only later this particular bacterium was reclassified as *Bifidobacterium bifidum* by Orla Jensen. However, because their morphological and physiological features are to a large degree similar to those of lactobacilli, they were classified as members of the genus *Lactobacillus* during a large part of the twentieth century.

Bifidobacteria are commensals that play a key role in human–microbe interactions in the gastrointestinal tract (GIT) of new-borns, and are believed to play an important role in maintaining a healthy GIT. For the latter reason bifidobacteria have become economically important and they are added in high numbers as live bacteria to numerous so-called functional food preparations carrying various health-related claims. However, despite the generally accepted importance of bifidobacteria in GIT well-being, the underlying molecular mechanisms by which these commensal bacteria function as health-promoting or probiotic microorganisms is far from understood. Recent genome sequencing has unravelled the genetic make-up of some members of the genus *Bifidobacterium*, but the availability of the full genomic sequence of complete bifidobacterial sequences merely represents the first step in moving towards a better understanding of the biology and metabolic capabilities of these bacteria.

Ecological distribution of bifidobacteria

Bifidobacteria are Gram-positive microorganisms with a high GC-content genome, and belong to the *Actinobacteria* phylum. In particular, the order *Bifidobacteriales* constitutes a distinct taxon between this class and includes the family *Bifidobacteriaceae* which comprises the following six genera: *Aeriscardovia, Alloscardovia, Bifidobacterium, Gardnerella, Parascardovia* and *Scardovia,* (Masco *et al.*, 2007; Stackebrandt *et al.*, 1997). Members of the genus *Aeriscardovia* are found in the pig caecum (Simpson *et al.*, 2004), whereas representatives of all other genera are commonly found in the human body. In particular, members of the genera *Scardovia* and *Parascardovia* are commonly located in the human oral cavity (Jian and Dong, 2002), those belonging to *Alloscardovia* are isolated from various human clinical samples (e.g. urine, blood, urethra, oral cavity, tonsil and abscesses of lung and aortic valve) (Masco *et al.*, 2007), while those of the genus *Gardnerella* constitute part of the normal vaginal flora (Aroutcheva *et al.*, 2001; Jian *et al.*, 2001). The genus *Bifidobacterium* comprises 32 species (two of them are recently discovered, i.e. *Bifidobacterium tsurumiense* and *Bifidobacterium mongoliense* (Okamoto *et al.*, 2008; Watanabe *et al.*, 2009), with four species further subdivided into subspecies. The different species exhibit more than 93% identity across the length of their 16S rRNA-encoding sequences (Miyake *et al.*, 1998), and are encountered in six ecological niches, including the human intestine, oral cavity, insect and animal intestine, sewage, blood and food. All these niches are directly or indirectly associated with the human/animal intestinal environment, while the last three may be considered the result of contaminations from their original natural source (Ventura *et al.*, 2008). Taking their different ecological niches into account and combining this information with a comparative analysis of their 16S rRNA sequences and other conserved genes, the currently known *Bifidobacterium* species can be clustered into six different groups: *Bifidobacterium adolescentis, Bifidobacterium asteroides, Bifidobacterium boum, Bifidobaterium longum, Bifidobacterium pullorum,* and *Bifidobacterium pseudolongum* (Ventura *et al.*, 2009a) (Fig 3.1). Interestingly, *B. asteroides*, isolated from the insect gut, is assumed to represent the species that is most closely related to the ancestral progenitor of the *Bifidobacterium* genus (Ventura *et al.*, 2009a).

The combined data on ecological origin and genomic sequences represents an important starting point for understanding the physiological and metabolic behaviour of microorganisms. In fact, it is widely accepted that niche-derived selective pressure is the driving force for improving the ecological fitness that allows an organism to survive and compete in a particular environment. In the case of bifidobacteria, the genomic era has provided valuable information about how these microorganisms are genetically adapted to colonize the mammalian gut (Ventura *et al.*, 2007a). Because complex carbohydrates, being derived from the diet as well as from host cells (e.g. mucin), are thought to represent the major carbon sources in the human colon, the acquired and genetically embedded capacity to metabolize such carbohydrates constitutes a critical adaptation of bifidobacteria to this highly competitive ecological niche. The currently available bifidobacterial genomes have revealed a very extensive array of genes encoding glycosyl hydrolases as well as glycosyl transferases, which collectively form the so-called glycobiome of a given microorganism.

An ecological study of bifidobacterial populations associated with animal faeces has revealed a wide distribution of bifidobacteria in the gut of a wide variety of animals (e.g. birds, ungulates, lagomorphs and rodents) (Lamendella *et al.*, 2008). According to Lamendella *et al.* (2008) various bifidobacteria, such as *Bifidobacterium animalis, B. adolescentis, Bifidobacterium dentium* and *Bifidobacterium catenulatum*, possess a cosmopolitan lifestyle, whereas other bifidobacterial species display a highly specialized ecological adaptation to a particular animal gut (e.g. *Bifidobacterium cuniculi* found in rabbits, *Bifidobacterium angulatum* in cows and *Bifidobacterium gallinarum* in chicken). Such a host-specific ecological adaptation has also been noticed for certain human-derived bifidobacteria (Ruiz *et al.*, 2009).

Recently, the distribution of bifidobacteria in the gut of human subjects of varying age showed that certain bifidobacterial species were exclusively present as part of the adult human gut microbiota (i.e. *B. pseudolongum*) whereas others were found to be widely distributed irrespective of host age (i.e. *B. longum, Bifidobacterium breve, Bifidobacterium pseudocatenulatum* and *B. adolescentis*) (Turroni *et al.*, 2009).

Bifidobacteria living in the human intestine are subject to growing interest as a result of their health-promoting or so-called probiotic features. These enteric microorganisms, which are characterized as saccharoclastic, are able to hydrolyse a large variety of oligosaccharides, many of which are not digested by their human host. Such substrates can improve the number and/or activity of

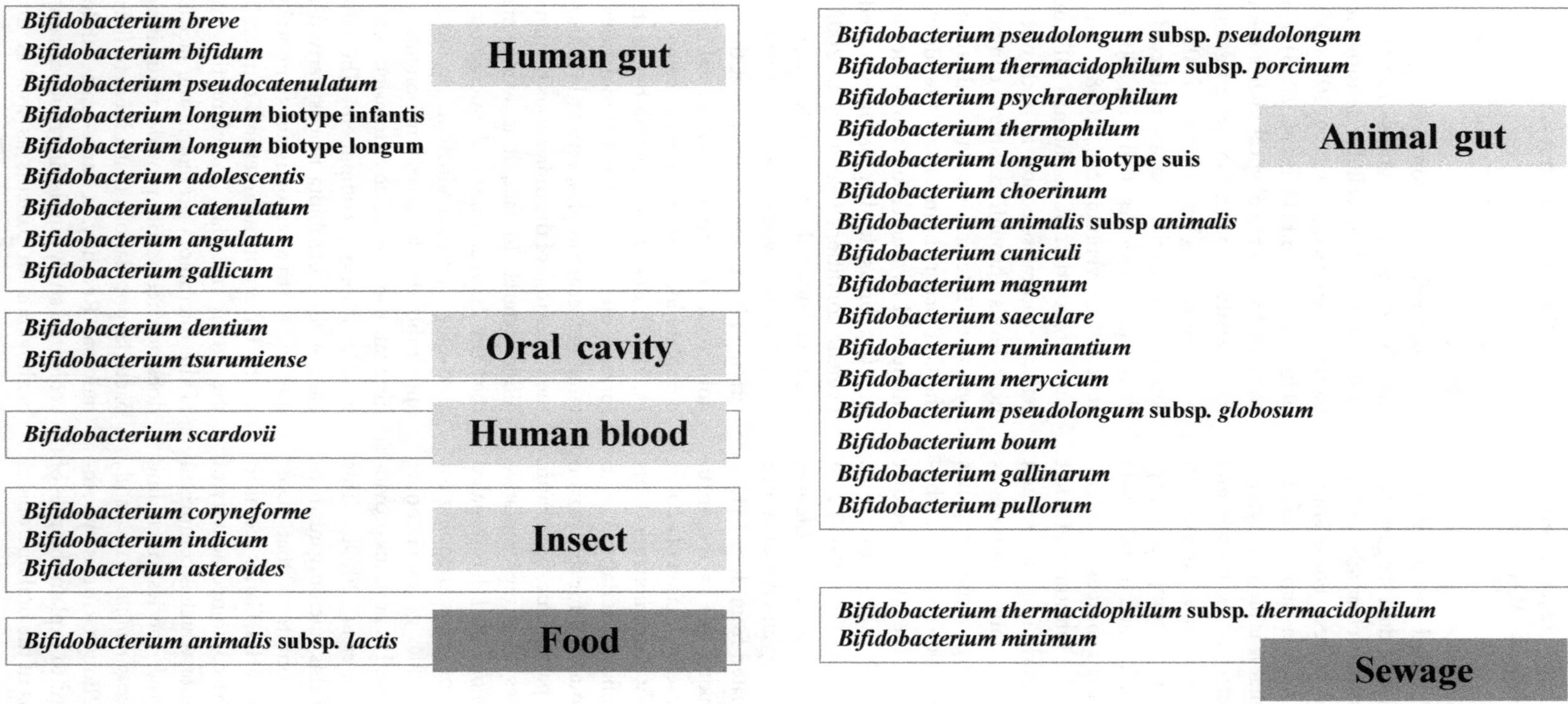

Figure 3.1 Ecological distribution of bifidobacterial species in different ecological niches.

probiotic enteric microorganisms in the GIT, and some of these carbohydrates are commercially exploited as so-called prebiotics (Vernazza *et al.*, 2006).

How these different ecological specializations are genetically determined is still not fully understood, although genomic data suggests that bifidobacteria may possess particular adaptations to explain their ecological success. For example, genome analysis of *B. longum* subsp. *infantis* ATCC15697 has shown that it is an archetypical human milk-utilizing bacterium as its genome features genes encoding enzymes involved in the breakdown of human milk component (e.g. human milk oligosaccharides and urea) (Sela *et al.*, 2008). In contrast, the genomes of enteric bifidobacteria residing in the intestines of adult humans do not appear to harbour genes related to the utilization of human milk components and they instead contain a large arsenal of genes dedicated to the metabolism of complex carbohydrates commonly found in the adult-type diet (in particular plant-derived fibres) (Ventura *et al.*, 2007a,b). Among the bifidobacteria residing in the mammalian GIT, only one species, *B. dentium*, resides in the human oral cavity, and this organism has in particular been isolated from deep lesions found in tooth caries, linking this bacterium to the development of this disease (Aas *et al.*, 2008; Becker *et al.*, 2002). Notably, *B. dentium* represents the only so far described bifidobacterial species to display cariogenic behaviour, and for this reason should be considered a pathogenic bacterium. Recently, genome analyses of a representative strain of this species, *B. dentium* Bd1, has revealed the presence of large sets of genes encoding for putative virulence factors, like genes encoding adhesins and proteases, as well as proteins providing acid tolerance (Ventura *et al.*, 2009b). This suggests an adaptation of this microorganism to tolerate the harsh acidic conditions that occur during the development of caries. Caries seems to be the final outcome of the combined activity of the overall plaque microbiota, which is comprised of many different bacterial species, including *B. dentium* and *Streptococcus mutans* (Aas *et al.*, 2008; Theilade, 1986). Although various hypotheses have been put forward to explain the role of individual microorganisms of the oral microbiota in caries development, the precise mechanisms causing this disease are not yet fully understood.

Bifidobacteria and the human gut microbiota

The bacterial population residing in the human gut, also known as the gut microbiota, represents one of the highest cellular densities found in a natural ecosystem, reaching 10^{11}–10^{12} cells/ml of luminal content (Gill *et al.*, 2006). However, in contrast to other ecological niches the microbial biodiversity found in the human gut microbiota is relatively low (Ventura *et al.*, 2008), where the dominant bacterial taxa are restricted to *Bacteroidetes* and *Firmicutes* (Gill *et al.*, 2006; Rezzonico *et al.*, 2003; Ventura *et al.*, 2008). The structure and composition of the human GIT microbiota reflects natural selection at both microbial and host levels through complex, competitive and symbiotic interactions. Bacterial colonization of the GIT occurs immediately after delivery and is influenced by numerous factors including infant diet (e.g. breast feeding or formula feeding) and hygiene conditions (Fanaro *et al.*, 2003). One of the main bacterial colonizers is represented by the bifidobacteria, whose numbers decline with age, although it has been claimed that the native bifidobacterial microbiota remains stable during the complete life span of a human being (Ventura *et al.*, 2008). Besides a large number of studies performed on bifidobacterial populations encountered in faecal samples, very little is known about the real extent of the indigenous human intestinal bifidobacteria. In fact, it is worth mentioning that the human intestinal microbiota includes a so-called mucosa-adherent, or indigenous or autochthonous, bacterial component, as well as a luminal, also referred to as transient, contaminant or allochthonous component. The latter is the part of the microbiota that is derived from the diet and for that reason is just temporally present in the lumen of the intestine, thus ending up in the faecal material. Therefore the exploration of the microbiota present in the faecal material does not necessarily provide a true image of the bacterial populations that adhere to the intestinal mucosa. Recently, a large-scale bifidobacterial ecological study employing mucosa samples from healthy human patients as well as from faecal samples was performed (Turroni *et al.*, 2009). In this study a polyphasic approach involving isolation of bifidobacteria using selective media was coupled to molecular analysis of the rRNA gene sequences [16S rRNA gene and Internally transcribed spacer (ITS) 16S–23S spacer sequences]. Notably, the culturable bifidobacterial population from intestinal as well as faecal samples include six main phylogenetic taxa, i.e. *B. longum*, *B. pseudocatenulatum*, *Bifidobacterium adolescentis*, *B. pseudolongum*, *B. breve* and *B. bifidum*, and two species mostly detected in faecal samples, i.e. *B. dentium* and *B. animalis* subsp. *lactis*. These results suggest that the latter two species do not belong to the human intestinal mucosa-adherent bifidobacterial population. Finally, these analyses revealed a significant intersubject variability and compositional differences between faecal and mucosa-adherent bifidobacterial communities. In contrast, a modest diversification of bifidobacterial populations was noticed between different intestinal regions within the same individual (intrasubject variability) (Turroni *et al.*, 2009).

Recent approaches, that were directed to explore the biodiversity of the intestinal microbiota such as metagenomic and microbiomic analyses, have shown that the GIT microbiota consists of many novel and, as yet, unculturable bacterial components (Gill *et al.*, 2006; Kurokawa *et al.*, 2007; Palmer *et al.*, 2007; Wang *et al.*, 2005). Surprisingly, relatively few bifidobacterial rRNA gene sequences were identified in the above mentioned metagenomic studies and it should be pointed out that the results achieved in these ecological investigations may have been skewed by (i) the efficacy of the protocols used for extracting DNA directly from the environmental samples (e.g. faecal or human tissues), (ii) the accuracy of the PCR primers used, as well as (iii) the bias introduced by the applied PCR conditions. In fact, the low level of detection of 16S rRNA sequences belonging to bifidobacteria may be explained by the fact that bifidobacteria, being high G+C Gram-positive bacteria, are less susceptible to cell lysis and that it is therefore more difficult to isolate chromosomal DNA from these bacteria as compared to Gram-negative bacteria (Hopkins *et al.*, 2005; Kauffmann *et al.*, 2004; Ventura *et al.*, 2008). Recently, the molecular complexity and biodiversity of the human mucosa-adherent bifidobacterial population was assessed through a diversity analysis of bifidobacteria-specific 16S rRNA-encoding sequences (Turroni *et al.*, 2009). Surprisingly, 13% of the total sequences (representing nearly 1300 clone-derived sequences) acquired in the above mentioned study had not been previously identified and some of these sequences showed greater than 3% sequence divergence from publicly available bifidobacterial 16S rRNA gene sequences, suggesting that these sequences represent novel bifidobacterial species (Turroni *et al.*, 2009).

The impact of genomics on bifidobacterial taxonomy

In the last decade prokaryotic taxonomy has benefited from advances in microbial genomics and ecology, and in particular from the ease with which sequence data can be obtained. Bacterial species are currently characterized using a polyphasic approach that incorporates genotypic and phenotypic (including chemotaxonomic) properties (Stackebrandt *et al.*, 1997; Vandamme *et al.*, 1996). Since the 1970s, DNA–DNA hybridization methods have been considered to be the cornerstone of bacterial taxonomy and at present a bacterial species is 'a category that circumscribes a genomically coherent group of individual isolates/strains sharing many unconnected features, comparatively tested under highly standardized conditions' (Coenye *et al.*, 2005). In this context, the genetic similarity between bacteria is evaluated by the extent with which their genomes hybridize under well defined conditions. Bacteria that display a DNA–DNA hybridization value of more than 70%, while exhibiting less than 5% difference in their melting temperature, are considered to belong to the same species. Conversely, isolates, sharing a level of genomic DNA hybridization that is less than 50%, do not belong to the same species. Comparative sequence analysis of the 16S rRNA gene is commonly used to determine the phylogenetic position of novel isolates. Strains that show over 97% of 16S rRNA sequence identity are considered to belong to the same species, which is in most cases consistent with a minimal genomic hybridization level of 70% (Vandamme *et al.*, 1996).

As suggested by the ad-hoc committee for the re-evaluation of the species definition in bacteriology (Stackebrandt *et al.*, 1997), the introduction of novel taxonomic tools is needed to provide increased sensitivity, consistency and robustness to the taxonomy of prokaryotes. Among these novel tools, a particularly interesting development is the analysis of complete genome sequences. In fact, with the availability of complete genome sequences it has become possible to reconstruct phylogenies on the basis of a much larger data set per species, allowing a more reliable and representative inference of the tree of life.

In recent years, bifidobacterial taxonomy has advanced through the introduction of comparative sequence analysis of the 16S rRNA gene (Matsuki *et al.*, 1998, 1999; Ventura *et al.*, 2001a,b, 2003b; Ventura and Zink, 2002), or housekeeping genes such as *rec*A (Kullen *et al.*, 1997; Ventura and Zink, 2003), *tuf*A (Ventura *et al.*, 2003a), *atp*D (Ventura *et al.*, 2004), *gro*EL (Jian *et al.*, 2001; Ventura *et al.*, 2004), *dna*K (Ventura *et al.*, 2005c), *grp*E (Ventura *et al.*, 2005c), *clp*P (Ventura *et al.*, 2005b), and *hrc*A (Ventura *et al.*, 2005a). However, several studies raised concerns that single-gene trees may not adequately reconstruct the true extent of phylogenetic relationships that exist between bacteria. This could be caused by horizontal gene transfer (HGT), anomalous mutation rates and variable rates of recombination (Ventura *et al.*, 2007a). In contrast, a phylogenetic tree based on a combined alignment of a set of conserved orthologous proteins (so-called supertrees (Bininda-Emonds, 2004)] is currently considered to produce the most reliable representation of the evolutionary relationships between bacteria (Brown *et al.*, 2001).

When such an approach was applied to analyse the evolutionary history of the genus *Bifidobacterium*, six different phylogenetic clusters, i.e. *B. longum*, *B. adolescentis*, *B. pullorum*, *B. pseudolongum*, *B. boum* and *B. asteroides*, were identified (Ventura *et al.*, 2009a), which substantiates the findings based on 16S rRNA classification. Noticeably, such a phylogenetic analysis of known members of the genus *Bifidobacterium*, as based on the concatenation of six gene sequences, i.e. *clp*C, *xfp*, *dna*J1, *rpo*C, *dna*B, *dna*G and *pur*F, generated a phylogenetic tree of high discriminatory power and robustness (Ventura *et al.*, 2006b). Moreover, this analysis provided genome-based evidence that *B. asteroides* is most closely related to the projected ancestor of all currently recognized bifidobacterial groups (Fig. 3.2). Such finding was not previously identified by the 16S rRNA sequence analysis, suggesting the high discriminatory power of this method.

Genome variability in bifidobacteria

Analysis of bacterial genome data provides useful information for the understanding of processes underlying speciation and evolution in prokaryotes. In fact, the increasing number of available bacterial genome sequences has enabled comparative genome analyses to precisely define genetic differences among such fully sequenced microorganisms. However, genome sequencing is still a laborious and expensive technique. In contrast, DNA microarray technology provides a useful adjunct to current tools for assessment of differences and changes in the bacterial genetic content. Hybridization experiments by means of a DNA array based on a reference genome sequence and using chromosomal DNA isolated from a test strain provides information on genome variability. This technique is also known as Comparative Genomic Hybridization (CGH) or DNA genomotyping (Clewley, 2002). So far, DNA microarrays have been used to compare interspecific and intraspecific variations in bacteria at the genomic level for various pathogenic bacteria, such as *Mycobacterium tuberculosis*, *Mycobacterium jejuni* (Mahairas *et al.*, 1996), *Vibrio cholerae*, *Staphylococcus aureus* (Fitzgerald *et al.*, 2001), *Salmonella enterica* (Ventura *et al.*, 2007b), *Streptococcus pyogenes* (Smoot *et al.*, 2002), *Xylella fastidiosa* (Nunes *et al.*, 2003), *Escherichia coli* (Dobrindt *et al.*, 2003; Fraser *et al.*, 1995), and *Bacillus anthracis*, as well as non-pathogenic or opportunistic pathogens like members belonging to the genus *Sewanella*, *Lactobacillus johnsonii* (Schell *et al.*, 2002) and *Pseudomonas aeruginosa* (Wolfgang *et al.*, 2003).

When CGH was used to assess the genome variation within the enteric *B. longum* species, using a *B. longum* NCC2705-based spotted DNA microarray, a large genetic variability at intraspecific and interspecific level was noticed (Rezzonico *et al.*, 2003). Such CGH experiments depicted seven large genome regions of variability

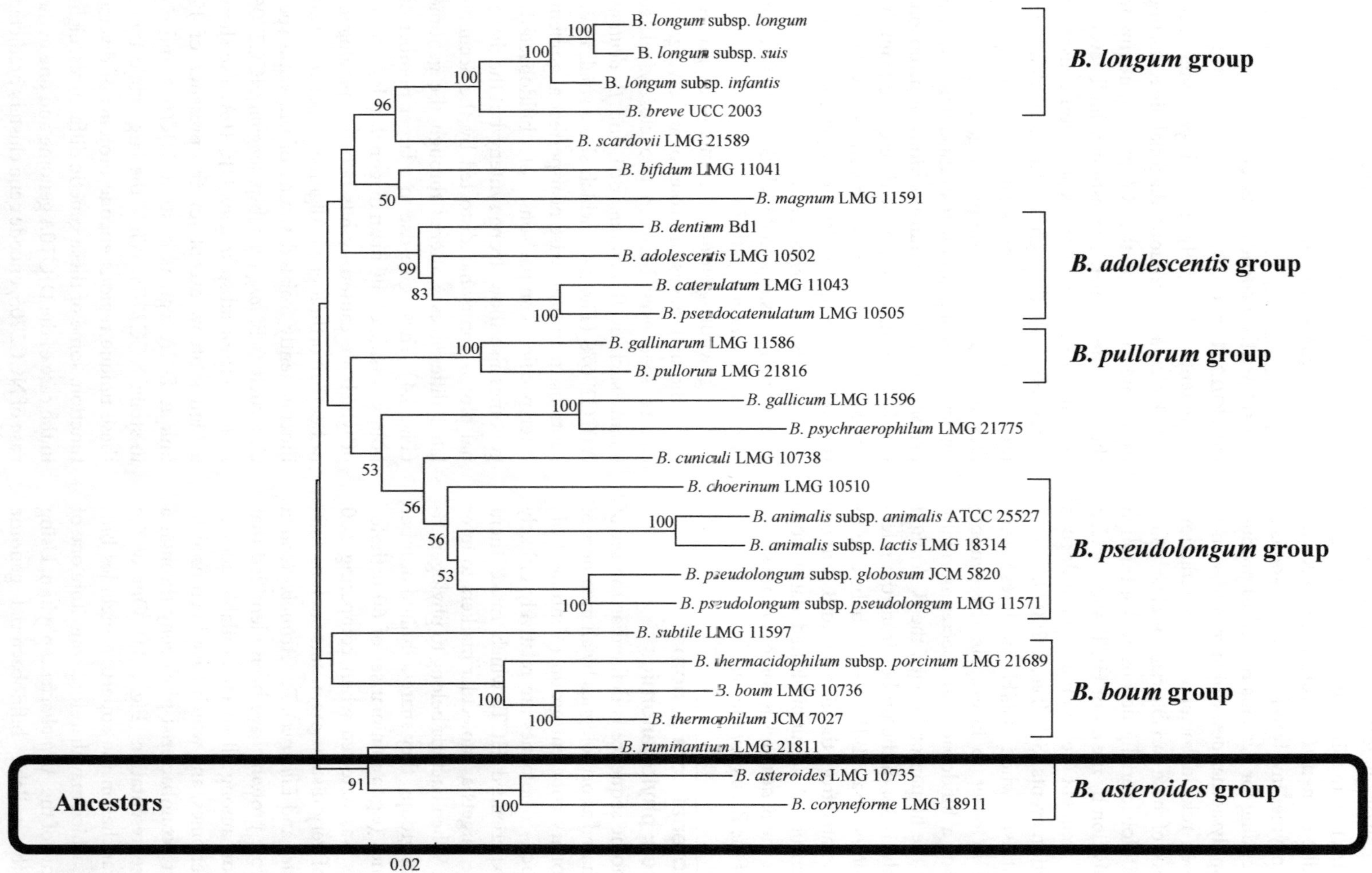

Figure 3.2 Phylogenetic tree of the genus *Bifidobacterium* constructed from the concatenation of *clp*C, *dna*B, *dna*G, *dna*J1, *pur*F, *rpo*C and *xfp* gene sequences. The phylogenetic groups are highlighted as well as the species that may represent the ancestor of the genus.

often associated with deviating G+C content relative to the average G+C content of the genome. Sequence analysis indicates that these regions include restriction-modification systems, the Bl-1 prophage like-element (Ventura *et al.*, 2006), a cluster of enzymes involved in sugar metabolism, and genes coding for proteins related to production of exopolysaccharides, which may play an as yet unknown role in bacteria–host interaction. The presence of these variable regions clearly suggest that bifidobacterial genomes evolve through DNA acquisition by means of HGT, and are also shaped by frequent deletion and homologous recombination events (Schell *et al.*, 2002).

In contrast, when CGH analyses were performed in order to investigate the genome variability of a bifidobacterial species normally residing in the human oral cavity, the *B. dentium* species, only 1–3% of the probes from the reference *B. dentium* strain Bd1 failed to hybridize to the DNA isolated from the nine tested *B. dentium* strains (Ventura *et al.*, unpublished data), suggesting that the *B. dentium* genome is more stable than that of the *B. longum* species.

Bifidobacterial core genome sequences: phylogenomics

A core genome sequence is defined as the set of all genes shared as orthologues by all members of an evolutionary coherent group (Ventura *et al.*, 2007b). Such orthologues are relatively unlikely to have experienced HGT, which make them appropriate targets as molecular markers to infer phylogeny. When bacteria belong to phylogenetically distant groups, they share a limited number of core genes (e.g. a minimal set of 60 orthologous genes was detected when comparing 100 genomes of very distant organisms such as *Bacteria*, *Archaea* and *Eukarya*). For *Actinobacteria*, a set of 74 core proteins have been identified that are common among all so far available actinobacterial genomes, and which have been used to reconstruct actinobacterial phylogeny through a supertree representation (Fig. 3.3a). One of the deepest branches in the supertree includes bifidobacteria, which may therefore constitute one of the ancestors of the *Actinobacteria* phylum. Using all currently available bifidobacterial genome sequences, it was possible to identify 696 orthologous proteins that have been used to analyse the speciation process in the genus *Bifidobacterium* (Fig. 3.3b). The evolutionary development of bifidobacteria using this phylogenomics approach is consistent with that derived using the 16S rRNA gene sequences (Ventura *et al.*, 2007b).

Comparative genomics of bifidobacteria

Alignments of closely related genomes usually depict a continuous diagonal line starting from one corner of the XY plot figure up to the opposite corner, suggesting a high level of synteny of the conserved genes. Non-conserved genes, obtained through insertions or deletions, lead to gaps in these plots rather than to signals. Deviations from the simple diagonal plot may reflect events of genomic rearrangements, such as genome deletions or transpositions or inversions that have occurred since the divergence from the last common ancestor that was shared by the two strains being compared. The most obvious of such deviations is the formation of an X-shaped plot, suggesting inversions that include the replication origin (Ventura *et al.*, 2007b).

Using whole-genome alignments at protein level, the first stages in evolutionary divergence can be seen even between two closely related strains such as *B. dentium* Bd1 and *B. dentium* ATCC27678 (Fig. 3.4), which show small interruptions in an otherwise near-perfect alignment. Interspecific comparisons of bifidobacterial genomes also show strong synteny, but the diagonal plot is somewhat distorted by the accumulated differences scattered through the genome (Fig. 3.4). These give rise to a larger number of points outside of the main diagonal, at least some of which are caused by hits between paralogues rather than orthologues. Alignments between two independently isolated strains of the same species, such as *B. longum* subsp. *longum* NCC2705 and *B. longum* subsp. *longum* DJO10A, also show a high synteny, except for the presence of 17 unique DNA regions in strain DJO10A and six in strain NCC2705. While the large majority of these unique regions encode proteins of diverse function, some of these genetic differences, eight in the case of the DJO10A genome and one in the case of NCC2705, encode gene clusters predicted to be involved in traits related to the human intestinal environment, such as oligosaccharide

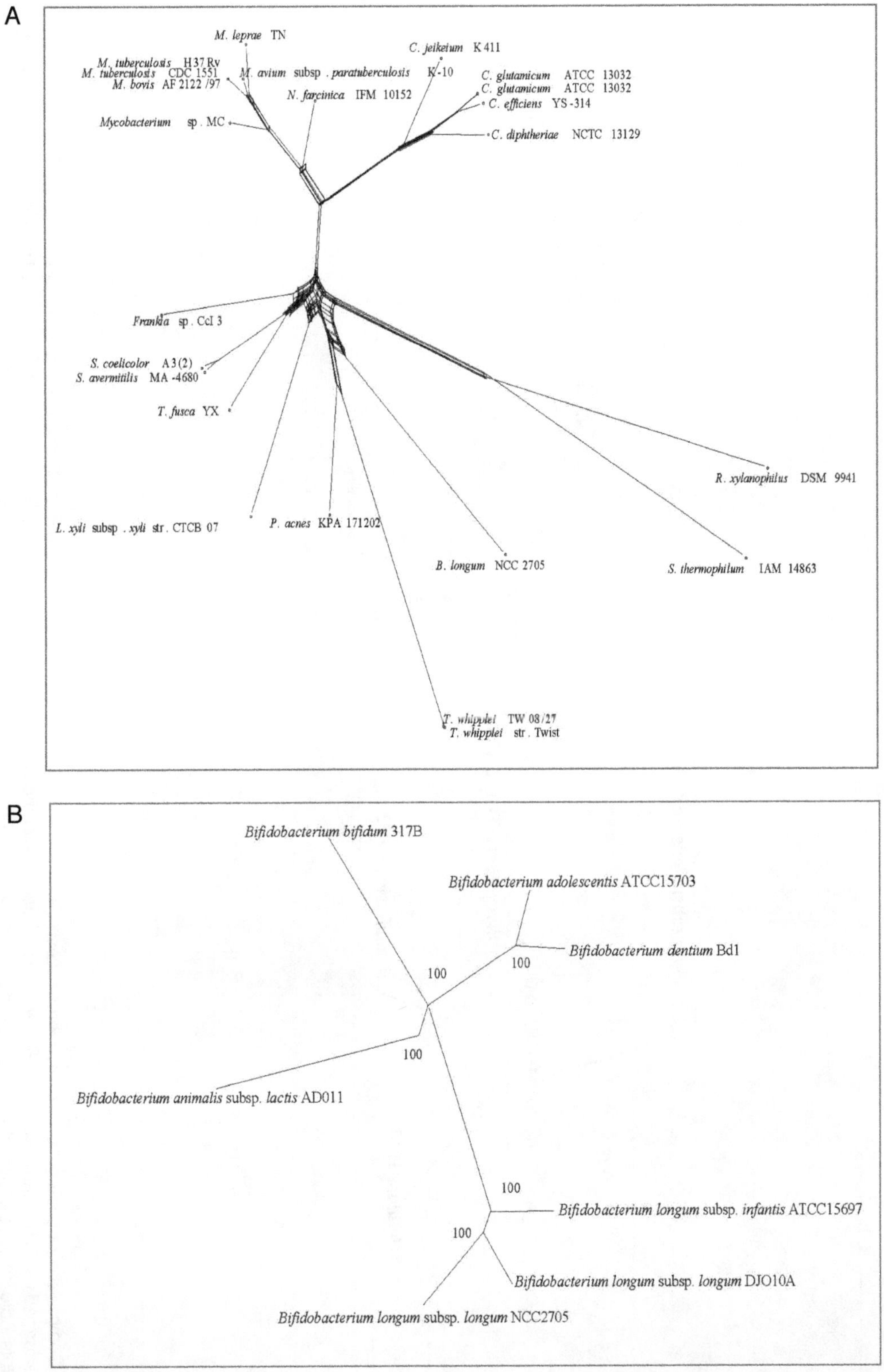

Figure 3.3 Phylogenomic trees of the *Actinobacteria* phylum (a) and of representative species of the genus *Bifidobacterium* (b). Phylogenomic trees were computed based on the minimal core gene sequences of *Actinobacteria* and *Bifidobacterium*.

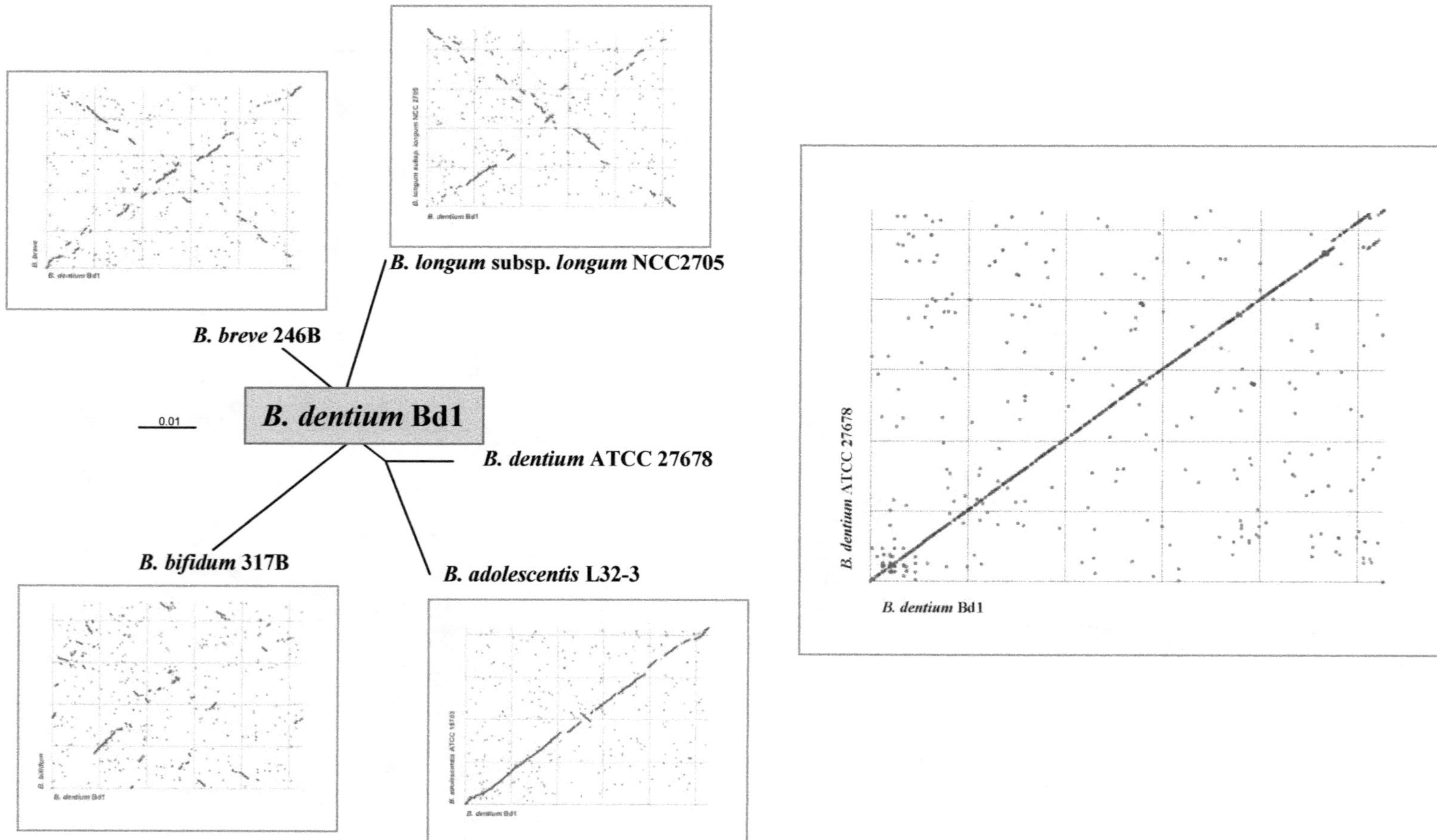

Figure 3.4 Comparison of gene order conservation between two genome pairs, illustrating different forms of *Bifidobacterium* genome evolution. The representation has been performed using the chromosome of *B. dentium* Bd1 as the reference genome in each comparison. X and Y axes represent the linearized chromosomes. Grey and black dots indicate pairs of homologous genes that are in the same orientation in both genomes, or in the opposite orientation relative to each other, respectively. The right most part of the figure shows the comparison of *B. dentium* Bd1 with *B. dentium* ATCC27678.

and polyol utilization, or related to resistance to xenobiotics (such as arsenic resistance) or competition against other bacteria (such as lantibiotic production) (Lee *et al.*, 2008). Moreover, there is an apparent genome reduction in strain NCC2705, consistent with previous observations for bacteria residing in a stable environment without HGT, in which redundant genes accumulate mutations before a subsequent deletion (Nilsson *et al.*, 2005). Finally, the comparison between the DJO10A and NCC2705 genomes revealed 248 unique sequences larger than 10 bp with the majority of them being short and encoding few if any genes (Lee *et al.*, 2008). However, these unique sequences are not clustered around the *ori*C and *ter*C, which in other bacterial genomes were shown to be the regions with the highest intraspecies genome variation (Bron *et al.*, 2006; Schell *et al.*, 2002).

Genome-wide comparison between phylogenetic distant bifidobacterial strains such as *B. dentium* Bd1 with *B. longum* subsp. *longum* as well as *B. breve* UCC2003 produced defined alignments (Fig. 3.4), which display an X-shaped distribution.

Comparative genome analyses of two recently sequenced *B. animalis* subsp. *lactis* strains, i.e. *B. animalis* subsp. *lactis* DSM10140 and *B. animalis* subsp. *lactis* Bl-04, revealed a nearly perfect alignment. In fact, only 47 short nucleotide sequence differences were detected between these strains, 39 of which were present within coding regions and eight within intergenic regions (Barrangou *et al.*, 2009).

Comparative genome analysis of bifidobacteria at intraspecific level revealed a very different core and pangenome, i.e. core genome plus unique genes, between the different *Bifidobacterium* species. In the case of the *B. dentium* and *B. animalis* subsp. *lactis* species the core and pangenome sequences appear to be practically identical, whereas in the *B. longum* species the number of unique genes are much more abundant, suggesting that the latter species has a greater reservoir of genes that are important to ensure survival and efficient adaptation to diverse ecological niches.

Assignment of annotated bifidobacterial ORFs into Clustered Orthologue Gene (COG) categories reveals a high overall conservation of COG categories and associated frequencies between different *Bifidobacterium* phylogenetic groups (e.g. *B. longum*, *B. adolescentis* and *B. animalis*) (Barrangou *et al.*, 2009). These include housekeeping functions, such as transcription and translation, but in particular those involved in carbohydrate metabolism and amino acid metabolism (Barrangou *et al.*, 2009). This reinforces the notion that carbohydrate metabolism has to a large degree shaped the bifidobacterial genome (Ventura *et al.*, 2007a).

Conclusions

Access to a large set of bifidobacterial genome sequences has significantly improved our understanding of the evolutionary path followed by this group of bacteria. Moreover, genomic and comparative genomic analyses of bifidobacteria have revealed key genes of this group of microorganisms (e.g. carbohydrate metabolism-encoding genes) that are worthy of continued investigation for their potential roles in colonization of the human gut. In this context, in depth ecological investigations using novel culture independent approaches coupled with the understanding of the metabolic capabilities and nutrient requirements of bifidobacteria will surely allow the identification of new bifidobacterial species from the human gut. An important issue is represented by the future impact of the availability of a constantly expanding number of bifidobacterial genome sequences on bifidobacterial taxonomy and bifidobacterial evolution. Looking at this from a broad perspective, it can be expected that the availability of sequenced genomes will facilitate the development of a universal genome sequence analysis scheme, which will allow the adoption of a more natural species concept.

References

Aas, J.A., Griffen, A.L., Dardis, S.R., Lee, A.M., Olsen, I., Dewhirst, F.E., Leys, E.J., and Paster, B.J. (2008). Bacteria of dental caries in primary and permanent teeth in children and young adults. J. Clin. Microbiol. *46*, 1407–1417.

Aroutcheva, A.A., Simoes, J.A., Behbakht, K., and Faro, S. (2001). *Gardnerella vaginalis* isolated from patients with bacterial vaginosis and from patients with healthy vaginal ecosystems. Clin. Infect. Dis. 33, 1022–1027.

Barrangou, R., Briczinski, E.P., Traeger, L.L., Loquasto, J.R., Richards, M., Horvath, P., Coute-Monvoisin, A.C., Leyer, G., Rendulic, S., Steele, J.L., *et al.* (2009). Comparison of the complete genome sequences of

Bifidobacterium animalis subsp. *lactis* DSM 10140 and Bl-04. J. Bacteriol. *191*, 4144–4151.

Becker, M.R., Paster, B.J., Leys, E.J., Moeschberger, M.L., Kenyon, S.G., Galvin, J.L., Boches, S.K., Dewhirst, F.E., and Griffen, A.L. (2002). Molecular analysis of bacterial species associated with childhood caries. J. Clin. Microbiol. *40*, 1001–1009.

Bininda-Emonds, O.R.P. (2004). The evolution of supertrees. Trends Ecol. Evol. *19*, 315–322.

Bron, P.A., Molenaar, D., de Vos, W.M., and Kleerebezem, M. (2006). DNA micro-array-based identification of bile-responsive genes in *Lactobacillus plantarum*. J. Appl. Microbiol. *100*, 728–738.

Brown, J.R., Douady, C.J., Italia, M.J., Marshall, W.E., and Stanhope, M.J. (2001). Universal trees based on large combined protein sequence data sets. Nat. Genet. *28*, 281–285.

Clewley, J.P. (2002). Genomotyping: comparative bacterial genomics using arrays. Commun. Dis. Public Health *5*, 258–259.

Coenye, T., Gevers, D., Van de Peer, Y., Vandamme, P., and Swings, J. (2005). Towards a prokaryotic genomic taxonomy. FEMS Microbiol. Rev. *29*, 147–167.

Dobrindt, U., Agerer, F., Michaelis, K., Janka, A., Buchrieser, C., Samuelson, M., Svanborg, C., Gottschalk, G., Karch, H., and Hacker, J. (2003). Analysis of genome plasticity in pathogenic and commensal *Escherichia coli* isolates by use of DNA arrays. J. Bacteriol. *185*, 1831–1840.

Fanaro, S., Chierici, R., Guerrini, P., and Vigi, V. (2003). Intestinal microflora in early infancy: composition and development. Acta Paediatr. Suppl. *91*, 48–55.

Fitzgerald, J.R., Sturdevant, D.E., Mackie, S.M., Gill, S.R., and Musser, J.M. (2001). Evolutionary genomics of *Staphylococcus aureus*: insights into the origin of methicillin-resistant strains and the toxic shock syndrome epidemic. Proc. Natl. Acad. Sci. U.S.A *98*, 8821–8826.

Fraser, C.M., Gocayne, J.D., White, O., Adams, M.D., Clayton, R.A., Fleischmann, R.D., Bult, C.J., Kerlavage, A.R., Sutton, G., Kelley, J.M., *et al.* (1995). The minimal gene complement of *Mycoplasma genitalium*. Science *270*, 397–403.

Gill, S.R., Pop, M., Deboy, R.T., Eckburg, P.B., Turnbaugh, P.J., Samuel, B.S., Gordon, J.I., Relman, D.A., Fraser-Liggett, C.M., and Nelson, K.E. (2006). Metagenomic analysis of the human distal gut microbiome. Science *312*, 1355–1359.

Hopkins, M.J., Macfarlane, G.T., Furrie, E., Fite, A., and Macfarlane, S. (2005). Characterisation of intestinal bacteria in infant stools using real-time PCR and northern hybridisation analyses. FEMS Microbiol. Ecol. *54*, 77–85.

Jian, W., and Dong, X. (2002). Transfer of *Bifidobacterium inopinatum* and *Bifidobacterium denticolens* to *Scardovia inopinata* gen. nov., comb. nov., and *Parascardovia denticolens* gen. nov., comb. nov., respectively. Int. J. Syst. Evol. Microbiol. *52*, 809–812.

Jian, W., Zhu, L., and Dong, X. (2001). New approach to phylogenetic analysis of the genus *Bifidobacterium* based on partial *HSP60* gene sequences. Int. J. Syst. Evol. Microbiol. *51*, 1633–1638.

Kauffmann, I.M., Schmitt, J., and Schmid, R.D. (2004). DNA isolation from soil samples for cloning in different hosts. Appl. Microbiol. Biotechnol. *64*, 665–670.

Kullen, M.J., Brady, L.J., and O'Sullivan, D.J. (1997). Evaluation of using a short region of the *recA* gene for rapid and sensitive speciation of dominant bifidobacteria in the human large intestine. FEMS Microbiol. Lett. *154*, 377–383.

Kurokawa, K., Itoh, T., Kuwahara, T., Oshima, K., Toh, H., Toyoda, A., Takami, H., Morita, H., Sharma, V.K., Srivastava, T.P., et al. (2007). Comparative metagenomics revealed commonly enriched gene sets in human gut microbiomes. DNA Res. *14*, 169–181.

Lamendella, R., Santo Domingo, J.W., Kelty, C., and Oerther, D.B. (2008). Bifidobacteria in feces and environmental waters. Appl. Environ. Microbiol. *74*, 575–584.

Lee, J.H., Karamychev, V.N., Kozyavkin, S.A., Mills, D., Pavlov, A.R., Pavlova, N.V., Polouchine, N.N., Richardson, P.M., Shakhova, V.V., Slesarev, A.I., et al. (2008). Comparative genomic analysis of the gut bacterium *Bifidobacterium longum* reveals loci susceptible to deletion during pure culture growth. BMC Genomics *9*, 247.

Mahairas, G.G., Sabo, P.J., Hickey, M.J., Singh, D.C., and Stover, C.K. (1996). Molecular analysis of genetic differences between *Mycobacterium bovis* BCG and virulent *M. bovis*. J. Bacteriol. *178*, 1274–1282.

Masco, L., Crockaert, C., Van Hoorde, K., Swings, J., and Huys, G. (2007). *In vitro* assessment of the gastrointestinal transit tolerance of taxonomic reference strains from human origin and probiotic product isolates of *Bifidobacterium*. J. Dairy Sci. *90*, 3572–3578.

Matsuki, T., Watanabe, K., Tanaka, R., Fukuda, M., and Oyaizu, H. (1999). Distribution of bifidobacterial species in human intestinal microflora examined with 16S rRNA-gene-targeted species-specific primers. Appl. Environ. Microbiol. *65*, 4506–4512.

Matsuki, T., Watanabe, K., Tanaka, R., and Oyaizu, H. (1998). Rapid identification of human intestinal bifidobacteria by 16S rRNA-targeted species- and group-specific primers. FEMS Microbiol. Lett. *167*, 113–121.

Miyake, T., Watanabe, K., Watanabe, T., and Oyaizu, H. (1998). Phylogenetic analysis of the genus *Bifidobacterium* and related genera based on 16S rDNA sequences. Microbiol. Immunol. *42*, 661–667.

Nilsson, A.I., Koskiniemi, S., Eriksson, S., Kugelberg, E., Hinton, J.C., and Andersson, D.I. (2005). Bacterial genome size reduction by experimental evolution. Proc. Natl. Acad. Sci. U.S.A *102*, 12112–12116.

Nunes, L.R., Rosato, Y.B., Muto, N.H., Yanai, G.M., da Silva, V.S., Leite, D.B., Goncalves, E.R., de Souza, A.A., Coletta-Filho, H.D., Machado, M.A., *et al.* (2003). Microarray analyses of *Xylella fastidiosa* provide evidence of coordinated transcription control of laterally transferred elements. Genome Res. *13*, 570–578.

Okamoto, M., Benno, Y., Leung, K.P., and Maeda, N. (2008). *Bifidobacterium tsurumiense* sp. nov., from hamster dental plaque. Int. J. Syst. Evol. Microbiol. *58*, 144–148.

Palmer, C., Bik, E.M., DiGiulio, D.B., Relman, D.A., and Brown, P.O. (2007). Development of the human infant intestinal microbiota. PLoS Biol. *5*, e177.

Rezzonico E.V.M., Cuonoud, G., Pessi, G., Giliberti, G., and Arigoni, F. (2003). DNA-array based analysis of genome variation and gene expression in bifidobacteria, 4th Conference on Functional Genomics of Gram-positive Microorganisms, Baveno, Italy.

Ruiz, L., Sanchez, B., de Los Reyes-Gavilan, C.G., Gueimonde, M., and Margolles, A. (2009). Coculture of *Bifidobacterium longum* and *Bifidobacterium breve* alters their protein expression profiles and enzymatic activities. Int. J. Food Microbiol. *133*, 148–153.

Schell, M.A., Karmirantzou, M., Snel, B., Vilanova, D., Berger, B., Pessi, G., Zwahlen, M.C., Desiere, F., Bork, P., Delley, M., *et al.* (2002). The genome sequence of *Bifidobacterium longum* reflects its adaptation to the human gastrointestinal tract. Proc. Natl. Acad. Sci. U.S.A *99*, 14422–14427.

Sela, D.A., Chapman, J., Adeuya, A., Kim, J.H., Chen, F., Whitehead, T.R., Lapidus, A., Rokhsar, D.S., Lebrilla, C.B., German, J.B., *et al.* (2008). The genome sequence of *Bifidobacterium longum* subsp. *infantis* reveals adaptations for milk utilization within the infant microbiome. Proc. Natl. Acad. Sci. U.S.A *105*, 18964–18969.

Simpson, P.J., Ross, R.P., Fitzgerald, G.F., and Stanton, C. (2004). *Bifidobacterium psychraerophilum* sp. nov. and *Aeriscardovia aeriphila* gen. nov., sp. nov., isolated from a porcine caecum. Int. J. Syst. Evol. Microbiol. *54*, 401–406.

Smoot, J.C., Barbian, K.D., Van Gompel, J.J., Smoot, L.M., Chaussee, M.S., Sylva, G.L., Sturdevant, D.E., Ricklefs, S.M., Porcella, S.F., Parkins, L.D., *et al.* (2002). Genome sequence and comparative microarray analysis of serotype M18 group A *Streptococcus* strains associated with acute rheumatic fever outbreaks. Proc. Natl. Acad. Sci. U.S.A *99*, 4668–4673.

Stackebrandt, E., Sproer, C., Rainey, F.A., Burghardt, J., Pauker, O., and Hippe, H. (1997). Phylogenetic analysis of the genus *Desulfotomaculum*: evidence for the misclassification of *Desulfotomaculum guttoideum* and description of *Desulfotomaculum orientis* as *Desulfosporosinus orientis* gen. nov., comb. nov. Int. J. Syst. Bacteriol. *47*, 1134–1139.

Theilade, E. (1986). The non-specific theory in microbial etiology of inflammatory periodontal diseases. J. Clin. Periodontol. *13*, 905–911.

Tissier, H. (1900). Recherchers sur la flora intestinale normale et pathologique du nourisson, Thesis, University of Paris, Paris, France.

Turroni, F., Foroni, E., Pizzetti, P., Giubellini, V., Ribbera, A., Merusi, P., Cagnasso, P., Bizzarri, B., de'Angelis, G.L., Shanahan, F., *et al.* (2009). Exploring the diversity of the bifidobacterial population in the human intestinal tract. Appl. Environ. Microbiol. *75*, 1534–1545.

Vandamme, P., Pot, B., Gillis, M., de Vos, P., Kersters, K., and Swings, J. (1996). Polyphasic taxonomy, a consensus approach to bacterial systematics. Microbiol. Rev. *60*, 407–438.

Ventura, M., Elli, M., Reniero, R., and Zink, R. (2001a). Molecular microbial analysis of *Bifidobacterium* isolates from different environments by the species-specific amplified ribosomal DNA restriction analysis (ARDRA). FEMS Microbiol. Ecol. *36*, 113–121.

Ventura, M., Reniero, R., and Zink, R. (2001b). Specific identification and targeted characterization of *Bifidobacterium lactis* from different environmental isolates by a combined multiplex-PCR approach. Appl. Environ. Microbiol. *67*, 2760–2765.

Ventura, M., and Zink, R. (2002). Rapid identification, differentiation, and proposed new taxonomic classification of *Bifidobacterium lactis*. Appl. Environ. Microbiol. *68*, 6429–6434.

Ventura, M., and Zink, R. (2003). Comparative sequence analysis of the tuf and recA genes and restriction fragment length polymorphism of the internal transcribed spacer region sequences supply additional tools for discriminating *Bifidobacterium lactis* from *Bifidobacterium animalis*. Appl. Environ. Microbiol. *69*, 7517–7522.

Ventura, M., Canchaya, C., Meylan, V., Klaenhammer, T.R., and Zink, R. (2003a). Analysis, characterization, and loci of the *tuf* genes in *Lactobacillus* and *Bifidobacterium* species and their direct application for species identification. Appl. Environ. Microbiol. *69*, 6908–6922.

Ventura, M., Meylan, V., and Zink, R. (2003b). Identification and tracing of *Bifidobacterium* species by use of enterobacterial repetitive intergenic consensus sequences. Appl. Environ. Microbiol. *69*, 4296–4301.

Ventura, M., Canchaya, C., Zink, R., Fitzgerald, G.F., and van Sinderen, D. (2004). Characterization of the *groEL* and *groES* loci in *Bifidobacterium breve* UCC 2003: genetic, transcriptional, and phylogenetic analyses. Appl. Environ. Microbiol. *70*, 6197–6209.

Ventura, M., Canchaya, C., Bernini, V., Del Casale, A., Dellaglio, F., Neviani, E., Fitzgerald, G.F., and van Sinderen, D. (2005a). Genetic characterization of the *Bifidobacterium breve* UCC 2003 *hrcA* locus. Appl. Environ. Microbiol. *71*, 8998–9007.

Ventura, M., Kenny, J.G., Zhang, Z., Fitzgerald, G.F., and van Sinderen, D. (2005b). The *clpB* gene of *Bifidobacterium breve* UCC 2003: transcriptional analysis and first insights into stress induction. Microbiology *151*, 2861–2872.

Ventura, M., Zink, R., Fitzgerald, G.F., and van Sinderen, D. (2005c). Gene structure and transcriptional organization of the *dnaK* operon of *Bifidobacterium breve* UCC 2003 and application of the operon in bifidobacterial tracing. Appl. Environ. Microbiol. *71*, 487–500.

Ventura, M., Canchaya, C., Bernini, V., Altermann, E., Barrangou, R., McGrath, S., Claesson, M.J., Li, Y., Leahy, S., Walker, C.D., *et al.* (2006). Comparative genomics and transcriptional analysis of prophages identified in the genomes of *Lactobacillus gasseri*, *Lactobacillus salivarius*, and *Lactobacillus casei*. Appl. Environ. Microbiol. 72, 3130–3146.

Ventura, M., Canchaya, C., Del Casale, A., Dellaglio, *F56*, 2783–2792.

Ventura, M., Canchaya, C., Fitzgerald, G.F., Gupta, R.S., and van Sinderen, D. (2007a). Genomics as a means to understand bacterial phylogeny and ecological adaptation: the case of bifidobacteria. Antonie van Leeuwenhoek *91*, 351–372.

Ventura, M., Canchaya, C., Tauch, A., Chandra, G., Fitzgerald, G.F., Chater, K.F., and van Sinderen, D. (2007b). Genomics of *Actinobacteria*: tracing the evolutionary history of an ancient phylum. Microbiol. Mol. Biol. Rev. *71*, 495–548.

Ventura, M., O'Flaherty, S., Claesson, M.J., Turroni, F., Klaenhammer, T.R., van Sinderen, D., and O'Toole, P.W. (2009a). Genome-scale analyses of health-promoting bacteria: probiogenomics. Nat. Rev. Microbiol. *7*, 61–71.

Ventura, M., Turroni, F., Zomer, A., Foroni, E., Giubellini, V., Canchaya, C., Bottachini, F., Claesson, M.J., Zhang, Z., Mantzourani, M., *et al.* (2009b). The *Bifidobacterium dentium* Bd1 genome sequence reflects its genetic adaptation to the human oral cavity. Plos. Genetics *12*, el000785.

Vernazza, C.L., Gibson, G.R., and Rastall, R.A. (2006). Carbohydrate preference, acid tolerance and bile tolerance in five strains of *Bifidobacterium*. J. Appl. Microbiol. *100*, 846–853.

Wang, M., Ahrne, S., Jeppsson, B., and Molin, G. (2005). Comparison of bacterial diversity along the human intestinal tract by direct cloning and sequencing of 16S rRNA genes. FEMS Microbiol. Ecol. *54*, 219–231.

Watanabe, K., Makino, H., Sasamoto, M., Kudo, Y., Fujimoto, J., and Demberel, S. (2009). *Bifidobacterium mongoliense* sp. nov., from airag, a traditional fermented mare's milk product from Mongolia. Int. J. Syst. Evol. Microbiol. *59*, 1535–1540.

Wolfgang, M.C., Kulasekara, B.R., Liang, X., Boyd, D., Wu, K., Yang, Q., Miyada, C.G., and Lory, S. (2003). Conservation of genome content and virulence determinants among clinical and environmental isolates of *Pseudomonas aeruginosa*. Proc. Natl. Acad. Sci. U.S.A *100*, 8484–8489.

Metabolism of Bifidobacteria

4

David A. Sela, Neil P. J. Price and David A. Mills

Abstract

The genus *Bifidobacterium* possesses a unique fructose-6-phosphate phosphoketolase pathway employed to ferment carbohydrates. Much metabolic research on bifidobacteria has focused on oligosaccharide metabolism as these carbohydrate polymers are available in their otherwise nutrient-limited habitats. Interestingly, infant-associated bifidobacterial phylotypes appear to have evolved the ability to ferment milk oligosaccharides, whereas adult-associated species utilize plant oligosaccharides, consistent with what they encounter in their respective environments. As breast-fed infants often harbour a bifidobacteria dominated gut microbiota, there have been numerous applications to mimic the bifidogenic properties of milk oligosaccharides. These are broadly classified as plant-derived fructo-oligosaccharides or dairy-derived galacto-oligosaccharides, which are differentially metabolized and distinct from milk oligosaccharide catabolism. In the following chapter we review bifidobacterial carbohydrate metabolism, as well as several other topics including nitrogen and iron metabolism.

Introduction

Members of the genus *Bifidobacterium* are Gram-positive bacteria possessing high chromosomal G+C content as do other *Actinobacteria*. Moreover, bifidobacteria are non-motile, non-sporulating, generally non-gas producing, catalase-negative (with a few exceptions) saccharoclastic anaerobes. Bifidobacteria have been isolated from the human intestine/faeces, oral cavity, reproductive tract in addition to the insect gastrointestinal tract (GIT) and sewage. This chapter reviews what is currently known about bifidobacterial metabolism with an emphasis on oligosaccharide utilization in these bacteria.

As a defining characteristic of life, central metabolism furnishes energy and precursor metabolites to an interconnected web of biochemical networks. Thus, exogenous matter is acquired and transformed to achieve two primary homeostatic objectives, energy generation and biosynthesis of cellular components. Energy generation in the bifidobacteria, as with other heterotrophic anaerobes deficient for respiration, proceeds by fermentation of carbohydrates encountered in their local environment. Of particular relevance to understanding the biology of the bifidobacteria is their capacity to extract energy from soluble and conjugated oligosaccharides (Gibson *et al.*, 1995; Schell *et al.*, 2002). Hydrolytic enzymes degrade sugar polymers while incomplete oxidation of their constituent monosaccharides is coupled to adenosine triphosphate (ATP) generation primarily through substrate-level phosphorylation. Like other fermentative microorganisms, maintenance of a favourable redox balance is achieved by recycling reduced equivalents through the export of organic acid end products. Subsequently, anabolic operations utilize ATP to assemble scavenged or endogenously synthesized monomers into macromolecules.

The bifidobacteria have evolved a relatively simple metabolic circuitry, particularly in fluxes observed in axenic laboratory cultures – although pure culture models do not fully reflect the lifestyle or environmental experiences that confront the bifidobacteria within their animal hosts. Fitness within these competitive niches mandates participation in highly evolved interactions with phylogenetically disparate microbes. This is clearly evident in the complex syntrophic interactions between bifidobacteria and other symbionts within the lower GIT (Sonnenburg *et al.*, 2006; Xu and Gordon, 2003). Accordingly, recent technological advances have enabled scientists to systematically address the evolution of biochemical co-dependence and resource partitioning within microbial consortia. Needless to say, access to genome sequences has elevated our knowledge of the biochemical capabilities within the genus *Bifidobacterium* and presented several hypotheses to be experimentally tested. Genome comparisons of bifidobacteria isolated from dissimilar habitats (e.g. adult human vs. infant human vs. honeybee) have revealed potential adaptive outcomes that are intimately linked to the divergence of these species (Sela *et al.*, 2008; M. Ventura, personal communication).

In general, metabolic studies on bifidobacteria have centred on the physiological implications of oligosaccharide utilization in order to direct nutritional enrichment of *in vivo* cell densities. While additional features of bifidobacterial metabolism have been previously explored, genomics has inspired renewed efforts to characterize and contextualize the influence of bifidobacterial metabolites on other organisms that reside in their niche.

Carbohydrate metabolism of the bifidobacteria

Species of the genus *Bifidobacterium* are believed to degrade hexoses exclusively by the fructose-6-phosphate phosphoketolase (F6PK) pathway, alternatively referred to as the 'bifid shunt' (Fig. 4.1) (de Vries and Stouthamer, 1967; Scardovi, 1986; Veerkamp, 1969). Very few challenges to the primacy of this pathway have arisen to date. Indeed it was the identification of this fermentative pathway that prompted de Vries and Stouthamer to conclude that the

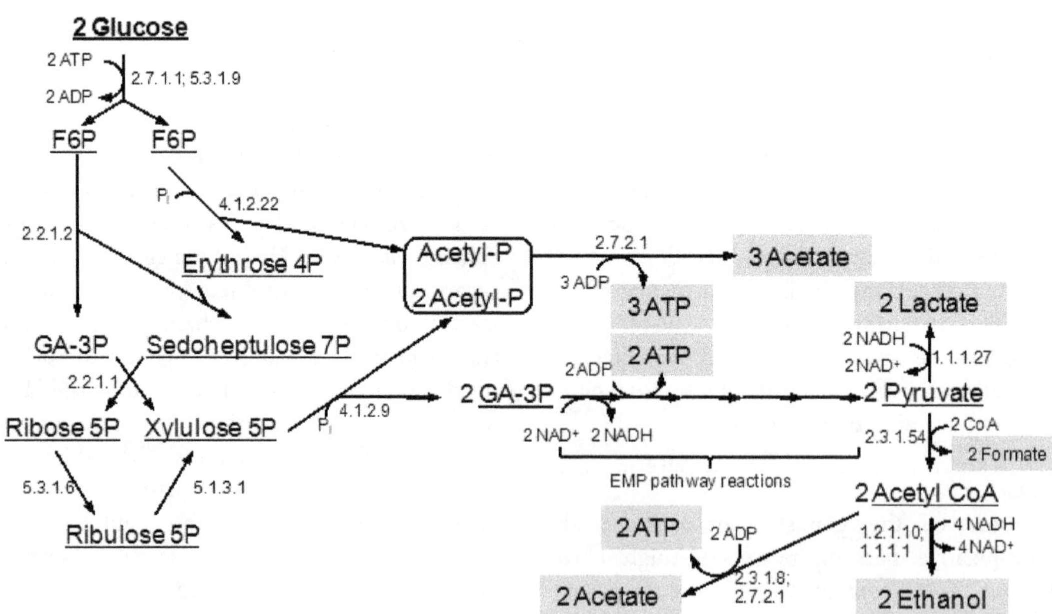

Figure 4.1 Bifidobacterial carbohydrate fermentation via the fructose-6-phosphate phosphoketolase (F6PK) pathway. Stoichiometric coefficients represent the utilization of two glucose molecules. Shaded boxes indicate end-products generated during hexose fermentation. Numbering refers to the enzyme commission (EC) number of the catabolic reaction. Pi, phosphate; F6P, fructose 6-P; GA-3P, glyceradehyde 3-phosphate.

bifidobacteria comprise a taxonomic entity distinct from the Lactic Acid Bacteria (LAB) (de Vries and Stouthamer, 1967). This finding was corroborated by their high genomic G+C content common to members of the phylum *Actinobacteria*, their characteristic bifid morphology and the lack of a classical heterofermentative pathway. In contrast to heterofermentation exhibited by certain LAB, CO_2 is not produced during fermentation by bifidobacteria except in strains capable of utilizing gluconate (Asano *et al.*, 2005). The absence of a 6-phosphogluconate intermediate distinguishes LAB heterofermentation from the bifidobacterial F6PK pathway (Kandler, 1983). Furthermore, bifidobacteria lack several genes for glycolysis and gluconeogenesis, including the key regulator phosphofructokinase (PFK). PFK typically plays a central role in controlling glycolysis as it catalyses one of the pathway's rate determining reactions and is the major flux-controlling enzyme of glycolysis. Results of isotopic labelling experiments with ^{13}C-fructose confirm that the F6PK pathway is predominant for the degradation of carbohydrates by *Bifidobacterium* evidently bypassing glycolysis. The lack of gluconeogenesis may be a consequence of the clade's ability to harvest environmental carbohydrates in the gut.

The F6PK pathway terminates with a theoretical yield of 1.5 mols acetate and 1 mol lactate secreted per mol glucose consumed (de Vries and Stouthamer, 1967; Wolin *et al.*, 1998). This yield is variable in actively growing cells, dependent on the particular culture conditions and substrates consumed (Amaretti *et al.*, 2007; Ruas-Madiedo *et al.*, 2005). Giving rise to its name, the key enzymatic step in the F6PK pathway is catalysed by fructose-6 phosphate phosphoketolase (EC 4.1.2.22) that cleaves hexose phosphate to erythrose-4-phosphate and acetyl phosphate. Additional carbon interconversions catalysed by a transketolase (EC 2.2.1.1) and transaldolase (EC 2.2.1.2) yield pentose phosphates from tetrose and hexose phosphates to be further degraded to acetyl phosphate and glyceraldehyde-3-phosphate. Enzymes participating in the Emden-Meyerhoff-Parnas pathway oxidize glyceraldehyde-3-phosphate to pyruvate that in turn is reduced to lactate through lactate dehydrogenase activity (EC 1.1.1.27). Acetyl-phosphates donate their phosphates to ADP by way of acetate kinase (EC 2.7.2.12) in the energy-generation step of the pathway. It has been postulated that formate, along with acetate, could be produced subsequent to phosphoroclastic cleavage of pyruvate (van der Meulen *et al.*, 2004). Pyruvate formate lyase (PFL; EC 2.3.1.54) probably degrades pyruvate to formate and acetyl-CoA as genes that encode the PFL heterodimer have been identified in all bifidobacterial genomes sequenced to date (Zhu and Shimizu, 2004). Interestingly, the formate acetyltransferase gene appears in a conserved cluster flanked by NAD^+ synthase (EC 6.3.5.1) and other genes potentially involved in cofactor recycling or formate metabolism (Markowitz *et al.*, 2008).

Formate production, along with the reduction of acetaldehyde to ethanol, dictates the ratio of secreted acetate to lactate to a certain extent. van der Meulen and colleagues determined that end product ratios are directly proportional to the specific rate of sugar consumption. In instances where the fermentation rate increases, the balance of end products favours lactic acid and lower concentrations of acetic acid, formic acid and ethanol (van der Meulen *et al.*, 2006a). The converse was observed for slowly consumed substrates such as oligofructose. In addition, bifidobacteria produce millimolar concentrations of succinic acid while actively growing, ceasing when nutrients are depleted. When substrates are slowly fermented, the main route for NAD^+ recycling is ethanol production, as 2 mol of NAD^+ is regenerated per mol ethanol produced. It is likely that succinic acid is produced by converting phosphoenolpyruvate (PEP) into oxaloacetate although it is currently unclear how CO_2 is incorporated (van der Meulen *et al.*, 2006a). Partial reduction of acetyl-CoA is favourable as it provides more carbon for ATP production as was observed in *Bifidobacterium animalis* subsp. *lactis* and *Bifidobacterium longum*. Strains of these species were observed to convert more than 50% of the acetyl-CoA pool into acetic acid (Ruas-Madiedo *et al.*, 2005). The redox status of the cell would be deleteriously out of equilibrium if it were not for an additional mechanism to regenerate NAD^+. Additional electron sinks such as mannitol, erythritol or glycerol were not detected in the bifidobacteria in contrast with certain LAB species (Richter *et al.*, 2003; van der Meulen *et al.*, 2006a). Aside from substrate identity, there are additional factors that influence carbohydrate consumption rate and thus dictates whether the end product balance is shifted towards lactate production. This necessarily includes the availability and regulation of transporters and intracellular enzymes that incorporate sugars into the F6PK pathway (Caescu *et al.*, 2004; Krzewinski *et al.*, 1996).

Fructose-6-phosphate phosphoketolase

The characteristic reaction of bifidobacterial central metabolism is catalysed by the thiamin diphosphate-dependent fructose-6-phosphate phosphoketolase (F6PK; EC 4.1.2.22), which performs the phosphoroclastic cleavage of D-fructose-6-phosphate into erythrose-4-phosphate and acetyl-1-phosphate (Scardovi and Trovatelli, 1965). The mechanism of F6PK has been investigated using chirally-labelled substrates (Fig. 4.2) (Merkler and Retey, 1981). Carbon atoms 1 and 2 of fructose-6P are transferred to the coenzyme to form 1,2-dihydroxyethyl-thiamine pyrophosphate. Elimination of water from this intermediate generates an enol which is then stereo-selectively protonated with inversion to generate 2-acetylthiamin pyrophosphate (acetyl-TPP). The last step is a phospholysis to regenerate the TPP coenzyme to produce acetyl phosphate. Thus carbon-1 of the original fructose-6P gives rise to the methyl group of the acetyl phosphate product, while carbons 3–6 are retained in the erythrose-4-phosphate. Acetyl phosphate may be hydrolysed to produce acetic acid or may be converted into acetyl-CoA

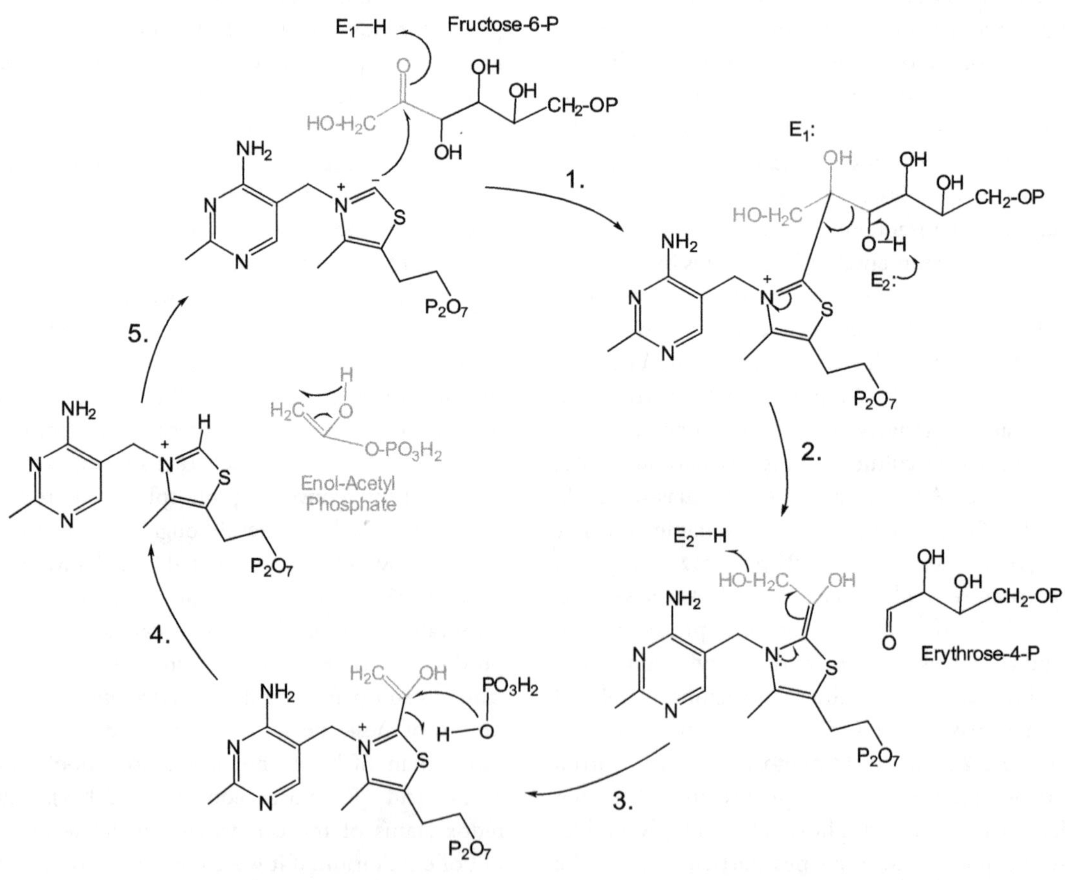

Figure 4.2 Thiamine Pyrophosphate (TPP) Mechanism for the *Bifidobacterium* Phosphoketolase (F6PK). 1. Carbon atoms 1 and 2 of fructose-6P (shown in grey) are transferred to TPP. 2. Release of erythrose-4-P (shown in black) and formation of 1,2-dihydroxyethyl-thiamine pyrophosphate. 3. Elimination of water from this intermediate generates an enol which is then stereo-selectively protonated with inversion to generate acetyl-TPP. 4. Phospholysis to release the acetyl phosphate enol product (shown in grey) and to regenerate the TPP cofactor.

by phosphotransacetylase activity. Notably, a phosphotransacetylase gene has been localized proximal to the F6PK gene in bifidobacterial genomes sequenced to date.

The F6PK biochemical reaction is commonly assayed to identify environmental isolates belonging to the genus *Bifidobacterium* (Orban and Patterson, 2000). There are two known fructose-6-phosphate phosphoketolase enzymes. One is specific only for fructose-6-phosphate while a second type is also active on D-xylulose-5-phosphate (EC 4.1.2.9) (Grill *et al.*, 1995; Sgorbati *et al.*, 1976). It was previously held that the presence of one or both of these enzymes is correlated to whether the bifidobacteria descends from a human-colonizing lineage; however, this has not been convincingly demonstrated (Grill *et al.*, 1995; Yin *et al.*, 2005).

The bifunctional F6PK type plays a dual role in central metabolism as it catalyses a later step in the pathway to yield glyceraldehyde-3-phosphate and acetyl-1-phosphate from D-xylulose-5-phosphate. Accordingly, this enzyme is encoded by the *xfp* gene first described in *B. animalis* subsp. *lactis* (Meile *et al.*, 2001). Genome sequencing and studies purposely targeting the *xfp* gene have clarified the molecular evolution of this gene (Yin *et al.*, 2005). This has further informed species-level identification strategies based on variability within the 5′ region adjacent to *xfp*.

Aside from the closely related *Gardnerella vaginalis*, F6PK has not been biochemically characterized in anaerobic bacteria outside the bifidobacteria (Gavini *et al.*, 1996). Although sequences with high identity to the bifidobacterial F6PK do appear in public sequence databases (Markowitz *et al.*, 2008), it is not currently known if *G. vaginalis* or other microbial species employ the F6PK pathway as the sole means to ferment carbohydrates.

Galactose metabolism

Lacto-*N*-biose phosphorylase (LNBP or GLNBP; EC 2.4.1.211) is critical in the degradation of lacto-*N*-biose (LNB) and galacto-*N*-biose (GNB) disaccharides, which are repeating subunits in soluble milk oligosaccharides and mucin glycoproteins, respectively (Kitaoka *et al.*, 2005). First characterized in *Bifidobacterium longum* subsp. *longum* NCC2705, LNBP cleaves both LNB and GNB into D-galactose-1-phosphate and an *N*-acetylhexosamine with anomeric inversion. LNBP is encoded within a putative galactose operon conserved in several bifidobacterial genomes (Kitaoka *et al.*, 2005; Markowitz *et al.*, 2008). Furthermore, the crystal structure of the *B. longum* subsp. *longum* JCM1217 LNBP protein has been solved while complexed with five different substrates (Hidaka *et al.*, 2009).

Interestingly, the *Bifidobacterium bifidum* JCM1254 genome possesses two LNBP isozymes designated *lnpA1* and *lnpA2* exhibiting similar specific activity (Nishimoto and Kitaoka, 2007b). Non-bifidobacterial LNBP homologues have been described in *Clostridium perfringens* and *Vibrio vulnificus* that are active only on mucin GNB or type I LNB disaccharides, respectively (Nakajima and Kitaoka, 2008; Nakajima *et al.*, 2008). The *B. bifidum* galactose gene cluster includes a carbohydrate-specific ABC transporter that may import LNB and/or GNB. In addition, a proximal galactose-1-phosphate uridylyltransferase (EC 2.7.7.10) and UDP-glucose 4-epimerase (EC 5.1.3.2) are involved in catabolism of mucin glycoconjugates (Podolsky, 1985). Galactose is liberated from mucin and other glycans as a hexose-1-phosphate by LNBP, which is subsequently directed into the F6PK pathway by way of a modified Leloir pathway. The galactosyl phosphate is first converted to UDP-galactose and then to the corresponding glucose nucleotide prior to utilization in energy generating or biosynthetic reactions.

Notwithstanding acquisition of oligosaccharide-bound galactose, there are generally two modes of galactose metabolism exhibited by intestinal anaerobes. PEP-dependent phosphotransferase systems (PTS) translocates lactose concomitant with phosphorylation to yield intracellular lactose-6-phosphate. This molecule is cleaved by a 6-phospho-β-D-galactosidase (EC 3.2.1.85) to yield glucose and galactose-6-phosphate (Suzuki *et al.*, 1996). Alternatively, soluble galactose is actively transported across the cell envelope or liberated from lactose by β-galactosidase (EC 3.2.1.23). In this instance, free galactose is phosporylated by a galactokinase (EC 2.7.1.6) into galactose-1-phosphate which therefore significantly differs from the LNBP-dependent LNB utilization pathway. In *B.*

longum subsp. *longum* NCC2705, a galactokinase (BL1210) is encoded adjacent to a galactose-1-phosphate uridylyltransferase gene (BL1211), albeit in the conspicuous absence of recognizable β-galactosidase or UDP-glucose epimerase genes. This deficiency, along with the appearance of another UDP-glucose 4-epimerase gene (BL1671) unlinked to other galactose-related genes, led Kitaoka *et al.* to postulate that galactose utilization proceeds primarily through the LNBP pathway that processes LNB and GNB (Kitaoka *et al.*, 2005).

D-Fructose fermentation

Utilization of the D-fructose ketohexose has been experimentally determined to proceed through an inducible fructokinase (EC 2.7.1.4) in *B. longum* subsp. *longum* which, interestingly, is highly specific for fructose only (Caescu *et al.*, 2004). Bacterial fructokinases typically phosphorylate mannose to yield mannose-6-phosphate, thus are often referred to as mannofructokinases (Sproul *et al.*, 2001). In addition to the bifidobacteria, the actinobacteria *Streptomyces violaceoruber* also possesses a strict D-fructokinase (Sabater *et al.*, 1972). It is still unknown if the bifidobacteria, which generally ferment mannose, employ a dedicated kinase or isomerize mannose first to D-fructose. Furthermore, a PTS system that translocates fructose has been described in *Bifidobacterium breve* UCC2003 (Maze et al., 2007).

Pentose utilization in the bifidobacteria

Reflecting the composition of their host's diet, the ability of bifidobacteria to ferment plant-derived pentoses may be determined, in part, by whether they are primarily an infant or adult colonizing phylotype (Sela *et al.*, 2008). This phenomenon is clearly exhibited in *Bifidobacterium longum* subsp. *infantis* which is deficient for xylose and arabinose fermentation (Degnan and Macfarlane, 1991). This is particularly striking when considering that the phylogenetic near-neighbour *B. longum* subsp. *longum* does ferment arabinose and xylose (Sakata *et al.*, 2002). Deriving broad conclusions from fermentation profiling is confounded by inconsistent subspecies typing of *B. longum* strains. Nevertheless, on the whole, *B. longum* subsp. *infantis* typically isolated from infant faeces are incapable of fermenting xylose and arabinose whereas subspecies *longum* commonly isolated from adults do so readily.

Among those strains that ferment pentose sugars, xylose is probably isomerized (EC 5.3.1.5) and phosphorylated (EC 2.7.1.17) prior to entry into the F6PK pathway as xylulose-5-phosphate. This is less efficient than hexose fermentation, as it consumes one ATP and acetate, and is probably compensated by upregulating of xylose transporters and/or catabolic enzymes (Palframan *et al.*, 2003). Consistent with its xylose-negative phenotype, the *B. longum* subsp. *infantis* ATCC 15697 genome is devoid of a recognizable xylose isomerase in clear contrast to the *B. longum* subsp. *longum* NCC2705 genome that encodes this gene (Schell *et al.*, 2002; Sela *et al.*, 2008).

Even more intriguing is the divergence of arabinose metabolism between *B. longum* subsp. *infantis* and *B. longum* subsp. *longum*. The *B. longum* subsp. *infantis* chromosome contains severely degraded vestiges of arabinose catabolic alleles which are found intact in the *B. longum* subsp. *longum* genome (Sela *et al.*, 2008). These genes include arabinose isomerase (*araA*) and L-ribulose-5-phosphate 4-epimerase (*araD*), which are both inactivated in *B. longum* subsp. *infantis*. In *B. longum* subsp. *longum* these genes enable conversion of L-arabinose to L-xylulose-5-phosphate with the aid of a L-ribulokinase. A recent non-orthologous displacement appears responsible for arabinose gene decay in *B. longum* subsp. *infantis*. Specifically, genes specifying a putative fucosidase and permease (Blon_0425 and Blon_0426) displaced three genes predicted to encode arabinose isomerase, ribulose 5-phosphate epimerase and a sugar kinase, present in B. *longum* subsp. *longum* and other arabinose-positive bifidobacteria. This genome remodelling strongly suggests a reorientation towards interactions with mammalian glycans containing fucose in lieu of arabinose in bifidobacteria that have not interacted with plant-derived sugars from an evolutionary perspective.

Regulation of bifidobacterial carbohydrate metabolism

While poorly understood, the mechanisms governing physiological responses to carbohydrates constitute an area of active inquiry. Of particular importance, *B. longum* subsp. *longum* NCC2705 exhibits a clear preference for lactose over glucose contributing to an atypical reverse diauxy (Kim *et al.*, 2003; Parche *et al.*, 2006). This phenomenon is also known in *Streptococcus thermophilus*, although this *firmicute* does not ferment glucose well and possesses CcpA-dependent carbon catabolite repression in contrast to *B. longum* (van den Bogaard *et al.*, 2000). Lactose represses glucose uptake in *B. longum* subsp. *longum* through downregulating a glucose permease suggesting evolutionary tuning for lactose utilization. The molecular mechanisms underlying this repression are not currently well understood.

A model for global regulatory control has yet to be fully conceived for the bifidobacteria. Multiple studies of carbohydrate regulation have linked the induction of individual genes or operons, often glycosidases and transporters, to the presence of a particular carbohydrate. Accordingly, sucrose and raffinose was demonstrated to induce expression of sucrose phosphorylase in *B. animalis* subsp. *lactis* (Trindade *et al.*, 2003). Perrin and colleagues observed that a *B. longum* subsp. *infantis* fructokinase and a β-fructofuranosidase is upregulated in the presence of D-fructose to deconstruct fructan polymers (Perrin *et al.*, 2001). This induction by fructose is repressed in the presence of glucose by a still unknown mechanism. Similarly, *B. breve* UCC2003 possesses a β-fructofuranosidase found within a fructooligosaccharide operon which is induced in the presence of exogenous sucrose and short-chain fructo-oligosaccharides (FOS). Conversely the fructooligosaccharide operon is repressed while growing on mixtures of sucrose and glucose or sucrose and fructose. This repression is putatively linked to the LacI-like repressor encoded upstream of the FOS operon and the ROK family glucokinase present downstream (Ryan *et al.*, 2005).

Interestingly, certain bifidobacteria do not exhibit a consumption preference towards easily accessible, and thus energetically favourable monosaccharides, when presented with complex mixtures. Monosaccharides do not repress oligosaccharide utilization or the production of hydrolytic enzymes in *B. adolescentis* MB 239 (Amaretti *et al.*, 2006). Remarkably, *B. adolescentis* MB 239 simultaneously ferments mixed oligosaccharides or mixed monosaccharides. In fact, monosaccharides promoted β-fructofuranosidase production in this strain.

Bifidobacterial oligosaccharide metabolism

Most bifidobacterial metabolic research has focused on oligosaccharide metabolism. This is largely attributable to the pervasiveness of these polymers in an otherwise nutrient-limited habitat within their animal host. These fibres transit undigested though the animal GIT where they selectively promote the growth of bifidobacteria or other bacteria capable of utilizing oligosaccharides. This trait can be exploited to manipulate the microbial composition of the lower intestine to promote putative favourable health benefits (Macfarlane and Cummings, 1999; Moro *et al.*, 2006). Accordingly, the desirable characteristics associated with bifidobacteria have been widely reported and will be described in depth in Chapter 6, *Probiotic Properties of Bifidobacteria* (also reviewed in Parvez *et al.*, 2006). Briefly, specific bifidobacterial strains have been demonstrated to abate irritable bowel syndrome (O'Mahony *et al.*, 2005), inflammatory bowel diseases (Furrie *et al.*, 2005; O'Mahony *et al.*, 2008), prevent and manage diarrhoea (Chouraqui *et al.*, 2004), increase digestive efficiency of carbohydrate polymers (Flint *et al.*, 2008), modulate the immune system (Klein *et al.*, 2008) and erect a barrier to pathogenic colonization (Lievin *et al.*, 2000; Madsen *et al.*, 2001), among other health benefits.

In general, bifidobacteria encounter plant-derived oligosaccharides subsequent to colonizing mature mammals. However, it is their atypical capability to subside on milk oligosaccharides that has inspired much attention and partially explains the bifidobacteria's early dominance of the infant GIT consortium. Thus, milk oligosaccharides are another example of how evolution has refined milk to confer health benefits to the infant in addition to supplying nutrition (Ward and German, 2004). Therefore milk's role in the establishment of a protective microbiota is the model by which oligosaccharide-based bifidogenic interventions are conceived.

Human milk oligosaccharide structure and properties

Human milk contains a high concentration of soluble oligosaccharides (HMO) comprising the third most abundant class of molecules that, surprisingly, exceeds milk's protein content (Coppa *et al.*, 1993). The concentration of milk oligosaccharides varies by individual and declines over the course of lactation, but it is generally understood to be present at ≥4 g/L in milk with higher levels observed in colostrums (Asakuma *et al.*, 2008; Chaturvedi *et al.*, 2001). Milk oligosaccharides incorporate D-glucose, D-galactose, N-acetylglucosamine, L-fucose and sialic acid (Neu5Ac) monosaccharides assembled in various configurations. Lactose (Galβ1–4Glc) is present at the reducing end and the molecule is elongated from a β1–3 linkage (with additional β1–6 linkage in branched forms) with several repeating Galβ1–3/4GlcNAc units. Further complexity is provided by terminal fucosylation by way of α1–2/3/4 linkages and/or α2–3 and α2–6 sialylation. To date, almost 200 distinct molecular species have been identified in pooled breast milk samples (Ninonuevo *et al.*, 2006). When considering the immense combinatorial potential for these molecules, the identification of relatively few structures suggest that their synthesis within the mammary epithelia has been constrained by a functional imperative. Accordingly, HMO structures vary among maternal genotypes and is dependent on Lewis and secretor status (Kunz *et al.*, 2000). Further information on HMO structure and function is provided by recent reviews (Bode, 2006; German *et al.*, 2008).

The HMO molecules largely escape degradation by infant digestive enzymes as well the low pH encountered in the stomach (Engfer *et al.*, 2000; Gnoth *et al.*, 2000). Thus, they persist intact during passage to the distal infant GIT where they exhibit several bioactive functions including bifidogenic properties. In a notable non-bifidogenic role, HMOs complement milk's to passive immunity against pathogen colonization. Accordingly, pathogens often deploy lectins as adhesins with specific affinity for mucins or epithelial surface glycoconjugates. These same glycan motifs are represented in the HMO structure which flood the lower GIT and serve as soluble ligand mimics, thus protecting against pathogen colonization by titrating out preferred binding surfaces recognized by adhesins. This remarkable example of mother-originated participation in neonatal innate immunity is reviewed in Newburg (2009). HMO-mediated ligand mimicry occurs in the intestinal mucosa (Newburg *et al.*, 2004; Ruiz-Palacios *et al.*, 2003), in the laryngopharynx (Andersson *et al.*, 1986), and the urinary tract, as some HMOs are absorbed whole and excreted by the infant (Obermeier *et al.*, 1999).

In addition to providing fermentable substrates and mimicking pathogen binding sites, HMOs are believed to regulate expression of host glycosyltransferases to direct the epithelial glycan epitopes presented to microbes (Angeloni *et al.*, 2005; Vanmaele *et al.*, 1999). Furthermore, there is mounting evidence that HMO directly modulates immune responses (Eiwegger *et al.*, 2004; Velupillai and Harn, 1994).

Bifidobacterial consumption of the HMO core structure

Select bifidobacterial phylotypes are proficient at capturing and utilizing HMO as a sole carbon source (LoCascio *et al.*, 2007, 2009; Ward *et al.*, 2006, 2007). Often present in the neonate's lower GIT, *B. longum* subsp. *infantis* is particularly adept at metabolizing HMO and thus regarded as the archetypical HMO consumer (Sela *et al.*, 2008). Evidently this is a conserved phenotype among *B. longum* subsp. *infantis*, whereas other infant-associated bifidobacteria exhibit strain-specific variation in their capacity to consume HMO (unpublished observations). This phenotypic variability includes oligosaccharide depolymerization or metabolism of HMO constituent monosaccharides as observed in *B. bifidum* and *B. breve*, respectively (LoCascio *et al.*, 2009; Ward *et al.*, 2007). These phenotypes hint at niche partitioning among bifidobacterial phylotypes. Whereas particular bifidobacterial species may not directly access HMO carbon, they release oligosaccharide-bound monosaccharides (e.g. *B. bifidum*) for use by scavenging bacteria such as *B. breve*. Accordingly, a mixed bifidobacterial genomic microarray was used to assess the transcriptional response to breast feeding. Not surprisingly, faecal RNA exhibited enrichment for carbohydrate utilization transcripts attributable to the bifidobacteria (Gonzalez *et al.*, 2008).

Efforts to further characterize HMO consumption prompted the development of quantitative methods to glycoprofile (MS) oligosaccharide utilization at high-resolution (Ninonuevo *et al.*, 2006). Accordingly, mass-spectrometry (MS) analyses of spent culture supernatants revealed that most bifidobacteria generally utilize one HMO composition corresponding to Lacto-*N*-tetraose (LNT; Galβ1–3GlcNAc β1–3 Galβ1–4Glc) or Lacto-*N*-neotetraose (LNnT; Galβ1–4GlcNAc β1–3 Galβ1–4Glc) (LoCascio *et al.*, 2007, 2009). While *B. longum* subsp. *infantis* is able to fully utilize these tetrasaccharideisomers, *B. longum* subsp. *longum*, *B. bifidum* and *B. breve* exhibit incomplete degradation of this composition (LoCascio *et al.*, 2009). The adult-associated bifidobacteria, *B. adolescentis* and *B. animalis*, are unable to degrade LNT/LNnt or other detectable HMO species.

The HMO core structure in type I glycans comprises a repeating component referred to as lacto-*N*-biose (LNB; Galβ1–3GlcNAc). Kilogram quantities of LNB can now be synthesized in a one-pot reaction with recombinant enzymes cloned from *B. bifidum* and *B. longum* subsp. *longum* (Nishimoto and Kitaoka, 2007c). Whereas oligosaccharide purification from fluid milk remains problematic, *in vitro* synthesis represents a significant advance in the acquisition of specific substrates for HMO research. Accordingly, LNB enhances growth of several bifidobacterial species that also ferment this disaccharide as a sole carbon source (Kiyohara *et al.*, 2009). In these instances LNB serves as a proxy for HMO as it is not typically abundant in milk and lacks the full structural diversity, and likely function, of HMO.

Aside from *B. longum* subsp. *infantis*, infant-associated bifidobacteria consume glycan-bound LNB by mechanisms involving extracellular LNB-liberating enzymes or intracellular catabolic processes. Accordingly, an extracellular 1,2-α-L-fucosidase (EC 3.2.1.63) has been characterized in *B. bifidum* which cleaves terminal 1,2-α-fucosyl linkages to galactose residues of LNB, thus permitting further metabolism to proceed (Katayama *et al.*, 2004; Nagae *et al.*, 2007). Subsequent to de-fucosylation, a lacto-*N*-biosidase (EC 3.2.1.140) liberates LNB from lacto-*N*-tetraose and other HMO compositions lacking fucosylated or sialylated LNB residues (Wada *et al.*, 2008). Transport of LNB across the cell membrane is facilitated by an ABC transporter that includes a dedicated LNB binding subunit (Suzuki *et al.*, 2008; Wada *et al.*, 2007). Interestingly, *B. longum* subsp. *longum* possesses an endo-α-*N*-acetylgalactosaminidase (EC 3.2.1.97) which liberates galacto-*N*-biose (GNB) from *O*-linked mucin glycans (Fujita *et al.*, 2005). Indeed the presence of both this endo-α-N-acetylgalactosaminidase and fucosidase has been linked to the *B. bifidum* mucin degradation phenotype, with expression of both genes induced in the presence of porcine mucin (Ruas-Madiedo *et al.*, 2008). An intracellular phosphorylase cleaves LNB or GNB disaccharides derived from HMO or mucin glycans, respectively (Kitaoka *et al.*, 2005; Nishimoto and Kitaoka, 2007b). Finally, a modified Leloir pathway was determined to feed LNB or GNB degradation products into the F6PK pathway to generate ATP (Nishimoto and Kitaoka, 2007a).

B. longum subsp. *infantis* preferentially consumes short chain HMO

Primarily isolated from infants, *B. longum* subsp. *infantis* biology has been clearly influenced by the evolutionary and metabolic interactions emanating to and from mother's milk (Sela *et al.*, 2008). Modern investigations of HMO metabolism have focused on *B. longum* subsp. *infantis* competence for growth on purified human milk oligosaccharides as a sole carbon source (Ward *et al.*, 2006). Further inquiry has demonstrated that *B. longum* subsp. *infantis* preferentially consumes several small mass oligosaccharides that are most abundant early in lactation while other bifidobacteria are capable of consuming only a lacto-*N*-tetraose-like composition, albeit incompletely (LoCascio *et al.*, 2007; Ninonuevo *et al.*, 2008).

Accordingly, the completed genome sequence of *B. longum* subsp. *infantis* ATCC 15697 exhibits discernable features to enable this interesting consumption phenotype that is unique among infant-associated bifidobacteria (Sela *et al.*, 2008). This includes a remarkable lineage-specific expansion of the solute binding proteins of ABC-type transporters with affinity towards oligosaccharides. In addition, the *B. longum* subsp. *infantis* genome contains several catabolic clusters encoding glycosyl hydrolases that cleave milk oligosaccharides and/or mucin glycans. Notable are sialidases (EC 3.2.1.18) and fucosidases (EC 3.2.1.51) active on α-linked terminal sialic acid and α-linked, respectively. The most significant HMO-related gene cluster is 43 kb in length situated on the same strand for the most part. This large cluster includes four glycosyl hydrolases active on HMO linkages (i.e. α-fucosidase, α-sialidase, β-galactosidase and β-hexosaminidase) bounded by ABC-type transporters and oligosaccharide binding proteins (Fig. 4.3). Significantly, the oligosaccharide binding proteins of the large HMO gene cluster exhibit an extreme divergence in peptide sequence relative to homologous proteins encoded elsewhere on the ATCC 15697 chromosome or other bifidobacterial genomes. This haplotype is not unique to ATCC 15697 as all *B. longum* subsp. *infantis* apparently possess HMO catabolic clusters (Sela *et al.*, 2008, and unpublished observations). Furthermore, MS-based proteomics have detected and confirmed that many of these predicted HMO-active enzymes are expressed while ATCC 15697 ferments HMO (Sela *et al.*, 2008).

In addition, the *in vivo* carbon flux through the F6PK pathway while growing on HMO has been investigated. Acetate produced while fermenting carbohydrates is both secreted and used in *de novo* fatty acid synthesis. Fatty acid synthesis in the bifidobacteria is likely accomplished by the multi-domain fatty acid synthase I common to *Actinobacteria* and eukaryotes (Schweizer and Hofmann, 2004). Accordingly, metabolic flux through the F6PK pathway was monitored by detecting perturbations to isotopomer ratios in fatty acids with gas chromatography–mass spectrometry (Price *et al.*, 2006). This approach was used to quantify the carbon flux through the F6PPK pathway by measuring specific carbohydrate-derived metabolites. After uptake of various ^{13}C-labelled sugars by bifidobacteria, the fractional distribution of ^{13}C into carbohydrate and fatty acid metabolites could be analysed by mass spectrometry. This analysis verified that both sialic acid (Neu5Ac) and HMO are indeed catabolized through the F6PK pathway in *B. longum* subsp. *infantis* ATCC 15697 (Sela *et al.*, 2008).

Milk oligosaccharide mimetics

There have been numerous attempts to mimic milk oligosaccharides in applications seeking to enrich bifidobacteria. Milk oligosaccharide mimetics are broadly classified as plant-derived fructo-oligosaccharides or dairy-derived galacto-oligosaccharides. Both classes lack the complexity inherent to HMOs in terms of monosaccharide composition, linkage diversity and structural organization. Nevertheless, *in vitro* and clinical evidence supports their use as bifidogenic substrates (Gibson *et al.*, 1995; Kruse *et al.*, 1999; Roberfroid *et al.*, 1998). These oligosaccharides are commercially distributed, as their production is currently more viable than industrial purification or synthesis of HMOs. It remains unclear, however, if these milk oligosaccharide mimetics exhibit the full complement of immunological and pathogen deflection functions, and whether they selectively increase bifidobacterial populations to the exclusion of undesirable genera.

Plant-derived fructo-oligosaccharides

Inulin fructo-oligosaccharides (FOS) are highly abundant in plant material such as chicory, garlic and sunchokes. Inulin-type FOS represent a simple polymer (DP 2–60) of β(2 – 1) linked

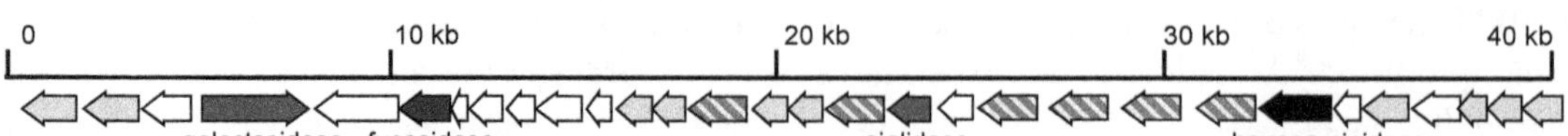

Figure 4.3 Milk oligosaccharide utilization cluster from *B. longum* subsp. *infantis* ATCC 15697. The 43 kbp gene cluster contains the four glycosyl hydrolases (black) that cleave HMO linkages, permeases (shaded in grey) as well as their associated oligosaccharide binding proteins (cross-hatched).

D-fructose residues with the degree of polymerization dependent on the purification regime and plant origin (Roberfroid, 2005; Wang *et al.*, 1999). A D-glucose moiety is often found at the terminus linked by an α(1–2) linkage as in sucrose (Niness, 1999). Conversely, oligofructose (OF) DPs range between 2 and 10, and is derived from chicory or synthesized by way of transfructosylation. OF extracted from plants possess a fraction retaining terminal glucosyl moieties whereas synthetic OF does not. Both inulin and OF are collectively referred to as fructo-oligosaccharides and differ primarily in degree of polymerization.

The FOS β(2 – 1) glycosidic linkage is recalcitrant to human digestion reducing their caloric value and the health benefits associated with high fibre consumption (Niness, 1999). Following transit to the colon, these fructans are hydrolysed by bacterial β-fructofuranosidases (EC 3.2.1.26) that are induced in the presence of its substrate in some bifidobacteria (Warchol *et al.*, 2002). This enzyme liberates fructose prior to phosphorylation and entry into the fructose-6-phosphate phosphoketolase pathway. Remarkably, *B. animalis* DN-173 010 preferentially consumes inulin-type FOS over monosaccharides and is deficient for glucose utilization altogether. Moreover, short chain oligofructose has been shown to be metabolized preferentially by DN-173 010 indicating an initial translocation of short-chain fructans, which, in turn, induces glyolytic enzymes, leaving long chain fructans intact (Perrin *et al.*, 2001; van der Meulen *et al.*, 2004). The preference for short chain fructo-oligosaccharides in the bifidobacteria has been observed in other studies and is reminiscent of their preference for short chain milk oligosaccharides (Bouhnik *et al.*, 1999; LoCascio *et al.*, 2007). Bifidobacterial utilization of specific oligosaccharides may be restricted by the inability translocate higher mass molecules, sterically limited by peptidoglycan pore diameter. This is of particular importance to bifidobacteria that primarily possess intracellular glycosyl hydrolases.

As with glucose, the theoretical ratio of acetic acid to lactic acid produced is 1.5 while growing on fructose, although bifidobacteria display an interesting fermentation profile while growing on inulin-type FOS. The production of lactic acid begins high and generally decreases in the later stages of fermentation. Conversely, formate is secreted towards the end of the fermentation, although diffuse at the initiation of growth. van der Meulen and colleagues postulate that carbohydrates found at high intracellular concentration are efficiently fermented and stimulate lactic acid production to equilibrate the redox balance. In the absence of these conditions, low intracellular sugars trigger formate production to generate more ATP (van der Meulen *et al.*, 2004). Formate secretion during slow fermentations has been observed in other bacteria such as *Lactococcus lactis* (Palmfeldt *et al.*, 2004).

In vitro co-cultures of bifidobacteria and other bacteria subsisting on oligofructose has also provided insight into the complex interactions that exist within the gut microbial consortium. Accordingly, co-cultures of *B. longum* subsp. *longum* BB536 and two acetate-converting, butyrate-producing bacteria, *Anaerostipes caccae* DSM 14662 or *Roseburia intestinalis* DSM 14610, were grown with oligofructose as the sole carbon source (Falony *et al.*, 2006). *A. caccae* DSM 14662 did not degrade oligofructose but grew well on fructose liberated by *B. longum* subsp. *longum* BB536. *A. caccae* also utilized lactate produced by bifidobacteria as was observed in co-culture fermentations with *B. adolescentis* indicating two potential cross-feeding pathways (Belenguer *et al.*, 2006). In a co-culture of both *R. intestinalis* and *B. longum,* initial oligofructose degradation concomitant with acetate production by the bifidobacterium prompted oligofructose utilization by *R. intestinalis* DSM 14610 (Falony *et al.*, 2006). This provided a link between the observed bifidogenic properties of oligofructose and butyrate production, which does not originate from bifidobacteria, but observed when providing oligofructose (Le Blay *et al.*, 1999; Tsukahara *et al.*, 2003). These intriguing observations potentially reflect an isolated syntrophic pathway present in the gut, although it is uncertain if butyrogenesis is due to a direct stimulation of non-bifidobacterial members of the gut consortium (Duncan *et al.*, 2003, 2004).

In a series of experiments, Falony *et al.* characterized four distinct inulin utilization phenotypes in bifidobacteria and subsequently performed competitive co-culture fermentations with *Bacteroides thetaiotamicron* LMG 11262 while supplying inulin as the sole carbohydrate (Falony *et al.*, 2009a,b). Not surprisingly, *B. breve*, which, unlike *B. thetaiotamicron*, does not ferment oligofructose or inulin, was outcompeted. *B. adolescentis* exhibited competitiveness towards the middle of the fermentation following partial inulin degradation. *B. angulatum* competed even earlier in the fermentation cycle, consuming short fractions of inulin immediately as they were produced. Finally, *B. longum* subsp. *longum* dominated the co-culture as it rapidly consumes oligofructose and degrades inulin.

Furthermore, the ability of *Bacteroides* to metabolize oligofructose was compared to that of *B. longum* subsp. *longum* BB536. Both *B. thetaiotamicron* LMG 11262 and *Bacteroides fragilis* LMG 10263 grew well on oligofructose, similar to fructose, whereas *B. longum* subsp. *longum* generates biomass slower on oligofructose than on fructose and produces formic acid and ethanol at the expense of lactate production. Kinetic analysis of *B. longum* subsp. *longum* fermentations indicated a preferential metabolism of the short OF fraction with larger chains consumed subsequent to the depletion of smaller molecules. This is in contrast to *Bacteroides* strains which did not exhibit preference for specific oligofructose oligomers, and degraded all oligofructose simultaneously releasing fructose into the culture medium. This observation is indicative of a differential oligofructose utilization mechanism probably associated with glycosyl hydrolase subcellular localization. Accordingly, *Bacteroides* secrete glycolytic enzymes into the culture medium or affix them to their extracellular surface (Xu *et al.*, 2003). Of note, a constitutively expressed extracellular *Bacteroides* fructanase has been observed (Blatch and Woods, 1993). Contrastingly, *Bifidobacterium* species often possess inducible intracellular glycosyl hydrolases including several characterized β-fructofuranosidases (Ehrmann *et al.*, 2003; Warchol *et al.*, 2002). Thus, it is likely that the bifidobacteria degrade oligofructose intracellularly with slower growth on larger OF attributable to inefficient translocation (van der Meulen *et al.*, 2006b). This is consistent with HMO metabolism in *B. longum* subsp. *infantis* and potentially constitutes an evolutionary stable strategy to compete for oligosaccharides with other members of the GIT consortia (Sela *et al.*, 2008). In addition, the emphasis on capturing and translocating oligosaccharides may explain how the bifidobacteria effectively compete with *Bacteroides* for oligosaccharides in the infant GIT (Kleessen *et al.*, 2001; Langlands *et al.*, 2004).

Little is known about the regulation of FOS utilization in the bifidobacteria. Ryan *et al.* examined the genetic structure of the FOS operon in *B. breve* UCC2003 that includes a novel glycosyl hydrolase active only on the fructosyl linkage to glucose and previously unknown in the bifidobacteria. This enzyme termed FosC exhibits low activity on the β(2–1) linkages between fructosyl moieties and potentially explains the poor growth of *B. breve* on inulin. The *B. breve* UCC2003 *fos* operon is induced by the presence of sucrose (a substrate of FosC) and short-chain FOS.

Dairy-derived galacto-oligosaccharides

Similar to plant-derived fructo-oligosaccharides, galacto-oligosaccharides are non-digestible carbohydrates that possess bifidogenic function (reviewed in Boehm and Stahl, 2007). Galacto-oligosaccharides (GOS) are more stable following thermal processing and at lower pHs compared to FOS, properties that favour this fibre for use in various food formulations. GOS is often manufactured in applications of LeChatelier's principle by manipulating galactose concentrations in the presence of β-galactosidase to prompt the reverse reaction or trans-galactosylation. This process is relatively inefficient and monosaccharides and disaccharides necessarily contaminate GOS preparations unless the carbohydrate mixture is further purified.

The basic structure of GOS incorporates lactose at the reducing end which is typically elongated with up to six galactose residues (Gal(β1-3/4/6)Gal(β1–4)Glc). Trans-galactosylation synthesis is primarily accomplished with whey-derived lactose as the substrate (Yanahira *et al.*, 1995). Like plant-derived FOS, the structural potential is markedly less varied than that of milk oligosaccharides albeit more complex than linear fructans.

Human milk possesses a large fraction of galactose bound in lactose, soluble oligosaccharides and glycoconjugates. This has led to confusion over the relative contribution of naturally occurring GOS to the human milk glycome (Boehm *et al.*, 2005; Macfarlane *et al.*, 2008). This may have been further perpetuated by misperceiving the synthetic GOS structure with for that of the HMO core, which contains *N*-acetylglucosamine. While some short-chain galactosyllactose (Gal(β1–3/4/6)Gal(β1–4) Glc) has been detected in millimolar concentrations in breast milk, this is several orders of magnitude lower than the typical HMO pool (Sumiyoshi *et al.*, 2004).

The requisite catabolic reaction in GOS utilization is β-galactosidase activity (EC 3.2.1.23) which cleaves terminal β-galactosyl linkages found in industrial or naturally occurring GOS preparations. β-galactosidases have been isolated and characterized from a variety of bifidobacteria, which typically encode multiple isozymes in their genomes. In general, bifidobacterial β-galactosidases are primarily classified into glycosyl hydrolase (GH) Family 42 and GH Family 2 along with a few exceptions. Several bifidobacterial β-galactosidases are found fused with other glycosidic domains. Biochemical characterization of purified or recombinant β-galactosidases has demonstrated that, in addition to lactose-degrading β-galactosidases (βgal-I), specific enzymes selectively degrade GOS (Hung *et al.*, 2001; Moller *et al.*, 2001). *B. adolescentis* DSM 20083 possesses one such GOS-inducible enzyme which hydrolyse oligosaccharide-bound galactose residues and not lactose (βgal-II) (van Laere *et al.*, 2000). βgal-II showed activity towards β-galactosyl residues linked through 1–4, 1–6, 1–3, and 1–1 bonds, with degree of polymerization (DP) values up to 8. For the most part, β-galactosidases are predicted to be expressed intracellularly. There are, however, notable exceptions as a few extracellular β-galactosidases have been studied in *B. adolescentis*, *B. longum* subsp. *infantis* and *B. bifidum* (Amaretti *et al.*, 2006; Moller *et al.*, 2001; Rabiu *et al.*, 2001).

In addition to β-galactosidases, endogalactanases (EC 3.2.1.89) also contribute towards GOS metabolism. Accordingly, the *B. longum* subsp. *longum* NCC2705 extracellular endogalactanase (*galA*) processively releases galactotrisaccharides from longer GOS or arabinogalactan oligomers (Hinz *et al.*, 2005a). Arabinogalactan is, as its name implies, a common arabinose-substituted galactan found in various plant materials (Hinz *et al.*, 2005b). Trisaccharide degradation products are believed to be translocated into the cytosol where they are further degraded by intracellular β-galactosidases. Accordingly, the endogalactanase appears in a gene cluster in *B. longum* subsp. *longum* NCC2705 with a potential oligosaccharide transporter (BL0260-BL0264), a β-galactosidase (BL0259) and an *lacI* family regulatory protein (BL0257) (Schell *et al.*, 2002). Interestingly, the expression of this β-galactosidase and components of the ABC transporter has been recently demonstrated to be upregulated while growing on GOS (Gonzalez *et al.*, 2008). This specific response to GOS provides further evidence that this locus is a primary contributor to GOS metabolism in *B. longum* subsp. *longum*. A homologue of this endogalactanase is absent from the *B. adolescentis* genome, congruent with previous studies where endoglycanase activity was absent in cell-free extract (van Laere *et al.*, 2000). However, the remaining GOS operon appears intact in *B. adolescentis* containing a duplicated β-galactosidase. Interestingly, *B. longum* subsp. *infantis* possesses a truncated endogalactanase gene (Blon_0440), which lacks the majority of its catalytic domain sequence and is located adjacent to a degraded β-galactosidase remnant in the absence of the sugar transporter (Schell *et al.*, 2002). It appears that these genes became expendable subsequent to the evolutionary divergence of the *infantis* and *longum* subspecies. This is consistent with the general remodelling of the subsp. *infantis* catabolic potential towards host-derived glycans at the expense of plant sugars such as type I arabinogalactans upon which this cluster is active (Sela *et al.*, 2008).

In addition to glycolytic enzymes, transport systems participate in oligosaccharide fermentation as translocation rate and specificity are limiting parameters for substrate utilization. Accordingly, bifidobacterial transport of GOS and derivatives may involve both known lactose PEP-dependent phosphotransferase systems to transport lactose and/or GOS, or other permeases with specificity for trisaccharides and tetrasaccharides that are induced by GOS (Gonzalez *et al.*, 2008; Gopal *et al.*, 2001; Parche *et al.*, 2007).

Amaretti *et al.* (2007) investigated *B. adolescentis* MB 239 fermentations of GOS, lactose, galactose and glucose in an effort to study its kinetics and metabolic implications. Accordingly, *B. adolescentis* MB 239 exhibits a stringent preference towards lactose and a DP 3 trisaccharide. Following consumption of these smaller carbohydrates, longer chain oligosaccharides were simultaneously utilized. As was previously demonstrated, *B. adolescentis* does not preferentially metabolize glucose before oligosaccharides (Amaretti *et al.*, 2006). Interestingly, *B. adolescentis* exhibits both intracellular and extracellular hydrolysis of β(1–4) galactosides as galactose accumulates in the medium at a faster rate than they are consumed. Lactose and GOS both induced intracellular and extracellular β-galactosidase activity, whereas galactose induced only secreted β-galactosidase. Formic acid was not produced while *B. adolescentis* ferments GOS, leading Amaretti *et al.* (2007) to postulate that the observed deviation from the theoretical ratio of lactate to acetate is due to the efflux of metabolites to anabolic operations. Ethanol production was not observed while growing on galactose, with moderate amounts generated during growth on lactose and GOS, and inexplicably the highest concentrations were observed during glucose fermentations.

B. bifidum NCIMB41171 has proved to be a model for bifidobacterial β-galactosidase studies as four isozymes have been characterized from this GOS producing strain (Goulas *et al.*, 2007, 2009a; Tzortzis *et al.*, 2005). As is the case with *B. adolescentis* MB239, *B. bifidum* NCIMB411171 is more efficient at utilizing lactose and GOS than simple monosaccharides which may be the result of an efficient transport system as is predicted in *B. longum* subsp. *infantis* ATCC15697 (Amaretti *et al.*, 2006; Goulas *et al.*, 2009b; Sela *et al.*, 2008). The four isozymes (termed BbgI, BbgII, BbgIII and BbgIV) were interpreted to act complementary to one another by localizing to different parts of the cell or by activity on distinct galactosyl linkages. Accordingly, BbgIII is believed to be expressed on the cell surface where it participates in the extracellular degradation of galactans. While BbgII exhibits a preference for a specific GOS fraction (DP 3 and 4) over lactose, in contrast to the other three isoenzymes. Aside from the inducible BbgIV, the other three characterized *B. bifidum* β-galactosidases are constitutively expressed.

Genome features that enable oligosaccharide metabolism

Aside from the central fermentative pathway genes of the F6PK pathway, there are two classes of genes and genomic features that directly contribute to oligosaccharide metabolism. These are glycosyl hydrolases that cleave the oligosaccharides into constituent subunits and membrane spanning transporters specific for soluble oligosaccharides or their degradation products. Table 4.1 lists bifidobacterial glycosyl hydrolases assigned an enzyme commission (EC) number in the Integrated Microbial Genomes Database (Markowitz *et al.*, 2008). High-throughput sequencing of bacterial genomes has provided a detailed survey of bifidobacterial physiology in general and the biology of oligosaccharide utilization specifically.

In bifidobacterial genomes, glycosyl hydrolases are often linked and clustered with genes for transport of an assortment of sugars most notably oligosaccharides. Based on currently available bifidobacterial genome sequences it has been estimated that ~5% of their gene content is devoted to carbohydrate transport (Ventura *et al.*, 2009). This includes ATP-binding cassette (ABC) transporters, permeases as well as PEP-dependent PTS (Maze *et al.*, 2007; Parche *et al.*, 2007; Schell *et al.*, 2002; Sela *et al.*, 2008). Transport systems are particularly critical to the competitive strategy employed to acquire oligosaccharide-bound carbon. Conversely, *Bacteroides* and other members of the colon microbial consortium utilize an arsenal of extracellular secreted glycosidases (Ramsay *et al.*, 2006; Sela *et al.*, 2008; Xu *et al.*, 2003). Analysis of the complete genome of *B. longum* subsp. *longum* NCC2705 revealed a genetic organization of oligosaccharide-active genes into catabolic clusters (Schell *et al.*, 2002). The predicted NCC2705 proteome consists of 8.5% proteins designated as carbohydrate transport/metabolism that is significantly more than bacterial genomes available at the time. Most striking was the presence of high-affinity oligosaccharide transporters that allow *B. longum* subsp. *longum* to compete for the structurally diverse oligosaccharides encountered in the distal GIT. The genetic

Table 4.1 Glycosyl Hydrolases of the sequenced bifidobacteria

		B. adolescentis ATCC 15703	B. adolescentis L2-32	B. angulatum DSM 20098	B. animalis lactis AD011	B. catenulatum DSM 16992	B. dentium ATCC 27678	B. gallicum DSM 20093	B. longum DJO10A	B. longum NCC2705	B. longum infantis ATCC 15697	B. pseudocatenulatum DSM 20438
	Sequencing status	Finished	Draft	Draft	Finished	Draft	Draft	Draft	Finished	Finished	Finished	Draft
	Genome size (Mbp)	2.1	2.4	2	1.9	2.1	2.6	2	2.4	2.3	2.8	2.3
Glycosyl Hydrolase												
EC:3.2.1.-	General Glycosidase	2	2	2	2	2	2	2	2	2	2	2
EC:3.2.1.1	Alpha-amylase.	1	1	0	1	0	0	0	0	0	0	0
EC:3.2.1.10	Oligo-1,6-glucosidase.	1	2	1	1	0	1	1	1	1	1	2
EC:3.2.1.14	Chitinase.	0	0	0	0	0	0	0	0	0	0	1
EC:3.2.1.18	Exo-alpha-sialidase.	0	0	0	0	0	0	0	0	0	2	0
EC:3.2.1.20	Alpha-glucosidase.	3	3	1	3	1	4	1	2	1	1	1
EC:3.2.1.21	Beta-glucosidase.	5	4	2	3	4	12	1	2	1	2	4
EC:3.2.1.22	Alpha-galactosidase.	2	4	2	2	3	5	0	1	1	1	2
EC:3.2.1.23	Beta-galactosidase.	5	5	4	4	3	4	1	3	3	4	5
EC:3.2.1.24	Alpha-mannosidase.	1	0	0	0	1	0	0	3	3	2	0
EC:3.2.1.25	Beta-mannosidase.	0	0	0	0	0	1	0	0	0	0	0
EC:3.2.1.26	Beta-fructofuranosidase.	1	1	1	1	1	2	1	1	1	1	1
EC:3.2.1.31	Beta-glucuronidase.	0	0	0	0	0	1	0	1	0	0	0
EC:3.2.1.37	Xylan 1,4-beta-xylosidase.	2	1	1	1	3	2	1	1	0	0	4
EC:3.2.1.4	Cellulase.	1	1	1	0	1	0	0	0	0	0	2
EC:3.2.1.41	Pullulanase.	1	1	1	0	0	1	1	0	0	0	1
EC:3.2.1.45	Glucosylceramidase.	1	1	0	2	1	1	0	0	0	0	1
EC:3.2.1.51	Alpha-L-fucosidase.	0	0	0	0	0	1	0	0	0	3	0
EC:3.2.1.52	Beta-N-acetylhexosaminidase.	1	1	1	1	1	0	1	2	2	4	2
EC:3.2.1.54	Cyclomaltodextrinase.	0	0	0	0	0	0	0	1	1	1	0
EC:3.2.1.55	Alpha-N-arabinofuranosidase.	2	0	1	1	3	3	0	3	3	0	3
EC:3.2.1.78	Mannan endo-1,4-beta-mannosidase.	0	0	0	1	0	1	0	0	0	0	0
EC:3.2.1.89	Arabinogalactan endo-1,4-beta-galactosidase.	0	0	0	0	0	0	0	1	1	0	0
EC:3.2.1.96	endo-beta-N-acetylglucosaminidase.	0	0	0	0	0	0	0	1	1	0	0

organization of glycosyl hydrolases and oligosaccharides in the bifidobacteria was confirmed in successive genome sequences (Barrangou *et al.*, 2009; Lee *et al.*, 2008; Sela *et al.*, 2008). These clusters exhibit a conserved modular architecture that consists of a regulatory element (*lac*I-type or otherwise), ABC transporters with an extracellular oligosaccharide binding protein and one or more genes encoding glycosyl hydrolases. Several of these gene clusters contain recent gene duplications and mobile elements that may reflect the selective pressure to innovate catabolic capabilities in the distal GIT.

The complete genome sequence of the closely related *B. longum* subsp. *infantis* ATCC15697, revealed a very different evolutionary strategy for this infant-linked phylotype (Sela *et al.*, 2008). While the genomic architecture of oligosaccharide-active loci remains conserved, the quantity and character of specific genes encoded within these clusters diverge appreciably from subspecies *longum*. The primary differences between the two subspecies include the presence of a suite of oligosaccharide binding proteins (Noll *et al.*, 2008; Sela *et al.*, 2008). In addition, there are several catabolic clusters dedicated to transport and metabolism of mammalian-type glycans that are fucosylated and/or sialylated. Moreover, and in contrast with subspecies *longum*, xylose and arabinose catabolism is absent or degraded in the subspecies *infantis* genome. Finally, the clustering of glycosyl hydrolases and oligosaccharide binding proteins in a large (~43 kbp) locus active on milk oligosaccharides and/or mucopolysaccharides is a significant genomic innovation that highlights the importance of the metabolism of these sugars in the *infantis* lineage.

As more bifidobacterial genomes are sequenced, it will become apparent if this general catabolic architecture is conserved across bifidobacteria or is a hallmark of *B. longum*-related phylotypes. Functional genomics performed on *B. breve* UCC2003 have shown this species possesses glycolytic operons for FOS utilization (Ryan *et al.*, 2005), α-glucosides (Pokusaeva *et al.*, 2009) and starches (O'Connell Motherway *et al.*, 2008), which are not always proximal to an ABC transporter. Nevertheless, it appears that many of the *B. longum* oligosaccharide catabolic clusters are conserved and are co-linear to *B. adolescentis* (unpublished observation). Interestingly, the smaller *B. animalis* subsp. *lactis* genome possesses a smaller repertoire of glycosidases with ABC transporters appearing in atypical non-operon structure (Barrangou *et al.*, 2009).

Species identification by metabolic profiles

The ability to ferment sugars has been historically used to delineate species boundaries. Early bifidobacteria classification schemes were based on the conversion of carbohydrate substrates into acid in conjunction with serological properties (Reuter, 1971). Assaying carbohydrate fermentation profiles may be easy to perform, although it often yields imprecise results. As such, many fermentation studies have been performed without the benefit of molecular strain typing, thus inconsistencies in species/strain identification confound linkage of phenotype to phylogeny. Currently, molecular typing of bifidobacterial isolates using conserved phylogenetic markers dominate laboratory standard operating procedures. Nevertheless, there are certain instances where it is desirable to assay fermentation phenotype, particularly in the identification of closely related subspecies. This is notably useful for differentiating subspecies belonging to the *B. longum* complex. For example, *B. longum* subsp. *longum* and subsp. *suis* are able to ferment L-arabinose, whereas subsp. *infantis* is incapable of utilizing this plant-derived pentose (Sakata *et al.*, 2002).

Bifidobacterial vitamin biosynthesis

In the most comprehensive study reported to date, Deguchi *et al.* (1985) tested the ability of 24 bifidobacterial strains encompassing five human-associated phylotypes to synthesize various water-soluble vitamins (thiamine, folic acid, nicotinic acid, pyridoxine, vitamin B12 and riboflavin). Many strains are competent for synthesis and excretion of thiamine, folate, nicotinate, pyridoxine and vitamin B12, although riboflavin synthesis is more variable. Moreover, intracellular accumulation of thiamine, nicotinate and folate varied widely among the tested bifidobacteria. Specifically, these three vitamins accumulated to high concentrations in all *B. bifidum* and *B. longum* subsp. *infantis* strains. While many of the

B. longum subsp. *longum* and *B. breve* strains tested accumulate these vitamins, they do so at a significantly lower concentration. Most *B. adolescentis* strains do not accumulate vitamins and require the addition of thiamine and nicotinate to the culture media to achieve optimal growth. Interestingly, the lower accumulating *B. longum* subsp. *longum* strains exhibited feedback repression of thiamine synthesis when exogenous thiamine was supplemented. In addition, the bifidobacteria produce biotin and are generally believed to require vitamin B12 and panthotenic acid for growth (Noda *et al.*, 1994; Tamura, 1983).

The genome sequence of *B. longum* subsp. *infantis* provided additional insight into how this phylotype may supplement host nutritional requirements (Sela *et al.*, 2008). Accordingly, anabolic pathways for riboflavin, thiamine and folate were reconstructed from the genome sequence, whereas the complete biosynthetic capability for biotin, pyridoxine, cobalamin or panthenate was not found, which is consistent with predictions from the *B. longum* subsp. *longum* NCC2705 genome (Schell *et al.*, 2002).

Nitrogen requirements

Bifidobacterial proteolytic systems are somewhat poorly understood in contrast to the extensive body of research generated on the lactic acid bacteria. However there have been a few studies of proteolytic metabolism in the bifidobacteria. Seo *et al.* demonstrated that *B. breve* cell extracts possess aminopeptidase and iminopeptidase activities, while *B. longum* subsp. *infantis*, *B. longum* subsp. *longum* and *B. adolescentis* exhibit amino-, di-, tri-, and carboxypeptidase activities (Seo *et al.*, 2007). Likewise, the *B. longum* subsp. *longum* NCC2705 genome encodes more than 20 peptidases as well as oligopeptide transporters (Schell *et al.*, 2002). Although widely distributed in the LAB, a cell wall-associated proteinase such as a PrtP has not been identified in the bifidobacteria (Janer *et al.*, 2005). Accordingly, bifidobacteria often grow poorly in pure milk as they lack extracellular proteolytic capacity often exhibited by lactic acid bacteria (Seo *et al.*, 2007; Shah, 2000).

In an additional study, *B. longum* BORI dipeptidase activity on Ala-Ala dipeptide was experimentally determined to be two times that of other tested *B. longum* strains (Folkertsma and Fox, 1992). Seo *et al.* (2007) purified the monomeric dipeptidase enzyme in addition to cloning and biochemically characterizing recombinant *pepD* dipeptidase. Similarly, a *B. animalis* subsp. *lactis* endopeptidase (*pepO*) active on purified α_{S1}-casein (f1–23) is likely to be instrumental in the degradation of whey peptides (Janer *et al.*, 2005). This may contribute to the known enhancement of bifidobacterial growth by providing glycomacropeptide and/or whey protein to the culture (Janer *et al.*, 2004).

Biosynthesis of monomers

Pathways for the biosynthesis of monomers that are further assembled into macromolecules have been identified in bifidobacterial genome sequences. Both *B. longum* subsp. *longum* NCC 2705 and *B. longum* subsp. *infantis* ATCC 15697 are predicted to synthesize at least 19 amino acids from NH_4^+ (Schell *et al.*, 2002; Sela *et al.*, 2008). Interestingly, Schell *et al.* postulate that cysteine biosynthesis and sulphur assimilation may be coupled in *B. longum* and proceed by way of an atypical pathway involving cystathionine γ-synthase, cystathionine β-synthetase and cystathionine γ-lyase to assimilate succinylhomoserine and H_2S or methanethiol produced by other colonic microbes (Schell *et al.*, 2002). The pathways for purine and pyrimidine biosynthesis are found intact in both subsp. *infantis* and subsp. *longum* as the bifidobacteria are not nearly as fastidious as certain lactic acid bacteria.

Utilization of inorganic nitrogen

Bifidobacteria are generally able to utilize NH_4^+ as their sole nitrogen source, with many species preferring inorganic nitrogen (Tanaka and Mutai, 1980). Ammonium is predicted to be translocated across the cell envelope through a dedicated transporter with glutamate dehydrogenase (EC 1.4.1.3) and glutamine synthetase (EC 6.3.1.2) probably used in assimilation (Sela *et al.*, 2008; Ventura *et al.*, 2004). To date, glutamine synthetase has been purified and characterized from several bifidobacterial species (Ventura *et al.*, 2004).

Urease

Several studies have characterized urease activity in the bifidobacteria. In one study, 414 strains representing 21 bifidobacterial phylotypes were assayed for the presence of ureolytic activity on Christiansen's agar (EC 3.3.1.5) (Christensen, 1946). The strongest urease activity was found in most *Bifidobacterium suis* (now *Bifidobacterium longum* subsp. *suis*) in addition to other species. High levels of urease are observed in the absence of exogenous urea suggesting that the ureolytic enzyme is not inducible, congruent with typical bacterial ureases (Crociani and Matteuzzi, 1982). A previous study conducted on human gut anaerobes reported that *B. longum* subsp. *infantis* invariably exhibits urease activity while only one strain out of 12 assayed *B. bifidum* isolates hydrolysed urea (Suzuki *et al.*, 1979). Recently, the first bifidobacterial urease genetic sequence has been described in *B. longum* subsp. *infantis* ATCC 15697 (Sela *et al.*, 2008). This intriguing discovery has prompted further questions regarding the interconnecting of bifidobacterial metabolism to that of its infant host. Accordingly, the provision of nitrogen to the distal GIT is limiting while the infant is nursing. Depending on the mother's nutritional status, the protein concentration of milk is often inadequate to satisfy the increased nitrogen demands necessitated by rapid infant growth, let alone the metabolic requirements of the resident microbiota. The urea content of human milk constitutes a substantial fraction (~15%) of the total available nitrogen and represents a potential reservoir for both the infant and its microbial inhabitants (Donovan and Lonnerdal, 1989). Urea nitrogen accession is believed to be catalysed by bacterial ureolytic activity. This reclamation of milk urea, in addition to urea produced by metabolic processes, is known as urea nitrogen salvaging (reviewed by Fuller and Reeds, 1998).

Iron

Gut microbiota require ferric iron for growth and are inhibited by divalent metal ion chelators. Animal hosts attempt to sequester free iron from pathogens by deploying iron-binding proteins such as lactoferrin (reviewed in Schaible and Kaufmann, 2004). Symbionts such as the bifidobacteria potentially benefit their host by restricting pathogen access to iron. To accomplish this, some bifidobacteria anchor a ferrireductase to their extracellular surface to convert ferric iron to ferrous iron. Fe^{2+} is subsequently transported into the cell by way of an ATP-dependent divalent metal permease that also relies on maintenance of a proton gradient (Bezkorovainy and Solberg, 1989). Under anaerobic conditions ferrous iron is available, although bifidobacteria may capture more than required for growth in competition with other microbes. A reduction in free iron likewise diminishes the concentration of free radicals within their host.

Bifidobacteria typically encounter lactoferrin from breast milk or secreted by the host mucosa. Iron is partitioned in breast milk with 35% bound to lactoferrin, 30% associated with the milk-fat globule membrane and the rest is chelated by small molecules such as citrate (Lonnerdal, 1989). It is not known if the bifidobacteria fulfil their metabolic requirement for iron from citrate-bound or xanthine oxidase ions (Bezkorovainy and Kot, 1998). Instead, lactoferrin-bound iron is probably preferred, as was observed for *B. breve* (Miller-Catchpole *et al.*, 1997). Lactoferrin binding proteins have been characterized in *B. bifidum* which correlates with observed lactoferrin-mediated bifidogenesis (Kim *et al.*, 2002). Moreover, it has been demonstrated that ferric iron adsorbs to the cell envelope in experiments conducted on *B. thermophlium* (Kot and Bezkorovainy, 1999). Although the bifidobacteria were previously not believed to produce siderophores, recently a genetic sequence has been annotated as such in *B. longum* subsp. *longum* DJO10A (Lee *et al.*, 2008; Miller-Catchpole *et al.*, 1997).

Bifidobacterial media

There have been numerous media developed for routine bifidobacteria isolation and enumeration that exploits aspects of their metabolism and intrinsic resistance to antimicrobial compounds. Common weakly- or non-selective growth media includes tryptone phytone yeast (TPY) medium (Scardovi, 1986), de Man, Rogosa and Sharpe (MRS) (De Man *et al.*, 1960) and reinforced clostridial medium (RCM) (Willis *et al.*, 1973) among others. A multitude of bifidobacteria selective media have been developed with a few highlighted here. A widely used formulation is *Bifidobacterium* iodoacetate medium which uses iodoacetate 25 mg/l (BIM-25) or 50 mg/l (BIM-50) (Munoa and Pares, 1988). Iodoacetate inhibits glyceraldehyde-3-phosphate dehydrogenase that is not essential for bifidobacterial fermentative metabolism thus restricting the growth of contaminants. In addition, the RCM-based BIM contains several antimicrobials including nalidixic acid, polymixin B, and kanamycin. Another widely used medium was developed by Beeren which combines low pH conditions with the addition of propionic acid to select for bifidobacteria (Beerens, 1990). Recently, several groups have been supplementing 50 mg/l mupirocin to MRS-derived media to produce *bifidobacterium* selective medium (BSM) (Simpson *et al.*, 2004; Turroni *et al.*, 2009).

An alternative or synergistic strategy is to use growth-enhancing compounds to promote and select for bifidobacterial growth. Accordingly, many groups have developed media with oligosaccharides as the sole carbohydrate source. Hartemink and colleagues added raffinose to the inhibitory agents lithium chloride and propionate (Hartemink *et al.*, 1996). This medium uses casein as a protein source and therefore does not select for bifidobacteria incapable of hydrolysing casein. Along those same lines, transoligosaccharides have been added to media preparations as well (Thitaram, 2005).

Future of bifidobacterial metabolism research

While whole genome sequencing has advanced bifidobacterial metabolism, there are many areas that remain underexplored. It is critical to understand carbohydrate flux in these saccharolytic organisms both mechanistically and at systems-level resolution. At present the molecular details of nitrogenous matter and micronutrient acquisition remain vague. In addition, the biosynthesis of cellular components such as fatty acids and cell wall components have been largely ignored recently and deserve to be pursued with renewed vigour. These studies have been hindered somewhat by the lack of dependable genetic tools in the bifidobacteria. Hopefully recent progress to generate a reverse-genetic system will now permit genome-informed physiological studies (O'Connell Motherway *et al.*, 2008).

In addition, there is much unknown regarding the metabolic relationship between bifidobacteria and milk components encountered in the infant GIT. Accordingly, milk glycopeptides have already been observed to stimulate bifidobacterial growth *in vitro* (Janer *et al.*, 2004; Liepke *et al.*, 2002). Unquestionably the diversity of these molecules demand a concerted effort to match specific glycopeptide structures with function to fully appreciate their bioactive properties as bifidogenic substrates or otherwise. Experiments performed in animal models or clinical studies will be critical to confirm *in vivo* activity of these bioactive components in milk.

Finally, the logical extension of applied metabolic research will require investigating the consequences to the gut microbial consortium in its entirety, following exposure to an assortment of oligosaccharide substrates. This is of particular importance to discriminate the influence of specific dietary molecules on the establishment and maintenance of microbial communities. There is strong evidence that the functional composition and ecological succession within human-associated microbiota influences the health status of their hosts (Turnbaugh *et al.*, 2006). Thus bifidobacteria-dominated consortia may have a large impact on host metabolism as they are abundant in both infants and adults. Clearly, important hypotheses remain to be tested in this regard when considering that bacterial metabolites have been detected in distal host tissue (Martin *et al.*, 2008).

Acknowledgements

DAS was supported by a pre-doctoral training grant (NIH-NIGMS T32-GM08799). Contents of this chapter are solely the responsibility of the authors and do not necessarily represent the official views of the NIGMS or NIH. We also acknowledge grant support from the University of California Discovery Grant Program, the California Dairy Research Foundation, and by USDA NRI-CSREES Award 2008-35200-18776 (DAM).

References

Amaretti, A., Bernardi, T., Tamburini, E., Zanoni, S., Lomma, M., Matteuzzi, D., and Rossi, M. (2007). Kinetics and metabolism of *Bifidobacterium adolescentis* MB 239 growing on glucose, galactose, lactose, and galactooligosaccharides. Appl. Environ. Microbiol. 73, 3637–3644.

Amaretti, A., Tamburini, E., Bernardi, T., Pompei, A., Zanoni, S., Vaccari, G., Matteuzzi, D., and Rossi,

M. (2006). Substrate preference of *Bifidobacterium adolescentis* MB 239: compared growth on single and mixed carbohydrates. Appl. Microbiol. Biotechnol. *73*, 654–662.

Andersson, B., Porras, O., Hanson, L.A., Lagergard, T., and Svanborg-Eden, C. (1986). Inhibition of attachment of *Streptococcus pneumoniae* and *Haemophilus influenzae* by human milk and receptor oligosaccharides. J. Infect. Dis. *153*, 232–237.

Angeloni, S., Ridet, J.L., Kusy, N., Gao, H., Crevoisier, F., Guinchard, S., Kochhar, S., Sigrist, H., and Sprenger, N. (2005). Glycoprofiling with micro-arrays of glycoconjugates and lectins. Glycobiology *15*, 31–41.

Asakuma, S., Urashima, T., Akahori, M., Obayashi, H., Nakamura, T., Kimura, K., Watanabe, Y., Arai, I., and Sanai, Y. (2008). Variation of major neutral oligosaccharides levels in human colostrum. Eur. J. Clin. Nutr. *62*, 488–494.

Asano, T., Yuasa, K., Nishiwaki, S., and Iino, H. (2005). Isolation rates of gluconate-fermenting bifidobacteria from human faeces. Microbial. Ecol. Health Dis. *17*, 114–120.

Barrangou, R., Briczinski, E.P., Traeger, L.L., Loquasto, J.R., Richards, M., Horvath, P., Coute-Monvoisin, A.C., Leyer, G., Rendulic, S., Steele, J.L., Broadbent, J.R., Oberg, T., Dudley, E.G. Schuster, S., Romero, D.A., and Roberts, R.F. (2009). Comparison of the complete genome sequences of *Bifidobacterium animalis* subsp. *lactis* DSM 10140 and Bl-04. J. Bacteriol. *191*, 4144–4151.

Beerens, H. (1990). An elective and selective isolation medium for *Bifidobacterium* spp. Lett. Appl. Microbiol. *11*, 155–157.

Belenguer, A., Duncan, S.H., Calder, A.G., Holtrop, G., Louis, P., Lobley, G.E., and Flint, H.J. (2006). Two routes of metabolic cross-feeding between *Bifidobacterium adolescentis* and butyrate-producing anaerobes from the human gut. Appl. Environ. Microbiol. 72, 3593–3599.

Bezkorovainy, A., and Kot, E. (1998). Interaction of bifidobacteria with ferric iron. Int. Dairy J. *8*, 507–512.

Bezkorovainy, A., and Solberg, L. (1989). Ferrous iron uptake by *Bifidobacterium breve*. Biol Trace Elem. Res. *20*, 251–267.

Blatch, G.L., and Woods, D.R. (1993). Molecular characterization of a fructanase produced by *Bacteroides fragilis* BF-1. J. Bacteriol. *175*, 3058–3066.

Bode, L. (2006). Recent advances on structure, metabolism, and function of human milk oligosaccharides. J. Nutr. *136*, 2127–2130.

Boehm, G., and Stahl, B. (2007). Oligosaccharides from milk. J. Nutr. *137*, 847S-849S.

Boehm, G., Stahl, B., Jelinek, J., Knol, J., Miniello, V., and Moro, G.E. (2005). Prebiotic carbohydrates in human milk and formulas. Acta. Paediatr. Suppl. *94*, 18–21.

Bouhnik, Y., Vahedi, K., Achour, L., Attar, A., Salfati, J., Pochart, P., Marteau, P., Flourie, B., Bornet, F., and Rambaud, J.C. (1999). Short-chain fructo-oligosaccharide administration dose-dependently increases fecal bifidobacteria in healthy humans. J. Nutr. *129*, 113–116.

Caescu, C.I., Vidal, O., Krzewinski, F., Artenie, V., and Bouquelet, S. (2004). *Bifidobacterium longum* requires a fructokinase (Frk; ATP:D-fructose 6-phosphotransferase, EC 2.7.1.4) for fructose catabolism. J. Bacteriol. *186*, 6515–6525.

Chaturvedi, P., Warren, C.D., Altaye, M., Morrow, A.L., Ruiz-Palacios, G., Pickering, L.K., and Newburg, D.S. (2001). Fucosylated human milk oligosaccharides vary between individuals and over the course of lactation. Glycobiology *11*, 365–372.

Chouraqui, J.P., van Egroo, L.D., and Fichot, M.C. (2004). Acidified milk formula supplemented with *Bifidobacterium lactis*: impact on infant diarrhea in residential care settings. J. Pediatr. Gastroenterol. Nutr. *38*, 288–292.

Christensen, W.B. (1946). Urea Decomposition as a Means of Differentiating Proteus and Paracolon Cultures from Each Other and from *Salmonella* and *Shigella* Types. J. Bacteriol. *52*, 461–466.

Coppa, G.V., Gabrielli, O., Pierani, P., Catassi, C., Carlucci, A., and Giorgi, P.L. (1993). Changes in carbohydrate composition in human milk over 4 months of lactation. Pediatrics *91*, 637–641.

Crociani, F., and Matteuzzi, D. (1982). Urease activity in the genus *Bifidobacterium*. Ann. Microbiol. *133*, 417–423.

Deguchi, Y., Morishita, T., and Mutai, M. (1985). Comparative studies on synthesis of water-soluble vitamins among human species of bifidobacteria. Agric. and Biol. Chem. *49*, 13–19.

De Man, J., Rogosa, M., and Sharpe, M. (1960). A medium for the cultivation of lactobacilli. J. Appl. Microbiol. *23*, 130–135.

de Vries, W., and Stouthamer, A.H. (1967). Pathway of glucose fermentation in relation to the taxonomy of bifidobacteria. J. Bacteriol. *93*, 574–576.

Degnan, B.A., and Macfarlane, G.T. (1991). Comparison of carbohydrate substrate preferences in eight species of bifidobacteria. FEMS Microbiol. Lett. *68*, 151–156.

Donovan, S.M., and Lonnerdal, B. (1989). Non-protein nitrogen and true protein in infant formulas. Acta. Paediatr. Scand. *78*, 497–504.

Duncan, S.H., Holtrop, G., Lobley, G.E., Calder, A.G., Stewart, C.S., and Flint, H.J. (2004). Contribution of acetate to butyrate formation by human faecal bacteria. Br. J. Nutr. *91*, 915–923.

Duncan, S.H., Scott, K.P., Ramsay, A.G., Harmsen, H.J., Welling, G.W., Stewart, C.S., and Flint, H.J. (2003). Effects of alternative dietary substrates on competition between human colonic bacteria in an anaerobic fermentor system. Appl. Environ. Microbiol. *69*, 1136–1142.

Ehrmann, M.A., Korakli, M., and Vogel, R.F. (2003). Identification of the gene for beta-fructofuranosidase of *Bifidobacterium lactis* DSM10140(T) and characterization of the enzyme expressed in *Escherichia coli*. Curr. Microbiol. *46*, 391–397.

Eiwegger, T., Stahl, B., Schmitt, J., Boehm, G., Gerstmayr, M., Pichler, J., Dehlink, E., Loibichler, C., Urbanek, R., and Szepfalusi, Z. (2004). Human milk-derived oligosaccharides and plant-derived oligosaccharides stimulate cytokine production of cord blood T-cells *in vitro*. Pediatr. Res. *56*, 536–540.

Engfer, M.B., Stahl, B., Finke, B., Sawatzki, G., and Daniel, H. (2000). Human milk oligosaccharides are resistant to enzymatic hydrolysis in the upper gastrointestinal tract. Am. J. Clin. Nutr. *71*, 1589–1596.

Falony, G., Calmeyn, T., Leroy, F., and de Vuyst, L. (2009a). Coculture fermentations of Bifidobacterium species and *Bacteroides thetaiotaomicron* reveal a mechanistic insight into the prebiotic effect of inulin-type fructans. Appl. Environ. Microbiol. *75*, 2312–2319.

Falony, G., Lazidou, K., Verschaeren, A., Weckx, S., Maes, D., and de Vuyst, L. (2009b). *In vitro* kinetic analysis of fermentation of prebiotic inulin-type fructans by *Bifidobacterium* species reveals four different phenotypes. Appl. Environ. Microbiol. *75*, 454–461.

Falony, G., Vlachou, A., Verbrugghe, K., and de Vuyst, L. (2006). Cross-feeding between *Bifidobacterium longum* BB536 and acetate-converting, butyrate-producing colon bacteria during growth on oligofructose. Appl. Environ. Microbiol. *72*, 7835–7841.

Flint, H.J., Bayer, E.A., Rincon, M.T., Lamed, R., and White, B.A. (2008). Polysaccharide utilization by gut bacteria: potential for new insights from genomic analysis. Nat. Rev. Microbiol. *6*, 121–131.

Folkertsma, B., and Fox, P.F. (1992). Use of the Cd-ninhydrin reagent to assess proteolysis in cheese during ripening. J. Dairy Res. *59*, 217–224.

Fujita, K., Oura, F., Nagamine, N., Katayama, T., Hiratake, J., Sakata, K., Kumagai, H., and Yamamoto, K. (2005). Identification and molecular cloning of a novel glycoside hydrolase family of core 1 type O-glycan-specific endo-alpha-N-acetylgalactosaminidase from *Bifidobacterium longum*. J. Biol. Chem. *280*, 37415–37422.

Fuller, M.F., and Reeds, P.J. (1998). Nitrogen cycling in the gut. Annu. Rev. Nutr. *18*, 385–411.

Furrie, E., Macfarlane, S., Kennedy, A., Cummings, J.H., Walsh, S.V., O'Neil D.A., and Macfarlane, G.T. (2005). Synbiotic therapy (*Bifidobacterium longum* /Synergy 1) initiates resolution of inflammation in patients with active ulcerative colitis: a randomised controlled pilot trial. Gut *54*, 242–249.

Gavini, F., Van Esbroeck, M., Touzel, J.P., Fourment, A., and Goossens, H. (1996). Detection of Fructose-6-phosphate Phosphoketolase (F6PPK), a Key Enzyme of the Bifid-Shunt, in *Gardnerella vaginalis*. Anaerobe *2*, 191–193.

German, J.B., Freeman, S.L., Lebrilla, C.B., and Mills, D.A. (2008). Human milk oligosaccharides: evolution, structures and bioselectivity as substrates for intestinal bacteria. Nestle Nutr. Workshop Ser. Pediatr. Program *62*, 205–222.

Gibson, G.R., Beatty, E.R., Wang, X., and Cummings, J.H. (1995). Selective stimulation of bifidobacteria in the human colon by oligofructose and inulin. Gastroenterology *108*, 975–982.

Gnoth, M.J., Kunz, C., Kinne-Saffran, E., and Rudloff, S. (2000). Human milk oligosaccharides are minimally digested *in vitro*. J. Nutr. *130*, 3014–3020.

Gonzalez, R., Klaassens, E.S., Malinen, E., de Vos, W.M., and Vaughan, E.E. (2008). Differential transcriptional response of *Bifidobacterium longum* to human milk, formula milk, and galactooligosaccharide. Appl. Environ. Microbiol. *74*, 4686–4694.

Gopal, P.K., Sullivan, P.A., and Smart, J.B. (2001). Utilisation of galacto-oligosaccharides as selective substrates for growth by lactic acid bacteria including *Bifidobacterium lactis* DR10 and *Lactobacillus rhamnosus* DR20. Int. Dairy J. *11*, 19–25.

Goulas, T., Goulas, A., Tzortzis, G., and Gibson, G.R. (2009a). Comparative analysis of four beta-galactosidases from *Bifidobacterium bifidum* NCIMB41171: purification and biochemical characterisation. Appl. Microbiol. Biotechnol. *82*, 1079–1088.

Goulas, T., Goulas, A., Tzortzis, G., and Gibson, G.R. (2009b). Expression of four beta-galactosidases from *Bifidobacterium bifidum* NCIMB41171 and their contribution on the hydrolysis and synthesis of galactooligosaccharides. Appl. Microbiol. Biotechnol. *84*, 899–907.

Goulas, T.K., Goulas, A.K., Tzortzis, G., and Gibson, G.R. (2007). Molecular cloning and comparative analysis of four beta-galactosidase genes from *Bifidobacterium bifidum* NCIMB41171. Appl. Microbiol. Biotechnol. *76*, 1365–1372.

Grill, J.P., Crociani, J., and Ballongue, J. (1995). Characterization of fructose-6-phosphate phosphoketolases purified from *Bifidobacterium* species. Curr. Microbiol. *31*, 49–54.

Hartemink, R., Kok, B.J., Weenk, G.H., and Rombouts, F.M. (1996). Raffinose-*Bifidobacterium* (RB) agar, a new selective medium for bifidobacteria. J. Microbiol. Meth. *27*, 33–43.

Hidaka, M., Nishimoto, M., Kitaoka, M., Wakagi, T., Shoun, H., and Fushinobu, S. (2009). The crystal structure of galacto-N-biose/lacto-N-biose I phosphorylase: a large deformation of a TIM barrel scaffold. J. Biol. Chem. *284*, 7273–7283.

Hinz, S.W., Pastink, M.I., van den Broek, L.A., Vincken, J.P., and Voragen, A.G. (2005a). *Bifidobacterium longum* endogalactanase liberates galactotriose from type I galactans. Appl. Environ. Microbiol. *71*, 5501–5510.

Hinz, S.W., Verhoef, R., Schols, H.A., Vincken, J.P., and Voragen, A.G. (2005b). Type I arabinogalactan contains beta-D-Galp-(1--> 3)-beta-D-Galp structural elements. Carbohydr. Res. *340*, 2135–2143.

Hung, M.N., Xia, Z., Hu, N.T., and Lee, B.H. (2001). Molecular and biochemical analysis of two beta-galactosidases from *Bifidobacterium infantis* HL96. Appl. Environ. Microbiol. *67*, 4256–4263.

Janer, C., Arigoni, F., Lee, B.H., Peláez, C., and Requena, T. (2005). Enzymatic ability of *Bifidobacterium animalis* subsp. *lactis* to hydrolyze milk proteins: identification and characterization of endopeptidase O. Appl. Environ. Microbiol. *71*, 8460–8465.

Janer, C., Peláez, C., and Requena, T. (2004). Caseinomacropeptide and whey protein concentrate enhance *Bifidobacterium lactis* growth in milk. Food. Chem. *86*, 263–267.

Kandler, O. (1983). Carbohydrate metabolism in lactic acid bacteria. Ant. Van Leeuwenhoek *49*, 209–224.

Katayama, T., Sakuma, A., Kimura, T., Makimura, Y., Hiratake, J., Sakata, K., Yamanoi, T., Kumagai, H., and Yamamoto, K. (2004). Molecular cloning and characterization of *Bifidobacterium bifidum* 1,2-alpha-L-fucosidase (AfcA), a novel inverting glycosidase (glycoside hydrolase family 95). J. Bacteriol. *186*, 4885–4893.

Kim, T.B., Song, S.H., Kang, S.C., and Oh, D.K. (2003). Quantitative comparison of lactose and glucose utilization in *Bifidobacterium longum* cultures. Biotechnol. Prog. *19*, 672–675.

Kim, W.S., Tanaka, T., Kumura, H., and Shimazaki, K. (2002). Lactoferrin-binding proteins in *Bifidobacterium bifidum*. Biochem. Cell Biol. *80*, 91–94.

Kitaoka, M., Tian, J., and Nishimoto, M. (2005). Novel putative galactose operon involving lacto-N-biose phosphorylase in *Bifidobacterium longum*. Appl. Environ. Microbiol. *71*, 3158–3162.

Kiyohara, M., Tachizawa, A., Nishimoto, M., Kitaoka, M., Ashida, H., and Yamamoto, K. (2009). Prebiotic effect of Lacto-N-biose I on bifidobacterial growth. Biosci. Biotechnol. Biochem. 73, 1175–1179.

Kleessen, B., Hartmann, L., and Blaut, M. (2001). Oligofructose and long-chain inulin: influence on the gut microbial ecology of rats associated with a human faecal flora. Br. J. Nutr. *86*, 291–300.

Klein, A., Friedrich, U., Vogelsang, H., and Jahreis, G. (2008). *Lactobacillus acidophilus* 74–2 and *Bifidobacterium animalis* subsp. *lactis* DGCC 420 modulate unspecific cellular immune response in healthy adults. Eur. J. Clin. Nutr. *62*, 584–593.

Kot, E., and Bezkorovainy, A. (1999). Binding of ferric iron to the cell walls and membranes of *Bifidobacterium thermophilum*: effect of free radicals. J. Agric. Food Chem. *47*, 4606–4610.

Kruse, H.P., Kleessen, B., and Blaut, M. (1999). Effects of inulin on faecal bifidobacteria in human subjects. Br. J. Nutr. *82*, 375–382.

Krzewinski, F., Brassart, C., Gavini, F., and Bouquelet, S. (1996). Characterization of the lactose transport system in the strain *Bifidobacterium bifidum* DSM 20082. Curr. Microbiol. 32, 301–307.

Kunz, C., Rudloff, S., Baier, W., Klein, N., and Strobel, S. (2000). Oligosaccharides in human milk: structural, functional, and metabolic aspects. Annu. Rev. Nutr. *20*, 699–722.

Langlands, S.J., Hopkins, M.J., Coleman, N., and Cummings, J.H. (2004). Prebiotic carbohydrates modify the mucosa associated microflora of the human large bowel. Gut 53, 1610–1616.

Le Blay, G., Michel, C., Blottiere, H.M., and Cherbut, C. (1999). Prolonged intake of fructo-oligosaccharides induces a short-term elevation of lactic acid-producing bacteria and a persistent increase in cecal butyrate in rats. J. Nutr. *129*, 2231–2235.

Lee, J.H., Karamychev, V.N., Kozyavkin, S.A., Mills, D., Pavlov, A.R., Pavlova, N.V., Polouchine, N.N., Richardson, P.M., Shakhova, V.V., Slesarev, A.I., Weimer, B.C., and O'Sullivan, D.J. (2008). Comparative genomic analysis of the gut bacterium *Bifidobacterium longum* reveals loci susceptible to deletion during pure culture growth. BMC Genomics *9*, 247.

Liepke, C., Adermann, K., Raida, M., Magert, H.J., Forssmann, W.G., and Zucht, H.D. (2002). Human milk provides peptides highly stimulating the growth of bifidobacteria. Eur. J. Biochem. *269*, 712–718.

Lievin, V., Peiffer, I., Hudault, S., Rochat, F., Brassart, D., Neeser, J.R., and Servin, A.L. (2000). *Bifidobacterium* strains from resident infant human gastrointestinal microflora exert antimicrobial activity. Gut *47*, 646–652.

LoCascio, R.G., Ninonuevo, M.R., Freeman, S.L., Sela, D.A., Grimm, R., Lebrilla, C.B., Mills, D.A., and German, J.B. (2007). Glycoprofiling of bifidobacterial consumption of human milk oligosaccharides demonstrates strain specific, preferential consumption of small chain glycans secreted in early human lactation. J. Agric. Food Chem. *55*, 8914–8919.

LoCascio, R.G., Niñonuevo, M., Kronewitter, S.R., Freeman, S.L., German, J.B., Lebrilla, C.B., and Mills, D.A. (2009). A versatile and scalable strategy for glycoprofiling bifidobacterial consumption of human milk oligosaccharides. Microbial Biotechnology *2*, 333–342.

Lonnerdal, B. (1989). Trace element absorption in infants as a foundation to setting upper limits for trace elements in infant formulas. J. Nutr. *119*, 1839–1844; discussion 1845.

Macfarlane, G.T., and Cummings, J.H. (1999). Probiotics and prebiotics: can regulating the activities of intestinal bacteria benefit health? BMJ *318*, 999–1003.

Macfarlane, G.T., Steed, H., and Macfarlane, S. (2008). Bacterial metabolism and health-related effects of galacto-oligosaccharides and other prebiotics. J. Appl. Microbiol. *104*, 305–344.

Madsen, K., Cornish, A., Soper, P., McKaigney, C., Jijon, H., Yachimec, C., Doyle, J., Jewell, L., and De Simone, C. (2001). Probiotic bacteria enhance murine and human intestinal epithelial barrier function. Gastroenterology *121*, 580–591.

Markowitz, V.M., Szeto, E., Palaniappan, K., Grechkin, Y., Chu, K., Chen, I.M., Dubchak, I., Anderson, I., Lykidis, A., Mavromatis, K., *et al.* (2008). The integrated microbial genomes (IMG) system in 2007: data content and analysis tool extensions. Nucleic Acids Res. *36*, D528–533.

Martin, F.P., Wang, Y., Sprenger, N., Yap, I.K., Rezzi, S., Ramadan, Z., Pere-Trepat, E., Rochat, F., Cherbut, C., van Bladeren, P., *et al.* (2008). Top-down systems biology integration of conditional prebiotic modulated transgenomic interactions in a humanized microbiome mouse model. Mol. Syst. Biol. *4*, 205.

Maze, A., O'Connell-Motherway, M., Fitzgerald, G.F., Deutscher, J., and van Sinderen, D. (2007). Identification and characterization of a fructose phosphotransferase system in *Bifidobacterium breve* UCC2003. Appl. Environ. Microbiol. 73, 545–553.

Meile, L., Rohr, L.M., Geissmann, T.A., Herensperger, M., and Teuber, M. (2001). Characterization of the D-xylulose 5-phosphate/D-fructose 6-phosphate phosphoketolase gene (xfp) from *Bifidobacterium lactis*. J. Bacteriol. *183*, 2929–2936.

Merkler, I., and Retey, J. (1981). Stereochemical investigation of the phosphoketolase reaction. The formation of chiral [2H1,3H]acetyl phosphate. Eur. J. Biochem. *120*, 593–597.

Miller-Catchpole, R., Kot, E., Haloftis, G., Furmanov, S., and Bezkorovainy, A. (1997). Lactoferrin can supply iron for the growth of *Bifidobacterium breve*. Nutr. Research *17*, 205–213.

Moller, P.L., Jorgensen, F., Hansen, O.C., Madsen, S.M., and Stougaard, P. (2001). Intra- and extracellular beta-galactosidases from *Bifidobacterium bifidum* and *B. infantis*: molecular cloning, heterologous expression, and comparative characterization. Appl. Environ. Microbiol. *67*, 2276–2283.

Moro, G., Arslanoglu, S., Stahl, B., Jelinek, J., Wahn, U., and Boehm, G. (2006). A mixture of prebiotic oligosaccharides reduces the incidence of atopic dermatitis during the first six months of age. Arch. Dis. Child *91*, 814–819.

Munoa, F.J., and Pares, R. (1988). Selective medium for isolation and enumeration of *Bifidobacterium* spp. Appl. Environ. Microbiol *54*, 1715–1718.

Nagae, M., Tsuchiya, A., Katayama, T., Yamamoto, K., Wakatsuki, S., and Kato, R. (2007). Structural basis of the catalytic reaction mechanism of novel 1,2-alpha-L-fucosidase from *Bifidobacterium bifidum*. J. Biol. Chem. *282*, 18497–18509.

Nakajima, M., and Kitaoka, M. (2008). Identification of lacto-N-Biose I phosphorylase from *Vibrio vulnificus* CMCP6. Appl. Environ. Microbiol. *74*, 6333–6337.

Nakajima, M., Nihira, T., Nishimoto, M., and Kitaoka, M. (2008). Identification of galacto-N-biose phosphorylase from *Clostridium perfringens* ATCC13124. Appl. Microbiol. Biotechnol. *78*, 465–471.

Newburg, D.S. (2009). Neonatal protection by an innate immune system of human milk consisting of oligosaccharides and glycans. J. Anim. Sci. *87*, 26–34.

Newburg, D.S., Ruiz-Palacios, G.M., Altaye, M., Chaturvedi, P., Meinzen-Derr, J., Guerrero, M.L., and Morrow, A.L. (2004). Innate protection conferred by fucosylated oligosaccharides of human milk against diarrhea in breastfed infants. Glycobiology *14*, 253–263.

Niness, K.R. (1999). Inulin and oligofructose: what are they? J. Nutr. *129*, 1402S-1406S.

Niñonuevo, M.R., Park, Y., Yin, H., Zhang, J., Ward, R.E., Clowers, B.H., German, J.B., Freeman, S.L., Killeen, K., Grimm, R., and Lebrilla, C.B. (2006). A strategy for annotating the human milk glycome. J. Agric. Food Chem. *54*, 7471–7480.

Niñonuevo, M.R., Perkins, P.D., Francis, J., Lamotte, L.M., LoCascio, R.G., Freeman, S.L., Mills, D.A., German, J.B., Grimm, R., and Lebrilla, C.B. (2008). Daily variations in oligosaccharides of human milk determined by microfluidic chips and mass spectrometry. J. Agric. Food Chem. *56*, 618–26.

Nishimoto, M., and Kitaoka, M. (2007a). Identification of N-acetylhexosamine 1-kinase in the complete lacto-N-biose I/galacto-N-biose metabolic pathway in *Bifidobacterium longum*. Appl. Environ. Microbiol. 73, 6444–6449.

Nishimoto, M., and Kitaoka, M. (2007b). Identification of the putative proton donor residue of lacto-N-biose phosphorylase (EC 2.4.1.211). Biosci. Biotechnol. Biochem. *71*, 1587–1591.

Nishimoto, M., and Kitaoka, M. (2007c). Practical preparation of lacto-N-biose I, a candidate for the bifidus factor in human milk. Biosci. Biotechnol. Biochem. *71*, 2101–2104.

Noda, H., Akasaka, N., and Ohsugi, M. (1994). Biotin production by bifidobacteria. J. Nutr. Sci. Vitaminol. *40*, 181–188.

Noll, K.M., Lapierre, P., Gogarten, J.P., and Nanavati, D.M. (2008). Evolution of mal ABC transporter operons in the *Thermococcales* and *Thermotogales*. BMC Evol. Biol. *8*, 7.

O'Connell Motherway, M., Fitzgerald, G.F., Neirynck, S., Ryan, S., Steidler, L., and van Sinderen, D. (2008). Characterization of ApuB, an extracellular type II amylopullulanase from *Bifidobacterium breve* UCC2003. Appl Environ Microbiol *74*, 6271–6279.

O'Mahony, C., Scully, P., O'Mahony, D., Murphy, S., O'Brien, F., Lyons, A., Sherlock, G., MacSharry, J., Kiely, B., Shanahan, F., and O'Mahony, L. (2008). Commensal-induced regulatory T cells mediate protection against pathogen-stimulated NF-kappa B activation. PLoS Pathog. *4*, e1000112.

O'Mahony, L., McCarthy, J., Kelly, P., Hurley, G., Luo, F., Chen, K., O'Sullivan, G.C., Kiely, B., Collins, J.K., Shanahan, F., and Quigley, E.M. (2005). *Lactobacillus* and *Bifidobacterium* in irritable bowel syndrome: symptom responses and relationship to cytokine profiles. Gastroenterology *128*, 541–551.

Obermeier, S., Rudloff, S., Pohlentz, G., Lentze, M.J., and Kunz, C. (1999). Secretion of 13C-labelled oligosaccharides into human milk and infant's urine after an oral [13C]galactose load. Isotopes Environ. Health Stud. *35*, 119–125.

Orban, J.I., and Patterson, J.A. (2000). Modification of the phosphoketolase assay for rapid identification of bifidobacteria. J. Microbiol. Methods *40*, 221–224.

Palframan, R.J., Gibson, G.R., and Rastall, R.A. (2003). Carbohydrate preferences of Bifidobacterium species isolated from the human gut. Curr. Issues Intest. Microbiol. *4*, 71–75.

Palmfeldt, J., Paese, M., Hahn-Hagerdal, B., and van Niel, E.W. (2004). The pool of ADP and ATP regulates anaerobic product formation in resting cells of *Lactococcus lactis*. Appl. Environ. Microbiol. *70*, 5477–5484.

Parche, S., Amon, J., Jankovic, I., Rezzonico, E., Beleut, M., Barutcu, H., Schendel, I., Eddy, M.P., Burkovski, A., Arigoni, F., and Titgemeyer, F. (2007). Sugar transport systems of *Bifidobacterium longum* NCC2705. J. Mol. Microbiol. Biotechnol. *12*, 9–19.

Parche, S., Beleut, M., Rezzonico, E., Jacobs, D., Arigoni, F., Titgemeyer, F., and Jankovic, I. (2006). Lactose-over-glucose preference in *Bifidobacterium longum* NCC2705: glcP, encoding a glucose transporter, is subject to lactose repression. J. Bacteriol. *188*, 1260–1265.

Parvez, S., Malik, K.A., Ah Kang, S., and Kim, H.Y. (2006). Probiotics and their fermented food products are beneficial for health. J. Appl. Microbiol. *100*, 1171–1185.

Perrin, S., Warchol, M., Grill, J.P., and Schneider, F. (2001). Fermentations of fructo-oligosaccharides and their components by *Bifidobacterium infantis* ATCC 15697 on batch culture in semi-synthetic medium. J. Appl. Microbiol. *90*, 859–865.

Podolsky, D.K. (1985). Oligosaccharide structures of human colonic mucin. J. Biol. Chem. *260*, 8262–8271.

Pokusaeva, K., O'Connell-Motherway, M., Zomer, A., Fitzgerald, G.F., and van Sinderen, D. (2009). Characterization of two novel alpha-glucosidases from *Bifidobacterium breve* UCC2003. Appl. Environ. Microbiol. *75*, 1135–1143.

Price, N.P., Whitehead, T.R., and Cote, G.L. (2006). Gas chromatography-mass spectrometry (GC-MS) techniques for metabolic flux analysis of the bifido shunt pathway. Biocatalysis and Biotransformation *24*, 95–98.

Rabiu, B.A., Jay, A.J., Gibson, G.R., and Rastall, R.A. (2001). Synthesis and fermentation properties of novel galacto-oligosaccharides by beta-galactosidases from *Bifidobacterium* species. Appl. Environ. Microbiol. *67*, 2526–2530.

Ramsay, A.G., Scott, K.P., Martin, J.C., Rincon, M.T., and Flint, H.J. (2006). Cell-associated alpha-amylases of butyrate-producing *Firmicute* bacteria from the human colon. Microbiology *152*, 3281–3290.

Reuter, G. (1971). Designation of type strains for *Bifidobacterium* species. Int. J. Syst. Bacteriol. *21*, 273–275.

Richter, H., Hamann, I., and Unden, G. (2003). Use of the mannitol pathway in fructose fermentation of *Oenococcus oeni* due to limiting redox regeneration capacity of the ethanol pathway. Arch. Microbiol. *179*, 227–233.

Roberfroid, M.B. (2005). Introducing inulin-type fructans. Br. J. Nutr. *93* Suppl. 1, S13–25.

Roberfroid, M.B., Van Loo, J.A., and Gibson, G.R. (1998). The bifidogenic nature of chicory inulin and its hydrolysis products. J. Nutr. *128*, 11–19.

Ruas-Madiedo, P., Gueimonde, M., Fernandez-Garcia, M., de los Reyes-Gavilán, C.G., and Margolles, A. (2008). Mucin degradation by *Bifidobacterium* strains isolated from the human intestinal microbiota. Appl. Environ. Microbiol. *74*, 1936–1940.

Ruas-Madiedo, P., Hernandez-Barranco, A., Margolles, A., and de los Reyes-Gavilán, C.G. (2005). A bile salt-resistant derivative of *Bifidobacterium animalis* has an altered fermentation pattern when grown on glucose and maltose. Appl. Environ. Microbiol. *71*, 6564–6570.

Ruiz-Palacios, G.M., Cervantes, L.E., Ramos, P., Chavez-Munguia, B., and Newburg, D.S. (2003). *Campylobacter jejuni* binds intestinal H(O) antigen (Fuc alpha 1,2Gal beta 1,4GlcNAc), and fucosyloligosaccharides of human milk inhibit its binding and infection. J. Biol. Chem. *278*, 14112–14120.

Ryan, S.M., Fitzgerald, G.F., and van Sinderen, D. (2005). Transcriptional regulation and characterization of a novel beta-fructofuranosidase-encoding gene from *Bifidobacterium breve* UCC2003. Appl. Environ. Microbiol. *71*, 3475–3482.

Sabater, B., Sebastian, J., and Asensio, C. (1972). Identification and properties of an inducible and highly specific fructokinase from *Streptomyces violaceoruber*. Biochem. Biophys. Acta *284*, 414–420.

Sakata, S., Kitahara, M., Sakamoto, M., Hayashi, H., Fukuyama, M., and Benno, Y. (2002). Unification of *Bifidobacterium infantis* and *Bifidobacterium suis* as *Bifidobacterium longum*. Int. J. Syst. Evol. Microbiol. *52*, 1945–1951.

Scardovi, V. (1986). Genus *Bifidobacterium*. In: Bergey's manual of systematic bacteriology, vol. 2. Sneath, P.H.A., Mair, N.S., Sharpe, M.E. and Holt, J.G., eds., (Baltimore, USA: Williams & Wilkins), pp. 1418–1434.

Scardovi, V., and Trovatelli, L.D. (1965). The fructose-6-phosphate shunt as peculiar pattern of hexose degradation in the genus *Bifidobacterium*. Annals of Microbiology *15*, 19–29.

Schaible, U.E., and Kaufmann, S.H. (2004). Iron and microbial infection. Nat. Rev. Microbiol 2, 946–953.

Schell, M.A., Karmirantzou, M., Snel, B., Vilanova, D., Berger, B., Pessi, G., Zwahlen, M.C., Desiere, F., Bork, P., Delley, M., *et al.* (2002). The genome sequence of *Bifidobacterium longum* reflects its adaptation to the human gastrointestinal tract. Proc. Natl. Acad. Sci. *99*, 14422–14427.

Schweizer, E., and Hofmann, J. (2004). Microbial type I fatty acid synthases (FAS): major players in a network of cellular FAS systems. Microbiol. Mol. Biol. Rev. *68*, 501–517.

Sela, D.A., Chapman, J., Adeuya, A., Kim, J.H., Chen, F., Whitehead, T.R., Lapidus, A., Rokhsar, D.S., Lebrilla, C.B., German, J.B., Price, N.P., Richardson, P.M., and Mills, D.A. (2008). The genome sequence of *Bifidobacterium longum* subsp. *infantis* reveals adaptations for milk utilization within the infant microbiome. Proc. Natl. Acad. Sci. U.S.A *105*, 18964–18969.

Seo, J.M., Ji, G.E., Cho, S.H., Park, M.S., and Lee, H.J. (2007). Characterization of a *Bifidobacterium longum* BORI dipeptidase belonging to the U34 family. Appl. Environ. Microbiol. 73, 5598–5606.

Sgorbati, B., Lenaz, G., and Casalicchio, F. (1976). Purification and properties of two fructose-6-phosphate phosphoketolases in *Bifidobacterium*. Antonie Van Leeuwenhoek *42*, 49–57.

Shah, N.P. (2000). Probiotic bacteria: selective enumeration and survival in dairy foods. J. Dairy Sci. *83*, 894–907.

Simpson, P.J., Fitzgerald, G.F., Stanton, C., and Ross, R.P. (2004). The evaluation of a mupirocin-based selective medium for the enumeration of bifidobacteria from probiotic animal feed. J. Microbiol. Methods *57*, 9–16.

Sonnenburg, J.L., Chen, C.T., and Gordon, J.I. (2006). Genomic and metabolic studies of the impact of probiotics on a model gut symbiont and host. PLoS Biol. *4*, e413.

Sproul, A.A., Lambourne, L.T., Jean-Jacques, D.J., and Kornberg, H.L. (2001). Genetic control of manno(fructo)kinase activity in *Escherichia coli*. Proc. Natl. Acad. Sci. *98*, 15257–15259.

Sumiyoshi, W., Urashima, T., Nakamura, T., Arai, I., Nagasawa, T., Saito, T., Tsumura, N., Wang, B., Brand-Miller, J., Watanabe, Y., and Kimura, K. (2004). Galactosyllactoses in the milk of Japanese women: changes in concentration during the course of lactation. J. Appl. Glycoscience *51*, 341–344.

Suzuki, K., Benno, Y., Mitsuoka, T., Takebe, S., Kobashi, K., and Hase, J. (1979). Urease-producing species of intestinal anaerobes and their activities. Appl. Environ. Microbiol. *37*, 379–382.

Suzuki, M., Saito, T., and Itoh, T. (1996). Purification and characterization of 6-phospho-beta-galactosidase from *Lactobacillus gasseri* JCM 1031. Biosci. Biotechnol. Biochem. *60*, 139–141.

Suzuki, R., Wada, J., Katayama, T., Fushinobu, S., Wakagi, T., Shoun, H., Sugimoto, H., Tanaka, A., Kumagai, H., Ashida, H., Kitaoka, M., and Yamamoto, K. (2008). Structural and thermodynamic analyses of solute-binding protein from *Bifidobacterium longum* specific for core 1 disaccharide and lacto-N-biose I. J. Biol. Chem. *283*, 13165–13173.

Tamura, Z. (1983). Nutriology of bifidobacteria. Bifid. Microflora *2*, 3–16.

Tanaka, R., and Mutai, M. (1980). Improved medium for selective isolation and enumeration of *Bifidobacterium*. Appl. Environ. Microbiol. *40*, 866–869.

Thitaram, S.N., Siragusa, G.R., and Hinton, A. (2005). *Bifidobacterium*-selective isolation and enumeration from chicken caeca by a modified oligosaccharide antibiotic-selective agar medium. Lett. Appl. Microbiol. *41*, 355–360.

Trindade, M.I., Abratt, V.R., and Reid, S.J. (2003). Induction of sucrose utilization genes from *Bifidobacterium lactis* by sucrose and raffinose. Appl. Environ. Microbiol. *69*, 24–32.

Tsukahara, T., Iwasaki, Y., Nakayama, K., and Ushida, K. (2003). Stimulation of butyrate production in the large intestine of weaning piglets by dietary fructo-oligosaccharides and its influence on the histological variables of the large intestinal mucosa. J. Nutr. Sci. Vitaminol. *49*, 414–421.

Turnbaugh, P.J., Ley, R.E., Mahowald, M.A., Magrini, V., Mardis, E.R., and Gordon, J.I. (2006). An obesity-associated gut microbiome with increased capacity for energy harvest. Nature *444*, 1027–1031.

Turroni, F., Foroni, E., Pizzetti, P., Giubellini, V., Ribbera, A., Merusi, P., Cagnasso, P., Bizzarri, B., de'Angelis, G.L., Shanahan, F., van Sinderen, D., and Ventura, M. (2009). Exploring the diversity of the bifidobacterial population in the human intestinal tract. Appl. Environ. Microbiol. *75*, 1534–1545.

Tzortzis, G., Goulas, A.K., and Gibson, G.R. (2005). Synthesis of prebiotic galactooligosaccharides using whole cells of a novel strain, *Bifidobacterium bifidum* NCIMB 41171. Appl. Microbiol. Biotechnol. *68*, 412–416.

van den Bogaard, P.T., Kleerebezem, M., Kuipers, O.P., and de Vos, W.M. (2000). Control of lactose transport, beta-galactosidase activity, and glycolysis by CcpA in *Streptococcus thermophilus*: evidence for carbon catabolite repression by a non-phosphoenolpyruvate-dependent phosphotransferase system sugar. J. Bacteriol. *182*, 5982–5989.

van der Meulen, R., Adriany, T., Verbrugghe, K., and de Vuyst, L. (2006a). Kinetic analysis of bifidobacterial metabolism reveals a minor role for succinic acid in the regeneration of NAD+ through its growth-associated production. Appl. Environ. Microbiol. *72*, 5204–5210.

van der Meulen, R., Avonts, L., and de Vuyst, L. (2004). Short fractions of oligofructose are preferentially metabolized by *Bifidobacterium animalis* DN-173, 010. Appl. Environ. Microbiol. *70*, 1923–1930.

van der Meulen, R., Makras, L., Verbrugghe, K., Adriany, T., and de Vuyst, L. (2006b). *In vitro* kinetic analysis of oligofructose consumption by *Bacteroides* and *Bifidobacterium* spp. indicates different degradation mechanisms. Appl. Environ. Microbiol. *72*, 1006–1012.

van Laere, K.M., Abee, T., Schols, H.A., Beldman, G., and Voragen, A.G. (2000). Characterization of a novel beta-galactosidase from *Bifidobacterium adolescentis* DSM 20083 active towards transgalactooligosaccharides. Appl. Environ. Microbiol. *66*, 1379–1384.

Vanmaele, R.P., Heerze, L.D., and Armstrong, G.D. (1999). Role of lactosyl glycan sequences in inhibiting enteropathogenic *Escherichia coli* attachment. Infect. Immun. *67*, 3302–3307.

Veerkamp, J.H. (1969). Catabolism of glucose and derivatives of 2-deoxy-2-amino-glucose in *Bifidobacterium bifidum* var. *pennsylvanicus*. Arch. Biochem. Biophys. *129*, 257–263.

Velupillai, P., and Harn, D.A. (1994). Oligosaccharide-specific induction of interleukin 10 production by B220+ cells from schistosome-infected mice: a mechanism for regulation of CD4+ T-cell subsets. Proc. Natl. Acad. Sci. U.S.A. *91*, 18–22.

Ventura, M., O'Flaherty, S., Claesson, M.J., Turroni, F., Klaenhammer, T.R., van Sinderen, D., and O'Toole, P.W. (2009). Genome-scale analyses of health-promoting bacteria: probiogenomics. Nat. Rev. Microbiol. *7*, 61–71.

Ventura, M., van Sinderen, D., Fitzgerald, G.F., and Zink, R. (2004). Insights into the taxonomy, genetics and physiology of bifidobacteria. Antonie van Leeuwenhoek *86*, 205–223.

Wada, J., Ando, T., Kiyohara, M., Ashida, H., Kitaoka, M., Yamaguchi, M., Kumagai, H., Katayama, T., and Yamamoto, K. (2008). *Bifidobacterium bifidum* lacto-N-biosidase, a critical enzyme for the degradation of human milk oligosaccharides with a type 1 structure. Appl. Environ. Microbiol. *74*, 3996–4004.

Wada, J., Suzuki, R., Fushinobu, S., Kitaoka, M., Wakagi, T., Shoun, H., Ashida, H., Kumagai, H., Katayama, T., and Yamamoto, K. (2007). Purification, crystallization and preliminary X-ray analysis of the galacto-N-biose-/lacto-N-biose I-binding protein (GL-BP) of the ABC transporter from *Bifidobacterium longum* JCM1217. Acta. Crystallogr. Sect. F Struct. Biol. Cryst. Commun. *63*, 751–753.

Wang, J., Sporns, P., and Low, N.H. (1999). Analysis of food oligosaccharides using MALDI-MS: quantification of fructo-oligosaccharides. J. Agric. Food Chem. *47*, 1549–1557.

Warchol, M., Perrin, S., Grill, J.P., and Schneider, F. (2002). Characterization of a purified beta-fructofuranosidase from *Bifidobacterium infantis* ATCC 15697. Lett. Appl. Microbiol. 35, 462–467.

Ward, R.E., and German, J.B. (2004). Understanding milk's bioactive components: a goal for the genomics toolbox. J. Nutr. *134,* 962S-967S.

Ward, R.E., Ninonuevo, M., Mills, D.A., Lebrilla, C.B., and German, J.B. (2006). *In vitro* fermentation of breast milk oligosaccharides by *Bifidobacterium infantis* and *Lactobacillus gasseri.* Appl. Environ. Microbiol. 72, 4497–4499.

Ward, R.E., Ninonuevo, M., Mills, D.A., Lebrilla, C.B., and German, J.B. (2007). *In vitro* fermentability of human milk oligosaccharides by several strains of bifidobacteria. Mol. Nutr. Food Res. *51,* 1398–1405.

Willis, A.T., Bullen, C.L., Williams, K., Fagg, C.G., Bourne, A., and Vignon, M. (1973). Breast milk substitute: a bacteriological study. Br. Med. J. *4,* 67–72.

Wolin, M.J., Zhang, Y., Bank, S., Yerry, S., and Miller, T.L. (1998). NMR detection of 13CH313COOH from 3-13C-glucose: a signature for *Bifidobacterium* fermentation in the intestinal tract. J. Nutr. *128,* 91–96.

Xu, J., Bjursell, M.K., Himrod, J., Deng, S., Carmichael, L.K., Chiang, H.C., Hooper, L.V., and Gordon, J.I. (2003). A genomic view of the human-*Bacteroides thetaiotaomicron* symbiosis. Science *299,* 2074–2076.

Xu, J., and Gordon, J.I. (2003). Inaugural Article: Honor thy symbionts. Proc. Natl. Acad. Sci. *100,* 10452–10459.

Yanahira, S., Kobayashi, T., Suguri, T., Nakakoshi, M., Miura, S., Ishikawa, H., and Nakajima, I. (1995). Formation of oligosaccharides from lactose by *Bacillus circulans* beta-galactosidase. Biosci. Biotechnol. Biochem. *59,* 1021–1026.

Yin, X., Chambers, J.R., Barlow, K., Park, A.S., and Wheatcroft, R. (2005). The gene encoding xylulose-5-phosphate/fructose-6-phosphate phosphoketolase (*xfp*) is conserved among *Bifidobacterium* species within a more variable region of the genome and both are useful for strain identification. FEMS Microbiol. Lett. *246,* 251–257.

Zhu, J., and Shimizu, K. (2004). The effect of *pfl* gene knockout on the metabolism for optically pure D-lactate production by *Escherichia coli.* Appl. Microbiol. Biotechnol. *64,* 367–375.

Acid and Bile Resistance and Stress Response in Bifidobacteria

5

Borja Sánchez, Lorena Ruiz, Douwe van Sinderen, Abelardo Margolles and Aldert L. Zomer

Abstract

Members of the genus *Bifidobacterium* are Gram-positive bacteria which are commonly found in the gastrointestinal tract (GIT) of mammals, including humans. Because of their probiotic properties they are frequently incorporated as functional ingredients in food products. From probiotic production to storage and GIT delivery, bifidobacteria encounter a plethora of stresses, which negatively affect their viability and consequently their probiotic activity. In order to cope with these environmental challenges they need to protect themselves through stress-induced adaptative responses. A range of publications have investigated stress response in bifidobacteria and, from these published results, it has emerged that these bacteria employ an interacting network to manage their response to various environmental stresses. This chapter summarizes the current knowledge on the molecular mechanisms used by bifidobacteria to tolerate the deleterious action of acid, heat and bile on their physiology.

Introduction

Bifidobacteria are strictly anaerobic microorganisms that exhibit a typical 'forked' (bifid) shape, from which they derived their name. From a phylogenetic perspective the genus *Bifidobacterium* belongs to the subclass *Actinobacteridae* of the phylum *Actinobacteria* of high GC-content, Gram-positive bacteria (Ventura *et al.*, 2007a). They represent one of the most common inhabitants of the mammalian and animal gastrointestinal tract (GIT) (Vaughan *et al.*, 2002), which they colonize immediately following birth of their host and where, at least in healthy breast-fed infants, they frequently constitute the dominant genus (Penders *et al.*, 2005). In healthy adults the proportion of *Bifidobacterium* species in the GIT remains relatively stable, about 3% (Vaughan *et al.*, 2002). In the human GIT their presence has been associated with various health-promoting or probiotic effects; for example, they are thought to play an essential role in the development of the immune system (Leahy *et al.*, 2005, and references therein). This has led to the widespread use of many bifidobacterial strains as functional components of probiotic foods.

From production of these probiotics to storage and passage through the oral cavity, stomach and small intestine, bifidobacteria encounter an abundance of different environmental stresses. In particular, exposure to oxygen or other oxygen-derived radicals, organic acids and bile, as well as osmotic, heat and cold stress have a major negative impact on bifidobacterial viability and, consequently, probiotic functionality. Although bifidobacteria do not generally take part in the fermentation, they are added as functional food ingredients or used as supplements. In the production process one of the major stress situations is spray drying or freeze drying the final product. During this process they will encounter mainly oxidative, osmotic and heat stress making this an important research subject. Subsequently, once bifidobacteria enters the GIT they must overcome biological barriers, the most prevalent of which being the acidic pH in the stomach and the presence of bile in the intestine. To cope with these life-threatening challenges bifidobacteria resort to a variety of survival strategies, described below.

Heat stress tolerance

Relatively little is known about heat stress tolerance in *Bifidobacterium* species, even though they are of increasing industrial relevance. An adaptive response was shown for *B. adolescentis* NCC251, as pre-exposure to moderate levels of heat stress increased their survival rate when the cells were subsequently exposed to severe heat stress regimes (Schmidt and Zink, 2000). Cross protection between other stresses (e.g. bile salt, osmotic stress) was also observed, although pre-conditioning the cells to the same stress condition generally ensured better survival than pre-conditioning using different stress conditions (Schmidt and Zink, 2000). Additionally, many *Bifidobacterium* strains coped better with industrial processes such as spray drying or freeze drying when they were encapsulated in a skimmed milk or soy milk matrix (Simpson *et al.*, 2005; Wang *et al.*, 2004). In contrast, microencapsulation with starch, a widely used method, did not increase survival rates (O'Riordan *et al.*, 2001). The particular methodology used in the rehydration process of dried cells has also been shown to be important: when rehydration of the spray dried cells was conducted at higher temperatures an increase in survival rate was shown for representatives of *Bifidobacterium* species (Wang *et al.*, 2004).

Heat-adaptation of *Bifidobacterium longum* NCC2705 is thought to involve the accumulation of certain chaperone proteins (ClpC, ClpB, DnaK, GrpE), as detected by label-free quantitative proteomics described by Guillaume *et al.* (2009). Surprisingly GroEL and GroES were found to be strongly downregulated in the heat-adapted strain, although the same authors note that these differences were not detected at the transcriptome level, suggesting that the differences are caused by protein stability, possibly a consequence of the upregulation of other chaperones. Generally, GroEL and GroES are upregulated under heat shock conditions (Narberhaus, 1999), as is the case for *Bifidobacterium breve* UCC2003 (Ventura *et al.*, 2004b). However, in the work described by Guillaume *et al.* (2009) heat tolerant mutants were analysed, and the reason for downregulation the GroELS chaperone machinery in these mutants is as yet unknown.

Protein misfolding and aggregation

Like other bacteria, bifidobacteria are capable of synthesizing a particular set of proteins protecting the cell from the deleterious effects caused by the accumulation of unfolded and/or misfolded proteins. Several of these protective proteins act as molecular chaperones, such as GroEL (Hsp60), DnaK (Hsp70) and ClpB (Hsp100) (Hartl and Hayer-Hartl, 2002; Schirmer *et al.*, 1996), or as proteases, such as ClpP and ClpC (Wickner *et al.*, 1999), and play key roles in several post-translational events to prevent protein denaturation, aggregation and misfolding. Traditionally, heat shock proteins (Hsps) have been classified according to their molecular mass in the following principal families, which are designated Hsp100, Hsp90, Hsp70, Hsp60, Hsp33 and small Hsp (or sHsp). Table 5.1 presents the chaperones and related proteins from high GC-content, Gram-positive species, their activities and their presence in bifidobacterial genomes.

Analysis of the *B. breve* UCC2003 genome sequence has revealed a conserved set of chaperone-encoding genes represented by *groEL*/*groES*, *dnaK*, *grpE*, *dnaJ1* and *dnaJ2*, as well as the Clp-ATPase specifying genes *clpB* and *clpC*, and the protease-encoding genes $clpP_1P_2$ (Ventura *et al.*, 2006). The Hsp20 protein appears to be encoded by only 7 of the 16 bifidobacterial genomes currently available. For a review on their exact functions see Ventura *et al.* (2006) and references therein.

Although many of the Hsps appear to be universally conserved among bacteria, the mechanisms underlying their regulated expression seem to be much more variable. In general, however, the heat-shock response in bacteria is controlled by a combination of three different control systems, i.e. selective promoter recognition by alternative σ factors, transcriptional repression and transcriptional activation. Alternative σ factors allow the RNA polymerase to recognize specific heat-shock promoters. However, despite the fact that bifidobacterial genomes contain at least one *rpoE* (σ^E) homologue, no information is currently available regarding their role in the physiological adjustment to different stress challenges. The

Table 5.1 Heat-inducible chaperone and charonin families in *Actinobacteria*

Chaperone family	Name	Mass	(Suggested) Function	Bifidobacterial homologue	References
Hsp100	ClpB	96	Solubilization of protein aggregates	ClpB	Schirmer *et al.* (1996)
	ClpC	95	Substrate recognition subunit of the proteolytic complex	ClpC	Schirmer *et al.* (1996)
	ClpX	46	Substrate recognition subunit of the proteolytic complex	ClpX	Schirmer *et al.* (1996)
ClpP	ClpP	25	Protein unfolding for proteolysis	ClpP	Wickner *et al.* (1999)
Hsp90	HtpG	71	Prevention of protein aggregation	absent	Dollins *et al.* (2007)
Hsp70	DnaK	69	Co- and post-translational protein folding	DnaK	Wickner *et al.* (1999)
DnaJ	DnaJ	35	Enhancing of ATPase activity	DnaJ1/DnaJ2	Wickner *et al.* (1999)
GrpE	GrpE	23	Nucleotide exchange factor, temperature sensor	GrpE	Wickner *et al.* (1999)
Hsp60	GroEL	57	Post-translational protein folding	GroEL	Wickner *et al.* (1999)
Charonin	Lon	87	Substrate sequestration for proteolysis	absent	
Hsp33	Hsp33	33	Protection during oxidative stress	absent	Jakob *et al.* (1999)
sHsps	Hsp20	20	Prevention of protein aggregation	Hsp20*	Wickner *et al.* (1999)
	Hsp18	18		absent	
	Hsp16	16		absent	
Hsp10	GroES	10	Post-translational protein folding	GroES	Wickner *et al.* (1999)

*Only present in certain *Bifidobacterium* species.

regulatory mechanisms based on transcriptional repression involve repressor proteins that are produced under non-stressed conditions, and that by binding to specific sequences positioned in the promoter regions of certain heat-shock genes prevent their transcription. Stress treatments inactivate such repressors through a variety of mechanisms, thereby allowing the transcriptional de-repression of their targeted heat-shock genes.

The most well known heat shock regulatory protein, HrcA, and its DNA binding region are highly conserved among eubacteria, including bifidobacteria. The HrcA protein binds to the consensus sequence TTAGCACTC-N9-GAGTGCTAA (Schulz and Schumann, 1996). It is known to regulate the transcription of the chaperone-encoding genes *dnaK*, *groEL*, *grpE*, *dnaJ*, *groEL* and *groES* in *Bacillus subtilis* (Schulz and Schumann, 1996), while in *Streptomyces* species it was found to regulate *groELS*, as well as its own expression and a second DnaJ-encoding gene (Grandvalet *et al.*, 1998). In *B. subtilis* and possibly others, a negative feedback loop exists between HrcA and GroEL, where the GroEL chaperone protein complex, when not sequestered by misfolded proteins, enhances HrcA repression of its target promoters (Mogk *et al.*, 1997). The *hrcA*, *groEL* and *groES* genes of *B. breve* UCC2003 are expressed upon osmotic shock, ethanol shock and moderate heat stress (Ventura *et al.*, 2004b and 2005a; A.L. Zomer *et al.*, unpublished). Upregulation was not found for all members of the HrcA regulon when bifidobacterial cells are exposed to a severe heat shock regime, as *hrcA* and *groEL* exhibited increased transcription, while *groES* and *dnaJ2* did not (Table 5.2). These differences could be caused by differences in mRNA stability, especially because of the drastically increased temperature (Δt13). In bifidobacterial genomes the CIRCE motif (controlling inverted repeat of chaperone expression) is found exclusively in the promoter regions of the genes encoding for GroEL and GroES, as well as in the promoter region of the *hrcA/dnaJ2* operon. Binding studies show that a his-tagged version of HrcA indeed forms complexes with CIRCE-containing promoters (Zomer *et al.*, unpublished).

In high GC-content, Gram-positives, including bifidobacteria, regulation of *dnaK*, *grpE*, *dnaJ* and *clpB* is mediated by HspR where it binds to the conserved sequence CTTGAGT-N7-ACTCAAG (Bucca *et al.*, 1997; Grandvalet *et al.*, 1999; Ventura *et al.*, 2005c,e). HspR-encoding genes are almost exclusively confined to members of the *Actinomycetales* order, with the exception of *Helicobacter, Campylobacter, Deinococcus* and related species, where this gene is also present, while HAIR sequences (for HspR associated inverted repeat) can be found in the promoter regions of various chaperone-encoding genes (Andersen *et al.*, 2005; Schmid *et al.*, 2005; Spohn and Scarlato, 1999). A negative feedback loop, similar to that described for HrcA, exists between HspR and DnaK. In *Streptomyces*, and possibly this applies to *Bifidobacterium* species as well, unsequestered DnaK forms a complex with HspR, which dramatically increases HspR's binding activity to its DNA targets (Bucca *et al.*, 2000 and 2003). The *B. breve* UCC2003 genome contains putative HAIR motifs in the promoter regions of the *dnaK* operon, the *clpB* gene, the *clgR* operon, as well as in the promoter regions of the gene encoding the principal sigma factor and that specifying endonuclease IV. Binding studies have shown that HspR forms complexes with most HAIR-containing regions except for the last two mentioned, although these two genes are upregulated under severe heat stress conditions (Table 5.2) (Zomer *et al.*, unpublished). Closer inspection of the functional binding sites revealed the presence of a larger inverted repeat structure around the general HAIR motif. Mutation of this structure resulted in a decreased binding efficiency of the six-histidine tagged (H6)-HspR protein, suggesting that the extension is a valid part of the bifidobacterial HAIR consensus sequence, AAAsTTGAGysw-N[5]-CTCAAsTTTT, which is therefore 8bp longer than the canonical HAIR motif TTGAGy-N[7]-ACTCAA (Bucca *et al.*, 2003).

The regulation of heat shock genes through transcriptional activation is controlled by proteins that positively regulate the expression of their gene targets. The only transcriptional activator of stress response described thus far in *Actinobacteria* is the Clp Gene Regulator (ClgR) (Bellier and Mazodier, 2004; Engels *et al.*, 2004; Ventura *et al.*, 2005d). The gene specifying this regulator was found in *B. breve* UCC2003 and its role in heat shock regulation has been investigated in considerable detail (Ventura *et al.*, 2005b,d). In various *Actinobacteria* it has been shown that ClgR regulates the expression of protease-encoding genes such as *clpP*$_1$*P*$_2$, *clpC* and *lon* (Bellier and Mazodier, 2004; Engels *et al.*, 2004; Ventura *et al.*, 2005b,d), while it has also been implicated in the response to DNA damage (Engels *et al.*, 2005). Its binding site, determined to be CGCT-4N-GCGNAC in *Streptomyces* and WNNWCGCYNANRGCGWWS in *Corynebacterium* suggests that it is not particularly well conserved. It has been shown previously that ClgR activity is controlled by degradation by the ClpCP protease in *Streptomyces* and in *Corynebacterium* species (Engels *et al.*, 2004). However, the suggested degradation signal, consisting of two alanine residues in the C-terminus of the regulator, is only present in ClgR encoded by *Streptomyces lividans* (Bellier *et al.*, 2006; Viala and Mazodier, 2002). Degradation of ClgR by ClpCP provides a feedback loop, analogous to what has been described for HrcA and HspR.

A purified histidine-tagged ClgR (H6-ClgR) from *B. breve* was shown to bind to the promoter regions of the *clpC* and *clpP1P2* genes. However, it was observed that the *clpP1* and *clpC* promoter regions were only bound by the purified ClgR

Table 5.2 Regulated genes members of the putative stress regulatory network in *B. breve* UCC2003.

Locus	Product	(Putative) function	42°C	44°C	47°C	50°C	Ethanol	Osmotic	Putative regulator/motif
Bbr_0076	Hsp20	Chaperone	19.4	26.7	51.8	119.3	1.8	5.4	
Bbr_0124	DnaK	Chaperone protein	3.5	7.0	10.2	29.3	3.6	16.7	HspR GAAGTTGAGTGATGTAGGCTCAAGTTTT & HspR AAACCTGAGCCTACATCGCTCAACTTTT
Bbr_0125	GrpE	GrpE protein	3.2	6.7	7.8	22.8	3.6	13.1	
Bbr_0126	DnaJ1	Chaperone protein	3.6	8.2	10.2	24.9	1.6	16.2	
Bbr_0127	HspR	Regulator	2.6	6.1	8.8	23.8	1.9	15.4	
Bbr_0216		Modification methylase	–1.3	1.3	5.3	3.1	–1.1	1.9	LexA TCGAACAAATGTTCTA
Bbr_0496	MutY	DNA glycosylase	1.1	2.3	2.5	-1.7	3.9	1.7	LexA TCGAACATCTTCTCGA
Bbr_0585	ImpB	DNA repair	1.3	4.3	33.0	18.3	1.9	4.7	LexA TCGAACAAATGTTCGA
Bbr_0794	ClpP1	Protease	1.0	1.2	1.0	0.7	3.4	–2.3	ClgR CTACGCTGGCCGCCAAA
Bbr_0795	ClpP2	Protease	–1.1	1.2	–1.4	–3.5	3.3	–1.5	
Bbr_1004	HrcA	Regulator	2.0	5.2	11.2	3.5	5.7	6.9	HrcA TTAGCACTCGAAACCAATGAGTGCTAA & ClgR CTTCGCCGCAAACGAAC
Bbr_1005	DnaJ2	Chaperone protein	1.4	2.4	2.0	–4.4	3.8	1.5	
Bbr_1013	RuvA	Holliday junction DNA helicase	1.1	2.0	5.7	6.7	2.6	1.5	LexA TCGAACAGTTGTTCGA
Bbr_1014	RuvB	Holliday junction DNA helicase	1.0	1.9	2.2	–2.4	2.4	1.0	
Bbr_1046	Nfo	Endonuclease IV	1.1	1.7	8.9	18.1	5.0	1.9	HspR ATGCTTAAGTCTCCTTCGGTCAGATGGT
Bbr_1171	HrdB	principal sigma factor	1.1	1.7	5.8	13.1	1.7	1.3	HspR TCCGTGGAGGGTGGTTTACTCACATAGG
Bbr_1180	RecX	Recombination regulator	1.2	2.9	17.7	10.9	3.5	2.6	
Bbr_1181	RecA	Recombinase	1.4	3.3	10.2	12.0	4.3	2.4	LexA TCGAACAAATGTTCGC
Bbr_1182	–	–	1.0	1.7	9.1	66.9	3.6	1.2	
Bbr_1183	ClgR	Regulator	1.6	3.6	57.2	416.3	7.1	3.3	HspR AAACTTGAGTCTATTCACAGCAAGTTTT
Bbr_1271	LexA	Regulator	1.1	1.9	2.8	7.0	3.4	6.8	LexA TCGAACAGATGTTTGA & TCGAACAAGTGTTCTA & LexA TCGAACATCTGTTCTT & TGGCACAAATGTTCGA
Bbr_1356	ClpC	Protease	–1.1	1.2	–1.2	–2.7	2.4	–1.6	ClgR CTCCTCTCACGGCGCAA
Bbr_1364	GroEL	Chaperone protein	2.1	3.6	3.8	1.1	3.8	1.1	HrcA TTGGCACTCGGCAAGGGCGAGTGCTAA & HrcA TTAGCACTCGGAGATCGAGAGTGATAA
Bbr_1668	GroES	Chaperone protein	1.5	2.3	3.3	3.3	3.3	1.8	HrcA TTAGCACTCGAGTTGGCAGAGTGCTAA
Bbr_1781	ClpB	Chaperone protein	4.2	12.2	27.3	94.6	4.4	8.0	HspR AAACTTGAGTCACTTACACTCAACTTTT

Numbers indicate the ratio of up or downregulation.

Underlined: first member of operon.

protein in the presence of a crude lysate of heat stressed *B. breve* UCC2003 cells (Ventura *et al.*, 2005b,d). This binding activity was lost upon proteolytic treatment of the crude lysate, thus indicating the existence of (a) proteinaceous co-factor(s) required for ClgR binding. DNase I footprinting experiments showed that the *clp*P1 promoter region protected by ClgR binding contains an imperfect palindromic motif (CGCT-N4-GCCNA), which may represent the operator site for the ClgR protein. This motif is similar to the operator site of the ClgR homologue in other *Actinobacteria* (Bellier and Mazodier, 2004; Engels *et al.*, 2004). In these latter bacteria, it was shown that the ClgR regulon includes heat shock genes, such as *clpC*, *clpP* and *lon* (Engels *et al.*, 2005). Despite these similarities, the bifidobacterial ClgR homologue seems to uniquely require a co-factor(s) as a prerequisite to act as a binding protein. In pull-down assays using whole cell extracts from heat stressed *B. breve* UCC2003 cultures, it was shown that a 56 kDa protein is co-purified with the H6-ClgR protein (Ventura *et al.*, 2005d). Moreover, the H6-ClgR-56 kDa protein co–eluate mixture was able to bind the *clpP1* promoter region without the assistance of any other co-factor. Taken together, these results would represent a novel type of co-factor-mediated transcriptional regulation in high G+C Gram-positive bacteria. Preliminary results identified the 56 kDa protein to be the *B. breve* UCC2003 GroEL chaperone. Another interesting and unique feature of the bifidobacterial ClgR protein is that it contains an N-terminal domain of around 90 amino acid residues which is absent in homologues from other members of the *Actinobacteria* group. A truncated version of the H6-ClgR protein was able to bind the promoter regions of *clpP1P2*, *clpC* and surprisingly, also with the promoter of *hrcA*. EMSA's with mutated regions of the clpC promoter showed that the ClgR binding site is **T**N**C**G**CT**NNN**GGC**GN**AA** instead of CGCT-N[4]-GCCNA with the bases most important for binding marked in bold.

Taken together, these findings, along with the fact that the ClgR protein from other *Actinobacteria* does not appear to require any cofactor for binding to the *clpP1* promoter region, suggest that this N-terminal extension interacts with the cofactor molecule.

Regulation of the small heat shock protein Hsp20 still remains a mystery. Although it is strongly upregulated under heat stress; no known binding site is present in the promoter region of this gene (Ventura *et al.*, 2007b). Additionally, binding studies using purified His-tagged ClgR, HspR or HrcA reveal no complex formation between the DNA and any of these regulators, indicative that these proteins are not involved in its regulation. It is possible that an alternative sigma factor is responsible for activating the expression of this gene, of which there are two candidates on the genome of *B. breve* UCC2003; however, these do not possess a high level of similarity with known stress sigma factors in other high GC-content Gram-positive species. *In silico* analysis of the upstream region of this gene using WebSIDD (Bi and Benham, 2004) does, however, reveal the presence of a so-called *Stress* Induced DNA Duplex Destabilization (SIDD) site at position 71 or 98 upstream of both the transcriptional start sites determined by Ventura *et al.* (Zomer *et al.*, unpublished; Ventura *et al.*, 2007b). SIDD sites in the promoter region of genes are assumed to be involved in the increased expression of these genes in responses to environmental changes which are known to transiently alter superhelicity, such as the heat stress or osmotic stress employed in this study (see Hatfield and Benham, 2002; Wang and Benham, 2008, and references therein).

Several other heat shock related regulators are known, such as RheA in *Streptomyces* (Servant *et al.*, 2000), CtsR in many Gram-positives (Kruger and Hecker, 1998) and many stress related sigma factors (see Ventura *et al.*, 2006, and references therein). However, we were not able to identify these in the genome sequences of the currently available *Bifidobacterium* species at the NCBI website (http://www. ncbi. nlm. nih. gov/).

SOS response

Induction of Hsps forms a significant part of the bacterial response to increased temperatures in order to deal with the inflicted protein damage. Another type of stress response occurs when DNA becomes damaged by these environmental stresses. Under normal conditions, expression of genes involved in DNA damage repair (so-called SOS genes) is very low as they are under tight negative control. Repression of these genes is managed by the LexA regulator, which binds to the SOS box in the promoter regions of these genes (Cheo *et al.*, 1991). Activation of the SOS genes occurs when DNA is damaged, which can happen as a result of heat or oxidative stress. This DNA damage results in single stranded DNA (ssDNA) accumulation at the site of (stalled) replication, which in turn activates RecA. LexA autocleavage is promoted by this activated form of RecA and derepression of the SOS genes takes place (Sassanfar and Roberts, 1990).

In *B. breve* UCC2003, SOS response occurs when the cells are subjected to both heat shock and oxidative stress (A.L. Zomer *et al.*, unpublished). Induction of RecA expression might be regulated by HspR under heat shock conditions because of the proximity of *recA* and *recX* to the HspR-controlled *clgR* and Bbr_1182 genes, which together may form a transcriptional unit. Induction of *recA* expression by HspR could activate LexA autoproteolysis when single stranded DNA is available from stalled replication forks and, consequently, result in rapid upregulation of the SOS genes (Fig. 5.1).

Reconstruction of the bifidobacterial stress gene regulatory network

A stress regulatory network of *B. breve* UCC2003 can be constructed using the current knowledge obtained from the research described in this chapter and other available literature data. Although simplified, this network shows the complexity of stress gene regulation in *B. breve* UCC2003 (Fig. 5.1). Interplay between the various regulators occurs at different levels and is indicative of a tightly controlled stress regulon.

Regulation of the HrcA regulon depends on the activity of the HrcA regulator, where binding is dependent on the availability of unsequestered GroEL. As was shown for the HrcA protein from *B. subtilis*, misfolded proteins that accumulate following stress sequester GroELS away from HrcA, which in turn relieves repression of the HrcA regulon (Mogk *et al.*, 1997). DnaJ2, a member of the HrcA regulon, could, in the same manner,

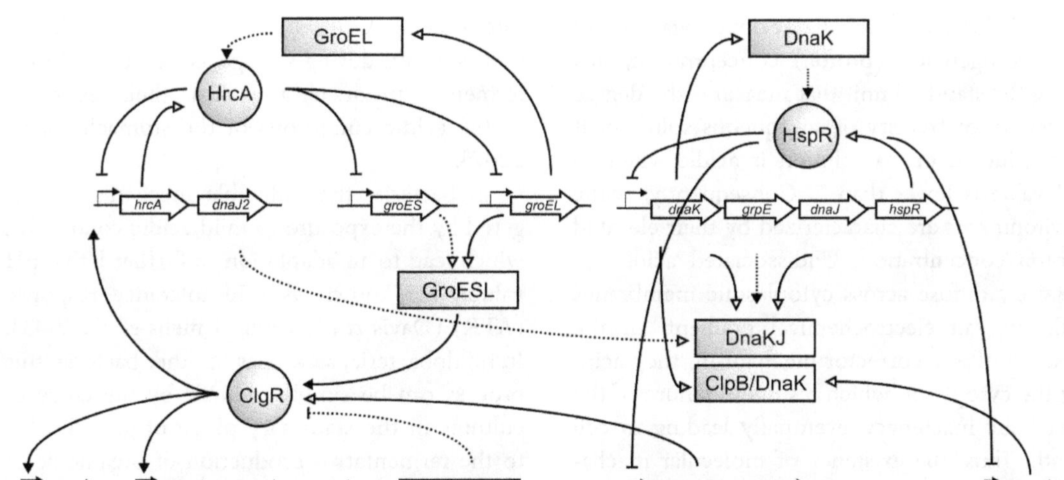

Figure 5.1 Major stress gene regulatory network of *B. breve* UCC 2003.

Dotted lines: Presumed interaction. Closed lines: proven interaction.

interact with DnaK of the HspR regulon, forming a DnaKJ2 complex, thereby relieving the binding of HspR to its target promoters, since its activity is dependent on unsequestered DnaK (Bucca *et al.*, 2000). Additionally, misfolded protein would also sequester DnaK by the formation of the DnaKJ complex. HspR in turn may activate the ClgR operon, thereby inducing SOS response by overproduction of RecA, and, if certain conditions are met, activation of the ClgR regulon, which contain *clpC*, *clpP* and *hrcA*, thereby making a full circle to HrcA. In addition, the binding activity of ClgR appears to be dependent on GroEL(S), and it is possible that GroEL(S) needs to be present in large quantities in an unsequestered form, i.e. when GroELS is no longer able to refold proteins when protein aggregates are formed under severe stress conditions. Induction of ClpC and ClpP would either ensure complete degradation of these large protein aggregates or facilitate chaperone access by breaking up the large complexes in smaller fragments.

Acid pH as stress factor for bifidobacteria

pH is defined as minus the decimal logarithm of the hydrogen ion (proton) concentration, this being the standard unit that measures the degree of acidity or basicity of an aqueous solution. It is considered that a solution is acidic when its pH value is lower than 7. Consequently, acidic environments are characterized by their elevated proton concentration. Undissociated acids can passively diffuse across cytoplasmic membranes following an electrochemical gradient. In the absence of any corrector mechanism, they acidify the cytoplasm, which results in failure of the metabolic machinery, eventually leading to cell death. Thus, the existence of molecular mechanisms allowing the maintenance of cytoplasmic pH at physiological values is crucial for bacterial survival in acidic environments.

Owing to their catabolic metabolism, bifidobacteria acidify their surroundings through fermentation of sugars, to mainly acetic and lactic acids (Vernazza *et al.*, 2006). Thus, their metabolic activity involves an acidic stress in the stationary phase of growth (Saarela *et al.*, 2004). In addition, probiotic bifidobacteria included in functional foods can face acidic conditions during product storage. This is particularly noteworthy in the case of fermented dairy products, in which post-acidification negatively affects their viability (Jayammanne and Adams, 2006). Finally, and after being ingested, bifidobacteria encounter severe acidic conditions when they arrive in the human stomach, where food stays for around 90 minutes at pH values close to 2.

Probiotic bifidobacteria must retain certain viability levels in order to exert their beneficial effects on the human host. It has been suggested that the optimal dose for a probiotic strain should comprise between 10^9–10^{10} CFUs per day (Tuomola *et al.*, 2001). In the case of bifidobacteria, exposure to the acidic pH of the stomach appears to be more harmful than other stresses, such as bile salts (Kheadr *et al.*, 2007). In relation to this, recent results suggest that mild acidic stress would rather induce cells to enter a viable, but not cultivable, state than decrease their viability (Lahtinen *et al.*, 2008). Thus, a good tolerance to acidic pH is a crucial trait for the probiotic success of a *Bifidobacterium* strain. Isolation of bifidobacterial strains showing good tolerance to acidic pH (among other characteristics) is therefore desirable, since it would potentially enhance their survival during storage as part of probiotic fermented products, as well as their resistance to the acidic conditions of the stomach (Sanz, 2007).

In bacteria, the inducible mechanisms triggered by the exposure to mild acidic conditions, which lead to an adaptation to further lethal pH values, are known as acid tolerance response (ATR) (Davis *et al.*, 1996; Samelis *et al.*, 2003). In bifidobacteria, as well as in other bacteria, this process can be simply induced by the entry of cultures in the stationary phase of growth, due to the fermentative production of organic acids (Buchanan and Edelson, 1996). In fact, this may be one of the reasons explaining why cells harvested at such a growth phase are more resistant to lethal acidic conditions (Samelis *et al.*, 2003).

Survival of bifidobacteria in harsh acidic environments can be enhanced if cells are previously exposed to sublethal pH values through ATR induction (Maus and Ingham, 2003). Furthermore, acid-pH resistant mutants with a stable phenotype can be obtained after long exposure to severe acidic conditions (Collado and Sanz, 2006). Isolation of strains showing high resistance to acid pH is very suitable for the food industry, as long as acid-resistant strains also possess cross-resistance to other stresses (Chung *et al.*, 1999; Collado and Sanz, 2007). For instance, several acid-resistant strains generated by Collado and co-workers showed cross-resistance to bile, NaCl and heat shock (Collado and Sanz, 2006). Reversibly, exposure of bifidobacteria to sublethal doses of other stresses, such as low-high temperature, may confer cross-resistance to acidic pH (Saarela *et al.*, 2004).

Viability of bifidobacteria at the pH values of fermented dairy products and synthetic gastric fluids is very variable (Maus and Ingham, 2003; Sanz, 2007). It can be considered that, in general, resistance of bifidobacteria to acidic pH is weak, with the exception of *Bifidobacterium animalis* (Masco *et al.*, 2007). This species is the most used *Bifidobacterium* spp. in fermented probiotic products, since they also satisfactorily tolerate the presence of oxygen (Vernazza *et al.*, 2006). Therefore, and on the contrary to other *Bifidobacterium* species, *B. animalis* is able to maintain its viability levels during storage in fermented milks (Jayamanne and Adams, 2006, 2009). The high membrane ATPase activity induction observed for this species might be responsible for its elevated acid resistance, this activity not being induced in non-resistant strains (Matsumoto *et al.*, 2004).

The acid resistance of *B. animalis* may be connected to its excellent survival through the human GIT (Alander *et al.*, 2001). In spite of the weak resistance of other *Bifidobacterium* spp. to acidic pH, certain strains are also able to induce an ATR. For instance, *B. longum* BL1 was selected for showing good resistance to exposure to artificial gastric juice (pH = 3). This resistance was correlated with the maintenance of its viability during storage in fermented milks. *B. longum* BL1 was able to maintain a higher cytoplasmic pH than other non-resistant strains, which again was shown to be dependent on the presence of an inducible membrane ATPase activity (Takahashi *et al.*, 2007).

Finally, acquisition of acid-resistance can induce changes in the surface properties of *Bifidobacterium*. In this sense, some acid-resistant strains presented a better ability to adhere to human mucin, and some others showed an enhanced ability to inhibit the adhesion of certain pathogens, or to displace previously adhered pathogens (Collado *et al.*, 2006). Care should be taken in the sense that acid-adaptation may cause deep changes in the antibiotic susceptibilities of the bifidobacterial strain (Collado and Sanz, 2007). In this way, resistance to aminoglycosides, chloramphenicol, erythromycin, and tetracycline increased in several *Bifidobacterium* strains that survived exposure to severe acid conditions (pH 2.0 for 60 min) (Kheadr *et al.*, 2007).

Molecular targets of acidic pH

Cytoplasmic acidification can be lethal if the pH falls under physiological values. Most metabolic enzymes have optimal activities at slightly acid pHs, so severe pH decrease will literally halt bacterial metabolism. In bacteria, ATR is characterized as being a complex process implying changes in the transcription of several genes and in the synthesis of a high number of proteins (Bore *et al.*, 2007; Davis *et al.*, 1996; Len *et al.*, 2004). In bifidobacteria, molecular studies of the ATR are very scarce, but currently it is known that this process is extremely dependent on the strain and on the species, and in general it is not easily inducible (Saarela *et al.*, 2004).

Two of the main mechanisms involved in pH resistance are the active extrusion of protons and the alkalinization of the cytoplasm. In anaerobic bacteria, it is known that the proton excess occurring in acidic environments is counteracted by the action of F_0F_1-ATPase in an energy-dependent manner (Cotter and Hill, 2003; Matsumoto *et al.*, 2004). This enzyme catalyses the hydrolysis of ATP, which enables the translocation of protons across the cytoplasmic membrane against an electrochemical gradient. Thus, the action of the F_0F_1-ATPase maintains cytoplasmic pH at physiological values when bifidobacteria encounter harsh acidic conditions. Besides F_0F_1-ATPase, ammonia formation is a common mechanism of tolerance to pH widely distributed among microorganisms since, as a basic compound, it can capture one proton, yielding an ammonium ion and therefore helping to buffer the cytoplasm at low pHs (Cotter and Hill, 2003).

As pointed out, F_0F_1-ATPase activity is directly related to the ATR in bifidobacteria, as in other anaerobes and facultative anaerobes

(Miwa *et al.*, 1997). This increase in the activity of the enzyme has been supported with molecular data, being shown that both the genes coding for the enzyme and the amount of protein are upregulated/overproduced in *B. animalis* and *B. longum* (Sánchez *et al.*, 2008). To date, the only proteomic data available on the response and adaptation of *B. longum* to acidic pH have been obtained through the isolation of an acid-resistant mutant from the acid pH-susceptible strain *B. longum* NCIMB8809 (Sánchez *et al.*, 2007a). Comparison of the 2-dimensional electrophoresis (2DE) patterns revealed that the F_0F_1-ATPase was upregulated in acidic conditions. This fact provided additional evidence of the involvement of this enzyme in the energy-dependent extrusion of the excess of protons generated at acidic pHs. In bifidobacteria, the eight subunits of this enzyme are encoded in an operon and arranged into two groups, one for each of the two complexes. These two complexes give the name to the enzyme, one being membrane-integral (F_0) and the other cytoplasmic (F_1). 2DE analysis showed that *B. longum* synthesizes higher amounts of the two main cytoplasmic subunits of the F_0F_1-ATPase, AtpA (α subunit) and AtpD (β subunit) (Sánchez *et al.*, 2007a). In this sense, it has also been shown that the membranes of *B. animalis* contained a greater amount of the β subunit of the F_0F_1-ATPase when grown at acidic pH (Sánchez *et al.*, 2006). The transcription of the F_0F_1-ATPase operon has been extensively studied in *B. animalis* subsp. *lactis* by Ventura and co-workers. The operon is transcribed into two mRNAs, one containing the information for all the subunits and other, smaller, coding for the cytoplasmic subunits (Ventura *et al.*, 2004a). Both mRNAs could be detected in higher amounts following exposure to acid pH, which suggests a specific regulation of F_0F_1-ATPase at the transcription level. On the other hand, the acid-adapted *B. longum* strain showed a deep reorganization in the relative amounts of certain glycolytic proteins, which was correlated with additional physiological data to an enhancement of the glycolytic pathway (Sánchez *et al.*, 2007a). These changes concerned proteins involved in the catabolism of complex carbohydrates, such as alpha-1,4-glucosidase, UDP-glucose 4-epimerase, and phosphoglucomutase, whose biological activity contributes to fuelling the bifid shunt (Table 5.3). Theoretically, this reorganization would increase the synthesis of ATP, which would promote the F_0F_1-ATPase activity.

The involvement of the F_0F_1-ATPase in the ATR has been corroborated with some physiological data. For instance, in a fermenter-scale study, Mättö *et al.* (2006) showed that the addition of pepsin to *B. animalis* cultures conferred a protective effect against low pH by facilitating the activity of the F_0F_1-ATPase. In fact, the presence of glucose in the acidic buffer slowed down the loss of viability of *B. animalis* cultures harvested at the stationary phase of growth, by serving as a metabolic fuel for the F_0F_1-ATPase (Sánchez *et al.*, 2006).

In addition to the clear involvement of the F_0-F_1 ATPase in the pH response of *Bifidobacterium*, the amount of enzymes responsible for the biosynthesis of branched-chain amino acids, as well as glutamine synthetase, was increased in *B. longum* grown under acidic conditions (Sánchez *et al.*, 2007a). These proteomic data supported the hypothesis formulated by Len *et al.* (2004), in which branched-chain amino acid synthesis would be coupled to the bioconversion of glutamate in oxoglutarate, glutamate being obtained by deamination of one molecule of glutamine. Ammonia released in the last reaction would act as a buffer in acidic conditions by capturing one proton (van de Guchte *et al.*, 2002). This fact was corroborated with physiological experiments, in which higher concentrations of the branched-chain amino acid valine and ammonium ion were measured in *B. longum* cells grown under acidic conditions (Sánchez *et al.*, 2007a).

It is also worth noting the down-regulation of bile salt hydrolase (BSH) as a response to acidic pH in *B. longum*. Moreover, BSH was constitutively underproduced in the acid-adapted strain. Lower BSH activities were measured in the cytoplasm of the acid-adapted strain and as a response to acidic pH (Sánchez *et al.*, 2007a). BSH catalyses the so-called bile salt deconjugation, in which the amino acid moiety of the bile salt is released, yielding a deconjugated bile acid. Its physiological role in intestinal bacteria is not well understood, mainly because deconjugated bile acids are more toxic for cells than bile salts (Grill *et al.*, 2000; Noriega *et al.*, 2006). It seems that

Table 5.3 Main protein or gene players involved in bile and pH response in *Bifidobacterium*

Protein or gene	Putative function	Physiological relevance	Reference
Proteins and genes involved in bile response			
B. breve			
Bbr_0838 ↑	Multidrug resistance protein	ND	Gueimonde *et al.* (2009)
B. longum			
Ctr	Sodium dependent bile acid transporter	Bile salt efflux	Price *et al.* (2006)
BL0920 ↑	Efflux transporter protein	Bile salt efflux	Gueimonde *et al.* (2009)
HtrA ↑	Protease	ND	Savijoki *et al.* (2005)
DnaK ↑	Molecular chaperone	ND	Savijoki *et al.* (2005); Sánchez *et al.* (2005)
GroEL ↑	Chaperonin	ND	Savijoki *et al.* (2005); Sánchez *et al.* (2005)
Xfp ↑	Fructose-6-P phosphoketolase	Increased glycolytic flux	Sánchez *et al.* (2005)
Gap ↑	Glyceraldehyde-3-phosphate dehydrogenase C	Increased glycolytic flux	Sánchez *et al.* (2005)
Pgk ↑	Phosphoglycerate kinase	Increased glycolytic flux	Sánchez *et al.* (2005)
GalE1 ↑	UDP-glucose 4-epimerase	Increased glycolytic flux	Sánchez *et al.* (2005)
IlvE ↑	Probable branched-chain amino acid aminotransferase	ND	Sánchez *et al.* (2005)
BL1000 ↑[b]	Response regulator of two-component system	ND	Sánchez *et al.* (2005)
OppA ↑	Oligopeptide-binding protein of ABC transporter	Shift on tripeptide uptake	Ruiz *et al.* (2009)
DppD ↑	ATP-binding protein of peptide ABC transporter	Shift on tripeptide uptake	Ruiz *et al.* (2009)
Eno ↑	Enolase	ND	Ruiz *et al.* (2009)
GluA1 ↓	Glutamine synthetase	ND	Ruiz *et al.* (2009)
PbpA ↑	Cell division protein FtsI	ND	Ruiz *et al.* (2009)
B. animalis			
gtf01207 ↑	Priming glycosyltransferase	EPS production	Ruas-Madiedo *et al.* (2009)

Table 5.3 Continued

Protein or gene	Putative function	Physiological relevance	Reference
AtpD[a] ↑	H(+)-ATPase, beta subunit	Regulation of intracellular pH	Sánchez *et al.* (2006)
BA41 ↑	Xylulose-5-phosphate/fructose-6-phosphate phosphoketolase	Shift on glucose consumption	Sánchez *et al.* (2007b)
BA38[a] ↓/BA53[a] ↑	Glyceraldehyde-3-phosphate dehydrogenase	Shift on glucose consumption	Sánchez *et al.* (2007b)
BA5[a] ↓	Methionine synthase	ND	Sánchez *et al.* (2007b)
BA40[a] ↓	O-Acetylhomoserine sulfhydrolase	ND	Sánchez *et al.* (2007b)
BA27[a] ↑	Ketol-acid reductoisomerase 1	ND	Sánchez *et al.* (2007b)
BA52[a] ↑	CTP synthase	ND	Sánchez *et al.* (2007b)
BA55[a] ↓	Long-chain-fatty-acid CoA ligase	Shift on fatty acid composition	Sánchez *et al.* (2007b); Ruiz *et al.* (2007)
BA8[a] ↑	Heat shock protein ClpB	ND	Sánchez *et al.* (2007b)
BA1 ↑	Trypsin-like serine protease homologue to HtrA	ND	Sánchez *et al.* (2007b)
BA22 ↑	Thioredoxin-dependent thiol peroxidase	ND	Sánchez *et al.* (2007b)
BA23 ↑	Co-chaperone GrpE	ND	Sánchez *et al.* (2007b)
BA26 ↑	Chaperone protein GroES	ND	Sánchez *et al.* (2007b)
BA30 ↑	Chaperone protein GroEL	ND	Sánchez *et al.* (2007b)
BA12 ↑	Chaperone protein DnaK	ND	Sánchez *et al.* (2007b)
BA21[a] ↑	DNA protection during starvation protein	ND	Sánchez *et al.* (2007b)
BA9[a]	Cell division initiation protein	ND	Sánchez *et al.* (2007b)
BA2[a]	Chromosome segregating ATPase	ND	Sánchez *et al.* (2007b)
BA47[a]	Bile salt hydrolase	Increased deconjugation activity of glyco- and tauro-deoxycholate	Sánchez *et al.* (2007b); Noriega *et al.* (2006)
Proteins and genes involved in the acid tolerance response			
B. animalis			
AtpD, *atpD* ↑	H(+)-ATPase, beta subunit	Regulation of intracellular pH	Ventura *et al.* (2004a); Sánchez *et al.*(2006)

Protein or gene	Putative function	Physiological relevance	Reference
atpA ↑	H(+)-ATPase, alpha subunit	Regulation of intracellular pH	Ventura *et al.* (2004a)
B. longum			
AtpA ↑	H(+)-ATPase, alfa subunit	Regulation of intracellular pH	Sánchez *et al.* (2007a)
AtpD ↑	H(+)-ATPase, beta subunit	Regulation of intracellular pH	Sánchez *et al.* (2007a)
MetE[c] ↑	Methionine synthase	ND	Sánchez *et al.* (2007a)
MetB[c] ↑	Cystathionine gamma-synthase	ND	Sánchez *et al.* (2007a)
CysD[c] ↑	O-Acetylhomoserine (thiol)-lyase	ND	Sánchez *et al.* (2007a)
BSH[c] ↓	Bile salt hydrolase	Decreased deconjugation activity of glyco- and tauro-deoxycholate	Sánchez *et al.* (2007a)
AglA ↑	Probable alpha-1,4-glucosidase	Modified glycolytic flux	Sánchez *et al.* (2007a)
Pgm ↑	Phosphoglucomutase	Modified glycolytic flux	Sánchez *et al.* (2007a)
GalE1 ↑	UDP-glucose 4-epimerase	Modified glycolytic flux	Sánchez *et al.* (2007a)
IlvE ↑	Probable BCAA aminotransferase	Increased valine synthesis	Sánchez *et al.* (2007a)
IlvD ↑	Dihydroxy acid dehydratase	Increased valine synthesis	Sánchez *et al.* (2007a)
IlvC2 ↑	Ketol-acid reductoisomerase	Increased valine synthesis	Sánchez *et al.* (2007a)
GlnA1 ↑	Glutamine synthetase 1	Increased ammonia synthesis	Sánchez *et al.* (2007a)
GroES ↑↓	General stress chaperone GroES	ND	Sánchez *et al.* (2007a)
DnaJ ↓	Chaperone protein	ND	Sánchez *et al.* (2007a)

(↑)Induced.

(↓)Repressed.

[a]Involved in bile adaptation.

[b]Only detected in the presence of bile.

[c]Involved in acid pH adaptation.

by down-regulating BSH activity, deconjugated bile acid production is slowed down, therefore decreasing the harmful effect of bile. Remarkably, the *B. longum* acid-adapted strains 8809dpH, with a constitutively diminished BSH activity, also acquired a bile resistance phenotype.

The acid-resistant strain *B. longum* 8809dpH showed deep and stable variations in the amount of enzymes responsible for some metabolic pathways (Sánchez *et al.*, 2007a). These principally concerned the over-expression of enzymes involved in the biosynthesis of certain intermediaries of the synthesis of the sulphur amino acids cysteine and methionine: methionine synthase, cystathionine gamma-synthase and O-acetylhomoserine (thiol)-lyase. It has been suggested that *B. longum* cysteine biosynthesis and sulphur assimilation may be accomplished by an atypical pathway (Schell *et al.*, 2002). It is not clear whether these changes in the sulphur metabolism are directly related to acidic pH resistance, but it opens an interesting research field to understand how bifidobacteria can become adapted to acidic conditions. Finally, this acid-resistant strain also showed an increase in the amount of the general stress response chaperone GroES under acidic conditions, as observed in other bacteria (Cotter and Hill, 2003; van de Gutche *et al.*, 2002).

Bile tolerance: clues for the adaptation of bifidobacteria to the intestinal environment

Bile flow exerts an important physiological role facilitating the digestion of lipophilic compounds from diet, but it is also a potent antimicrobial agent directly influencing the establishment of the intestinal microbiota. Bacteria passing through the GIT may be able to cope with bile stress in order to survive and colonize the intestinal environment. Bile mainly affects the lipid and protein composition of membranes, disturbing its functionality, but it also promotes oxidative damage to DNA and has a deep impact on general metabolic routes i.e. sugar metabolism. This section gives an overview of the current knowledge of the molecular mechanisms involved in the interaction of bile with bifidobacteria.

Bile physiology

Bile is a complex mixture of organic and inorganic compounds produced in the liver. The organic constituents of bile include bile salts, cholesterol, phospholipids and proteins, although many endogenous (vitamins, steroids and essential trace metals) and exogenous substances (antibiotics and drugs) can also be detected (Begley *et al.*, 2005; Kristiansen *et al.*, 2004). Bile is synthesized in the hepatocytes from cholesterol and leaves the liver by way of the common hepatic duct which joins the cystic duct from the gallbladder. The gallbladder is not essential for bile secretion but facilitates its storage and preparation for fat digestion (Begley *et al.*, 2005). During digestion bile is secreted into the duodenum in a hormone regulated process (Johnson, 1998). Once in the intestine bile flow carries out its main physiological function which is the emulsification of liposoluble substances (lipids, nutrients and vitamins) to facilitate their absorption.

Bile acids are the major organic constituents in bile, representing approximately 50% of this fraction (Begley *et al.*, 2002). All those bile acids that are synthesized from cholesterol in the hepatocyte are called primary bile acids: cholic acid and chenodeoxycholic acid being the two main ones. Before secretion, all bile acids are further processed by means of conjugation to glycine or taurine, generating conjugated bile acids. In human bile the typical ratio of glyco- to tauro-conjugates is usually around 3:1, although this value may vary depending on specific populations or diet habits (Hardison, 1978; Sjovall, 1959). As a consequence of the conjugation process the pKa of the acid decreases and at physiological pH conjugated bile acids are almost fully ionized (in contrast to the unconjugated ones) and are called bile salts. Conjugation also modifies solubility of these compounds, making them freely soluble over a wide range of ionic strengths and pH values (Begley *et al.*, 2005; Hofmann and Hagey, 2008; Ridlon *et al.*, 2006). Once in the intestine, where fat digestion and absorption takes place, bile salts can also be reabsorbed in the distal ileum through active transport and returned to the liver in a process known as the enterohepatic circulation. However, a small fraction of the bile present in the small intestine escapes this absorption process and passes to the colon, where it exerts a strong selective pressure on the local microbiota. The gradient of bile salt concentrations along the GIT goes from ~40 mM (its concentration in the human hepatic bile) to less than 1 mM in the distal part of the large bowel (Begley *et al.*, 2005; Ejderhamn *et al.*, 1991; Thomas *et al.*, 2001).

During intestinal transit bile properties can be modified by the action of gut microbiota, resulting in the production of secondary bile acids, litocholic acid and deoxycholic acid being the major ones. The most prevalent bacterial modifications of bile include deconjugation, 7α-dehydroxilation and oxidation of the hydroxyl groups at C-7 (Ridlon *et al.*, 2006). Deconjugation is carried out by bile salt hydrolases (BSH) which have been found in many members of the indigenous microbiota, including *Bifidobacterium*, but has not been detected in bacteria isolated from environments in which bile is absent (Tanaka *et al.*, 2002; Begley *et al.*, 2006). This activity releases the amino acid (glycine or taurine) from the conjugated bile acids, rendering the free bile acid (Begley *et al.*, 2006). 7α-dehydroxilation is restricted to free bile acids and is mainly carried out by members of the genus *Clostridium*, being absent in *Bifidobacterium*. Owing to this reaction, cholic acid is converted to deoxycholic acid, and chenodeoxycholic acid to litocholic acid, both called secondary bile acids. Furthermore, 7α-dehydrogenation in the GIT is carried out by hydrosteroid dehydrogenases rendering dihydroxy bile acids such as ursodeoxycholic acid (UDCA) (Bortolini *et al.*, 1997). Both deconjugation and dehydroxylation of bile acids by intestinal bacteria increase their relative hydrophobicity (Carey, 1984). As a consequence of bile salt biotransformations by human intestinal bacteria, faecal bile acid composition is significantly different to biliary salt composition, primary bile acids being predominant in the biliary bile and secondary bile acids predominant in human faeces (Ridlon *et al.*, 2006).

Antimicrobial activity of bile

Bile tolerance is an *in vitro* selection criterion for probiotic bacteria and is generally considered necessary to evaluate their ability to resist the effects of bile acids which further influences their ability to survive to GIT conditions and colonize the intestinal environment. The physicochemical characteristics of bile determine its physiological function in humans and its antimicrobial activity. Bile acids are amphipatic molecules with detergent properties that self-associate in both aqueous and organic environments to form micelles. This property is directly related to their capacity to emulsify and solubilize fats (Carey *et al.*, 1984; Hofman, 2009; Roda *et al.*, 1983). During passage through GIT, bile in particular represents a key challenge to bacteria that survive and transit the stomach entering the small intestine. As a consequence of their detergent-like properties, one of the main targets of bile action is the bacterial cell surface. Exposure to bile salts modulate the expression of membrane proteins in enteric bacteria (Bron *et al.*, 2006; Sánchez *et al.*, 2006) and they deeply affect the fatty acid composition, the phospholipid composition and the cell wall functionality in *Bifidobacterium* (Gómez-Zavaglia *et al.*, 2002; Ruiz *et al.*, 2007). Changes in the cell surface of LAB and bifidobacteria, after the exposure to bile salts, have also been shown through electron microscope analysis (Ruiz *et al.*, 2007; Taranto *et al.*, 2006).

The type and structure of bile salt interacting with the bacterial membrane is also correlated to its cytotoxicity. Hydrophobic bile salts are more toxic since they have a higher affinity for the membrane. Furthermore conjugated bile salts are strong acids that are fully ionized at physiological pH values, so they are unable to cross the membrane unless a transport system is available. On the contrary, free bile acids are weak acids that can freely diffuse through the lipid bilayer, promoted by the proton transmembrane gradient (ΔpH) and without the need for a specific transporter (Kurdi *et al.*, 2003). For this reason, unconjugated bile acids (cholate and deoxyholate) are much more inhibitory for the cells than the conjugated forms.

Another well known deleterious effect of bile salts in cells is related to oxidative damage caused by the generation of free radicals (Begley *et al.*, 2005; Bernstein *et al.*, 1999; Payne *et al.*, 1998). A very common response to bile salts exposure in bacteria is the induction of stress proteins playing roles in oxidative damage remediation (Bron *et al.*, 2006; Leverrieur *et al.*, 2003; Sánchez *et al.*, 2005, 2007b, 2008). Several studies have also shown the action of bile on macromolecules by promoting aberrant secondary structure formation in the

RNA, inducing DNA damage and modifying protein conformation and turnover (Begley *et al.*, 2005; Bernstein *et al.*, 1999; Sánchez *et al.*, 2005).

Molecular aspects of bile response

Bacteria passively accumulate cholic acid, which freely pass the lipid bilayer, and this situation has a deep impact on the general metabolism and homeostasis of the cell. Intracellular accumulation of unconjugated bile acids reduces the intracellular pH, leading, at certain concentrations, to the dissipation of ΔpH. This effect, together with the disturbing of membrane integrity by bile salts, can lead to leakage of ions, proton and other cellular components, eventually causing cell death. Envelope composition and structure of bacteria also has a strong influence on bile tolerance. In this regard, both adaptation and response of bifidobacteria to bile was correlated to significant alterations in the lipid composition of the cell membrane, these changes being considered as responsible for a shift of the physicochemical characteristics of the cell envelope (Ruiz *et al.*, 2007). Production of exopolymers is also one of the common responses to bile in some bacteria, causing a deep impact on their surface properties, thus acting as a protective coat against environmental conditions, and eventually affecting its intestinal colonization ability (Alp and Aslim, 2010; Crawford *et al.*, 2008; Denou *et al.*, 2008; Lebeer *et al.*, 2007; Ruas-Madiedo *et al.*, 2009).

Independently of its natural resistance levels to bile, bacteria can also acquire adaptative mutations as a reaction to bile exposure, and develop resistance phenotypes that are stable through generations (Sánchez *et al.*, 2008). Bile adaptation has mainly been studied in Gram-positive bacteria and the particular mechanisms involved in bile adaptation seem to be strain dependent (Begley *et al.*, 2005), although a common characteristic associated with bile adaptation in different species is related to a decrease in cell length (Leverrier *et al.*, 2003; Margolles *et al.*, 2003). Several *Bifidobacterium* strains have been stably-adapted to bile salts, through a progressive adaptation after growth in increasing concentrations of bile extracts (Noriega *et al.*, 2004). Furthermore, adaptation to bile has also been found to protect bacteria against other stressful factors, i.e. acid or heat, this being cross-resistance strain dependent (Saarela *et al.*, 2004; Sánchez *et al.*, 2006). Bile adaptation in bifidobacteria has been correlated to stable changes in carbohydrate fermentation profiles and some glycosidase activities (Margolles *et al.*, 2003; Noriega *et al.*, 2004; Ruas-Madiedo *et al.*, 2005), changes in fatty acid composition and membrane protein profiles (Margolles *et al.*, 2003; Ruiz *et al.*, 2007), in the ability to compete with enteropathogens in human intestinal mucus (Gueimonde *et al.*, 2005, 2007), and in the tetracycline resistance phenotype, among others (Kheadr *et al.*, 2007; Noriega *et al.*, 2005).

Among the members of the genus *Bifidobacterium*, molecular studies on bile stress have mainly been focused on the species *B. longum* and *B. breve*, commonly found in human intestinal faeces, and *B. animalis* subsp. *lactis*, typically isolated from functional foods (Gueimonde *et al.*, 2009; Price *et al.*, 2006; Sánchez *et al.*, 2005, 2007b; Savijoki *et al.*, 2005). Minimal inhibitory concentrations of bile have been described for *Bifidobacterium* species, ranging from 0.125% to 2% (Margolles *et al.*, 2003), although differences in the experimental conditions (bile origin, broth type, conditions assayed, etc.) make it very difficult to compare results from different studies. Remarkably, bile tolerance is a strain-specific trait that can be extremely variable within species or genus. Although the molecular mechanisms involved in bile tolerance in bifidobacteria are still poorly understood, the increasing appearance of *Bifidobacterium* genome projects (NCBI website: http://www.ncbi.nlm.nih.gov/) and the recent advances achieved in the field of molecular tools in this genus (O'Connell-Motherway *et al.*, 2008; Yasui *et al.*, 2008) have made possible more accurate knowledge of the molecular mechanisms of bifidobacterial responses and adaptation to bile. Some of these mechanisms are encompassed in the following paragraphs.

As mentioned before, unconjugated bile acids can passively accumulate in cells freely crossing the lipid bilayer. A well characterized mechanism of bile salt resistance in bacteria involves the action of efflux pumps, responsible for the efflux of bile salts and other toxic compounds (Gunn, 2000; Lin *et al.*, 2003). In this regard, 4 MDR efflux pumps have been described in bifidobacteria (Gueimonde *et al.*, 2009; Margolles *et al.*, 2005, 2006; Price *et al.*, 2006). The expression of two of them, ctr and BL0920 from *B. longum*, have been shown to confer bile resistance through a bile salt extrusion activity. The *ctr* gene of *B. longum* codes for a putative sodium-dependent bile acid transporter and it was shown to confer cholic acid and other antimicrobial compounds resistance, when cloned into an *E. coli* efflux-negative strain. The involvement of the *ctr* gene product in the efflux of radioactive labelled cholate (Price *et al.*, 2006) was also shown. In addition to this, the product of the BL0920 gene, which codes for a putative efflux pump, was over-expressed more than 30 fold when *B. longum* cells were exposed to subinhibitory concentrations of ox bile. The BL0920 product was shown to confer bile resistance when cloned into *E. coli* strain and to promote efflux of fluorescently labelled ursodeoxycholic acid (Gueimonde *et al.*, 2009). Both studies suggest that active efflux of bile acids is the mechanism underlying the bile resistance phenotype observed. Interestingly, the BL0920 homologue in *B. breve* UCC2003, Bbr_0838, is highly induced in the presence of bile as well, suggesting a similar function (Gueimonde *et al.*, 2009).

Exposure to bile salts also has a deep impact on central metabolic pathways of Gram-positive microorganisms, this situation having been related to higher energy requirements for the cells under stress conditions. Several works have studied the effect of bile on the sugar metabolism of *B. longum* and *B. animalis* subsp. *lactis*, showing that bacteria are able to re-route the metabolites to different branches, with the aim of modifying the production of energy and the redox equivalents of the cell (Begley *et al.*, 2005; Sánchez *et al.*, 2008), although the particular way to cope with bile stress is strain/species dependent (Sánchez *et al.*, 2005, 2007b). While *B. longum* showed an overexpression of most of the enzymes of the glycolitic pathway, in *B. animalis* subsp. *lactis* the only enzymes of the bifid shunt which were overexpressed were Xylulose-5-phosphate/fructose-6-phosphate phosphoketolase and Glyceraldehyde-3-phosphate dehydrogenase. This data points to a different glucose consumption of both strains in the presence of bile, which is in fact higher in *B. longum* than in *B. animalis*. In addition, *B. animalis* subsp. *lactis* seems to increase the acetic formation through the bifid shunt, which could theoretically allow bifidobacteria to synthesize more ATP by substrate level phosphorylation. In this regard, intracellular ATP concentration was shown to decrease as the bile salt concentration in the growth medium increases, suggesting that *Bifidobacterium* response to bile is ATP-dependent (Sánchez *et al.*, 2006). Bile was also shown to promote changes in the redox status of *B. animalis*, which is dependent on the metabolic reactions occurring at a given time (Sánchez *et al.*, 2007b). On the contrary, in *B. longum* the acetic/lactic ratio decreased in the presence of bile, thus enhancing the formation of lactic acid (Sánchez *et al.*, 2005).

Another well described effect of bile exposure is caused by the production of oxygen free radicals, which causes oxidative damage in cells. Bifidobacterial mechanisms to cope with oxidative stress promoted by bile salts could include the induction of a DPS protein and a thioredoxin-dependent thiol peroxidase in *B. animalis* (Sánchez *et al.*, 2007b), both enzymes being involved in SOS response (Comtois *et al.*, 2003; Nair and Finkel, 2004). Regarding oxidative stress, *B. animalis* was also shown to reduce the production of enzymes involved in the last steps of the methionine biosynthesis, whose sulphur group is susceptible to be oxidized: methionine synthase and O-acetylhomoserine sulfhydrolase. Interestingly, in a bile adapted *B. animalis* strain, these two enzymes were highly overproduced (Sánchez *et al.*, 2007b), suggesting that methionine metabolism is crucial for regulation of oxidative stress (Hondorp and Matthews, 2004; Sánchez *et al.*, 2008).

The effect of bile on protein conformation is counteracted through a chaperone/protease-mediated response which could conduct the correct folding of proteins and promote a quicker recycling of misfolding proteins (Kampinga, 2006). In this regard, *B. animalis* displayed a broader variety of these molecules to cope with this stress than *B. longum*. In *B. animalis* six putative chaperones and proteases were found to be overproduced in the presence of bile, as previously found in other Gram-positive bacteria (Duché *et al.*, 2002; Leverrier *et al.*, 2003): ClpB, a trypsin-like protease homologue to HtrA, GrpE, GroES, GroEL and DnaK (Sánchez *et al.*, 2007b). Three of these chaperones, HtrA, GroEL and

DnaK, were also found to be present in higher amounts in the cytoplasm of *B. longum* grown in the presence of bile (Sánchez *et al.*, 2005; Savijoki *et al.*, 2005). Thus, these three chaperones seem to constitute a common chaperone/protease bile-mediated response in *Bifidobacterium*, although they have also been related to some other stresses, DnaK to heat and osmotic shock response and GroEL to moderate heat shock in *B. breve* (De Dea Lindner *et al.*, 2007).

In relation to the effects of bile on the cell surface, previously referred to as the first target of bile action, bile has been shown to modify the expression of genes involved in EPS production, and modify the concentration of proteins involved in fatty acid synthesis, cell-wall binding proteins, cell division proteins, membrane embedded ATPases and cell wall 'moonlighting' proteins, with an important role in the colonization of the intestinal environment (Ruas-Madiedo *et al.*, 2009; Ruiz *et al.*, 2009; Sánchez *et al.*, 2006, 2007b). Bile induced the expression of *gtf01207* in *B. animalis*, which codes for a putative priming glycosyltransferase involved in EPS synthesis. The increased production of EPS was demonstrated in the presence of bile by cryo-scanning electron microscopy and quantification with lectin-conjugated probes (Ruas-Madiedo *et al.*, 2009). Exocellular polymers could coat bacterial surface with a protective layer against the action of bile salts and a correlation has been recently reported between EPS production and bile tolerance of *Bifidobacterium* spp. (Alp *et al.*, 2009). Production of exopolymers may also have an important effect on gastrointestinal colonization of bacteria. In this regard, a significant correlation has been reported between EPS production and intestinal colonization of *Lactobacillus* in an *in vivo* model (Denou *et al.*, 2008).

Remarkably, it is well known that bile induces changes in the lipidic composition of bifidobacterial membranes through inducing changes in the concentration of proteins involved in fatty acid synthesis. In this regard, long-chain-fatty-acid-CoA ligase was found to be highly underproduced in *B. animalis* in the presence of bile and constitutively underproduced in a *B. animalis* bile resistant mutant compared with the wild type (Sánchez *et al.*, 2007b). Transcriptomic analysis of *B. animalis* Bb-12 also showed down-regulation by bile of several genes involved in fatty acid biosynthesis (Garrigues *et al.*, 2005). These results are in agreement with studies which demonstrate that bile induces changes in the membrane composition in bifidobacteria (Kociubinski *et al.*, 2002; Kurdi *et al.*, 2006; Ruiz *et al.*, 2007); these changes could favour tolerance to bile decreasing the bile salt diffusion rate to the cytoplasm. Changes in the levels of surface proteins with important biological functions were also detected in *B. longum* in the presence of bile, such as enolase. Surface displayed enolase is able to capture human plasminogen, thus suggesting a role in the human gut colonization (Candela *et al.*, 2007; Knaust *et al.*, 2007). Also oligopeptide binding proteins were accumulated on the cell surface, which was reflected in a different tripeptide transport rate in cells grown with bile (Ruiz *et al.*, 2009). Also related to the bifidobacterial surfaces, it was shown that a bile resistant mutant of *B. animalis* subsp. *lactis* showed higher amounts of two proteins related to the cell division process (cell division initiation protein and chromosome segregation ATPase), both explaining one of the consequences of bile adaptation in bifidobacteria: a decrease in cell size (Margolles *et al.*, 2003; Sánchez *et al.*, 2007b). Interestingly, the cell division protein FtsI was also induced by bile in *B. longum* (Ruiz *et al.*, 2009).

In relation to bacterial modification of bile salts, adaptation to bile in *B. animalis* has also been related to an increase in the synthesis and activity of BSH (Noriega *et al.*, 2006; Sánchez *et al.*, 2007b). This activity, which seems to play a role in bile adaptation, is constitutively expressed in bifidobacteria although no induction was found in both *B. animalis* and *B. longum* as a response to bile (Grill *et al.*, 1995; Sánchez *et al.*, 2005, 2007b). However, *in vivo* assays have shown the intracellular accumulation of BSH in *B. longum* in the GIT of rabbits, suggesting that factors different from bile could modify the expression of this gene (Yuan *et al.*, 2008). All these observations suggest a possible relationship between BSH activity and bile resistance, although the role of BSHs in bacterial detoxification of bile salts and its contribution to bile resistance in probiotics is not yet clear, as some authors did not report any relationship between BSH activity and bile tolerance (Ahn *et al.*, 2003; Moser and Savage, 2001).

In addition, BSH enzymes have generally been found located intracellularly (Begley *et al.*, 2006), although intracellular accumulation of conjugated bile acids needs active transport mechanisms. In this regard, an operon-like structure containing the *bsh* gene and two more open reading frames, coding for a carboxylase and a transporter, was found in *B. animalis* (Kim and Lee, 2008), suggesting that the transporter could be responsible for the intake of conjugated bile acids.

Finally, response to bile has also been related to the increased concentration of a response regulator, belonging to a two-component regulatory system, in *B. longum* (Sánchez *et al.*, 2005). Also, overproduction of two enzymes involved in the biosynthesis of branched chain amino acids was found in *B. longum* and *B. animalis* as a response to bile (Sánchez *et al.*, 2005, 2007b), this metabolic route also being overproduced in response to acid in bifidobacteria and other Gram-positive bacteria (Len *et al.*, 2004; Sánchez *et al.*, 2005).

Cross-resistance mechanisms promoted by bile

In recent years it has been shown that the mechanisms of different stress responses overlap (van de Guchte *et al.*, 2002), even when the toxicity mechanism seems to be different. The acquisition of a stable resistance phenotype against one type of bile acid in *Bifidobacterium*, also conferred resistance to other bile salts. Also, adaptation to bile led to a higher tetracycline resistance *in vitro* and a better toleration of the acid pH (Noriega *et al.*, 2004, 2005). Remarkably, some MDR transporters have been shown to be involved in the acquisition of bile-cross resistance phenotypes. In *B. longum*, the MDR transporter named *ctr* was demonstrated to confer resistance to cholic acid, SDS and antibiotics (chloramphenicol, erythromycin and tetracycline), when cloned into *E. coli* (Price *et al.*, 2006).

Regarding the phenomena of cross-adaptation to GIT conditions, pH and bile, in *Bifidobacterium*, it can be concluded that protection against one of these two stresses, also confers protection against the other, this resistance being strain dependent (Noriega *et al.*, 2004; Saarela *et al.*, 2004; Sánchez *et al.*, 2006, 2007a,b). In light of proteomic studies under bile and acidic conditions in *B. animalis* subsp. *lactis*, it can be deduced that a common link between both resistance mechanisms is the F_0F_1-ATPase (Sánchez *et al.*, 2006). As previously mentioned, bile stress generates a proton excess, as a consequence of the intracellular accumulation of unconjugated bile acids, mimicking acid stress. A higher induction capacity of the F_0F_1-ATPase in the bile adapted mutant strain of *B. animalis* could be related to a tighter capacity to regulate the intracellular pH, imposed directly by acidic conditions or indirectly through bile exposure, by inducing proton pumping (Sánchez *et al.*, 2006, 2007b). Fig. 5.2 shows an overview of the responses to bile and pH in *Bifidobacterium*.

Future perspectives

Fundamental knowledge on environmental stress gene regulation in bifidobacteria is the key to their survival in commercial products and in their natural niche, and will thus impact on their interaction with the host. Despite the commercial and scientific interest, up until recently the genetics of bifidobacteria had not developed to a degree that was comparable to other high G+C content Gram-positive bacteria. The sequencing of several bifidobacterial genomes is an important step towards bridging this scientific gap; however, functional genomics experiments need to be performed to explore various aspects such as the mode of probiotic action, or the genetics and functional genomics of stress response. This latter area of investigation is considered highly important from an applied (bifidobacterial strains must demonstrate resilience to the adverse conditions encountered in industrial processes, whilst also surviving storage and passage through the stomach and small intestine), as well as academic (the mechanisms of protein/DNA damage repair, adaptive responses to hostile environments), perspective. Understanding adaptive responses in such biotechnologically important bacteria generates fundamental information that may ultimately improve their viability, delivery efficacy, intestinal colonization and consequently their health-promoting functionality when applied as probiotic bacteria.

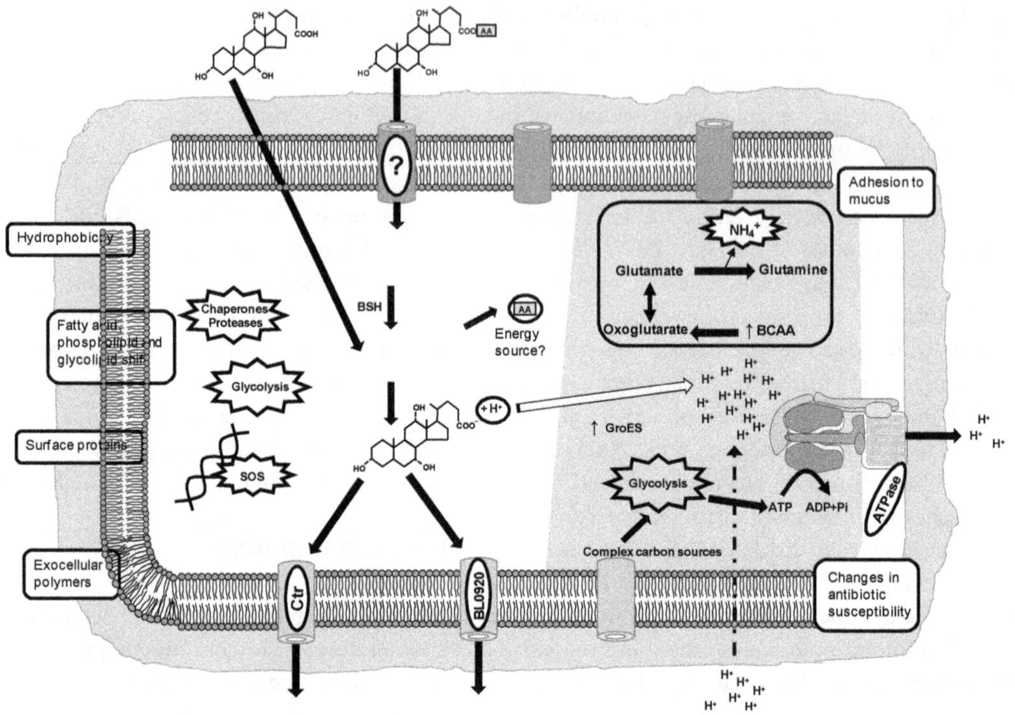

Figure 5.2 Mechanisms of bile response (white background) and acid pH response (grey background) in bifidobacteria. See text for details.

References

Ahn, Y.T., Kim, G.B., Lim, K.S., Baek, Y.J., and Kim, H.U. (2003). Deconjugation of bile salts by *Lactobacillus acidophilus* isolates. Int. Dairy J. *13*, 303–311.

Alander, M., Mättö, J., Kneifel, W., Johansson, M., Kögler, B., Crittenden, R., Mattila-Sandholm, T., and Saarela, M. (2001). Effect of galacto-oligosaccharide supplementation on human faecal microflora and on survival and persistence of *Bifidobacterium lactis* Bb-12 in the gastrointestinal tract. Int. Dairy J. *11*, 817–825.

Alp, G., and Aslim, B. (2009). Relationship between the resistance to bile salts and low pH with exopolysaccharide (EPS) production of *Bifidobacterium* spp. isolated from infant's feces and breast milk. Anaerobe (in press)

Andersen, M.T., Brondsted, L., Pearson, B.M., Mulholland, F., Parker, M., Pin, C., Wells, J.M., and Ingmer, H. (2005). Diverse roles for HspR in *Campylobacter jejuni* revealed by the proteome, transcriptome and phenotypic characterization of an *hspR* mutant. Microbiology-SGM *151*, 905–915.

Begley, M., Gahan, C.G., and Hill, C. (2002). Bile stress response in *Listeria monocytogenes* LO28: adaptation, cross-protection, and identification of genetic loci involved in bile resistance. Appl. Environ. Microbiol. *68*, 6005–6012.

Begley, M., Gahan, C.G., and Hill, C. (2005). The interaction between bacteria and bile. FEMS Microbiol. Rev. *29*, 625–651.

Begley, M., Hill, C., and Gahan, C.G. (2006). Bile salt hydrolase activity in probiotics. Appl. Environ. Microbiol. *72*, 1729–1738.

Bellier, A., and Mazodier, P. (2004). ClgR, a novel regulator of *clp* and *lon* expression in *Streptomyces*. J. Bacteriol. *186*, 3238–3248.

Bellier, A., Gominet, M., and Mazodier, P. (2006). Post-translational control of the *Streptomyces lividans* ClgR regulon by ClpP. Microbiology-SGM *152*, 1021–1027.

Bernstein, C., Bernstein, H., Payne, C.M., Beard, S.E., and Schneider, J. (1999). Bile salt activation of stress response promoters in *Escherichia coli*. Curr. Microbiol. *39*, 68–72.

Bi, C., and Benham, C.J. (2004). WebSIDD: server for predicting stress-induced duplex destabilized (SIDD) sites in superhelical DNA. Bioinformatics. *20*, 1477–1479.

Bore, E., Langsrud, S., Langsrud, O., Rode, T.M., and Holck, A. (2007). Acid-shock responses in *Staphylococcus aureus* investigated by global gene expression analysis. Microbiology-SGM *153*, 2289–2303.

Bortolini, O., Medici, A., and Poli, S. (1997). Biotransformations on steroid nucleus of bile acids. Steroids *62*, 564–577.

Bron, P.A., Molenaar, D., de Vos, W.M., and Kleerebezem, M. (2006). DNA micro-array-based identification of bile-responsive genes in *Lactobacillus plantarum*. J. Appl. Microbiol. *100*, 728–738.

Bucca, G., Brassington, A M., Hotchkiss, G., Mersinias, V., and Smith, C.P. (2003). Negative feedback regulation of *dnaK*., *clpB* and *lon* expression by the DnaK chaperone machine in *Streptomyces coelicolor*, identified by transcriptome and *in vivo* DnaK.-depletion analysis. Mol. Microbiol. *50*, 153–166Bucca, G., Brassington, A.M., Schonfeld, H.J., and Smith, C.P. (2000). The HspR regulon of *Streptomyces coelicolor*: a role for the DnaK chaperone as a transcriptional co-repressor. Mol. Microbiol. *38*, 1093–1103.

Bucca, G., Hindle, Z., and Smith, C.P. (1997). Regulation of the *dnaK* operon of *Streptomyces coelicolor* A3(2) is governed by HspR, an autoregulatory repressor protein. J. Bacteriol. *179*, 5999–6004.

Buchanan, R.L., and Edelson, S.G. (1996). Culturing enterohemorrhagic *Escherichia coli* in the presence and absence of glucose as a simple means of evaluating the acid tolerance of stationary-phase cells. Appl. Environ. Microbiol. *11*, 4009–4013.

Candela, M., Bergmann, S., Vici, M., Vitali, B., Turroni, S., Eikmanns, B.J., Hammerschmidt, S., and Brigidi, P. (2007). Binding of human plasminogen to *Bifidobacterium*. J. Bacteriol. *189*, 5929–5936.

Carey, M.C. (1984). Bile acids and bile salts: ionization and solubility properties. Hepatology *4*, 66S-71S.

Cheo, D.L., Bayles, K.W., and Yasbin, R.E. (1991). Cloning and characterization of DNA damage-inducible promoter regions from *Bacillus subtilis*. J. Bacteriol. *173*, 1696–1703.

Chung, H.S., Kim, Y.B., Chun, S.L., and Ji, G.E. (1999). Screening and selection of acid and bile resistant bifidobacteria. Int. J. Food Microbiol. *47*, 25–32.

Collado, M.C., and Sanz, Y. (2006). Method for direct selection of potentially probiotic *Bifidobacterium* strains from human feces based on their acid-adaptation ability. J. Microbiol. Methods *66*, 560–563.

Collado, M.C., and Sanz, Y. (2007). Induction of acid resistance in *Bifidobacterium*: a mechanism for improving desirable traits of potentially probiotic strains. J. Appl. Microbiol. *103*, 1147–1157.

Collado, M.C., Gueimonde, M., Sanz, Y., and Salminen, S. (2006). Adhesion properties and competitive pathogen exclusion ability of bifidobacteria with acquired acid resistance. J. Food Prot. *69*, 1675–1679.

Comtois, S.L., Gidley, M.D., and Kelly, D.J. (2003). Role of thioredoxin system and the thiol-peroxidases Tpx and Bcp in mediating resistance to oxidative and nitrosative stress in *Helicobacter pylori*. Microbiology-SGM *149*, 121–129.

Cotter, P.D., and Hill, C. (2003). Surviving the acid test: responses of gram-positive bacteria to low pH. Microbiol. Mol. Biol. Rev. *67*, 429–453.

Crawford, R.W., Gibson, D.L., Kay, W.W., and Gunn, J.S. (2008). Identification of a bile-inducible exopolysaccharide required for *Salmonella* biofilm formation on gallstone surfaces. Infect. Immun. *76*, 5341–5349.

Davis, M.J., Coote, P.J., and O'Byrne, C.P. (1996). Acid tolerance in *Listeria monocytogenes*: the adaptive acid tolerance response (ATR) and growth-phase-dependent acid resistance. Microbiology-SGM *142*, 2975–2982.

de Dea Linder, J., Canchaya, C., Zhang, Z., Neviani, E., Fitzgerald, G.F., van Sinderen, D., and Ventura, M. (2007). Exploiting *Bifidobacterium* genomes: the molecular basis of stress response. Int. J. Food Microbiol. *120*, 13–24.

Denou, E., Pridmore, R.D., Berger, B., Panoff, J.M., Arigoni, F., and Brüsso, H. (2008). Identification of genes associated with the long-gut-persistance phenotype of the probiotic *Lactobacillus johnsonii* strain NCC533 using a combination of genomics and transcriptome analysis. J. Bacteriol. *190*, 3161–3168.

Dollins, D.E., Warren, J.J., Immormino, R.M., and Gewirth, D.T. (2007). Structures of GRP94–nucleotide complexes reveal mechanistic differences between the *hsp90* chaperones. Mol. Cell. *28*, 41–56.

Duché, O., Tremoulet, F., Namane, A., and Labadie, J. (2002). A proteomic analysis of the salt stress response of *Listeria monocytogenes*. FEMS Microbiol. Lett. *215*, 183–188.

Ejderhamn, J., Rafter, J.J., and Strandvik, B. (1991). Faecal bile acid excretion in children with inflammatory bowel disease. Gut 32, 1346–1351.

Engels, S., Ludwig, C., Schweitzer, J.E., Mack, C., Bott, M., and Schaffer, S. (2005). The transcriptional activator ClgR controls transcription of genes involved in proteolysis and DNA repair in *Corynebacterium glutamicum*. Mol. Microbiol. *57*, 576–591.

Engels, S., Schweitzer, J.E., Ludwig, C., Bott, M., and Schaffer, S. (2004). *clpC* and *clpP1P2* gene expression in *Corynebacterium glutamicum* is controlled by a regulatory network involving the transcriptional regulators ClgR and HspR as well as the ECF sigma factor sigmaH. Mol. Microbiol. *52*, 285–302.

Garrigues, C., Stuer-Lauridsen, B., and Johansen, E. (2005). Characterisation of *Bifidobacterium animalis* subsp. *lactis* Bb-12 and other probiotic bacteria using genomics, transcriptomics and proteomics. Aust. J. Dairy Technol. *60*, 84–92.

Gómez-Zavaglia, A., Kociubinski, G., Pérez, P., Disalvo, E., and de Antoni, G. (2002). Effect of bile on the lipid composition and surface properties of bifidobacteria. J. Appl. Microbiol. *93*, 794–799.

Grandvalet, C., de Crecy-Lagard, V., and Mazodier, P. (1999). The ClpB ATPase of *Streptomyces albus* G belongs to the HspR heat shock regulon. Mol. Microbiol. *31*, 521–532.

Grandvalet, C., Rapoport, G., and Mazodier, P. (1998). *hrcA*, encoding the repressor of the *groEL* genes in *Streptomyces albus* G, is associated with a second *dnaJ* gene. J. Bacteriol. *180*, 5129–5134.

Grill, J., Perrin,S., and Schneider,F. (2000). Bile salt toxicity to some bifidobacteria strains: role of conjugated bile salt hydrolase and pH. Can. J. Microbiol. *46*, 878–884.

Grill, J., Schneider, F., Crociani, J., and Ballongue, J. (1995). Purification and characterization of conjugated bile salt hydrolase from *Bifidobacterium longum* BB536. Appl. Environ. Microbiol. *61*, 2577–2582.

Gueimonde, M., Garrigues, C., van Sinderen, D., de los Reyes-Gavilán, C.G., and Margolles, A. (2009). Bile-inducible efflux transporter from *Bifidobacterium longum* NCC2705, conferring bile resistance. Appl. Environ. Microbiol. *75*, 3153–3160.

Gueimonde, M., Margolles, A., de los Reyes-Gavilán, C.G., and Salminen, S. (2007). Competitive exclusion of enteropathogens from human intestinal mucus by *Bifidobacterium* strains with acquired resistance to bile-a preliminary study. Int. J. Food Microbiol. *113*, 228–232.

Gueimonde, M., Noriega, L., Margolles, A., de los Reyes-Gavilán, C.G., and Salminen, S. (2005). Ability of *Bifidobacerium* strains with acquired resistance to bile to adhere to human intestinal mucus. Int. J. Food Microbiol. *101*, 341–346.

Guillaume, E., Berger, B., Affolter, M., and Kussmann, M. (2009). Label-free quantitative proteomics of two *Bifidobacterium longum* strains. J. Proteomics *72*, 771–784.

Gunn, J.S. (2000). Mechanisms of bacterial resistance and response to bile. Microbes and Infection 2, 907–913.

Hardison, W.G. (1978). Hepatic taurine concentration and dietary taurine as regulators of bile acid conjugation with taurine. Gastroenterology 75, 71–75.

Hartl, F.U., and Hayer-Hartl, M. (2002). Molecular chaperones in the cytosol: from nascent chain to folded protein. Science *295*, 1852–1858.

Hatfield, G.W., and Benham, C.J. (2002). DNA topology-mediated control of global gene expression in *Escherichia coli*. Annu. Rev. Genet. *36*, 175–203.

Hofman, A.F., and Hagey, L.R. (2008). Bile acids: chemistry, pathochemistry, biology, pathobiology, and therapeutics. Cell Mol. Life Sci. *65*, 2461–2483.

Hofman, A.F. (2009). Bile acids: trying to understand their chemistry and biology with the hope of helping patients. Hepatology *49*, 1403–1418.

Hondorp, E.R., and Matthews, R.G. (2004). Oxidative stress inactivates cobalamin-independent methionine synthase (MetE) in *Escherichia coli*. PLos Biol. 2, 1738–1753.

Jakob, U., Muse, W., Eser, M., and Bardwell, J.C. (1999). Chaperone activity with a redox switch. Cell *96*, 341–352.

Jayamanne V.S., and Adams, M.R. (2006). Determination of survival, identity and stress resistance of probiotic bifidobacteria in bio-yoghurts. Lett. Appl. Microbiol. *42*, 189–194.

Jayamanne, V.S., and Adams, M.R. (2009). Modelling the effects of pH, storage temperature and redox potential (E-h) on the survival of bifidobacteria in fermented milk. Int. J. Food Sci. Tech. *44*, 1131–1138.

Jonhson, L.R. (1998). Bile secretion and gallbladder function. Essential Medical Physiology, Jonhson, L.R., ed., (Philadelphia: Lippincott-Raven), pp. 465–471.

Kampinga H.H. (2006). Chaperones in preventing protein denaturation in living cells and protecting against cellular stress. Handb. Exp. Pharmacol. *172*, 1–42.

Kheadr, E., Dabour, N., Le Lay, C., Lacroix, C., and Fliss, I. (2007). Antibiotic susceptibility profile of bifidobacteria as affected by oxgall, acid, and hydrogen peroxide stress. Antimicrob. Agents Chemother. *51*, 169–174.

Kim, G.B., and Lee, B.H. (2008). Genetic analysis of a bile salt hydrolase in *Bifidobacterium animalis* subsp. *lactis* KL612. J. Appl. Microbiol. *105*, 778–790.

Knaust, A., Weber, M.V., Hammerschmidt, S., Bergmann, S., Frosch, M., and Kurzai, O. (2007). Cytosolic proteins contribute to surface plasminogen recruitment of *Neisseria meningitides*. J. Bacteriol. *189*, 3246–3255.

Kociubinski, G., Zavaglia, A.G., Perez, P.F., Disalvo, E.A., and de Antoni, G.L. (2002). Effect of bile components on the surface properties of bifidobacteria. J. Dairy Res. *69*, 293–302.

Kristiansen, T.Z., Bunkenborg, J., Gronborg, M., Molina, H., Thuluvath, P.J., Argani, P., Goggins, M.G., Maitra, A., and Pandey, A. (2004). A proteomic analysis of human bile. Mol. Cell Proteomics 3, 715–728.

Kruger, E., and Hecker, M. (1998). The first gene of the *Bacillus subtilis* clpC operon, ctsR, encodes a negative regulator of its own operon and other class III heat shock genes. J. Bacteriol. *180*, 6681–6688.

Kurdi, P., Kawanishi, K., Mizutani, K., and Yokota, A. (2006). Mechanism of growth inhibition by free bile acids in lactobacilli and bifidobacteria. J. Bacteriol. *188*, 1979–1986.

Kurdi, P., Tanaka, H., Van Veen, H.W., Asano, K., Tomita, F., and Yokota, A. (2003). Cholic acid accumulation and its diminution by short-chain fatty acids in bifidobacteria. Microbiology-SGM *149*, 2031–2037.

Lahtinen, S.J., Ahokoski, H., Reinikainen, J.P., Gueimonde, M., Nurmi, J., Ouwehand, A.C., and Salminen, S.J. (2008). Degradation of 16S rRNA and attributes of viability of viable but nonculturable probiotic bacteria. Lett. Appl. Microbiol. *46*, 693–698.

Leahy, S.C., Higgins, D.G., Fitzgerald, G.F., and van Sinderen, D. (2005). Getting better with bifidobacteria. J. Appl. Microbiol. *98*, 1303–1315.

Lebeer, S., Verhoeven, T.L., Perez Vélez, M., Vanderleyden, J., and de Keersmaecker, S.C. (2007). Impact of environmental and genetic factos on biofilm formation by the probioic strain *Lactobacillus rhamnosus* GG. Appl. Environ. Microbiol. *73*, 6768–6775.

Len, A.C., Harty, D.W., and Jacques, N.A. (2004). Proteome analysis of *Streptococcus mutans* metabolic phenotype during acid tolerance. Microbiology-SGM *150*, 1353–1366.

Leverrier, P., Dimova, D., Pichereau, V., Auffray, Y., Boyaval, P., and Jan, G. (2003). Susceptibility and adaptive response to bile salts in *Propionibacterium freudenreichii:* physiological and proteomic analysis. Appl. Environ. Microbiol. *69*, 3809–3818.

Lin, J., Sahin, O., Michael, L.O., and Zhang, Q. (2003). Critical role of multidrug efflux pump CmeABC in bile resistance and *in vivo* colonization of *Campylobacter jejuni*. Infect. Immun. *71*, 4250–4259.

Margolles, A., Flórez, A.B., Moreno, J.A., van Sinderen, D., and de los Reyes-Gavilán, C.G. (2006). Two membrane proteins from *Bifidobacterium breve* UCC2003 constitute an ABC-type multidrug transporter. Microbiology-SGM *152*, 3497–3505.

Margolles, A., García, L., Sánchez, B., Gueimonde, M., and de los Reyes-Gavilán, C.G. (2003). Characterisation of a *Bifidobacterium* strain with acquired resistance to cholate-a preliminar study. Int. J. Food Microbiol. *82*, 191–198.

Margolles, A., Moreno, J.A., van Sinderen, D., and de los Reyes-Gavilán, C.G. (2005). Macrolide resistance mediated by a *Bifidobacterium breve* membrane protein. Antimicrob. Agents Chemother. *49*, 479–481.

Masco, L., Crockaert, C., van Hoorde, K., Swings, J., and Huys, G. (2007). *In vitro* assessment of the gastrointestinal transit tolerance of taxonomic reference strains from human origin and probiotic product isolates of *Bifidobacterium*. J. Dairy Sci. *90*, 3572–3578.

Matsumoto, M., Ohishi, H., and Benno, Y. (2004). H+-ATPase activity in *Bifidobacterium* with special reference to acid tolerance. Int. J. Food Microbiol. May *93*, 109–113.

Mättö, J., Alakomi, H.L., Vaari, A., Virkajärvi, I., and Saarela, M. (2006). Influence of processing conditions on *Bifidobacterium animalis* subsp *lactis* functionality with a special focus on acid tolerance and factors affecting it. Int. Dairy J. *16*, 1029–1037.

Maus, J.E., and Ingham, S.C. (2003). Employment of stressful conditions during culture production to enhance subsequent cold- and acid-tolerance of bifidobacteria. J. Appl. Microbiol. *95*, 146–154.

Miwa, T., Esaki, H., Umemori, J., and Hino, T. (1997). Activity of H(+)-ATPase in ruminal bacteria with special reference to acid tolerance. Appl. Environ. Microbiol. *63*, 2155–2158.

Mogk, A., Homuth, G., Scholz, C., Kim, L., Schmid, F.X., and Schumann, W. (1997). The GroE chaperonin machine is a major modulator of the CIRCE heat shock regulon of *Bacillus subtilis*. EMBO J. *16*, 4579–4590.

Moser, S.A., and Savage, D.C. (2001). Bile salt hydrolase activity and resistance to toxicity of conjugated bile salts are unrelated properties in lactobacilli. Appl. Environ. Microbiol. *67*, 3476–3480.

Nair, S., and Finkel, E. (2004). Dps protects cells against multiple stresses during stationary phase. J. Bacteriol. *186*, 4192–4198.

Narberhaus, F. (1999). Negative regulation of bacterial heat shock genes. Mol. Microbiol. *31*, 1–8.

Noriega, L., Cuevas, I., Margolles, A., and de los Reyes-Gavilán, C.G. (2006). Deconjugation and bile salts hydrolase activity by *Bifidobacterium* strains with acquired resistance to bile. Int. Dairy J. *16*, 850–855.

Noriega, L., Gueimonde, M., Sánchez, B., Margolles, A., and de los Reyes-Gavilán, C.G. (2004). Effect of the adaptation to high bile salts concentrations on glycosidic activity, survival at low pH and cross-resistance to bile salts in *Bifidobacterium*. Int. J. Food Microbiol. *94*, 79–86.

Noriega, L., de los Reyes-Gavilán, C.G., and Margolles, A. (2005). Acquisition of bile salt resistance promotes antibiotic susceptibility changes in *Bifidobacterium*. J. Food Prot. *68*, 1916–1919.

O'Connell Motherway, M., Fitzgerald, G.F., Neirynck, S., Ryan, S., Steidler, L., and van Sinderen, D. (2008). Characterization of ApuB, an extracellular type II amylopululanase from *Bifidobacterium breve* UCC2003. Appl. Environ. Microbiol. *74*, 6271–6279.

O'Riordan, K., Andrews, D., Buckle, K., and Conway, P. (2001). Evaluation of microencapsulation of a *Bifidobacterium* strain with starch as an approach to prolonging viability during storage. J. Appl. Microbiol. *91*, 1059–1066.

Payne, C.M., Crowley, C., Washo-Stultz, D., Briehl, M., Bernstein, H., Bernstein, C., Beard, S., Holubec, H., and Warneke, J. (1998). The stress-response proteins poly (ADP-ribose) polymerase and NF-kappaB protect against bile salt- induced apoptosis. Cell Death Differ. *5*, 623–636.

Penders, J., Vink, C., Driessen, C., London, N., Thijs, C., and Stobberingh, E.E. (2005). Quantification of *Bifidobacterium* spp., *Escherichia coli* and *Clostridium difficile* in faecal samples of breast-fed and formula-fed infants by real-time PCR. FEMS Microbiol. Lett. *243*, 141–147.

Price, C.E., Reid, S.J., Driessen, A.J., and Abratt, V.R. (2006). The *Bifidobacterium longum* NCIMB 702259T *ctr* gene codes for a novel cholate transporter. Appl. Environ. Microbiol. *72*, 923–926.

Ridlon, J.M., Kang, D.J., and Hylemon, P.B. (2006). Bile salt biotransformations by human intestinal bacteria. J. Lipid Res. *47*, 241–259.

Roda, A., Hofmann, A.F., and Mysels, K.J. (1983). The influence of bile salt structure on self-association in aqueous solutions. J. Biol. Chem. *258*, 6362–6370.

Ruas-Madiedo, P., Gueimonde, M., Arigoni, F., de los Reyes-Gavilán, C.G., and Margolles, A. (2009). Bile affects the synthesis of exopolysaccharides by *Bifidobacterium animalis*. Appl. Environ. Microbiol. *75*, 1204–1207.

Ruas-Madiedo, P., Hernández-Barranco, A., Margolles, A., and de los Reyes-Gavilán, C.G. (2005). A bile salt-resistant derivative of *Bifidobacterium animalis* has an altered fermentation pattern when grown on glucose and maltose. Appl. Environ. Microbiol. *71*, 6564–6570.

Ruiz, L., Couté, Y., Sánchez, B., de los Reyes- Gavilán, C.G., Sánchez, J.C., and Margolles, A. (2009). The cell-envelope proteome of *Bifidobacterium longum* in an *in vitro* bile environment. Microbiology-SGM *155*, 957–967.

Ruiz, L., Sánchez, B., Ruas-Madiedo, P., de los Reyes-Gavilán, C.G., and Margolles, A. (2007). Cell envelope changes in *Bifidobacterium animalis* subsp. *lactis* as a response to bile. FEMS Microbiol. Lett. *274*, 316–322.

Saarela, M., Rantala, M., Hallamaa, K., Nohynek, L., Virkajärvi, I., and Mättö, J. (2004). Stationary-phase acid and heat treatments for improvement of the viability of probiotic lactobacilli and bifidobacteria. J. Appl. Microbiol. *96*, 1205–1214.

Samelis, J., Ikeda, J.S., and Sofos, J.N. (2003). Evaluation of the pH-dependent, stationary-phase acid tolerance in *Listeria monocytogenes* and *Salmonella* Typhimurium DT104 induced by culturing in media with 1% glucose: a comparative study with *Escherichia coli* O157:H7. J. Appl. Microbiol. *95*, 563–575.

Sánchez, B., Champomier-Vergès, M.C., Anglade, P., Baraige, F., de los Reyes-Gavilán, C.G., Margolles, A., and Zagorec, M. (2005). Proteomic analysis of global changes in protein expression during bile salt exposure of *Bifidobacterium longum* NCIMB 8809. J. Bacteriol. *187*, 5799–5808.

Sánchez, B., Champomier-Vergès, M.C., Collado, M.C., Anglade, P., Baraige, F., Sanz, Y., de los Reyes-Gavilán, C.G., Margolles, A., and Zagorec, M. (2007a). Low-pH adaptation and the acid tolerance response of *Bifidobacterium longum* biotype *longum*. Appl. Environ. Microbiol. 73, 6450–9459.

Sánchez, B., Champomier-Vergès, M.C., Stuer-Lauridsen, B., Ruas-Madiedo, P., Anglade, P., Baraige, F., de los Reyes-Gavilán, C.G., Johansen, E., Zagorec, M., and Margolles, A. (2007b). Adaptation and response of *Bifidobacterium animalis* subsp. *lactis* to bile: a proteomic and physiological approach. Appl. Environ. Microbiol. 73, 6757–6767.

Sánchez, B., de los Reyes-Gavilán, C.G., and Margolles, A. (2006). The F1F0-ATPase of *Bifidobacterium animalis* is involved in bile tolerance. Environ. Microbiol. *8*, 1825–1833.

Sánchez, B., Ruiz, L., de los Reyes-Gavilán, C.G., and Margolles, A. (2008). Proteomics of stress response in *Bifidobacterium*. Front. Biosci. *13*, 6905–6919.

Sanz, Y. (2007). Ecological and functional implications of the acid-adaptation ability of *Bifidobacterium*: A way of selecting improved probiotic strains. Int. Dairy J. *17*, 1284–1289.

Sassanfar, M., and Roberts, J.W. (1990). Nature of the SOS-inducing signal in *Escherichia coli*. The involvement of DNA replication. J. Mol. Biol. *212*, 79–96.

Savijoki, K., Suokko, A., Palva, A., Valmu, L., Kalkkinen, N., and Varmanen, P. (2005). Effect of heat-shock and bile salts on protein synthesis of *Bifidobacterium longum* revealed by [35S]methionine labelling and two-dimensional gel electrophoresis. FEMS Microbiol. Lett. *248*, 207–215.

Schell, M.A., Karmirantzou, M., Snel, B., Vilanova, D., Berger, B., Pessi, G., Zwahlen, M.C., Desiere, F., Bork, P., Delley, M., Pridmore, P.D., and Arigoni, F. (2002). The genome sequence of *Bifidobacterium longum* reflects its adaptation to the human gastrointestinal tract. Proc. Natl. Acad. Sci. U.S.A *99*, 14422–14427.

Schirmer, E.C., Glover, J.R., Singer, M.A., and Lindquist, S. (1996). HSP100/Clp proteins: a common mechanism explains diverse functions. Trends Biochem. Sci. *21*, 289–296.

Schmid, A.K., Howell, H.A., Battista, J.R., Peterson, S.N., and Lidstrom, M.E. (2005). HspR is a global negative regulator of heat shock gene expression in *Deinococcus radiodurans*. Mol. Microbiol. *55*, 1579–1590.

Schmidt, G., and Zink, R. (2000). Basic features of the stress response in three species of bifidobacteria: *B. longum*, *B. adolescentis*, and *B. breve*. Int. J. Food Microbiol. *55*, 41–45.

Schulz, A., and Schumann, W. (1996). hrcA, the first gene of the *Bacillus subtilis dnaK* operon encodes a negative regulator of class I heat shock genes. J. Bacteriol. *178*, 1088–1093.

Servant, P., Grandvalet, C., and Mazodier, P. (2000). The RheA repressor is the thermosensor of the HSP18 heat shock response in *Streptomyces albus*. Proc. Natl. Acad. Sci. U.S.A *97*, 3538–3543.

Simpson, P.J., Stanton, C., Fitzgerald, G.F., and Ross, R.P. (2005). Intrinsic tolerance of *Bifidobacterium* species to heat and oxygen and survival following spray drying and storage. J. Appl. Microbiol. *99*, 493–501.

Sjovall, J. (1959). Dietary glycine and taurine on bile acid conjugation on man. Bile acids and steroids 75. Proc. Soc. Exp. Biol. Med. *100*, 676–678.

Spohn, G., and Scarlato, V. (1999). The autoregulatory HspR repressor protein governs chaperone gene transcription in *Helicobacter pylori*. Mol. Microbiol. *34*, 663–674.

Takahashi, N., Xiao, J.Z., Miyaji, K., and Iwatsuki, K. (2007). H+-ATPase in the acid tolerance of *Bifidobacterium longum*. Milk Sci. Int. *62*, 151–153.

Tanaka, H., Doesburg, K., Iwasaki, T., and Mierau, I. (1999). Screening of lactic acid bacteria for bile salt hydrolase activity. J. Dairy Sci. *82*, 2530–2535.

Taranto, M.P., Pérez-Martínez, G., and Font de Valdez, G. (2006). Effect of bile acid on the cell membrane functionality of lactic acid bacteria for oral administration. Res. Microbiol. *157*, 720–725.

Thomas, L.A., Veysey, M.J., French, G., Hylemon, P.B., Murphy, G.M., and Dowling, R.H. (2001). Bile acid metabolism by fresh human colonic contents: a comparison of caecal versus faecal samples. Gut *49*, 835–842.

Tuomola, E., Crittenden, R., Playne, M., Isolauri, E., and Salminen, S. (2001). Quality assurance criteria for probiotic bacteria. Am. J. Clin. Nutr. 73, 393S-398S.

van de Guchte, M., Serror, P., Chervaux, C., Smokvina, T., Ehrlich, S.D., and Maguin, E. (2002). Stress responses in lactic acid bacteria. Antonie van Leeuwenhoek *82*, 187–216.

Vaughan, E.E., de Vries, M.C., Zoetendal, E., Ben-Amor, K., Akkermans, A.D.L., and de Vos, W.M. (2002). The intestinal LABs. Antonie van Leeuwenhoek *82*, 341–352.

Ventura, M., Canchaya, C., van Sinderen, D., Fitzgerald, G.F., and Zink, R. (2004a). *Bifidobacterium lactis* DSM 10140: identification of the *atp* (*atpBEFHAGDC*) operon and analysis of its genetic structure, characteristics, and phylogeny. Appl. Environ. Microbiol. *70*, 3110–3121.

Ventura, M., Canchaya, C., Zink, R., Fitzgerald, G.F., and van Sinderen, D. (2004b). Characterization of the *groEL* and *groES* loci in *Bifidobacterium breve* UCC 2003: genetic, transcriptional, and phylogenetic analyses. Appl. Environ. Microbiol. *70*, 6197–6209.

Ventura, M., Canchaya, C., Bernini, V., Del, C.A., Dellaglio, F., Neviani, E., Fitzgerald, G.F., and van Sinderen, D. (2005a). Genetic characterization of the *Bifidobacterium breve* UCC 2003 *hrcA* locus. Appl. Environ. Microbiol. *71*, 8998–9007.

Ventura, M., Fitzgerald, G.F., and van Sinderen, D. (2005b). Genetic and transcriptional organization of the *clpC* locus in *Bifidobacterium breve* UCC 2003. Appl. Environ. Microbiol. *71*, 6282–6291.

Ventura, M., Kenny, J.G., Zhang, Z., Fitzgerald, G.F., and van Sinderen, D. (2005c). The *clpB* gene of *Bifidobacterium breve* UCC 2003:

transcriptional analysis and first insights into stress induction. Microbiology-SGM *151*, 2861–2872.

Ventura, M., Zhang, Z., Cronin, M., Canchaya, C., Kenny, J.G., Fitzgerald, G.F., and van Sinderen, D. (2005d). The ClgR protein regulates transcription of the *clpP* operon in *Bifidobacterium breve* UCC 2003. J. Bacteriol. *187*, 8411–8426.

Ventura, M., Zink, R., Fitzgerald, G.F., and van Sinderen, D. (2005e). Gene structure and transcriptional organization of the *dnaK* operon of *Bifidobacterium breve* UCC 2003 and application of the operon in bifidobacterial tracing. Appl. Environ. Microbiol. *71*, 487–500.

Ventura, M., Canchaya, C., Zhang, Z., Bernini, V., Fitzgerald, G.F., and van Sinderen, D. (2006). How high G+C Gram-positive bacteria and in particular bifidobacteria cope with heat stress: protein players and regulators. FEMS Microbiol. Rev. *30*, 734–759.

Ventura, M., Canchaya, C., Tauch, A., Chandra, G., Fitzgerald, G.F., Chater, K.F., and van Sinderen. (2007a). Genomics of *Actinobacteria*: tracing the evolutionary history of an ancient phylum. Microbiol. Mol. Biol. Rev. *71*, 495–548.

Ventura, M., Canchaya, C., Zhang, Z., Fitzgerald, G.F., and van Sinderen, D. (2007b). Molecular characterization of *hsp20*, encoding a small heat shock protein of *Bifidobacterium breve* UCC2003. Appl. Environ. Microbiol. *73*, 4695–4703.

Vernazza, C.L., Gibson, G.R., and Rastall, R.A. (2006). Carbohydrate preference, acid tolerance and bile tolerance in five strains of *Bifidobacterium*. J. Appl. Microbiol. *100*, 846 – 853.

Viala, J., and Mazodier, P. (2002). ClpP-dependent degradation of PopR allows tightly regulated expression of the *clpP3 clpP4* operon in *Streptomyces lividans*. Mol. Microbiol. *44*, 633–643.

Wang, H., and Benham, C.J. (2008). Superhelical destabilization in regulatory regions of stress response genes. PLoS. Comput. Biol *4*, e17.

Wang, Y.C., Yu, R.C., and Chou, C.C. (2004). Viability of lactic acid bacteria and bifidobacteria in fermented soymilk after drying, subsequent rehydration and storage. Int. J. Food Microbiol. *93*, 209–217.

Wickner, S., Maurizi, M.R., and Gottesman, S. (1999). Posttranslational quality control: folding, refolding, and degrading proteins. Science *286*, 1888–1893.

Yasui, K., Kano, Y., Tanaka, K., Watanabe, K., Shimizu-Kadota, M., Yoshikawa, H., and Suzuki, T. (2008). Improvement of bacterial transformation efficiency using plasmid artificial modification. Nucleic Acids Res. *37*, e3.

Yuan, J., Wang, B., Sun, Z., Bo, X., Yuan, X., He, X., Zhao, H., Du, X., Wang, F., Jiang, Z., *et al.* (2008). Analysis of host-inducing proteome changes in *Bifidobacterium longum* NCC2705 grown *in vivo*. J. Proteome Res. *7*, 375–385.

Probiotic Properties of Bifidobacteria

6

Maddalena Rossi and Alberto Amaretti

Abstract

Bifidobacteria are major components of the indigenous bacterial population present in the human gut and are arguably most relevant to the health-promoting properties that have been attributed to elements of this microbiota. They exert a range of beneficial health effects, including the regulation of intestinal microbial homeostasis, the inhibition of pathogens and harmful bacteria that colonize and/or infect the gut mucosa, the modulation of local and systemic immune responses, the repression of procarcinogenic enzymatic activities within the microbiota, the production of vitamins, and the bioconversion of a number of dietary compounds into bioactive molecules.

This chapter summarizes the reported health-promoting properties of members of the genus *Bifidobacterium* and discusses what research is still necessary for an in depth understanding of the probiotic function. In fact, although experimental evidence of the probiotic effectiveness of bifidobacteria has a long history, little information is available on the molecular mechanisms underlying the health-promoting claims, especially on such complex phenomena as anticarcinogenic and anti-inflammatory effects.

Introduction

Probiotics are defined as 'live microbes which when administered in adequate amounts confer a health benefit to the host' (FAO/WHO working group, 2001). Another definition recognizes probiotic as 'a preparation of or a product containing viable, defined microorganisms in sufficient numbers, which alter the microflora (by implantation or colonization) in a compartment of the host and by that exert beneficial health effects in this host' (Schrezenmeir and de Vrese, 2001). These non-pathogenic organisms do not necessarily possess a phylogenetic relation to one another and are therefore best defined functionally rather than taxonomically. Probiotic microorganisms have been identified within different phyla of bacteria and yeasts, since a variety of strains of bifidobacteria, lactobacilli, enterococci, streptococci, propionibacteria, *Bacillus* sp., *E. coli* and yeasts have been claimed to promote beneficial effects on human health (de Vrese and Schrezenmeir, 2008). However, the majority of probiotics in use today belong to the genera *Lactobacillus* and *Bifidobacterium*.

The primary habitat of bifidobacteria is the gastrointestinal tract of mammals, birds and insects. Undoubtedly, they have certain unique qualities that make them among the first gut colonizers and one of the most promising microorganisms to be utilized as probiotics. They are considered one of the most important health-promoting bacteria, having beneficial health effects through a variety of different mechanisms. For instance, they participate in the regulation of intestinal microbial homeostasis, interfere with the ability of pathogens to colonize and infect the mucosa, modulate local and systemic immune responses, stabilize and preserve the gastrointestinal barrier function, repress procarcinogenic enzymatic activities and promote the bioconversion of a number of dietary compounds into bioactive healthy molecules (Servin, 2004; Marotti *et al.*, 2007; Hord, 2008; Kim *et al.*, 2008). Although several reports have provided insights into metabolic, trophic and protective functions that support the functional claims of bifidobacteria, few have addressed the causal relationship between the mechanisms of action and the beneficial effects. So far, the molecular mechanisms through which these activities are achieved remain largely unknown, and it is likely that more than one is at work simultaneously (Boirivant and Strober, 2007; Hord, 2008).

Evidence of the diverse ability to induce anti-inflammatory and cytokine responses suggests that some probiotic effects are associated with different capacities to express particular surface molecules (glycolipids, strain-associated molecular patterns) or to secrete products that interact with various components of epithelial cells (Hord, 2008). It is neither the objective of this chapter to review probiotic function in depth regarding host cell interaction, nor to summarize the effects that cause inhibition of pathogens. Nevertheless, it is evident that many of the mechanisms that explain probiotic activities, as will be described in this chapter, are also involved in mutual cross-talk between bifidobacteria and the gut, in the immune response, and in the barrier effect against proliferation of (opportunistic) pathogens. This chapter will focus on current advances in the physiology of bifidobacteria with respect to conjugated fatty acids and vitamin production, biotransformation of plant precursors in health-promoting molecules, detoxification of harmful compounds, antioxidant properties and the scientific data supporting anticarcinogenic action.

Production of vitamins

Vitamins cannot be synthesized by mammals, therefore they must be obtained through intestinal absorption from exogenous sources, such as the diet and the gut microbiota. It is well established that the ruminal microbiota provide a rich source of vitamins for ruminants and the faecal bacteria for coprophagic rodents (Stevens and Hume, 1998). Evidence that commensal colonic bacteria are a significant source of a range of vitamins to humans has been presented as well (Hill, 1997). In particular, the microbiota of the human colon is known to produce vitamin K (menaquinones) and most of the water-soluble vitamins of group B, including biotin, nicotinic acid, folates, riboflavin, thiamine, pyridoxine, panthotenic acid and cobalamin (Hill, 1997). Not surprisingly, the microbiome of the human distal gut revealed a variety of COGs which are involved in synthesis of several essential micronutrients (Gill *et al.*, 2006). Unlike dietary vitamins, which are mainly absorbed in the proximal part of the small intestine, the uptake of the microbial ones predominantly occurs in the colon. Colonocytes appear to be able to absorb biotin, thiamin, folates, riboflavin, panthotenic acid and menaquinones, indicating that microbiota-produced vitamins can contribute to systemic vitamin levels and especially to the homeostasis of the vitamins in localized epithelial cells (Said and Mohamed, 2006).

Among intestinal bacteria, bifidobacteria are generally considered to synthesize several vitamins of the group B, including folate, biotin, thiamine, nicotinic acid, pyridoxine, riboflavin and B12, but have not been reported to produce vitamin K. The production of group B vitamins could therefore be a potential mechanism for probiotic activity of bifidobacteria. Nonetheless, the capability of bifidobacteria to produce vitamins and to release them extracellularly, so that they can improve the vitamin status of the host, has never been explored in depth, with the sole exception of folate (Pompei *et al.*, 2007b; Strozzi and Mogna, 2008). Some studies, focusing on vitamin auxotrophies, demonstrated the production of folate, biotin, thiamine, nicotinic acid, pyridoxine and B12 by diverse strains of bifidobacteria, with wide differences among species and strains. The strains which are capable of growth in the absence of a certain vitamin, and which are likely to produce it, exhibited diverse levels of vitamin production in both cells and supernatants (Deguchi *et al.*, 1985; Noda *et al.*, 1994; Pompei *et al.*, 2007a). However, extensive screening of *Bifidobacterium* strains for auxotrophies, synthesis, and accumulation of vitamins has not been accomplished yet. The lack of suitable analytical methods with sufficient sensitivity and of a general synthetic medium, where most *Bifidobacterium* spp. abundantly grow, has hampered such a survey.

Based on the genome sequences of bifidobacteria (*B. longum* subsp. *longum* NCC2705, *B. longum* subsp. *longum* DJ010A, *B. longum* subsp. *infantis* ATCC 15697, *B. adolescentis* ATCC 15703 and *B. animalis* subsp. *lactis* AD011), diverse capabilities of vitamin production have been predicted. Complete pathways for the synthesis of folate, thiamine and nicotinate are present in *B. longum* subsp. *longum* NCC2705, whereas pathways for riboflavin, biotin, cobalamin, pantothenate, pyridoxine and menaquinone are missing (Schell *et al.*, 2002). The analysis of predicted pathways on the KEGG database (www.genome.jp/kegg) suggests that complete biosynthetic processes for biotin, pantothenate, pyridoxine, cobalamin and menaquinone are lacking in the currently sequenced bifidobacteria. The pathway for *de novo* synthesis of riboflavin is present but interrupted in *B. longum* subsp. *infantis* ATCC 15697, and the one for folate is present but interrupted in *B. adolescentis* ATCC 15703, while the complete pathway for nicotinate is present in *B. longum* subsp. *longum* DJ010A and *B. adolescentis* ATCC 15703. However, the actual ability to produce vitamins has not been established for the sequenced strains.

Production of folates

Folate is a water-soluble vitamin (vitamin B9 or B11) which is plentiful in green leafy vegetables in the form of mono- and polyglutamate. It is a carrier of one-carbon units and is involved in a number of metabolic pathways, including the biosynthesis of nucleotides and amino acids. Folate actively impacts on DNA replication since it is necessary for DNA methylation and repair. For that reason, it is essential for rapidly proliferating cells which undergo continuous renewal, such as erythrocytes, leucocytes and enterocytes. Because deficiency of folate causes several harmful effects in humans, such as birth defects, increased homocysteine levels with a possible risk of cardiac disease, anaemia and aberrant DNA synthesis causing increased cancer risk, it is frequently administered as dietary supplement. The daily recommended intake is between 170 and 300 μg, varying on the basis of sex, age and specific physiologic state (Shane, 1995).

Folates are found in food in the form of polyglutamates. They are first hydrolysed to the monoglutamate form, then absorbed mainly in the proximal part of the small intestine through an acidic pH-dependent carrier. Folates can be produced by the microbiota and absorbed across the large intestine (Camilo *et al.*, 1996; Kim *et al.*, 2004). Following absorption, they undergo hydrolysis to methyl-tetrahydrofoalte, which is the predominant form of folate in plasma (Shane, 1995).

Several strains of bifidobacteria have been screened for their ability to produce folate in low-folate or folate-free media. Twenty four strains of *B. bifidum*, *B. infantis*, *B. breve*, *B. longum* and *B. adolescentis* were cultured in a low-folate semisynthetic medium and significant differences in vitamin accumulation were found among the species tested. All *B. bifidum* and *B. infantis* strains were classified as high folate accumulators, while *B. breve*, *B. longum* and *B. adolescentis* were low accumulators; for all the strains, extracellular folate accounted for most of the vitamin (Deguchi *et al.*, 1984). Furthermore, 76 animal or human strains of *Bifidobacterium*, belonging to 15 different species, were screened in a folate-free semisynthetic medium (Pompei *et al.*, 2007a). The ability to produce folate was found only in 17 strains belonging to nine different species (*B. adolescentis*, *B. breve*, *B. pseudocatenulatum*, *B. animalis*, *B. bifidum*, *B. catenulatum*, *B. dentium*, *B. infantis* and *B. longum*), which produced mainly extracellular vitamin. Unlike Deguchi *et al.*, (1984), the highest folate levels were produced by four strains of *B. adolescentis* and two of *B. pseudocatenulatum*. Such an apparent discrepancy was interpreted as the result of both strain to strain variability and different experimental design. Even if the annotated genome of *B. longum* NCC2705 contains all the genes for folate biosynthesis, only one out of 15 *B. longum* strains was capable to grow without the vitamin. Likely, the presence of genes does not necessarily ensure that the corresponding enzymes are actually functional. This may be because of the presence of inactivated, remnant genes, where evolutionary pressures have caused loss of function, inactivating mutations, elimination of transcription/translation and/or post-translational processing. Thus, folate synthesis is regarded as a strain-specific trait.

In folate-depleted rats, the administration of diets containing bifidogenic ingredients (e.g. human milk solids or prebiotics) has been shown to increase the folate concentration in caecum, colon, plasma and colonic tissue (Krause *et al.*, 1996; Thoma *et al.*, 2003). These studies indicated that the number of commensal bifidobacteria in the large intestine correlated with the vitamin level, supporting evidence that bifidobacteria are capable of producing folate in the gut and that the vitamin synthetized in the large intestine can be absorbed and utilized by the host. The exploitation of folate-producing bifidobacteria as probiotic supplements has been recently investigated *in vivo* in animal and human trials. Three strains (*B. adolescentis* DSMZ 18350, *B. adolescentis* DSMZ 18352 and *B. pseudocatenulatum* DSMZ 18535) were specifically selected for their ability to produce high amounts of folate (Pompei *et al.*, 2007a). A mixture of freeze dried biomass of these strains was administered as a probiotic or synbiotic supplement to folate-depleted rats for 14 days producing a significant increase of the vitamin in the plasma and liver (Pompei *et al.*, 2007b). The same strains were given to 23 healthy volunteers in a pilot human study, causing a significant increase of folate concentration in the faeces of the subjects (Strozzi and Mogna, 2008). These results demonstrate that folate-producing probiotic strains may represent an endogenous source of the vitamin, preventing its deficiency in the colon. Localized folate production in the large intestine may provide the proliferating enterocytes with this vitamin, potentially reducing colonic carcinogenesis. Therefore, the trophic effects on colonocytes of folate-producing strains deserves to be evaluated. Furthermore, administration of folate-producing *B. longum* caused beneficial effects in lowering hyperhomocysteinaemia of haemodialysis patients and it was hypothesized that reduction in the serum level of homocysteine may be attributable to the supply of folate (Taki *et al.*, 2005).

Besides their exploitation as an endogenous source of vitamins, folate-producing bifidobacteria may also be used to fortify fermented dairy products, as milk is a poor source of folate. Seven strains of *Bifidobacterium* were evaluated for the capacity to enhance the folate concentration of reconstituted skimmed milk, resulting in a strain of *B. breve* being selected as the most promising (Crittenden *et al.*, 2003). Moreover, mixed culture fermentations of reconstituted skimmed milk were successfully carried out using folate-producing strains of *Bifidobacterium* in conjunction with strains of *Streptococcus thermophilus* and/or *Lactobacillus delbrueckii* subsp. *bulgaricus* from conventional yoghurt, demonstrating that it is possible to increase folate levels in fermented milk products through appropriate selection of bacterial strains.

Production of other B vitamins

Apart from folate, diverse strains of *Bifidobacterium* have been demonstrated to produce other B vitamins. Much of the current information on B vitamin production comes from a study that compared 24 bifidobacterial strains (belonging to the species *B. bifidum*, *B. infantis*, *B. breve*, *B. longum* and *B. adolescentis*) for the synthesis of water-soluble vitamins (Deguchi *et al.*, 1984). Pyridoxine was produced by all strains, although some variability among strains was observed. In contrast, riboflavin was not produced by any of the tested strains. The production of thiamine and nicotinic acid was observed with a pattern similar to folate, since all strains of *B. bifidum* and *B. infantis* and many strains of *B. breve* were classified as high producers, while most strains of *B. adolescentis* and some of *B. longum* and *B. breve* required these vitamins for growth and, consequently, were considered non-producers. Small amounts of vitamin B12 were found in all the strains (Deguchi *et al.*, 1984). A comparative study carried out to investigate biotin production in five strains of *B. bifidum*, *B. adolescentis*, *B. longum*, *B. breve*, and *B. infantis*, showed that all strains tested were capable of biotin biosynthesis (Noda *et al.*, 1993).

Despite the finding that a diverse range of bifidobacterial strains can produce the above water-soluble vitamins, information about the potential applications of vitamin-producing strains is still scarce. One study showed that the fermentation of soymilk with two strains of *B. infantis* and *B. longum* enhanced the content of riboflavin and thiamine, but negatively affected niacin concentration (Hou *et al.*, 2000). Moreover, even if diverse vitamins of group B can be absorbed in the colon, it is currently unclear whether commensal bifidobacteria can give any significant contribution to the vitamin status of the host, at either local or systemic level. Therefore, with the exception of folate, no strains of *Bifidobacterium* have been proposed yet as probiotic supplements to provide the host with vitamins of an endogenous origin.

Activation of healthy dietary compounds

Edible plants are dietary sources of hundreds of non-nutritional phytochemicals. These molecules are typically secondary metabolites that occur in small quantities in plant tissues, and, therefore, can be encountered in plant-derived foods. The major classes of bioactive phytochemicals are polyphenols (which include phenolic acids, flavonoids, stilbenes, lignans, and tannins), organosulphur compounds (which include glucosinolates and isothiocyanates), carotenoids, plant sterols, monoterpenes and alkaloids (Kris-Etherton *et al.*, 2002). Considerable attention has been given to such a broad variety of compounds and numerous epidemiological, clinical, and experimental studies have been conducted to identify them and to evaluate their effects on human health. The benefits of many dietary polyphenols are well documented and seem to be because of their ability to inactivate reactive oxygen species (Holst and Williamson, 2008). Therefore, many plant-derived compounds are active in the prevention of various diseases associated with oxidative stress and are responsible for the protective effect of fruits and vegetables against cardiovascular and neurodegenerative diseases, (van der Schouw *et al.*, 2000; Calabrese *et al.*, 2008), and the initiation and progression of cancer (de Kok *et al.*, 2008). Furthermore, beneficial activity of diverse phytochemicals arises from oestrogenic-like properties. Phytoestrogens are a broad group of non-steroidal compounds of different structure, falling into three main classes: isoflavones, coumestans and lignans. They have a common diphenolic group which has similarity in structure to the human female hormone 17-β-oestradiol, and have been shown to bind to alpha and beta oestrogen receptors (Turner *et al.*, 2007). Phytoestrogens mimic the action of oestrogens on target organs and exert many health benefits against hormone-dependent diseases, including protection against breast and prostate cancers, prevention and treatment of osteoporosis, lowering of haematic cholesterol and lipids, cardiovascular protection, antioxidant activity and alleviation of menopausal symptoms (Setchell and Cassidy, 1999; Branca and Lorenzetti, 2005).

Even though a normal diet provides a human being with a number of phytochemicals that are considered to be beneficial, it should be kept in mind that their biological activity depends on several factors which are not necessarily mutually related. These include abundance in the diet, effective bioavailability and intrinsic activity. The level of phytochemicals within the body is significantly affected by diverse phenomena, such as the digestive transformations of native compounds, the absorption in the intestine, the hepatic activity, and the biliary or urinary elimination (Scalbert and Williamson, 2000). The phytochemicals which are not absorbed in the small intestine reach the colon, where they may undergo extensive biotransformation by the resident microbiota (Manach *et al.*, 2004). Microbial transformation includes the hydrolysis of glycosides into the respective aglycones, which may lead to their degradation. Besides, the specific conversion of diverse molecules into bioactive metabolites can be accomplished by the microbiota, such as the conversion of lignans into enterolactone and enterodiol and the conversion of soy isoflavones into S-equol. During the absorption process, phytochemicals are conjugated in the intestine and later in the liver, being subjected to methylation, sulfation and β-glucuronidation. Then, they can be secreted in the duodenum with the bile as hydrophilic conjugates. Elements of the colonic microbiota are responsible for the enterohepatic recycling since those phytochemicals can be deconjugated by microbial enzymes, especially β-glucuronidase, in the colon. Here, they can be reabsorbed, leading to a longer presence in the body (Zhang *et al.*, 2007).

Within the human intestinal microbiota, bifidobacteria can produce a number of diverse glycosyl hydrolases which are necessary to hydrolyse oligo- and polysaccharides into fermentable carbohydrates in the colon (Schell *et al.*, 2002). Several strains of the genus *Bifidobacterium* are able to hydrolyse the glycosilated forms of some phytochemicals (including flavonoids, glucosinolates and ginsenosides) and take part in the deconjugation step transforming these molecules into bioactive compounds (Tsangalis *et al.*, 2002; Cheng *et al.*, 2004a; Bae *et al.*, 2004; Marotti *et al.*, 2007; Dabek *et al.*, 2008; Raimondi *et al.*, 2009).

Polyphenols

Polyphenols are effective in protecting against cancer and cardiovascular diseases. These effects arise from antioxidant properties, scavenging of free radicals, nitric oxide regulation, apoptosis induction, inhibition of cell proliferation and angiogenesis and oestrogenic activity (de Kok *et al.*, 2008). Polyphenols constitute a family of thousands of different molecules widely distributed in plants, occurring in high amounts in the human diet. They are classified as phenolic acids, flavonoids, stilbenes, and lignans (Fig. 6.1), based on the number of phenol rings and structural elements. Within flavonoids, six classes can be distinguished: flavonols, flavones, isoflavones, flavanones, anthocyanidins and flavanols. Most polyphenols are present in food in the form of esters, glycoconjugates or polymers, which have to undergo several chemical transformations before the bioactive compound is generated and absorbed.

Polyphenols, and especially those bound to hydrophilic moieties, are generally too polar to be absorbed through passive diffusion by enterocytes in the small intestine (Manach *et al.*, 2005). Unlike quercetin glucosides, which seem to be absorbed through a sodium-dependent glucose transporter, the polyphenols bound to different sugar moieties cannot be absorbed in their native form (Wiseman, 1999). The conjugated polyphenols which are not hydrolysed by human enzymes in the small intestine and escape absorption, reach the colon. Here they may be hydrolysed by the resident microbiota and their degradation products may subsequently be absorbed by colonocytes, even if absorption occurs less readily and efficiently in the colon than in the small intestine (Manach, 2004).

Isoflavones

Isoflavones occur abundantly in soybean and are increasingly being administered as functional dietary supplements (Wang and Murphy, 1994; Song *et al.*, 2007). Isoflavones can occur in the form of aglycones (daidzein, genistein and glycitein), glucosides (daidzin, genistin and glycitin), acetylglucosides (acetyldaidzin, acetylgenistin and acetylglycitin) and malonylglucosides (malonyldaidzin, malonylgenistin and malonylglycitin). The chemical form in which isoflavones occur affects availability, biological activity and, therefore, their physiological effects. Compared with the corresponding aglycones, glucose-conjugated isoflavones are poorly absorbed by the intestinal epithelium and have weaker biological activities (Setchell *et al.*, 2002). The ability to specifically release isoflavone aglycones (Fig. 6.2a) depends on the enzyme β-glucosidase and is carried out by mucosal and/or microbial enzymes (Day *et al.*, 1998; Garro *et al.*, 2006). Different microbial groups of the colon can transform glycosylated isoflavones into aglycones, such as lactobacilli, bifidobacteria, *E. coli* and *Bacteroides* spp., and probably many others (Hur *et al.*, 2000).

Members of the genus *Bifidobacterium* are known to produce β-glucosidases as well as many glycosyl-hydrolases. A wide diversity in β-glucosidase activity was found among *Bifidobacterium* species and among strains of the same species, with β-glucosidase activity mostly cell-associated (Tsangalis *et al.*, 2002; Marotti *et al.*, 2007; Dabek *et al.*, 2008). A quantitative correlation between β-glucosidase activity and the yield of daidzin hydrolysis could not be established, but the absence of β-glucosidase activity in a few strains correlated with the lack of daidzein production (Raimondi *et al.*, 2009). These observations indicate that the production of β-glucosidase may be considered as a potential probiotic property, since it may improve the *in vivo* availability of health-promoting soy isoflavones. The ability of bifidobacteria to produce β-glucosidase that transform the soybean isoflavone glycosides into their corresponding aglycones has been exploited to improve the biological activity of soymilk. In fact the production of bioactive isoflavones by bifidobacteria during soymilk fermentation is documented in various publications (Tsangalis *et al.*, 2002, 2005; Chien *et al.*, 2006; Otieno et Shah, 2007a,b; Pham and Shah, 2007; Tsangalis *et al.*, 2007; Wei *et al.*, 2007).

Daidzein, which is released from daidzin, can undergo a reductive metabolic pathway through dihydrodaidzein and tetrahydrodaidzein or dehydroequol, being transformed into S-equol by the gut microbiota in the colon (Fig. 6.2a) (Atkinson *et al.*, 2005; Yuan *et al.*, 2007). S-equol is the most effective compound in stimulating an estrogenic response among all isoflavone derivatives. It has

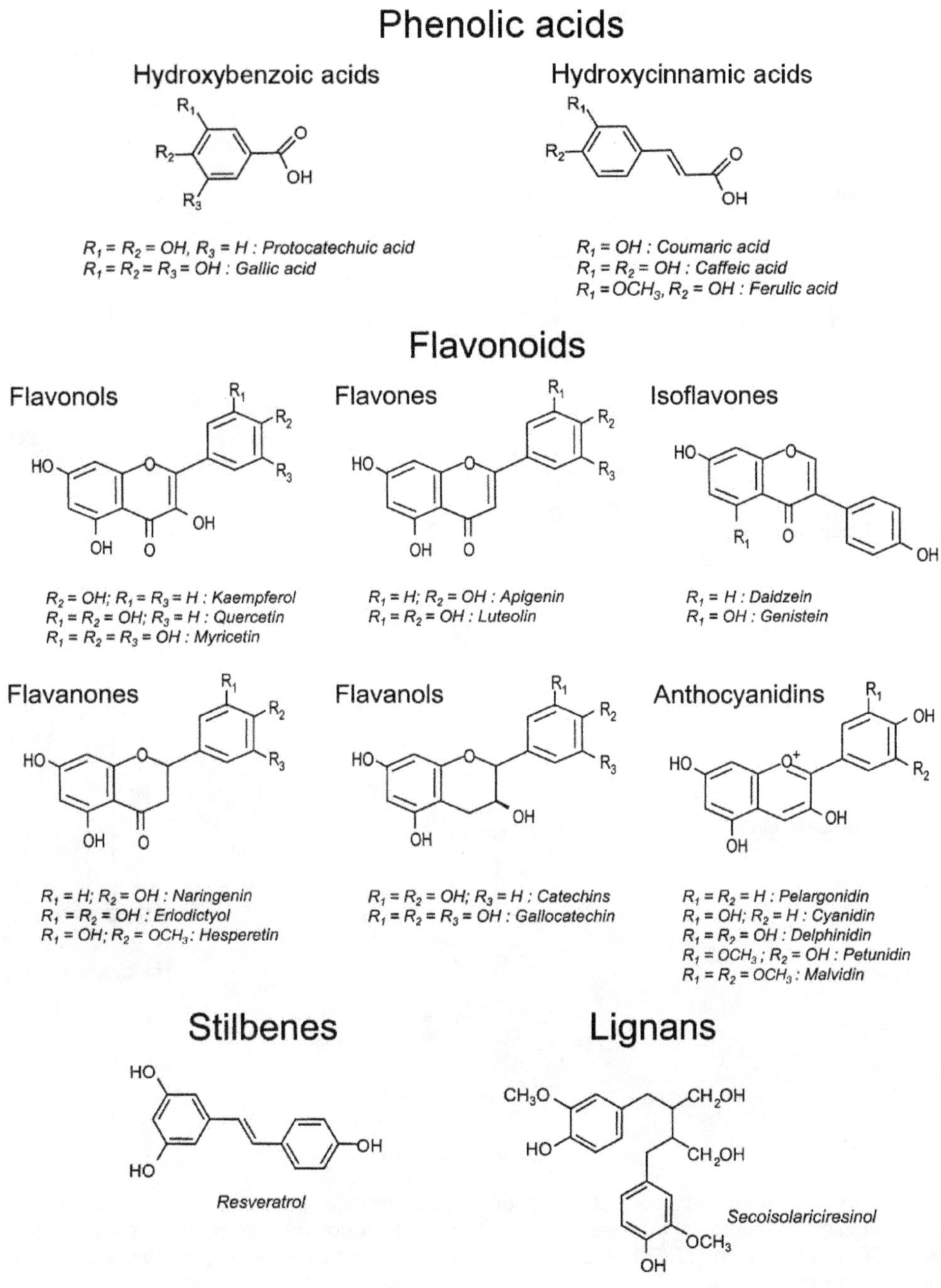

Figure 6.1 Chemical structure of polyphenols, adapted from Manach *et al.* (2004).

a longer half life and a greater biological activity, as well as superior antioxidant activity. Among humans, a large interindividual variability exists in the capacity to produce S-equol, since the production of S-equol occurs in only 30–40% of people. Recently, extensive research has looked for intestinal bacteria capable of producing S-equol from daidzein. The conversion of daidzein to S-equol was shown to be carried out *in vitro* by mixed cultures of the anaerobic intestinal bacteria *Lactobacillus* sp. Niu-O16 and *Eggerthella* sp. Julong 732. The former was isolated from bovine rumen and reduced daidzein to dihydrodaidzein, while the latter was isolated from human faeces

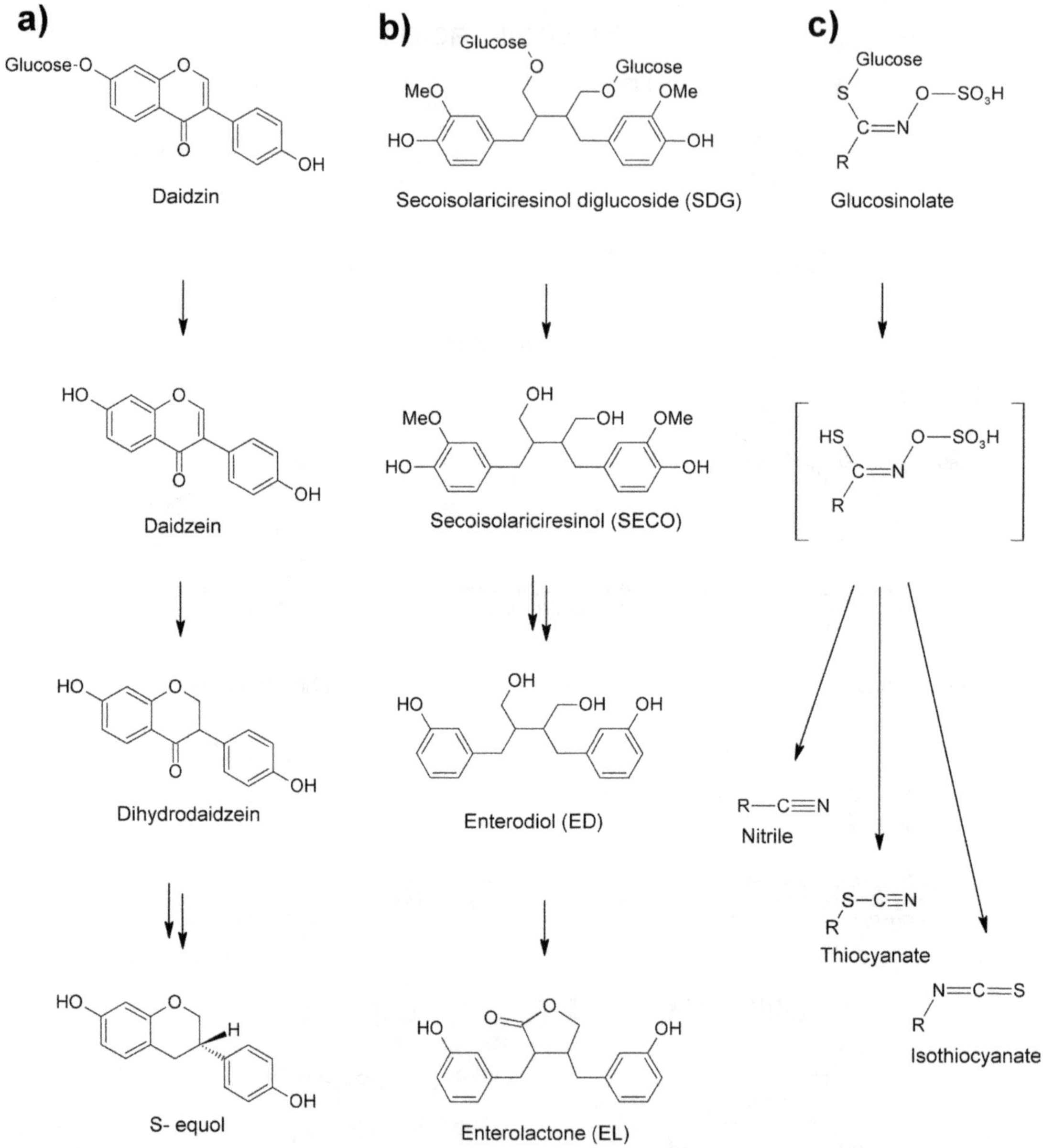

Figure 6.2 Examples of conversion pathways of phytochemicals into bioactive compounds by the commensal colonic microbiota. (a) Conversion of daidzin into S-equol; (b) conversion of secoisolariciresinol diglucoside (SDG) into enterodiol (ED) and enterolactone (EL); conversion of glucosinolates into nitriles, thiocyanates, and isothiocyanates.

and was shown to convert dihydrodaidzein and tetrahydrodaidzein to S-equol (Wang *et al.*, 2005, 2007). However, the gut microbiota consists of a multitude of species and it is likely that many microbial groups carry out the reductive steps of daidzein towards S-equol. Single strains performing the whole bioconversion of daidzein to S-equol have recently been isolated from human and animal faeces. Phylogenetic analysis has indicated that they belong to the species *Adlercreutzia equolifaciens* (Maruo *et al.*, 2008), and to the taxa *Coriobacteriaceae* (Matthies *et al.*, 2008), such as *Asaccharobacter celatus* (Minamida *et al.*, 2008), and *Clostridiales* (Tamura *et al.*, 2007; Yu *et al.*, 2008).

No bifidobacteria have so far been reported that are capable of performing a complete conversion of daidzein to S-equol, with the exception of a study by Tsangalis *et al.* (2002), a result which has, as yet, not been reproduced. When the involvement of bifidobacteria in the reductive reactions of the aglycone daidzein towards the production of S-equol was explored, none of the 22 strains belonging to eight species of human origin was capable of transforming daidzein into any reduced metabolite (Raimondi *et al.*, 2009).

Flavonols and flavanones

Flavonols are very ubiquitous in foods (such as citrus fruits, berries, onions, leafy vegetables and legumes) where they are commonly present at low concentrations. The main representatives of flavonols are quercetin and kaempferol which generally occur as 7-O-glycosylated and methylated forms. The most common sugar moieties are glucose, rhamnose and rutinose, but other sugars may occur, such as galactose, arabinose, xylose and glucuronic acid (Manach *et al.*, 2004). Rutin, the rutinoside of quercetin, is one of the major glycosides of quercetin (Fig. 6.1). Flavanones are found in some plants, and are present in high concentration in the albedo of citrus fruit. The main flavanones are naringenin, eriodictyol, and hesperidin, which are generally 3-O-glycosylated with rutinose or nehoesperidose. Similar to other polyphenols, flavonols and flavanones are antioxidants, and exert many biological activities which include anti-inflammatory effects and anti-proliferative activities against breast and gastrointestinal cancer (Graf *et al.*, 2005).

The sugar moiety of dietary glycoconjugated flavonols and flavanones is a major determinant affecting absorption in the small intestine. Glucose-conjugated quercetin is rapidly absorbed in the small intestine, probably through a sodium-dependent glucose transporter, while the other glyco-conjugates are not (Hollman *et al.*, 1995). Rutinosides, rhamnosides and other glycoconjugates reach the large intestine where they are hydrolysed by microbial glycosyl-hydrolases, after which their deconjugated products can be absorbed. Substantial person-to-person variation exists in bioavailability and absorption of active forms, and probably reflects the impact of the colonic microbiota. The involvement of intestinal bacteria in the deconjugation and metabolism of glycoconjugated flavonols and flavanones has been demonstrated with faecal fermentations (Olthof *et al.*, 2003; Rechner *et al.*, 2004). The latter report showed that the initial cleavage of the glycosidic bond is the result of efficient β-glucosidase activity that originates from elements of the microbiota.

Apart from isoflavones, information on the hydrolysis of other glycoconjugated flavonoids by bifidobacteria is scarce. Five selected strains of *B. pseudocatenulatum*, *B. infantis*, *B. adolescentis*, *B. catenulatum* and *B. breve* hydrolysed the glucosides of soy isoflavones and both the 3-O-glucoside and the 3-O-xylosylglucoside of the flavonol kaempferol (Marotti *et al.*, 2007). These strains were selected from a panel of 38 strains belonging to *B. longum*, *B. bifidum*, *B. pseudocatenulatum*, *B. infantis*, *B. adolescentis*, *B. catenulatum* and *B. breve*, because they produced the highest β-glucosidase level within their species. These observations suggest that bifidobacteria take part in the hydrolysis of a wide range of gluco-conjugated flavonoids through the production of β-glucosidase, thus enhancing the availability of flavonoids different from soy isoflavones, although this hypothesis needs to be confirmed. The hydrolysis of quercetin glycoconjugates by pure cultures of bifidobacteria has never been reported, and information on the hydrolysis of flavanones is also lacking. Flavonol and flavonone aglycons generated by the microbiota in the colon can undergo further microbial transformations, such as cleavage and/or dehydroxylation, but the involvement of bifidobacteria in these reactions, if any, has yet to be investigated.

Lignans

Lignans are found in high concentration in flaxseeds, berries, rye and a wide range of fruits and vegetables, and occur abundantly in western diets. They exert both antioxidant and phytoestrogenic activities and are protective against cancers of the breast, colon, and prostate, as well as atherosclerosis and diabetes (Adlercreutz, 2007). Secoisolariciresinol diglucoside (SDG) is a major dietary lignan (Fig. 6.1), which activation by the colonic microbiota is necessary to produce enterodiol (ED) and enterolactone (EL), with superior oestrogen-dependent and independent activity. In the colon, SDG is hydrolysed by microbial β-glucosidases to give the aglycone secoisolariciresinol (SECO). Then, SECO undergoes demethylation and dehydroxylation yielding the bioactive enterolignans ED and EL (Fig. 6.2b). Efforts have been carried out to characterize the intestinal bacteria which are responsible for the production of enterolignans. The conversion of SDG into ED and EL is the result of the interaction between dominant and subdominant intestinal anaerobes. The SDG hydrolysis can be carried out by some strains of the genera *Bacteroides* and *Clostridium*; demethylation of SECO can be carried out by strains of the genera *Butyrobacterium*, *Eubacterium* and *Peptostreptococcus*; dehydroxylation can be carried out by strains of the genera *Clostridium* and *Eggerthella* (Clavel *et al.*, 2006; Eeckhaut, 2008). As far as we know, bifidobacteria have never been tested for their ability to transform SDG and enterolignans. However, since many strains produce β-glucosidases capable of hydrolysing glucosides of other polyphenols, they could still be involved in the removal of glucose moieties from SDG.

Hydroxycinnamates

The free forms of hydroxycinnamates, such as caffeic, ferulic and *p*-coumaric acids (Fig. 6.1), are antioxidant, nitrite scavenging, and inhibit carcinogenesis *in vitro* and in animal models. Hydroxycinnamates are found in dietary plants as ester conjugates (Manach *et al.*, 2004). In coffee, fruits and many vegetables, caffeic acid occurs as an ester linked to quinic acid, resulting in chlorogenic acid. Ferulic and *p*-coumaric acids are found as esters of polysaccharides (e.g. pectin and arabinoxylans) in cereals and vegetables. Such esters cannot be hydrolysed or absorbed in the upper gastrointestinal tract, but caffeic, ferulic, and *p*-coumaric acids can be released in their free form in the colon (Bourne and Rice-Evans, 1998; Gonthier *et al.*, 2003). Interindividual differences in excretion profiles of hydroxycinnamates were observed, indicating that the composition of the microbiota may dictate their bioavailability (Rechner *et al.*, 2004).

The hydrolysis of the hydroxycinnamates esters is possible using faecal cultures that produce feruloyl esterases (Gonthier *et al.*, 2003). Nonetheless, an extensive investigation of the colonic species which are capable to produce cinnamoyl esterase and to release hydroxycinnamates from the respective esters has not been accomplished yet. Three isolates of *E. coli*, *Lactobacillus gasseri* and *B. lactis* produced cell-bound cinnamoyl esterase activity against chlorogenic acid and ethyl ferulic acid (Couteau *et al.*, 2001). Moreover, feruloyl oligosaccharides were utilized as carbon sources for the growth of one strain of *B. bifidum* (Yuan *et al.*, 2005). Therefore, the role of bifidobacteria in the release of caffeic and ferulic acids in the colon can be considered as a healthy effect. Likewise flavonols and flavanones, hydroxycinnamates can be subjected to microbial reductions and dehydroxylations that may lead to diverse end products (e.g. cinnamic or phenylpropionc acid), depending on the composition of the microbiota. Any potential role of bifidobacteria in such reactions has not been investigated in any detail.

Glucosinolates

Cruciferous vegetables (such as broccoli, Brussels sprouts, cabbage, cauliflower, kale, swede and turnip) contain glucosinolates in high amounts. These phytochemicals are primarily responsible for the reduced incidence of cancer conferred by eating significant quantities of these vegetables (Hayes *et al.*, 2008). Glucosinolates are a class of organic compounds that contain sulphur and nitrogen and are derived from glucose and an amino acid (IARC, 2004). They are β-thioglucoside N-hydroxysulfates containing a central carbon atom which is bound to the thioglucose group through a sulphur atom, to a sulphate group via a nitrogen atom, and to a side group (Fig. 6.2c). Glucosinolates are formed by the plant from one of eight different amino acids (alanine, valine, leucine, isoleucine, phenylalanine, methionine, tyrosine and tryptophan) yielding different side groups. Moreover, the side group may be substituted by different sized alkyl groups, so that more than one hundred natural glusinolates have been recognized, presenting diverse beneficial biological effects. The cancer chemopreventing activity of glucosinolates is not due to these compounds themselves, but to their hydrolysis end products, especially isothiocyanates (ITCs) (Valgimigli and

Iori, 2009). A number of distinct mechanisms for cancer prevention have been proposed as associated to different glucosinolate-derived ITCs. These include induction of detoxification phase II genes, enhancement of antioxidant defence, inhibition of pro-inflammatory reactions through the repression of NF-κB, inhibition of cytochrome P450 activity, inhibition of histone deacetylase, stimulation of cell-cycle arrest and apoptosis (Clarke *et al.*, 2008; Nian *et al.*, 2009).

Upon the disruption of plant tissues (owing to harvesting, food preparation, chewing, etc.), glucosinolates come into contact with plant myrosinase (Bones and Rossiter, 1996). Myrosinase cleaves off glucose, giving a transient unstable aglycone, which spontaneously rearranges into the corresponding ITC, nitrile, or thiocynate. At high or neutral pH, the hydrolysis of glucosinolates give primarily rise to ITCs (such as sulforaphane) while the formation of nitriles is favoured at low pH. Specific plant factors, in the presence of ferrous ions, promote the formation of epithionitriles from glucosinolates with a side chain containing a double bond. The indolyl glucosinolates, which are derived from tryptophan, generate an unstable ITC that can give rise to diverse indoles which have potent pharmacological effects (Bonnesen *et al.*, 2001).

After ingestion of cruciferous vegetables containing active plant myrosinase, the glucosinolates are hydrolysed in the upper intestine and ITCs are absorbed there. If vegetables are cooked, myrosinase is inactivated and intact glucosinolates can reach the colon where they are subjected to microbial transformation by the resident microbiota releasing isothiocyanates along with other metabolites (Krul *et al.*, 2002). As demonstrated with faecal cultures and *in vivo* experiments, the human intestinal microbiota is capable of extensive transformation of glucosinolates. In particular, intestinal bacteria belonging to the *Lactobacillus* genus, *Bacteroides thetaiotaomicron*, *E. coli* and some species of bifidobacteria posses myrosinase activity and are capable of glucosinolate hydrolysis (Nugon-Baudon *et al.*, 1990; Elfoul *et al.*, 2001; Krul *et al.*, 2002; Cheng *et al.*, 2004a). Microbial myrosinases release ITCs which can be absorbed through the colon. The absorption of ITCs is less efficient in the colon than in the small intestine, therefore they undergo a diverse fate, being further metabolized to amines or other compounds by the microbiota.

The ability of three strains of *B. adolescentis*, *B. pseudocateunulatum* and *B. longum* to hydrolyse the glucosinolates sinigrin (SNG) and glucotropaeolin (GTL) has been established *in vitro*. The above strains presented cell-associated myrosinase-like activity and utilized the glucose moiety of glucosinolates as a carbon source for growth. Since bifidobacteria produce acetic and lactic acids causing a pH decrease, nitriles were the major products of SNG and GTL in cultures of *B. adolescentis* JCM 7045. In buffered medium, with the pH kept in the range of faecal pH (6.1 to 6.6), lower amounts of nitriles were found. On the other hand, ITCs did not accumulate at significant amounts, because they were transformed into unidentified products (Cheng *et al.*, 2004a). Thus, it remains to be established whether the predominant production of nitriles by bifidobacteria actually occurs *in vivo* and whether bifidobacteria can effectively contribute to the release of healthy ITCs.

Ginsenosides

Ginseng is the root of the species within the genus *Panax*, e.g. *Panax ginseng*. It is not strictly regarded as a food ingredient, but it is used worldwide due to adaptogen, stimulant, antidiabetic and antitumor properties. Ginseng saponins (ginsenosides) are the main components of ginseng roots having pharmacological and biological activities, and are viewed as the active compounds behind the claims of ginseng's efficacy (Christensen, 2009). Ginsenosides contain a triterpene steroid-like aglycone moiety that is glycosylated with glucose, arbinose, rhamnose, and xylose moieties at diverse positions. After ingestion, ginsenosides are transformed through passage in the stomach, due to acidic pH, and pass through the small intestine to reach the large intestine, where they undergo microbial hydrolysis, forming compounds with superior biological activity.

Through acid treatment and incubation with the human intestinal microbiota or intestinal bacteria isolated from human faeces, the ginsenosides are mainly transformed into the aglycone or into other active compounds (Bae *et al.*, 2000, 2002, 2004), effective in inhibiting both NF-κB and JNK pathways and exerting potent anticarcinogenic, anti-proliferative, and anti-inflammatory effects (Choi *et al.*, 2007). Diverse bacterial isolates from human faeces have been shown to be capable of carrying out the above hydrolysis. They belonged to the genera *Prevotella, Fusobacterium, Bacteroides, Eubacterium, Streptococcus* and *Bifidobacterium* and transformed complex ginsenosides through many different hydrolytic steps. The transformation pathway, the intermediates, and the degree of conversion strongly differed, due to the diversity of the microbiota composition among individuals (Bae *et al.*, 2000, 2002, 2004).

Members of the genus *Bifidobacterium* were active in the hydrolysis of arabinose, xylose and glucose moieties of ginsenosides. In particular, some strains of *Bifidobacterium* (e.g. *B. breve* K-110 and a strain designated as *Bifidobacterium cholerium* K-103) were shown to produce β-glucosidase activity that removes diverse glucose moieties from ginsenosides (Bae *et al.*, 2000, 2002, 2004). Moreover, a β-xylosidase capable of hydrolysing a xylose moiety from the ginsenosides, was purified and characterized in *B. breve* K-110 (Shin *et al.*, 2003a). Likewise, in this same strain an α-arabinofuranosidase and an α-arabinopyranosidase, producing a diverse ginsenoside, were purified and characterized (Shin *et al.*, 2003b). Taken together, these results suggest that bifidobacteria can carry out some hydrolytic steps leading to bioactive healthy compounds from ginseng saponins, but their impact in activating ginsenosides *in vivo* has yet to be proved. In spite of this, some bifidobacteria strains have been proposed to be used along with other food microorganisms (e.g. *Lactobacillus delbrueckii*) or enzymes, to enhance the concentration of specific ginsenosides in ginseng-based products through fermentation and/or enzymatic digestion (Chi *et al.*, 2005; Chi and Ji, 2005).

Production of conjugated linoleic acid

Biological effects of CLA

Conjugated linoleic acids (CLA) are a group of positional and geometric isomers of linoleic acid (LA, *cis*9, *cis*12-C18:2) that contain conjugated double bonds. A total of 28 CLA isomers have been identified with diverse position (ranging from $\Delta^{7,9}$ to $\Delta^{12,14}$) and *cis/trans* geometry (Banni, 2002). *cis*9, *trans*11-CLA and *trans*10, *cis*12-CLA are the major isomers occurring in human diet, and, consequently, these are the most studied. The former isomer is mostly contained in ruminant meat and dairy products, while the latter is present in hydrogenated oils. They are naturally present in ruminant food products because of bacterial biohydrogenation of LA carried out by the rumen microbiota (Jenkins *et al.*, 2008). Since their discovery (Pariza and Hargraves, 1985), CLA are purported to have several beneficial effects. Considerable research efforts have unveiled many potential health-promoting properties of CLA, including anti-carcinogenesis, anti-atherosclerosis, anti-diabetes, anti-obesity, anti-allergy and enhancement of immune functions (Kelley and Erickson, 2003; Bhattacharya *et al.*, 2006; Moloney *et al.*, 2007; Mitchell and McLeod, 2008; Turpeinen *et al.*, 2008). It is hard to be conclusive about this argument, since these important health benefits which have been observed using animal models, have often not been similarly translated in human clinical trials (Plourde *et al.*, 2008). Moreover, several studies have raised concerns about the potential safety of CLA for humans (Pariza, 2004).

Many efforts have been made to unravel the molecular mechanisms responsible for such a wide range of biological activities. Several studies have demonstrated that various CLA isomers possess different biological activities, and often more than a single biological activity is due to a particular isomer (Gebauer *et al.*, 2007). Given that certain isomers exert individual effects on numerous biological functions, it is not surprising that some effects can be induced and/or enhanced synergistically by mixtures of CLA isomers. *cis*9, *trans*11-CLA presents the highest biological activity in most cases (e.g. anticarcinogenic, anti-diabetes, anti-allergy, immune-stimulating effects), while *trans*10, *cis*12-CLA is the most effective isomer to affect blood lipids and reduce body fat, as well as being the most cytotoxic for human breast cancer cells (Pariza, 2004; Kelley *et al.*, 2007).

CLA regulate a wide spectrum of processes, including proliferation, differentiation and development. They accumulate into tissues that are particularly rich of neutral lipids, such as mammary and adipose tissue, and inhibit the production of arachidonic acid from LA. Because arachidonic acid is the substrate for the cyclooxygenase and lipooxygenase pathways of eicosanoid biosynthesis, this results in lower levels of prostaglandins and thromboxans, which in turn are endogenous mediators of inflammation. In particular, formation of prostanoids from arachidonic acid is carried out by cyclooxygenase-1 (COX-1) and cyclooxygenase-2 (COX-2). *trans*10, *cis*12-CLA inhibits COX-2 activity through a regulatory cascade of nuclear factor kappa B (NF-κB), whereas COX-1 inhibition results from competition of CLA with arachidonic acid (Li *et al.*, 2006; Stachowska *et al.*, 2009).

Fatty acids and their metabolites modulate the activity of peroxisome proliferator-activated receptors (PPARs) (Marion-Letellier *et al.*, 2009). PPARs heterodimerize with the retinoid X receptor and act as transcription factors in a process where they translate different stimuli into changes in gene expression. PPARγ is the master adipogenic transcription factor, which induces anabolic processes such as triglyceride synthesis, glucose uptake and fatty acid uptake, playing a key role in adipogenesis and insulin sensitization (Tontonoz and Spiegelman, 2008). Inhibition of PPARγ activity, at both pre-translational and post-translational levels, by the tumour necrosis factor alpha (TNF-α) is involved in pathogenesis of insulin resistance, atherosclerosis, inflammation, and cancer cachexia (Ye, 2008). CLA behave as agonists of PPARγ, increasing the DNA-binding activity. In reporter assays using the mouse macrophage cell line RAW264.7, CLA activated PPARγ and inhibited interferon-γ (INF-γ) induced proinflammatory mediators, such as inducible nitric oxide synthase and COX-2 (Yu *et al.*, 2002).

Effects of CLA on human health

The attention on the health benefits of CLA has markedly increased since they were discovered as potent anticarcinogenic compounds (Pariza and Hargraves, 1985). The anti-cancer effects of CLA arise from the interaction of the different isomers with a number of pathways of signal transduction that affect apoptotic activity, cell cycle progression, proliferation, and angiogenesis (Ip *et al.*, 1999; Majumder *et al.*, 2002; Masso-Welch *et al.*, 2002; Kemp *et al.*, 2003; Lim *et al.*, 2005). Above all, NF-κB-induced signalling, PPAR pathways, and eicosanoids synthesis and signal transduction have been demonstrated to be a target of CLA isomers, thus exerting anti-cancer effects (Cheng *et al.*, 2004b).

The two most prominent isomers *cis*9, *trans*11-CLA and *trans*10, *cis*12-CLA have been shown to inhibit the initiation and the development of tumours in animals (Ip *et al.*, 2002) and repress the proliferation of different cancer cells (Majumder *et al.*, 2002; Kemp *et al.*, 2003; Maggiora *et al.*, 2004). Besides, *cis*9, *trans*11-CLA and *trans*10, *cis*12-CLA were shown to exhibit distinct effects on tumorigenesis in rats and mice and differences in potency against growth of human cancer cells (Maggiora *et al.*, 2004). Thus, *cis*9, *trans*11-CLA and/or *trans*10, *cis*12-CLA are claimed as chemopreventive agents against many of the most common cancer types, such as intestine, breast and prostate cancer, even though they may target diverse cancer cells and the mechanisms of their anti-tumour activity may be distinct (Hubbard *et al.*, 2003; Palombo *et al.*, 2002).

In several animal models, the dietary supplementation with CLA has been shown to decrease the development of atherosclerosis and to cause the regression of atherosclerotic lesions (Toomey *et al.*, 2006). In fact, *cis*9, *trans*11-CLA reduces the pro-inflammatory cytokines and decreases the atherogenesis related genes (Lee *et al.*, 2008). Moreover, CLA isomers inhibit cytokine-induced eicosanoid release from vascular smooth muscle cells, lower cholesterol accumulation, stimulate acceptor-dependent cholesterol efflux, and increase the expression of genes involved in cholesterol homeostasis (Ringseis *et al.*, 2008). Compared to studies in animal models, only a small number human trials have demonstrated the effects of CLA or single isomers on atherosclerosis. Therefore, clinical trials are not as conclusive as animal models and cell line results, although one study involving healthy subjects did support the role of CLA in anti-inflammatory and in anti-atheroscleroctic activities, showing that a dietary CLA supplement can reduce IgE, TNF-α and IL-1β, and increase IgA and IgM (Song *et al.*, 2005).

The potential use of CLA to control body fat has received significant attention and remains one of the major health claims which led to the widespread promotion and sale as dietary supplements. In animals, CLA intake induces a decrease in body fat mass and an increase in lean body mass, with *trans*10, *cis*12-CLA being pivotal in fat reduction (Evans *et al.*, 2002). *trans*10, *cis*12-CLA, but not *cis*9, *trans*11-CLA, altered the expression of a number of genes involved in lipid transport, fatty acid and steroid metabolism, lipolysis, beta-oxidation, cholesterol biosynthesis, membrane lipid metabolism, gluconeogenesis and the citrate cycle (Murphy *et al.*, 2007). Human trials suggest that CLA tend to increase the lean body mass (Whigham *et al.*, 2007). Nevertheless, the effect was much lower in humans than observed in animal studies, especially those involving mice (Tricon *et al.*, 2005). Conversely, no significant effects on body weight and composition were observed in other studies (Tricon and Yaqoob, 2006). Moreover, it raised the question as to whether fat loss in humans was accompanied by impairment in insulin sensitivity (Salas-Salvadò *et al.*, 2006). The incongruity between human and animal data has been ascribed to methodological differences in the experimental design, including age, gender, CLA dose and isomers. Still, the relatively obscure metabolic fate of CLA in humans may also be a contributing factor to the lack of consistency for CLA efficacy across studies (Plourde *et al.*, 2008).

Bacterial synthesis of conjugated linoleic acids

CLA are intermediates of the biohydrogenation of unsaturated fatty acids, a microbial reductive process occurring in the rumen which has important effects on fatty acids composition of meat, milk and dairy products (Jenkins *et al.*, 2008). The lipids which enter the rumen with grass and feedstuff are triacylglycerols, phospholipids and galactolipids, and are rich in unsaturated fatty acids such as α-linolenic (ALA, *cis*9, *cis*12, *cis*15-octadecatrienoic acid) and linoleic (LA) acids. In the rumen, lipids are efficiently hydrolysed by lipolytic enzymes of the microbiota. In the latter process, unsaturated fatty acids undergo the progressive saturation through microbial biohydrogenation, resulting in a low concentration of ALA and LA in meat and milk. The complete saturation of LA is carried out by microbial enzymes giving stearic acid as the end product. In the rumen, such a transformation occurs mostly through three steps, the first of which is the rapid conversion of *cis*9, *cis*12-LA into *cis*9, *trans*11-CLA, also known as rumenic acid (RA). Isomerization of LA into *cis*9, *trans*11-CLA is carried out by LA-isomerases of ruminal bacteria, and is followed by its slow reduction to *trans*11-C18:1 (vaccenic acid, VA) which can then be reduced to stearic acid by microbial activity. In animal tissues, VA can be converted back to *cis*9, *trans*11-CLA by the Δ9 desaturase enzyme, thus providing an additional mechanism for CLA production (Jenkins *et al.*, 2008).

Many efforts have been carried in order to identify the bacteria involved in the hydrogenation of unsaturated fatty acids. The first ruminal bacterium reported to produce CLA from LA was *Butyrivibrio fibrisolvens* (Wilde and Dawson, 1966). Further studies demonstrated that a single microbial group among ruminal bacteria, the *Butyrivibrio*, was involved in biohydrogenation of fatty acids and presented the highest capacity to convert LA into *cis*9, *trans*11-CLA (Jenkins *et al.*, 2008). The genus *Butyrivibrio* is phylogenetically placed within Cluster XIVa of the *Clostridium* subphylum, together with a vast array of low GC firmicutes, including members of *Clostridium*, *Coprococcus*, *Lachnospira*, *Eubacterium*, *Ruminococcus*, *Syntrophococcus*, *Roseburia* and *Acetitomaculum* (Kopecný *et al.*, 2001).

With the investigation of lipid biohydrogenation by non-ruminal bacteria and with the improvement of analytic techniques, which have enabled separation and identification of minor CLA isomers and intermediates, several bacterial species have been reported to convert LA into a variety of CLA isomers aside from *cis*9, *trans*11-CLA (Devillard *et al.*, 2007; McIntosh *et al.*, 2009). Among non-ruminal bacteria, the ability to transform LA into CLA has been found in diverse microbial genera, including species of lactic acid bacteria. Non-ruminal bacteria may also produce *cis*9, *trans*11-CLA as their main CLA isomer, but the specific activity is several fold lower than that of *Butyrivibrio*-related bacteria. In addition, only a small number of bacterial species are known to synthesize *trans*9, *trans*11-CLA and/or *trans*10, *cis*12-CLA as a minor and/or trace product of LA metabolism.

CLA and bifidobacteria

During the last decade, attention has focused on LA metabolism by the microbiota of the human colon. The ability to perform LA hydrogenation has been found in diverse isolates from the human faeces, indicating that CLA and VA production can occur in the human gut (Devillard *et al.*, 2007; McIntosh *et al.*, 2009). Nonetheless, the microbial community involved in LA transformation and the metabolic pathway seem to be more complex in the human gut than in the rumen (Fig. 6.3). Human colonic species involved in LA hydrogenation belong to clostridial Cluster IV and Cluster XIVa, and to species of *Bifidobacterium*; they all transform LA into CLA and VA (Devillard *et al.*, 2007). Because *Clostridium*-like bacteria are very abundant in the colonic microbiota of an adult, it is likely that they are quantitatively more important than bifidobacteria for LA metabolism in the human gut.

Some species of bifidobacteria (especially *B. breve* and *B. dentium*) can transform LA into *cis*9, *trans*11-CLA, while producing lower amounts of *trans*9, *trans*11-CLA and *trans*10, *cis*12-CLA (Coakley *et al.*, 2003). Various strains of *Roseburia*, a genus belonging to Cluster XIVa, have been shown to produce *cis*9, *trans*11-CLA and/or VA. In addition, several isolates of Cluster IV and Cluster XIVa and one of *Bifidobacterium* are known to produce 10-hydroxy-C18:1 (HFA), indicating the presence of an additional route of *cis*9, *trans*11-CLA production in the human intestine (Devillard *et al.*, 2007). The activity of transforming HFA into *cis*9, *trans*11-CLA and *trans*9, *trans*11-CLA by the faecal microbiota has also been observed, but the responsible bacteria are still unknown. The ability of various bifidobacterial isolates to transform LA into *cis*9, *trans*11-CLA raised the question as to whether CLA production may be regarded as a mechanism by which bifidobacteria exert (some of) their beneficial properties (Ewaschuk *et al.*, 2006). In the course of several investigations, more than one hundred strains of *Bifidobacterium* were screened for LA conversion into CLA or VA. All these studies have indicated that none of these

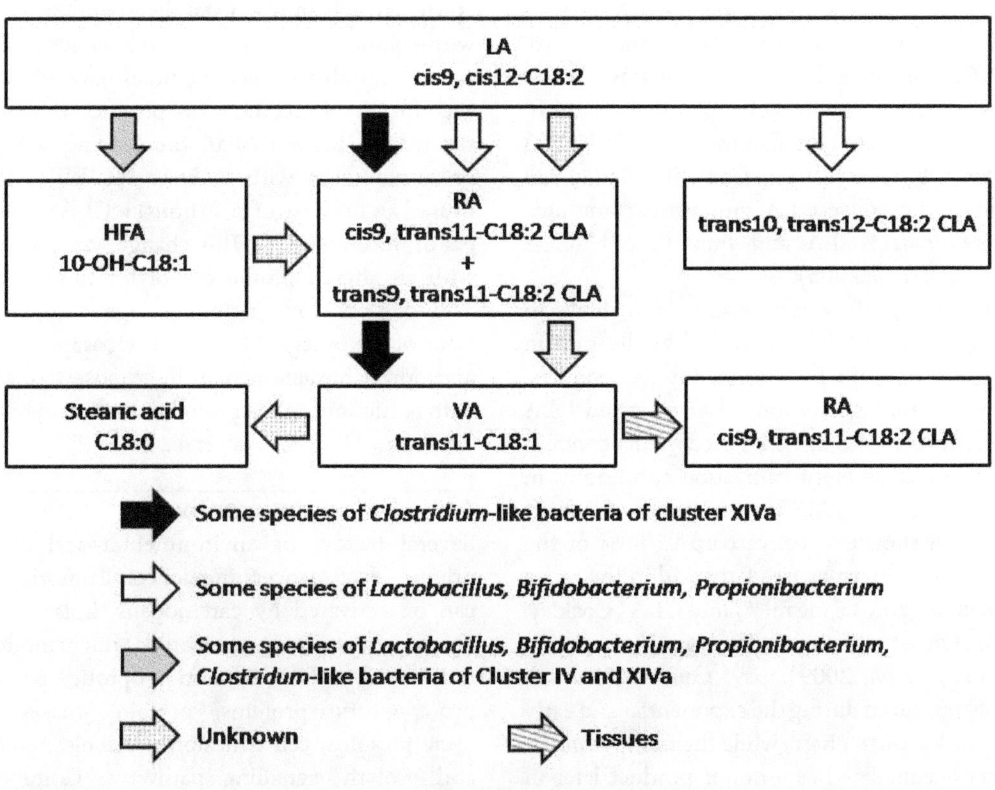

Figure 6.3 LA metabolism by the microbiota of the human gut, adapted from Devillard *et al.* (2007).

tested strains produce VA or stearic acid, while considerable interspecies differences in the ability to produce CLA were observed. Among the active CLA producers, the most efficient isolates were strains of *B. breve*, followed by strains of *B. dentium* and *B. lactis*, whereas low or negligible activity was found in strains of *B. adolescentis*, *B. pseudocatenulatum*, *B. angulatum*, and *B. infantis* (Coakley *et al.*, 2003; Oh *et al.*, 2003; Devillard *et al.*, 2007; Choi *et al.*, 2008; Chung *et al.*, 2008). However, data concerning the conversion of LA to CLA by *B. longum* are quite inconsistent, since most of the strains do not perform this reaction or possess only weak activity (Rosberg-Cody *et al.*, 2004; Xu *et al.*, 2008), while just a small number are good converters (Hennessy *et al.*, 2009). Furthermore, Rosberg-Cody *et al.* (2004) isolated a *B. bifidum* strain exhibiting considerable LA conversion, with production of *cis*9, *trans*11-CLA as the major isomer. A similar production level was reported for *B. bifidum* CRL 1399 (Van Nieuwenhove *et al.*, 2007).

*cis*9, *trans*11-CLA is the major isomer formed by all CLA-producing *Bifidobacterium* strains, generally accounting for more than 80% of CLA production. *trans*9, *trans*11-CLA is usually produced as a minor component, and *trans*10, *cis*12-CLA is generally produced at trace levels. In contrast, evidence has emerged that strains of *Bifidobacterium* contained in the probiotic VSL#3 mixture (*B. breve*, *B. longum*, and *B. infantis*) can all efficiently convert LA into similar amounts of *cis*9, *trans*11-CLA and *trans*10, *cis*12-CLA (Ewaschuk *et al.*, 2006).

The ability of certain strains of *B. breve* to convert LA into CLA could find application in the production of CLA-enriched dairy products, and some studies have indeed investigated CLA production in MRS or milk-based media containing LA under different cultivation conditions. In cysteine-containing MRS broth, strains of *B. breve* have been shown to convert up to 80% of the available LA (given as free fatty acid in the range between 0.1 and 2.8 mg ml^{-1}) into CLA (Coakley, 2003; Oh *et al.*, 2003; Chung *et al.*, 2008; Hennessy *et al.*, 2009). *cis*9, *trans*11-CLA was mainly produced during the exponential and early stationary growth phase, while the *trans*9, *trans*11 isomer accumulated as a minor product later in the stationary phase. In milk and reconstituted skim milk, the conversion yield of LA into CLA was lower than in cysteine-MRS broth, but it was increased through the addition of supplements which are commonly used to enhance the growth of bifidobacteria in milk, such as cysteine, sodium salts of short-chain fatty acids, yeast extract, and enzymatic digests of casein (Chung *et al.*, 2008; Hennessy *et al.*, 2009).

Even if bifidobacteria can be successfully exploited to produce CLA-enriched dairy products, scarce information is still available about the effectiveness of bifidobacteria in producing CLA isomers *in vivo* and affecting the systemic or local levels of CLA, and also the contribution of the intestinal microbiota to the CLA level in non-ruminants is still not completely understood. The amount of LA available for CLA production in the colon varies and is dependent on the amount ingested and the efficacy of absorption in the small intestine. Studies have shown that humans generally excrete some LA in faeces (Edionwe and Kies, 2001), suggesting that the substrate for microbial production of CLA is available. CLA produced by the commensal and/or administered bacteria have the potential to be absorbed by epithelial cells in the distal ileum and to interact with colonocytes, causing local beneficial effects. A small number of recent studies have addressed the efficacy of colonic CLA production by non-ruminants. In particular, the feeding with LA in combination with *B. breve* NCIMB 702258 caused an increase of *cis*9, *trans*11-CLA in the livers of mice and pigs. This change was associated with an altered profile of polyunsaturated fatty acid composition, including higher concentrations of the omega-3 fatty acids eicosapentaenoic and docosahexaenoic acid in adipose tissue and with reductions in the proinflammatory cytokines TNF-α and INF-γ (Wall *et al.*, 2009).

Anticarcinogenic activity

Several factors of environmental and genetic origin affect cancer incidence. Tumorigenesis can be activated by carcinogens, inflammatory agents, and tumour promoters which modulate transcription factors, anti-apoptotic proteins, pro-apoptotic proteins, protein kinases, cell cycle proteins, cell adhesion molecules, COX-2, and growth signalling pathways. Commensal bacteria may impact on carcinogenesis since the site of colorectal cancer coincides with the colonic region in which bacterial concentration is the highest (10^{11} bacteria per gram of luminal content). Germ-free animals exhibited lower incidence of colorectal cancer, thus implicating the involvement of the gut microbiota in its aetiology (Hope *et al.*, 2005; Vannucci *et al.*, 2008). Chemically-induced tumours occurred at a lower incidence in germ-free rats as compared to rats that harboured a normal microbiota (Reddy *et al.*, 1975, 1976b). In double-knockout mice, deficient in T-cell receptor β (TCRβ) and p53 genes, adenocarcinoma of the colon is known to develop spontaneously and early in life with a notable frequency (Funabashi *et al.*, 2001), but it was completely absent in germ-free mice (Kado *et al.*, 2001). Furthermore, faeces may contain mutagenic or genotoxic molecules which originate from bacterial biotransformation of pro-carcinogens (Rowland, 1999; Huycke and Gaskins, 2004). Such experimental data confirm that the intestinal microbiota influences the incidence of malignant transformation and that cancer is at least partially due to bacterial metabolism in the colon. Thus, it is conceivable that factors modulating composition and activity of the microbiota can affect the development of colorectal cancer.

Experimental evidence of the anticarcinogenic activity of bifidobacteria has a long history, as documented by several studies of their anti-genotoxic effects *in vitro* and *in vivo.* Feeding bifidobacteria effectively reduced chemically-induced DNA damage in the gastrointestinal mucosa of animals. Several studies exploiting different animal models demonstrated that bifidobacteria prevented putative preneoplastic lesions or tumours due to a number of chemical mutagens (e.g. heterocyclic aromatic amines, 1,2-dimethylhydrazine, azoxymethane *N'*-nitro-*N*-nitrosoguanidine, 1,2-dimethylhydrazine, 2-amino-3-methylimidazo(4,5-f)quinoline) and diverse strains belonging to different *Bifidobacterium* species exerted chemopreventive activity (Reddy and Rivenson, 1993; Pool-Zobel *et al.*, 1996; Arimochi *et al.*, 1997; Singh *et al.*, 1997; Rowland *et al.*, 1998; Zsivkovits *et al.*, 2003).

Besides animal trials, many studies have reported on the efficacy of diverse strains of *Bifidobacterium* (in the form of cells, cell free extracts, or cellular fractions) in inhibiting the growth of colon cancer cell lines. Bifidobacteria increased the activity of differentiation-related enzymes in a human colon cancer cell line, causing a decreased cellular growth rate and an increased differentiation (Baricault *et al.*, 1995). Furthermore, whole cells, cytoplasm extracts, polysaccharide- and peptidoglycan-containing fractions of bifidobacteria inhibited the growth of colon cancer cell lines (Sekine *et al.*, 1995; You *et al.*, 2004; Kim *et al.*, 2008; Lee et *al.*, 2008).

Several mechanisms of *Bifidobacterium* anticarcinogenic activity have been proposed. They include the enhancement of the immune response, the removal of potential carcinogens, the repression of the microbial population and/or activities producing putative carcinogens, the production of anti-tumorigenic or anti-mutagenic compounds in the colon, and the modification of metabolic activities of the microbiota (Commane *et al.*, 2005; Picard *et al.*, 2005; Geier *et al.*, 2006; Fotiadis *et al.*, 2008). Besides, a number of health benefits exerted by bifidobacteria (e.g. energy recovery, antagonism against pathogens, activation of dietary compounds into bioactive molecules, and production of CLA and vitamins) confer protection against cancer as well. However, some bifidobacteria-induced mechanisms which regulate the signal transcription pathways of the host have only recently been deciphered.

Removal of carcinogens

Bifidobacteria can reduce the incidence of cancer through binding and/or detoxifying harmful compounds (e.g. heterocyclic aromatic amines, polycyclic aromatic hydrocarbons, mycotoxins) and reactive oxygen species (Knasmuller *et al.*, 2001; Fuchs *et al.*, 2008). Mice which were fed a probiotic mixture of *L. acidophilus* and *B. longum* exhibited a lower level of the mutagen Trp-P-2 in most of the tissues (Orrhage *et al.*, 2002). A strain of *B. longum* efficiently bound amino acid pyrolysates, exerting a strong anti-mutagenic effect (Sreekumar and Hosono, 1998). Some bifidobacterial strains were shown to detoxify certain mycotoxines, including aflatoxin B1 (Peltonen *et al.*, 2001), ochratoxin A (Skrinjar *et al.*, 1996), and patulin (Fuchs *et al.*, 2008). Binding of harmful compounds is a physical phenomenon in which cell-wall peptidoglycans and polysaccharides play a major role, mostly due to a cation-exchange mechanism (Zhang and Otha, 1991).

Enzymes

Colonic bacteria can produce or release enzymes that transform pro-carcinogens into toxic compounds. β-glucuronidases, azoreductases, and nitroreductases produced by colonic bacteria are involved in the metabolism of a wide range of such compounds (Mastromarino *et al.*, 1976; McBain and Macfarlane, 1998). Microbial β-glucuronidases release carcinogens from hepatic-derived glucuronate-conjugates, which are responsible for the enterohepatic circulation of dietary molecules, drugs, and other foreign compounds (Hawksworth, 1971). In fact, after absorption, many carcinogenic molecules are metabolized in the liver and are conjugated to glucuronic acid before being delivered with the bile to the small intestine. Reaching the colon, they can be hydrolysed by bacterial β-glucuronidases, which release the parent compound or its activated hepatic metabolite. Azoreductases catalyse the reductive cleavage of azo-bonds converting molecules that contain an azo-group to the corresponding amino compound, some of which are toxic, mutagenic, or carcinogenic (e.g. aromatic amines). Likewise, nitroreductases participate in the conversion of aromatic nitro compounds to the corresponding amines, which are reactive intermediates whose end products are known mutagens and carcinogens (Nakamura *et al.*, 2002). In animal models, bifidobacteria were shown to antagonize various bacterial enzymes which activate non-absorbable pro-carcinogens that occur in the diet and/or arise from the microbiota metabolism or the environment, including β-glucosidase, β-glucuronidase, tryptophanase, and urease (Rowland *et al.*, 1998; Kim *et al.*, 2008). Human trials were consistent with such observations, since the administration of bifidobacteria resulted in lower pH, ammonia concentration, and β-glucuronidase activity in faeces (Benno and Mitsuoka, 1992; Bouhnik *et al.*, 1996).

The intestinal microbiota may promote carcinogenesis through the metabolism of bile salts, which are secreted in the small intestine and can be deconjugated by bacterial bile salt hydrolases (BSH) into the corresponding primary bile acids (e.g. chenodeoxycholic and cholic acids). Then, microbial hydrosteroid dehydrogenases can oxidize or epimerize the hydroxyl groups at positions 3, 7 or 12. Of these activities, 7-α-dehydroxilation represents the major bacterial biotransformation of primary bile acids, yielding secondary bile acids (e.g. deoxycholic and lithocholic). The secondary bile acids are more hydrophobic than the primary ones and are absorbed through passive diffusion by colonic epithelial cells (Ridlon *et al.*, 2006). They promote colon tumours (McGarr *et al.*, 2005) and are associated to colon carcinogenesis in a number of laboratory animal models and human epidemiological studies (Reddy *et al.*, 1976a; Bernstein *et al.*, 2005). Most bifidobacteria produce BSH (Begley *et al.*, 2006), but it has been shown that the consumption of a BSH-positive strain of *B. animalis* does not increase the production of secondary bile acids in pigs (Lepercq *et al.*, 2005). This is attributable to the fermentative metabolism of bifidobacteria which reduces the pH of faeces, inhibiting 7-α-dehydrogenase and the growth and/or the activity of the bacteria that can perform the production of secondary bile salts, such as clostridia and eubacteria (Thomas *et al.*, 2001).

Anti-inflammatory activity

Many events occurring in the early-onset of cancer are common to what happens in inflammation. Proinflammatory cytokines act as tumour promoters by regulating cytokines, chemokines, adhesion molecules, and angiogenesis (Yang and Pei, 2006). The development of intestinal inflammation is a consequence of an increased immune response to the commensal bacteria which is often triggered by alterations of the intestinal microbiota (Neish, 2009). In fact, colitogenic bacteria are responsible for the development of chronic intestinal inflammation leading to tumorigenesis (Uronis *et al.*, 2009). Bifidobacteria have an immune-modulatory effect and elicit different responses depending on whether they interact with the host in physiological or pathological conditions (Neish, 2009). Although they strengthen specific and innate immune responses in healthy individuals (Kekkonen *et al.*, 2008), they downregulate the secretion of proinflammatory cytokines in pathological conditions, such as inflammatory bowel diseases, inducing a strong anti-inflammatory response during inflammation through the negative modulation of IL-1β and TNF-α (Vanderpool *et al.*, 2008; Looijer-van Langen and Dieleman, 2009).

NF-kB is a transcription factor that is normally found in the cytoplasm as part of an inactive complex with inhibitory proteins of the IkB family. Inflammatory cytokines (e.g. TNF-α and IL-1β) and other stress-derived stimuli (e.g. radicals, carcinogens, tumour promoters, endotoxins and radiations) trigger a kinase cascade, causing NF-κB to enter the nucleus and alter the transcription of more than 200 genes. NF-κB signalling pathway is central in inflammatory response, and affect apoptosis and proliferation. Its activation is critical for tumorigenesis, being associated with cell survival and proliferation in many types of malignancies (Shen and Tergaonkar, 2009; Vallabhapurapu and Karin, 2009). The attenuation of NF-κB activation is important for bifidobacterial anti-inflammatory and anti-cancerogenic activity. Bifidobacteria contributed to the attenuation of LPS-induced NF-κB activation and limited the inflammatory response in cell cultures and in mice (Riedel *et al.*, 2006; O'Mahoney *et al.*, 2008). Feeding mice with a probiotic *B. infantis* strain, followed by an LPS challenge or infection with *Salmonella typhimurium*, resulted in induction of regulatory T cells, reduction of pro-inflammatory cytokines and T cell proliferation, and attenuation of NF-κB activation (O'Mahony *et al.*, 2008). Moreover, a *B. breve* strain released a low molecular weight (< 3kDa) component in the culture medium which inhibited the activation of NF-κB cascade (Heuvelin *et al.*, 2009). Such active metabolite inhibited TNFα-induced secretion of inflammatory chemokines (IL-8, GRO and MIP-1b) by intestinal epithelial cells.

Stimulation of macrophages, which are involved in the host defence against infection and tumour formation, contribute to anticarcinogenic activity of bifidobacteria. Viable or heat-killed *Bifidobacterium* species, as well as certain cell components, were capable of stimulating macrophage cell lines to produce hydrogen peroxide, nitric oxide, and cytokines (e.g. IL-6 and TNF-α), which mediate killing and growth inhibition of tumour cells (Miettinen *et al.*, 1996; Lee do *et al.*, 2008; Okada *et al.*, 2009).

Antioxidative activity

Both laboratory and clinical studies have demonstrated that oxidative stresses play a key role in all stages of carcinogenesis and that reactive oxygen species (ROS), particularly H_2O_2, are involved in tumour promotion (Valko *et al.*, 2007). Moreover, several other oxidants and free-radical generators are known tumour promoters, and many tumour promoters generate ROS (Paz-Elizur *et al.*, 2008). Bifidobacteria exert antioxidative activity, as a strain-specific feature, through mechanisms which have not been investigated. Both intact cells and cell-free extracts seem to exhibit such antioxidant properties (Lin and Yen, 1999a,b; Lin and Chang, 2000; Zanoni *et al.*, 2008). A functional food based on fermented soy milk containing lactic acid bacteria, including *Bifidobacterium*, exhibited enhanced antioxidative properties, with the level of antioxidative activity being dependent on the starter organism employed (Wang *et al.*, 2006).

Conclusions and future prospects

The intestinal microbiota plays a key role in normal gut function, human health and well being. Bifidobacteria constitute a significant portion of the intestinal population and have beneficial effects on their host, thus manipulation of the gut microbiota is often accomplished by oral administration of probiotic formulations containing bifidobacteria, aimed at increasing their number in the colon. Bifidobacteria acidify the large intestine, restrict putrefactive and potentially pathogenic bacteria, cause immune-stimulation, exert anticarcinogenic activity, produce vitamins, activate biologically active compounds, and remove carcinogens and harmful molecules. However, it is evident that the health benefits conferred by ingestion of bifidobacteria involve several mechanisms of action whose details are largely unknown.

Nowadays, the selection of probiotic bifidobacteria does not arise from the ability to confer specific health benefits. Most of the probiotic strains have been selected on the basis of technological features (e.g. stability, industrial-scale biomass productivity and recovery) and physiological properties, such as bile and gastric juice resistance as markers of transit tolerance in the upper gastrointestinal tract. The search for *in vitro* and *in vivo* evidence that support health-promoting claims usually follows the identification of a technologically valuable strain. Moreover, many current claims of health-promoting properties are based on strain-specific features.

A major issue is the absence of conclusive studies that disclose the variability in activation of cellular and systemic responses among phylogenetically-similar strains. The task of unravelling the molecular mechanisms that cause these different responses is daunting. To address this issue, a continuous and coordinated work in genetics, physiology and metabolism of bifidobacteria, and a defined model system for studying the causal relationships between their effects on health and mechanisms of action are necessary. Recent advances in omic approaches, allowing the simultaneous analysis of huge numbers of genes and proteins, are now available to compare different strains and the precise effectors by which probiotics influence human health. The more detailed knowledge of how bifidobacteria actively participate in the maintenance of health is expected to significantly broaden the judicious selection and utilization of probiotics.

References

Adlercreutz, H. (2007). Lignans and human health. Crit. Rev. Clin. Lab. *44*, 483–525.

Arimochi, H., Kinouchi, T., Kataoka, K., Kuwahara, T., and Ohnishi, Y. (1997). Effect of intestinal bacteria on formation of azoxymethane-induced aberrant crypt foci in the rat colon. Biochem. Biophys. Res. Commun. *238*, 753–757.

Atkinson, C., Frankenfeld, C.L., and Lampe, J.W. (2005). Gut bacterial metabolism of the soy isoflavone daidzein: exploring the relevance to human health. Exp. Biol. Med. *230*, 155–170.

Bae, E.A., Han, M.J., Choo, M.K., Park, S.Y., and Kim, D.H. (2002). Metabolism of 20(S)- and 20(R)-ginsenoside Rg3 by human intestinal bacteria and its relation to *in vitro* biological activities. Biol. Pharm. Bull. *25*, 58–63.

Bae, E.A., Han, M.J., Kim, E.J., and Kim, D.H. (2004). Transformation of ginseng saponins to ginsenoside Rh2 by acids and human intestinal bacteria and biological activities of their transformants. Arch. Pharm. Res. *27*, 61–67.

Bae, E.A., Park, S.Y., and Kim, D.H. (2000). Constitutive beta-glucosidases hydrolyzing ginsenoside Rb1 and Rb2 from human intestinal bacteria. Biol. Pharm. Bull. *23*, 1481–1485.

Banni, S. (2002). Conjugated linoleic acid metabolism. Curr. Opin. Lipidol. *13*, 261–266.

Baricault, L., Denariaz, G., Houri, J.J., Bouley, C., Sapin, C., and Trugnan, G. (1995). Use of HT-29, a cultured human colon cancer cell line, to study the effect of fermented milks on colon cancer cell growth and differentiation. Carcinogenesis *16*, 245–252.

Begley, M., Hill, C., and Gahan, C.G. (2006). Bile salt hydrolase activity in probiotics. Appl. Environ. Microbiol. *72*, 1729–1738.

Benno, Y., and Mitsuoka, T. (1992). Impact of *Bifidobacterium longum* on human fecal microflora. Microbiol. Immunol. *36*, 683–694.

Bernstein, H., Bernstein, C., Payne, C.M., Dvorakova, K., and Garewal, H. (2005). Bile acids as carcinogens in human gastrointestinal cancers. Mutat. Res. *589*, 47–65.

Bhattacharya, A., Banu, J., Rahman, M., Causey, J., and Fernandes, G. (2006). Biological effects of conjugated linoleic acids in health and disease. J. Nutr. Biochem. *17*, 789–810.

Boirivant, M., and Strober, W. (2007). The mechanism of action of probiotics. Curr. Opin. Gastroenterol. *23*, 679–692.

Bones, A.M., and Rossiter, J.T. (1996). The myrosinase-glucosinolate system, its organisation and biochemistry. Phys. Plant. *97*, 194–208.

Bonnesen, C., Eggleston, I.M., and Hayes, J.D. (2001). Dietary indoles and isothiocyanates that are generated from cruciferous vegetables can both stimulate apoptosis and confer protection against DNA damage in human colon cell lines. Cancer Res. *61*, 6120–6130.

Bouhnik, Y., Flourie, B., Andrieux, C., Bisetti, N., Briet, F., and Rambaud, J.C. (1996). Effects of *Bifidobacterium* sp. fermented milk ingested with or without inulin on colonic bifidobacteria and enzymatic activities in healthy humans. Eur. J. Clin. Nutr. *50*, 269–273.

Bourne, L.C., and Rice-Evans, C. (1998). Bioavailability of ferulic acid. Biochem. Biophys. Res. Commun. *253*, 222–227.

Branca, F., and Lorenzetti, S. (2005). Health effects of phytoestrogens. Forum Nutr. *57*, 100–111.

Calabrese, V., Cornelius, C., Mancuso, C., Pennisi, G., Calafato, S., Bellia, F., Bates, T.E., Giuffrida Stella, A.M., Schapira, T., Dinkova Kostova, A.T., and Rizzarelli, E. (2008). Cellular stress response: a novel target for chemoprevention and nutritional neuroprotection in aging, neurodegenerative disorders and longevity. Neurochem. Res. 33, 2444–2471.

Camilo, E., Zimmerman, J., Mason, J.B., Golner, B., Russell, R., Selhub, J., and Rosenberg, I.H. (1996). Folate synthesized by bacteria in the human upper small intestine is assimilated by the host. Gastroenterol. *110*, 991–998.

Cheng, D.L., Hashimoto, K., and Uda, Y. (2004a). *In vitro* digestion of sinigrin and glucotropaeolin by single strains of *Bifidobacterium* and identification of the digestive products. Food Chem. Toxicol. 42, 351–357.

Cheng, W.L., Lii, C.K., Chen, H.W., Lin, T.H., and Liu, Contribution, K.L. of conjugated linoleic acid to the suppression of inflammatory responses through the regulation of the N.F.-kappaB pathway. (2004b). J. Agric. Food Chem. *14*, 71–78.

Chi, H., and Ji, G.E. (2005). Transformation of ginsenosides Rb1 and Re from Panax ginseng by food microorganisms. Biotechnol. Lett. *27*, 765–771.

Chi, H., Kim, D.H., and Ji, G.E. (2005). Transformation of ginsenosides Rb2 and Rc *from Panax ginseng* by food microorganisms. Biol. Pharm. Bull. *28*, 2102–2105.

Chien, H.L., Huang, H.Y., and Chou, C.C. (2006). Transformation of isoflavone phytestrogens during the fermentation of soymilk with lactic acid bacteria and bifidobacteria. Food Microbiol. *23*, 772–778.

Choi, K., Kim, M., Ryu, J., and Choi, C. (2007). Ginsenosides compound K and Rh(2) inhibit tumor necrosis factor-alpha-induced activation of the NF-kappaB and JNK pathways in human astroglial cells. Neurosci. Lett. *421*, 37–41.

Choi, N.J., Park, H.G., Kim, Y.J., Kim, I.H., Kang, H.S., Yoon, C.S., Yoon, H.G., Park, S.I., Lee, J.W., and Chung, S.H. (2008). Utilization of monolinolein as a substrate for conjugated linoleic acid production by *Bifidobacterium breve* LMC 520 of human neonatal origin. J. Agric. Food Chem. *56*, 10908–10912.

Christensen, L.P. (2009). Ginsenosides chemistry, biosynthesis, analysis, and potential health effects. Adv. Food Nutr. Res. *55*, 1–99.

Chung, S.H., Kim, I.H., Park, H.G., Kang, H.S., Yoon, C.S., Jeong, H.Y., Choi, N.J., Kwon, E.G. and Kim, Y.J. (2008). Synthesis of conjugated linoleic acid by human-derived *Bifidobacterium breve* LMC 017: utilization as a functional starter culture for milk fermentation. J. Agric. Food. Chem. *56*, 3311–3316.

Clarke, J.D., Dashwood, R.H., and Ho, E. (2008). Multi-targeted prevention of cancer by sulforaphane. Cancer Lett. *269*, 291–304.

Clavel, T., Henderson, G., Engst, W., Doré, J., and Blaut, M. (2006). Phylogeny of human intestinal bacteria that activate the dietary lignan secoisolariciresinol diglucoside. FEMS Microbiol. Ecol. *55*, 471–478.

Coakley, M., Ross, R.P., Nordgren, M., Fitzgerald, G., Devery, R., and Stanton, C. (2003). Conjugated linoleic acid biosynthesis by human-derived *Bifidobacterium* species. J. Appl. Microbiol. *94*, 138–145.

Commane, D., Hughes, R., Shortt, C., and Rowland, I. (2005). The potential mechanisms involved in the anticarcinogenic action of probiotics. Mutat. Res. *591*, 276–289.

Couteau, D., McCartney, A.L., Gibson, G.R., Williamson, G., and Faulds, C.B. (2001). Isolation and characterization of human colonic bacteria able to hydrolyse chlorogenic acid. J. Appl. Microbiol. *90*, 873–881.

Crittenden, R.G., Martinez, N.R., and Playne, M.J. (2003). Synthesys and utilization of folate by yoghurt starter cultures and probiotic bacteria. Int. J. Food Microbiol. *80*, 217–220.

Dabek, M., McCrae, S.I., Stevens, V.J., Duncan, S.H., and Louis, P. (2008). Distribution of beta-glucosidase and beta-glucuronidase activity and of beta-glucuronidase gene *gus* in human colonic bacteria. FEMS Microbiol. Ecol. *66*, 487–495.

Day, A.J., DuPont, M.S., Ridley, S., Rhodes, M., Rhodes, M.J., Morgan, M.R., and Williamson, G. (1998). Deglycosylation of flavonoid and isoflavonoid glycosides by human small intestine and liver and beta glucosidase activity. FEBS Lett. *436*, 71–75.

de Kok, T.M., van Breda, S.G., and Manson, M.M. (2008). Mechanisms of combined action of different chemopreventive dietary compounds: a review. Eur. J. Nutr. *47*, S51–S59.

de Vrese, M., and Schrezenmeir, J. (2008). Probiotics, prebiotics, and synbiotics. Adv. Biochem. Eng. Biotechnol. *111*, 1–66.

Deguchi, Y., Morishita, T., and Mutai, M. (1985). Comparative studies on synthesis of water-soluble vitamins among human species of bifidobacteria. Agric. Biol. Chem. *49*, 13–19.

Devillard, E., McIntosh, F.M., Duncan, S.H., and Wallace, R.J. (2007). Metabolism of linoleic acid by human gut bacteria: different routes for biosynthesis of conjugated linoleic acid. J. Bacteriol. *189*, 2566–2570.

Edionwe, A.O., and Kies, C. (2001). Comparison of palm and mixtures of refined palm and soybean oils on serum lipids and fecal fat and fatty acid excretions of adult humans. Plant Foods Hum. Nutr. *56*, 157–165.

Eeckhaut, E., Struijs, K., Possemiers, S., Vincken, J.P., Keukeleire, D.D., and Verstraete, W. (2008). Metabolism of the lignan macromolecule into enterolignans in the gastrointestinal lumen as determined in the simulator of the human intestinal microbial ecosystem. J. Agric. Food Chem. *56*, 4806–4812.

Elfoul, L., Rabot, S., Khelifa, N., Quinsac, A., Duguay, A., and Rimbault, A. (2001). Formation of allyl isothiocyanate from sinigrin in the digestive tract of rats monoassociated with a human colonic strain of *Bacteroides thetaiotaomicron*. FEMS Microbiol. Lett. *197*, 99–103.

Evans, M.E., Brown, J.M., and McIntosh, M.K. (2002). Isomer-specific effects of conjugated linoleic acid (CLA) on adiposity and lipid metabolism. J. Nutr. Biochem. *13*, 508–516.

Ewaschuk, J.B., Walker, J.W., Diaz, H., and Madsen, K.L. (2006). Bioproduction of conjugated linoleic acid by probiotic bacteria occurs *in vitro* and *in vivo* in mice. J. Nutr. *136*, 1483–1487.

FAO/WHO working group. (2001). Report of a joint FAO/WHO expert consultation on evaluation of health and nutritional properties of probiotics in food including powder milk with live lactic acid bacteria. Córdoba, Argentina (October 1–4, 2001).

Fotiadis, C.I., Stoidis. C.N., Spyropoulos, B.G., and Zografos, E.D. (2008). Role of probiotics, prebiotics and synbiotics in chemoprevention for colorectal cancer. World J. Gastroenterol. *14*, 6453–6457.

Fuchs, S., Sontag, G., Stidl, R., Ehrlich, V., Kundi, M., and Knasmüller, S. (2008). Detoxification of patulin and ochratoxin A, two abundant mycotoxins, by lactic acid bacteria. Food Chem. Toxicol. *46*, 1398–1407.

Funabashi, H., Uchida, K., Kado, S., Matsuoka, Y., and Ohwaki, M. (2001). Establishment of a Tcrb and Trp53 genes deficient mouse strain as an animal model for spontaneous colorectal cancer. Exp. Anim. *50*, 41–47.

Garro, M.S., Aguirre, L., and Savoy de Giori, G. (2006). Biological activity of *Bifidobacterium longum* in response to environmental pH. Appl. Microbiol. Biotechnol. *70*, 612–617.

Gebauer, S.K., Psota, T.L., and Kris-Etherton, P.M. (2007). The diversity of health effects of individual trans fatty acid isomers. Lipids *42*, 787–799.

Geier, M.S., Butler, R.N., and Howarth, G.S. (2006). Probiotics, prebiotics and synbiotics: a role in chemoprevention for colorectal cancer? Cancer Biol. Ther. *5*, 1265–1269.

Gill, S.R., Pop, M., Deboy, R.T., Eckburg, P.B., Turnbaugh, P.J., Samuel, B.S., Gordon, J.I., Relman, D.A., Fraser-Liggett, C.M., and Nelson, K.E. (2006). Metagenomic analysis of the human distal gut microbiome. Science *312*, 1355–1359.

Gonthier, M.P., Verny, M.A., Besson, C., Remesey, C., and Scalbert, A. (2003). Chlorogenic acid bioavailability largely depends on its metabolism by the gut microflora in rats. J. Nutr. *133*, 1853–1859.

Graf, B.A., Milbury, P.E., and Blumberg, J.B. (2005). Flavonols, flavones, flavanones, and human health: epidemiological evidence. J. Med. Food *8*, 281–290.

Hawksworth, G., Drasar, B.S., and Hill, M.J. (1971). Intestinal bacteria and the hydrolysis of glycosidic bonds. J. Med. Microbiol. *4*, 451–459.

Hayes, J.D., Kelleher, M.O., and Eggleston, I.M. (2008). The cancer chemopreventive actions of phytochemicals derived from glucosinolates. Eur. J. Nutr. *47*, S73–S88.

Hennessy, A.A., Ross, R.P., Devery, R., and Stanton, C. (2009). Optimization of a reconstituted skim milk based medium for enhanced CLA production by bifidobacteria. J. Appl. Microbiol. *106*, 1315–1327.

Heuvelin, E., Lebreton, C., Grangette, C., Pot, B., Cerf-Bensussan, N., and Heyman, M. (2009). Mechanisms involved in alleviation of intestinal inflammation by *Bifidobacterium breve* soluble factors. PLoS One *4*, e5184.

Hill, M.J. (1997). Intestinal flora and endogenous vitamin synthesis. Eur. J. Cancer Prev. *6*, S43–S45.

Hollman, P.C.H., Devries, J.H.M., Vanleeuwen, S.D., Menglers, M.J.B., and Katan, M.B. (1995). Absorption of dietary qurecetin glycosides and quercetin in healthy ileostomy volunteers. Am. J. Clin. Nutr. *62*, 1276–1282.

Holst, B., and Williamson, G. (2008). Nutrients and phytochemicals: from bioavailability to bioefficacy beyond antioxidants. Curr. Opin. Biotechnol. *19*, 73–82.

Hope, M.E., Hold, G.L., Kain, R., and El-Omar, E.M. (2005). Sporadic colorectal cancer-role of the commensal microbiota. FEMS Microbiol. Lett. *244*, 1–7.

Hord, N.G. (2008). Eukaryotic–microbiota crosstalk: potential mechanisms for health benefits of prebiotics and probiotics. Annu. Rev. Nutr. *28*, 215–231.

Hou, J.W., Yu, R.C., and Chou, C.C. (2000). Changes in some components of soymilk during fermentation with bifidobacteria. Food Res. Int. 33, 393–397.

Hubbard, N.E., Lim, D., and Erickson, K.L. (2003). Effect of separate conjugated linoleic acid isomers on murine mammary tumorigenesis. Cancer Lett. *190*, 13–19.

Hur, H.G., Lay Jr., J.O., Beger, R.D., Freeman, J.P., and Rafii, F. (2000). Isolation of human intestinal bacteria metabolizing the natural isoflavone glycosides daidzin and genistin. Arch. Microbiol. *174*, 422–428.

Huycke, M.M., and Gaskins, H.R. (2004). Commensal bacteria, redox stress, and colorectal cancer: mechanisms and models. Exp. Biol. Med. (Maywood). *229*, 586–597.

I.A.R.C – International Agency for Research on Cancer Workgroup. (2004). Cruciferous vegetables, isothiocyanates and indoles, Handbooks of cancer prevention, vol 9. (Lyon: IARC Press).

Ip, C., Dong, Y., Ip, M.M., Banni, S., Carta, G., Angioni, E., Murru, E., Spada, S., Melis, M.P., and Saebo, A. (2002). Conjugated linoleic acid isomers and mammary cancer prevention. Nutr. Cancer *43*, 52–58.

Ip, M.M., Masso-Welch, P.A., Shoemaker, S.F., Shea-Eaton, W.K., and Ip, C. (1999). Conjugated linoleic acid inhibits proliferation and induces apoptosis of normal rat mammary epithelial cells in primary culture. Exp. Cell Res. *250*, 22–34.

Jenkins, T.C., Wallace, R.J., Moate, P.J., and Mosley, E.E. (2008). Board-invited review: Recent advances in biohydrogenation of unsaturated fatty acids within the rumen microbial ecosystem. J. Anim. Sci. *86*, 397–412.

Kado, S., Uchida, K., Funabashi, H., Iwata, S., Nagata, Y., Ando, M., Onoue, M., Matsuoka, Y., Ohwaki, M., and Morotomi, M. (2001). Intestinal microflora are necessary for development of spontaneous adenocarcinoma of the large intestine in T-cell receptor beta chain and p53 double-knockout mice. Cancer Res. *61*, 2395–2398.

Kekkonen, R.A., Lummela, N., Karjalainen, H., Latvala, S., Tynkkynen, S., Jarvenpaa, S., Kautiainen, H., Julkunen, I., Vapaatalo, H., and Korpela, R. (2008). Probiotic intervention has strain-specific anti-inflammatory effects in healthy adults. World J. Gastroenterol. *14*, 2029–2036.

Kelley, D.S., and Erickson, K.L. (2003). Modulation of body composition and immune cell functions by conjugated linoleic acid in humans and animal models: benefits vs. risks. Lipids *38*, 377–386.

Kelley, N.S., Hubbard, N.E., and Erickson, K.L. (2007). Conjugated linoleic acid isomers and cancer. J Nutr. *137*, 2599–2607.

Kemp, M.Q., Jeffy, B.D., and Romagnolo, D.F. (2003). Conjugated linoleic acid inhibits cell proliferation through a p53-dependent mechanism: effects on the expression of G1-restriction points in breast and colon cancer cells. J. Nutr. *133*, 3670–3677.

Kim, T.H., Yang, J., Darling, P.B., and O'Connor, D.L. (2004). A large pool of available folate exists in the large intestine of infants and piglets. J. Nutr. *134*, 1389–1394.

Kim, Y., Lee, D., Kim, D., Cho, J., Yang, J., Chung, M., Kim, K., and Ha, N. (2008). Inhibition of proliferation in colon cancer cell lines and harmful enzyme activity of colon bacteria by *Bifidobacterium adolescentis* SPM0212. Arch. Pharm. Res. *31*, 468–473.

Knasmüller, S., Steinkellner, H., Hirschl, A.M., Rabot, S., Nobis, E.C., and Kassie, F. (2001). Impact of bacteria in dairy products and of the intestinal microflora on the genotoxic and carcinogenic effects of heterocyclic aromatic amines. Mutat. Res. 480–481, 129–138.

Kopecný, J., Logar, R.M., and Kobayashi, Y. (2001). Phenotypic and genetic data supporting reclassification of *Butyrivibrio fibrisolvens* isolates. Folia Microbiol. (Praha). *46*, 45–48.

Krause, L.J., Forsberg, C.W., and O'Connor, D.L. (1996). Feeding human milk to rats increases *Bifidobacterium* in the cecum and colon which correlates with enhanced folate status. J. Nutr. *126*, 1505–1511.

Kris-Etherton, P.M., Hecker, K.D., Bonanome, A., Coval, S.M., Binkoski, A.E., Hilpert, K.F., Griel, A.E., and Etherton, T.D. (2002). Bioactive compounds in foods: their role in the prevention of cardiovascular disease and cancer. Am. J. Med. *113*, 71S-88S.

Krul, C., Humblot, C., Philippe, C., Vermeulen, M., van Nuenen, M., Havenaar, R., and Rabot, S. (2002). Metabolism of sinigrin (2-propenyl glucosinolate) by the human colonic microflora in a dynamic *in vitro* large-intestinal model. Carcinogenesis 23, 1009–1016.

Lee do, K., Jang, S., Kim, M.J., Kim, J.H., Chung, M.J., Kim, K.J., and Ha, N.J. (2008). Anti-proliferative effects of *Bifidobacterium adolescentis* SPM0212 extract on human colon cancer cell lines. BMC Cancer *8*, 310–317.

Lee, Y., Thompson, J.T., de Lera, A.R., and Vanden Heuvel, J.P. (2008). Isomer-specific effects of conjugated linoleic acid on gene expression in RAW 264.7. J. Nutr. Biochem. (in press).

Lepercq, P., Hermier, D., David, O., Michelin, R., Gibard, C., Beguet, F., Relano, P., Cayuela, C., and Juste, C. (2005). Increasing ursodeoxycholic acid in the enterohepatic circulation of pigs through the administration of living bacteria. Br. J. Nutr. *93*, 457–469.

Li, G., Dong, B., Butz, D.E., Park, Y., Pariza, M.W., and Cook, M.E. (2006). NF-kappaB independent inhibition of lipopolysaccharide-induced cyclooxygenase by a conjugated linoleic acid cognate, conjugated nonadecadienoic acid. Biochem. Biophys. Acta *1761*, 969–972.

Lim, D.Y., Tyner, A.L., Park, J.B., Lee, J.Y., Choi, Y.H., and Park, J.H. (2005). Inhibition of colon cancer cell proliferation by the dietary compound conjugated linoleic acid is mediated by the CDK inhibitor p21CIP1/WAF1. J. Cell Physiol. *205*, 107–113.

Lin, M.Y., and Chang, F.J. (2000). Antioxidative effect of intestinal bacteria *Bifidobacterium longum* ATCC 15708 and *Lactobacillus acidophilus* ATCC 4356. Dig. Dis. Sci. *45*, 1617–1622.

Lin, M.Y., and Yen, C.L. (1999a). Antioxidative ability of lactic acid bacteria. J. Agric. Food Chem. *47*, 1460–1466.

Lin, M.Y., and Yen, C.L. (1999b). Inhibition of lipid peroxidation by *Lactobacillus acidophilus* and *Bifidobacterium longum*. J. Agric. Food Chem. *47*, 3661–3664.

Looijer-van Langen, M.A., and Dieleman, L.A. (2009). Prebiotics in chronic intestinal inflammation. Inflamm. Bowel Dis. *15*, 454–462.

Maggiora, M., Bologna, M., Cerù, M.P., Possati, L., Angelucci, A., Cimini, A., Miglietta, A., Bozzo, F., Margiotta, C., Muzio, G., and Canuto, R.A. (2004). An overview of the effect of linoleic and conjugated-linoleic acids on the growth of several human tumor cell lines. Int. J. Cancer *112*, 909–919.

Majumder, B., Wahle, K.W., Moir, S., Schofield, A., Choe, S.N., Farquharson, A., Grant, I., and Heys, S.D. (2002). Conjugated linoleic acids (CLAs) regulate the expression of key apoptotic genes in human breast cancer cells. FASEB J. *16*, 1447–1449.

Manach C., Williamson, G., Morand, C., Scalbert, A., and Rémésy, C. (2005). Bioavailability and bioefficacy of polyphenols in humans. I. Review of 97 bioavailability studies. Am. J. Clin. Nutr. *81*, 230S-242S.

Manach, C., Scalbert, A., Morand, C., Rémésy, C., and Jiménez, L. (2004). Polyphenols: food sources and bioavailability. Am. J. Clin. Nutr. *79*, 727–747.

Marion-Letellier, R., Déchelotte, P., Iacucci, M., and Ghosh, S. (2009). Dietary modulation of peroxisome proliferator-activated receptor gamma. Gut *58*, 586–593.

Marotti, I., Bonetti, A., Biavati, B., Catizone, P., and Dinelli, G. (2007). Biotransformation of common bean (*Phaseolus vulgaris* L.) flavonoid glycosides by *Bifidobacterium* species from human intestinal origin. J. Agric. Food Chem. *55*, 3913–3919.

Maruo, T., Sakamoto, M., Ito, C., Toda, T., and Benno, Y. (2008). *Adlercreutzia equolifaciens* gen. nov., sp. nov., an equol-producing bacterium isolated from human faeces, and emended description of the genus *Eggerthella*. Int. J. Syst. Evol. Microbiol. *58*, 1221–1227.

Masso-Welch, P.A., Zangani, D., Ip, C., Vaughan, M.M., Shoemaker, S., Ramirez, R.A., and Ip, M.M. (2002). Inhibition of angiogenesis by the cancer chemopreventive agent conjugated linoleic acid. Cancer Res. *62*, 4383–4389.

Mastromarino, A., Reddy, B.S., and Wynder, E.L. (1976). Metabolic epidemiology of colon cancer: enzymic activity of fecal flora. Am. J. Clin. Nutr. *29*, 1455–1460.

Matthies, A., Clavel, T., Gütschow, M., Engst, W., Haller, D., Blaut, M., and Braune, A. (2008). Conversion of daidzein and genistein by a newly isolated anaerobic bacterium from mouse intestine. Appl. Environ. Microbiol. *74*, 4847–4852.

McBain, A.J., and Macfarlane, G.T. (1998). Ecological and physiological studies on large intestinal bacteria in relation to production of hydrolytic and reductive enzymes involved in formation of genotoxic metabolites. J. Med. Microbiol. *47*, 407–416.

McGarr, S.E., Ridlon, J.M., and Hylemon, P.B. (2005). Diet, anaerobic bacterial metabolism and colon cancer risk: a review of the literature. J. Clin. Gastroenterol. *39*, 98–109.

McIntosh, F.M., Shingfield, K.J., Devillard, E., Russell, W.R., and Wallace, R.J. (2009). Mechanism of conjugated linoleic acid and vaccenic acid formation in human faecal suspensions and pure cultures of intestinal bacteria. Microbiology *155*, 285–294.

Miettinen, M., Vuopio-Varkila, J., and Varkila, K. (1996). Production of human tumor necrosis factor alpha, interleukin-6, and interleukin-10 is induced by lactic acid bacteria. Infect. Immun. *64*, 5403–5405.

Minamida, K., Ota, K., Nishimukai, M., Tanaka, M., Abe, A., Sone, T., Tomita, F., Hara, H., and Asano,

K. (2008). *Asaccharobacter celatus* gen. nov., sp. nov., isolated from rat caecum. Int. J. Syst. Evol. Microbiol. *58*, 1238–1240.

Mitchell, P.L., and McLeod, R.S. (2008). Conjugated linoleic acid and atherosclerosis: studies in animal models. Biochem. Cell. Biol. *86*, 293–301.

Moloney, F., Toomey, S., Noone, E., Nugent. A., Allan, B., Loscher, C.E., and Roche, H.M. (2007). Antidiabetic effects of *cis*-9, *trans*-11-conjugated linoleic acid may be mediated via anti-inflammatory effects in white adipose tissue. Diabetes *56*, 574–582.

Murphy, E.F., Hooiveld, G.J., Muller, M., Calogero, R.A., and Cashman, K.D. (2007). Conjugated linoleic acid alters global gene expression in human intestinal-like Caco-2 cells in an isomer-specific manner. J. Nutr. *137*, 2359–2365.

Nakamura, J., Kubota, Y., Miyaoka, M., Saitoh, T., Mizuno, F., and Benno, Y. (2002). Comparison of four microbial enzymes in Clostridia and *Bacteroides* isolated from human feces. Microbiol. Immunol. *46*, 487–490.

Neish, A.S. (2009). Microbes in gastrointestinal health and disease. Gastroenterol. *136*, 65–80.

Nian, H., Delage, B., Ho, E., and Dashwood, R.H. (2009). Modulation of histone deacetylase activity by dietary isothiocyanates and allyl sulfides: studies with sulforaphane and garlic organosulfur compounds. Environ. Mol. Mutagen. *50*, 213–221.

Noda, H., Akasaka, N., and Ohsugi, M. (1994). Biotin production by bifidobacteria. J. Nutr. Sci. Vitaminol. *40*, 181–188.

Nugon-Baudon, L., Rabot, S., Wal, J.M., and Szylit, O. (1990). Interactions of the intestinal microflora with glucosinolates in rapeseed meal toxicity – 1st evidence of an intestinal lactobacillus possessing a myrosinase-like activity *in vivo*. J. Sci. Food Agric. *52*, 547–559.

Oh, D.K., Hong, G.H., Lee, Y., Min, S., Sin, H.S., and Cho, S.K. (2003). Production of conjugated linoleic acid by isolated *Bifidobacterium* strains. World J. Microbiol. Biotechnol. *9*, 907–912.

Okada, Y., Tsuzuki, Y., Hokari, R., Komoto, S., Kurihara, C., Kawaguchi, A., Nagao, S., and Miura, S. (2009). Anti-inflammatory effects of the genus *Bifidobacterium* on macrophages by modification of phospho-I kappaB and *SOCS* gene expression. Int. J. Exp. Pathol. *90*, 131–140.

Olthof, M.R., Hollman, P.C., Buijsman, M.N., van Amelsvoort, J.M., and Katan, M.B. (2003). Chlorogenic acid, quercetin-3-rutinoside and black tea phenols are extensively metabolized in humans. J. Nutr. *133*, 1806–1814.

O'Mahony, C., Scully, P., O'Mahony, D., Murphy, S., O'Brien, F., Lyons, A., Sherlock, G., MacSharry, J., Kiely, B., Shanahan, F., and O'Mahony, L. (2008). Commensal-induced regulatory T cells mediate protection against pathogen-stimulated NF-kappaB activation. PLoS Pathog. *4*, e1000112.

Orrhage, K.M., Annas, A., Nord, C.E., Brittebo, E.B., and Rafter, J. (2002). Effects of lactic acid bacteria on the uptake and distribution of the food mutagen Trp-P-2 in mice. J. Scand. J. Gastroenterol. *37*, 215–221.

Otieno, D.O., and Shah, N.P. (2007a). A comparison of changes in the transformation of isoflavones in soymilk using varying concentrations of exogenous and probiotic-derived endogenous beta-glucosidases. J. Appl. Microbiol. *103*, 601–612.

Otieno, D.O., and Shah, N.P. (2007b). Endogenous beta-glucosidase and beta-galactosidase activities from selected probiotic microorganisms and their role in isoflavone biotransformation in soymilk. J. Appl. Microbiol. *103*, 910–917.

Palombo, J.D., Ganguly, A., Bistrian, B.R., and Menard, M.P. (2002). The antiproliferative effects of biologically active isomers of conjugated linoleic acid on human colorectal and prostatic cancer cells. Cancer Lett. *177*, 163–172.

Pariza, M.W. (2004). Perspective on the safety and effectiveness of conjugated linoleic acid. Am. J. Clin. Nutr. *79*, 1132S-1136S.

Pariza, M.W., and Hargraves, W.A. (1985). A beef-derived mutagenesis modulator inhibits initiation of mouse epidermal tumors by 7,12-dimethylbenz[a]anthracene. Carcinogenesis *6*, 591–593.

Paz-Elizur, T., Sevilya, Z., Leitner-Dagan, Y., Elinger, D., Roisman, L.C., and Livneh, Z. (2008). DNA repair of oxidative DNA damage in human carcinogenesis: potential application for cancer risk assessment and prevention. Cancer Lett. *266*, 60–72.

Peltonen, K., el-Nezami, H., Haskard, C., Ahokas, J., and Salminen, S. (2001). Aflatoxin B1 binding by dairy strains of lactic acid bacteria and bifidobacteria. J. Dairy Sci. *84*, 2152–2166.

Pham, T.T., and Shah, N.P. (2007). Biotransformation of isoflavone glycosides by *Bifidobacterium animalis* in soymilk supplemented with skim milk powder. J. Food Sci. *72*, 316–324.

Picard, C., Fioramonti, J., Francois, A., Robinson, T., Neant, F., and Matuchansky, C. (2005). Review article: bifidobacteria as probiotic agents – physiological effects and clinical benefits. Aliment. Pharmacol. Ther. *22*, 495–512.

Plourde, M., Jew, S., Cunnane, S.C., and Jones, P.J. (2008). Conjugated linoleic acids: why the discrepancy between animal and human studies? Nutr. Rev. *66*, 415–421.

Pompei, A., Cordisco, L., Amaretti, A., Zanoni, S., Matteuzzi, D., and Rossi, M. (2007a). Folate production by bifidobacteria as a potential probiotic property. Appl. Environ. Microbiol. 73, 179–185.

Pompei, A., Cordisco, L., Amaretti, A., Zanoni, S., Raimondi, S., Matteuzzi, D., and Rossi, M. (2007b). Administration of folate-producing bifidobacteria enhances folate status in Wistar rats. J. Nutr. *137*, 2742–2746.

Pool-Zobel, B.L., Neudecker, C., Domizlaff, I., Ji, S., Schillinger, U., Rumney, C., Moretti, M., Vilarini, I., Scassellati-Sforzolini, R., and Rowland, I. (1996). *Lactobacillus*- and *Bifidobacterium* mediated antigenotoxicity in the colon of rats. Nutr. Cancer *26*, 365–380.

Raimondi, S., Roncaglia, L., De Lucia, M., Amaretti, A., Leonardi, A., Pagnoni, U.M., and Rossi, M. (2009). Bioconversion of soy isoflavones daidzin and daidzein by *Bifidobacterium* strains. Appl. Microbiol. Biotechnol. *81*, 943–950.

Rechner, A.R., Smith, M.A., Kuhnle, G., Gibson, G.R., Debnam, E.S., Srai, S.K., Moore, K.P., and Rice-Evans, C.A. (2004). Colonic metabolism of dietary polyphenols: influence of structure on microbial fermentation products. Free Radic. Biol. Med. *36*, 212–225.

Reddy, B.S., and Rivenson, A. (1993). Inhibitory effect of *Bifidobacterium longum* on colon, mammary, and liver carcinogenesis induced by 2-amino-3-methylimidazo[4,5-f]quinoline, a food mutagen. Cancer Res. *53*, 3914–3918.

Reddy, B.S., Narasawa, T., Weisburger, J.H., and Wynder, E.L. (1976a). Promoting effect of sodium deoxycholate on colon adenocarcinomas in germfree rats. J. Natl. Cancer Inst. *56*, 441–442.

Reddy, B.S., Narisawa, T., and Weisburger, J.H. (1976b). Colon carcinogenesis in germ-free rats with intrarectal 1,2-dimethylhydrazine and subcutaneous azoxymethane. Cancer Res. *36*, 2874–2876.

Reddy, B.S., Narisawa, T., Wright, P., Vukusich, D., Weisburger, J.H., and Wynder, E.L. (1975). Colon carcinogenesis with azoxymethane and dimethylhydrazine in germ-free rats. Cancer Res. *35*, 287–290.

Ridlon, J.M., Kang, D.J., and Hylemon, P.B. (2006). Bile salt biotransformations by human intestinal bacteria. J. Lipid Res. *47*, 241–259.

Riedel, C.U., Foata, F., Philippe, D., Adolfsson, O., Eikmanns, B.J., and Blum, S. (2006). Anti-inflammatory effects of bifidobacteria by inhibition of LPS-induced NF-kappaB activation. World. J. Gastroenterol. *12*, 3729–3735.

Ringseis, R., Wen, G., Saal, D., and Eder, K. (2008). Conjugated linoleic acid isomers reduce cholesterol accumulation in acetylated LDL-induced mouse RAW264.7 macrophage-derived foam cells. Lipids *43*, 913–923.

Rosberg-Cody, E., Ross, R.P., Hussey, S., Ryan, C.A., Murphy, B.P., Fitzgerald, G.F., Devery, R., and Stanton, C. (2004). Mining the microbiota of the neonatal gastrointestinal tract for conjugated linoleic acid-producing bifidobacteria. Appl. Environ. Microbiol. *70*, 4635–4641.

Rowland, I.R. (1999). The role of the gastrointestinal microbiota in colorectal cancer. Curr. Pharm. Des. *15*, 1524–1527.

Rowland, I.R., Rumney, C.J., Coutts, J.T., and Lievense, L.C. (1998). Effect of *Bifidobacterium longum* and inulin on gut bacterial metabolism and carcinogen-induced aberrant crypt foci in rats. Carcinogenesis *19*, 281–285.

Said, H.M., and Mohammed, Z.M. (2006). Intestinal absorption of water-soluble vitamins: an update. Curr. Opin. Gastroenterol. *22*, 140–146.

Salas-Salvadó, J., Márquez-Sandoval, F., and Bulló, M. (2006). Conjugated linoleic acid intake in humans: a systematic review focusing on its effect on body composition, glucose, and lipid metabolism. Crit. Rev. Food Sci. Nutr. *46*, 479–488.

Scalbert, A., and Williamson, G. (2000). Dietary intake and bioavailability of polyphenols. J. Nutr. *130*, 2073S-8205S.

Schell, M.A., Karmirantzou, M., Snel, B., Vilanova, D., Berger, B., Pessi, G., Zwahlen, M.C., Desiere, F., Bork, P., Delley, M., Pridmore, R.D., and Arigoni, F. (2002). The genome sequence of *Bifidobacterium longum* reflects its adaptation to the human gastrointestinal tract. Proc. Natl. Acad. Sci. U.S.A *99*, 14422–14427.

Schrezenmeir, J., and de Vrese, M. (2001). Probiotics, prebiotics, and synbiotics-approaching a definition. Am. J. Clin. Nutr. *73*, 361S-364S.

Sekine, K., Ohta, J., Onishi, M., Tatsuki, T., Shimokawa, Y., Toida, T., Kawashima, T., and Hashimoto, Y. (1995). Analysis of antitumor properties of effector cells stimulated with a cell wall preparation (WPG) of *Bifidobacterium infantis*. Biol. Pharm. Bull. *18*, 148–153.

Servin, A.L. (2004). Antagonistic activities of lactobacilli and bifidobacteria against microbial pathogens. FEMS Microbiol. Rev. *28*, 405–440.

Setchell, K.D.R., and Cassidy, A. (1999). Dietary Isoflavones: Biological Effects and Relevance to Human Health. J. Nutr. *129*, 758S-767S.

Setchell, K.D.R., Brown, N.B., Zimmer-Nechemias, L., Brashear, W.T., Wolfe, B., Kirscher, A.S., and Heubi, J.E. (2002). Evidence for lack of absorption of soy isoflavone glycosides in humans, supporting the crucial role of intestinal metabolism for bioavailability. Am. J. Clin. Nutr. *76*, 447–453.

Shane, B. (1995). Folate chemistry and metabolism. In Folate in Health and Disease, Bailey, L.B., ed. (New York: Marcel Dekker, Inc.), pp 1–22.

Shen, H.M., and Tergaonkar, V. (2009). NFkappaB signaling in carcinogenesis and as a potential molecular target for cancer therapy. Apoptosis *14*, 348–363.

Shin, H.Y., Lee, J.H., Lee, J.Y., Han, Y.O., Han, M.J., and Kim, D.H. (2003a). Purification and characterization of ginsenoside Ra-hydrolyzing beta-D-xylosidase from *Bifidobacterium breve* K-110, a human intestinal anaerobic bacterium. Biol Pharm Bull. *26*, 1170–1173.

Shin, H.Y., Park, S.Y., Sung, J.H., and Kim, D.H. (2003b). Purification and characterization of alpha-L-arabinopyranosidase and alpha-L-arabinofuranosidase from *Bifidobacterium breve* K-110, a human intestinal anaerobic bacterium metabolizing ginsenoside Rb2 and Rc. Appl. Environ. Microbiol. *69*, 7116–7123.

Singh, J., Rivenson, A., Tomita, M., Shimamura, S., Ishibashi, N., and Reddy, B.S. (1997). *Bifidobacterium longum*, a lactic acidproducing intestinal bacterium inhibits colon cancer and modulates the intermediate biomarkers of colon carcinogenesis. Carcinogenesis *18*, 833–841.

Skrinjar, M., Rasic, J.L., and Stojicic, V. (1996). Lowering of ochratoxin A level in milk by yoghurt bacteria and bifidobacteria. Folia Microbiol. (Praha) *41*, 26–28.

Song, H.J., Grant, I., Rotondo, D., Mohede, I., Sattar, N., Heys, S.D., and Wahle, K.W. (2005). Effect of CLA supplementation on immune function in young healthy volunteers. Eur. J. Clin. Nutr. *59*, 508–517.

Song, W.O., Chun, O.K., Hwang, I., Shin, H.S., Kim, B.G., Kim, K.S., Lee, S.Y., Shin, D., and Lee, S.G. (2007). Soy isoflavones as safe functional ingredients. J. Med. Food *10*, 571–580.

Sreekumar, O., and Hosono, A. (1998). Antimutagenicity and the influence of physical factors in binding

Lactobacillus gasseri and *Bifidobacterium longum* cells to amino acid pyrolysates. J. Dairy Sci. *81*, 1508–1516.

Stachowska, E., Dolegowska, B., Dziedziejko, V., Rybicka, M., Kaczmarczyk, M., Bober, J., Rac, M., Machalinski, B., and Chlubek, D. (2009). Prostaglandin E2 (PGE2) and thromboxane A2 (TXA2) synthesis is regulated by conjugated linoleic acids (CLA) in human macrophages. J. Physiol. Pharmacol. *60*, 77–85.

Stevens, C.E., and Hume, I.D. (1998). Contributions of microbes in vertebrate gastrointestinal tract to production and conservation of nutrients. Physiol. Rev. *78*, 393–427.

Strozzi, G.P., and Mogna, L. (2008). Quantification of folic acid in human feces after administration of *Bifidobacterium* probiotic strains. J. Clin. Gastroenterol. *42*, S179–184.

Taki, K., Takayama, F., and Niwa, T. (2005). Beneficial effects of Bifidobacteria in a gastroresistant seamless capsule on hyperhomocysteinemia in hemodialysis patients. J. Ren. Nutr. *15*, 77–80.

Tamura, M., Tsushida, T., and Shinohara, K. (2007). Isolation of an isoflavone metabolizing, *Clostridium*-like bacterium, strain TM-40, from human faeces. Anaerobe *13*, 32–35.

Thoma, C., Green, T.J., and Ferguson, L. (2003). Citrus pectin and oligofructose improve folate status and lower serum total homocysteine in rats. Int. J. Vitam. Nutr. Res. 73, 403–409.

Thomas, L.A., Veysey, M.J., French, G., Hylemon, P.B., Murphy, G.M., and Dowling, R.H. (2001). Bile acid metabolism by fresh human colonic contents: a comparison of caecal versus faecal samples. Gut *49*, 835–842.

Tontonoz, P., and Spiegelman, B.M. (2008). Fat and beyond: the diverse biology of PPARgamma. Annu. Rev. Biochem. *77*, 289–312.

Toomey, S., Harhen, B., Roche, H.M., Fitzgerald, D., and Belton, O. (2006). Profound resolution of early atherosclerosis with conjugated linoleic acid. Atherosclerosis *187*, 40–49.

Tricon, S., and Yaqoob, P. (2006). Conjugated linoleic acid and human health: a critical evaluation of the evidence. Curr. Opin. Clin. Nutr. Metab. Care *9*, 105–110.

Tricon, S., Burdge, G.C., Williams, C.M., Calder, P.C., and Yaqoob, P. (2005). The effects of conjugated linoleic acid on human health-related outcomes. Proc. Nutr. Soc. *64*, 171–182.

Tsangalis, D., Ashton, J.F., McGill, A.E.J., and Shah, N.P. (2002). Enzymic transformation of isoflavone phytoestrogens in soymilk by β-glucosidase-producing bifidobacteria. J. Food Sci. *67*, 3104–3113.

Tsangalis, D., Wilcox, G., Shah, N.P., and Stojanovska, L. (2005). Bioavailability of isoflavone phytoestrogens in postmenopausal women consuming soya milk fermented with probiotic bifidobacteria. Br. J. Nutr. *93*, 867–877.

Tsangalis, D., Wilcox, G., Shah, N.P., McGill, A.E., and Stojanovska, L. (2007). Urinary excretion of equol by postmenopausal women consuming soymilk fermented by probiotic bifidobacteria. Eur. J. Clin. Nutr. *61*, 438–441.

Turner, J.V., Agatonovic-Kustrin, S., and Glass, B.D. (2007). Molecular aspects of phytoestrogen selective binding at estrogen receptors. J. Pharm. Sci. *96*, 1879–1885.

Turpeinen, A.M., Ylönen, N., von Willebrand, E., Basu, S., and Aro, A. (2008). Immunological and metabolic effects of *cis*-9, *trans*-11-conjugated linoleic acid in subjects with birch pollen allergy. Br. J. Nutr. *100*, 112–119.

Uronis, J.M., Mühlbauer, M., Herfarth, H.H., Rubinas, T.C., Jones, G.S., and Jobin, C. (2009). Modulation of the intestinal microbiota alters colitis-associated colorectal cancer susceptibility. PLoS One. *4*, e6026.

Valgimigli, L., and Iori, R. (2009). Antioxidant and pro-oxidant capacities of ITCs. Environ. Mol. Mutagen. *50*, 222–237.

Valko, M., Leibfritz, D., Moncol, J., Cronin, M.T., Mazur, M., and Telser, J. (2007). Free radicals and antioxidants in normal physiological functions and human disease. Int. J. Biochem. Cell. Biol. *39*, 44–84.

Vallabhapurapu, S., and Karin, M. (2009). Regulation and function of NF-kappaB transcription factors in the immune system. Annu. Rev. Immunol. *27*, 693–733.

van der Schouw, Y.T., de Kleijn, M.J., Peeters, P.H., and Grobbee, D.E. (2000). Phyto-oestrogens and cardiovascular disease risk. Nutr. Metab. Cardiovasc. Dis. *10*, 154–167.

Van Nieuwenhove, C.P., Oliszewski, R., González, S.N., and Pérez Chaia, A.B. (2007). Conjugated linoleic acid conversion by dairy bacteria cultured in MRS broth and buffalo milk. Lett. Appl. Microbiol. *44*, 467–474.

Vanderpool, C., Yan, F., and Polk, D.B. (2008). Mechanisms of probiotic action: Implications for therapeutic applications in inflammatory bowel diseases. Inflamm. Bowel Dis. *14*, 1585–1596.

Vannucci, L., Stepankova, R., Kozakova, H., Fiserova, A., Rossmann, P., and Tlaskalova-Hogenova, H. (2008). Colorectal carcinogenesis in germ-free and conventionally reared rats: different intestinal environments affect the systemic immunity. Int. J. Oncol. *32*, 609–617.

Wall, R., Ross, R.P., Shanahan, F., O'Mahony, L., O'Mahony, C., Coakley, M., Hart, O., Lawlor, P., Quigley, E.M., Kiely, B., Fitzgerald, G.F., and Stanton, C. (2009). Metabolic activity of the enteric microbiota influences the fatty acid composition of murine and porcine liver and adipose tissues. Am. J. Clin. Nutr. *89*, 1393–1401.

Wang, H., and Murphy, P.A. (1994). Isoflavone content in commercial soybean foods. J. Agric. Food Chem. *42*, 1666–1673.

Wang, X.L., Hur,H.G., Lee, J.H., Kim, K.T., and Kim, S.I. (2005). Enantioselective synthesis of S-equol from dihydrodaidzein by a newly isolated anaerobic human intestinal bacterium. Appl. Environ. Microbiol. *71*, 214–219.

Wang, X.L., Kim, H.J., Kang, S.I., Kim, S.I., and Hur, H.G. (2007). Production of phytoestrogen S-equol from daidzein in mixed culture of two anaerobic bacteria. Arch. Microbiol. *187*, 155–160.

Wang, Y.C., Yu, R.C., and Chou, C.C. (2006). Antioxidative activities of soymilk fermented with lactic acid bacteria and bifidobacteria. Food Microbiol. *23*, 128–135.

Wei, Q.K., Chen, T.R., and Chen, J.T. (2007). Using of *Lactobacillus* and *Bifidobacterium* to product the isoflavone aglycones in fermented soymilk. Int. J. Food Microbiol. *117*, 120–124.

Whigham, L.D., Watras, A.C., and Schoeller, D.A. (2007). Efficacy of conjugated linoleic acid for reducing fat mass: a meta-analysis in humans. Am. J. Clin. Nutr. *85*, 1203–1211.

Wilde, P.F., and Dawson, R.M. (1966). The biohydrogenation of alpha-linolenic acid and oleic acid by rumen microorganisms. Biochem. J. *98*, 469–475.

Wiseman, H. (1999). The bioavailability of non-nutrient plant factors: dietary flavonoids and phyto-oestrogens. Proc. Nutr. Soc. *58*, 139–146.

Xu, H., Lee, H.Y., Hwang, B., Nam, J.H., Kang, H.Y., and Ahn, J. (2008). Kinetics of microbial hydrogenation of free linoleic acid to conjugated linoleic acids. J. Appl. Microbiol. *105*, 2239–2247.

Yang, L., and Pei, Z. (2006). Bacteria, inflammation, and colon cancer. World J. Gastroenterol. *12*, 6741–6746.

Ye, J. (2008). Regulation of PPARgamma function by TNF-alpha. Biochem. Biophys. Res. Commun. *374*, 405–408.

You, H.J., Oh, D.K., and Ji, G.E. (2004). Anticancerogenic effect of a novel chiroinositol-containing polysaccharide from *Bifidobacterium bifidum* BGN4. FEMS Microbiol. Lett. *240*, 131–136.

Yu, Y., Correll, P.H., and Vanden Heuvel, J.P. (2002). Conjugated linoleic acid decreases production of pro-inflammatory products in macrophages: evidence for a PPAR gamma-dependent mechanism. Biochem. Biophys. Acta *1581*, 89–99.

Yu, Z.T., Yao, W., and Zhu, W.Y. (2008). Isolation and identification of equol producing bacterial strains from cultures of pig faeces. FEMS Microbiol. Lett. *282*, 73–80.

Yuan, J.P., Wang, J.H., and Liu, X. (2007). Metabolism of dietary soy isoflavones to equol by human intestinal microflora-implications for health. Mol. Nutr. Food Res. *51*, 765–781.

Yuan, X., Wang, J., and Yao, H. (2005). Feruloyl oligosaccharides stimulate the growth of *Bifidobacterium bifidum*. Anaerobe *11*, 225–229.

Zanoni, S., Pompei, A., Cordisco, L., Amaretti, A., Rossi, M., and Matteuzzi, D. (2008). Growth kinetics on oligo- and polysaccharides and promising features of three antioxidative potential probiotic strains. J. Appl. Microbiol. *105*, 1266–1276.

Zhang, L., Zuo, Z., and Lin, G. (2007). Intestinal and hepatic glucuronidation of flavonoids. Mol. Pharm. *4*, 833–845.

Zhang, X.B., and Ohta, Y. (1991). Binding of mutagens by fractions of the cell wall skeleton of lactic acid bacteria on mutagens. J. Dairy Sci. *74*, 1477–1481.

Zsivkovits, M., Fekadu, K., Sontag, G., Nabinger, U., Huber, W.W., Kundi, M., Chakraborty, A., Foissy, H., and Knasmüller, S. (2003). Prevention of heterocyclic amine-induced DNA damage in colon and liver of rats by different lactobacillus strains. Carcinogenesis *24*, 1913–1918.

Antimicrobial Activity of the Genus *Bifidobacterium*

7

Ismail Fliss, Arthur C. Ouwehand, Ehab Kheadr, Sampo Lahtinen and Stephen J. Davids

Abstract

In view of their relative abundance in the large intestine, their totally inoffensive nature and their reputation as symbionts, bifidobacteria are viewed by many scientists as probable antagonists of many would-be colonizers of the large intestine. While science cannot at this point confirm that the beneficial effects of probiotics such as bifidobacteria necessarily include specific antimicrobial mechanisms that eliminate pathogens or prevent other intestinal bacteria from reaching numbers that lead to health problems, quite a few high-quality studies show how bifidobacteria might have such effects. Among the most likely antagonistic mechanisms, prodigious production of lactic and acetic acids, which in addition to lowering the ambient pH may act as metabolic inhibitors, should be considered the most significant. Another plausible mechanism is competition for adhesion sites on the intestinal epithelium. There is evidence that bifidobacteria secrete molecules of large molecular mass with the specific function of interfering with the attachment of other organisms. Other mechanisms, such as hydrogen peroxide or bacteriocin production are more speculative. The effectiveness of bifidobacteria as antagonists of pathogens or as competitors for epithelial adhesion sites varies considerably and appears to be strain-dependent.

Introduction

The intestinal microbiota of humans is a complex mixture of microbial organisms, mostly bacteria, comprising at least 80 genera and around 1800 phylotypes (Rajilić-Stojanović *et al.*, 2009). Their numbers in the large intestine are typically in the range of 10^{13}–10^{14} cells, which corresponds to a wet mass of less than 1 kg, or up to half of the faecal mass. Long considered only as commensal organisms, at least some genera are now being considered as symbionts by increasing numbers of scientists and medical practitioners. Less than 20% of the intestinal flora has been characterized taxonomically to any significant degree. Scientists thus distinguish between intestinal organisms that may be grown on laboratory media, those identified and quantified by nucleic acid amplification and the remaining anonymous majority. There is general agreement that bifidobacteria are major intestinal bacteria, making up as much as 2% of the total microbiota of the colon of an adult human being, although other bacteria, notably *Bacteroides*, *Eubacterium* *Clostridium*, faecal coliforms and faecal cocci make up the majority of the total gut microbiota. Flora metabolism is fermentative, with or without gas production and the vast majority of the organisms are either anaerobic or have limited tolerance to oxygen, although this latter trait is variable, even within species of anaerobes such as *Bifidobacterium*. What is known for sure is that perturbations of intestinal floral composition, due either to infection or to antibiotic therapy, generally results in diarrhoea and hence exposure of the subject to additional and often more dangerous infections (e.g. *Clostridium difficile*).

The term 'probiotic' was defined at least 20 years ago and refers to live microbial organisms that when fed to an animal are benign and colonize the large intestine, where they may persist for some unspecified but prolonged period of time and in doing so, 'stabilize' perturbed flora or make the intestinal environment refractory to invasion by pathogenic organisms. They thus confer a health benefit to the animal or human consumer of the probiotic. The widely accepted definition is as follows: 'live microbial organisms that confer a health benefit to the host when consumed in adequate amounts'. The vagueness of the expression 'health benefit' is deliberate. It is intended to include not only anti-diarrheal therapeutic effects known for at least a century but also possible prophylactic effects yet to be demonstrated experimentally, such as helping to maintain the intestinal epithelium in a state that is more refractory to cancer development, as well as effects that have been demonstrated but of which the long-term impact is only presumed, such as shortening gastrointestinal transit time (Bouvier *et al.*, 2001). Of course, the definition is intended to include especially the prophylactic benefit of keeping the host safe from enteric pathogens and there is much evidence that probiotics indeed do this.

The science of the intestinal flora and probiotics remains dogged by confusion surrounding at least three questions. One question is whether there is a relationship between the aptitude of a bacterial species to colonize the intestine and its aptitude to confer beneficial effects, in particular protection against pathogens. A second question concerns the relationship between intestinal flora and diet. Does the intestine function as a mixed chemostat (or can it be made to do so) in the sense that a bacterial population can be stabilized and made to persist therein by consuming a diet that provides the optimal mixture of nutrients? Of course, the inclusion of substances called 'prebiotics' in the diet has been proposed for this purpose, these being materials (primarily polysaccharides from bran and other plant fibres) that remain undigested until they reach the large intestine, where they break down and nourish probiotic organisms. A third question concerns the different microenvironments in the large intestine and the distribution of the various genera among them. Based on surface area (less than $1\,m^2$), the number of microbial cells in a confluent monolayer on the mucosal epithelium could not exceed 10^{11}, which is just one-hundredth to one-thousandth of the total microbial population of the intestine. Large numbers of them could also reside for long times in crevices and be involved in processes entirely different from those of either surface-attached or freely suspended genera. In spite of making up only a small fraction of the total flora, surface-associated bacteria might be by far the most significant in terms of host health and consist of a relatively small number of species. However, scientists appear increasingly partial to the idea that the intestinal flora is primarily a bio-film.

The persistence of an organism in the large intestine is not a criterion for meeting the definition of probiotic. In fact, probiotics owe their commercial viability to frequent consumption by healthy persons whose intestinal flora is normal and presumably stable, that is, well established, and perhaps even forming a bio-film. In the case of the therapeutic use of a probiotic following a perturbation of the intestinal flora, for example by antibiotic therapy, it is quite likely that the probiotic strain establishes itself and persists as a component of the flora for considerable time. It is generally acknowledged that the microbes that have become the most commercially successful probiotics (*Saccharomyces boulardii*, strains of *Lactobacillus* and especially *Bifidobacterium*) do not necessarily become permanent intestinal residents, making it necessary to ingest them on a regular basis as additives in food products (primarily dairy) if one hopes to enjoy the prophylactic benefits attributed to them. The ability of probiotic organisms to colonize the intestine for a long enough time to confer a benefit (e.g. protection against pathogens) varies considerably among strains within a specific species.

The factors involved in the success of an organism as an intestinal colonizer are associated primarily with nutrition, that is, the ability to utilize the available carbon sources faster than the competitors. These sources are generally 'low grade', that is, pentose sugars derived from plant fibre, peptides and other undigested nitrogenous compounds (e.g. Maillard reaction products), undigested oligosaccharides and disaccharides (melibiose and other fructosides, α-galactosides, fructans and β-glucans) and short-chain fatty acids, which are also significant carbon sources for intestinal epithelial cells. The material passing through the intestine normally contains very little 'easy substrate' such as glucose, fructose, sucrose or lactose. Successful utilization of substrates generally results in rapid production of organic acids and therefore inhibition of many other bacteria. There is currently no indication that any probiotic bacteria produce significant amounts of neutral compounds (acetone, diacetyl, butanol, etc.) instead of acids.

Probiotics appear to exert their protective effects by at least five mechanisms:

- They produce lactic and acetic acids, which are inhibitory to many undesirable bacteria.
- They utilize the available carbon sources faster than the undesirable competitors.
- They attach to the intestinal wall and thereby prevent pathogens from doing so and possibly displace attached pathogens.
- They appear to produce and excrete bacteriocins or bacteriocin-like substances, which exert more specific and more potent antibacterial effects than organic acids.
- They are believed to produce or shed substances that bring about local stimulation of the immune system, making it more responsive to invasion by pathogens.

There is no consensus concerning the relative significance of the postulated mechanisms. Bacteriocins are polypeptides of molecular mass less than 10 kDa, generally hydrophobic but water-soluble. They are highly disruptive to cell walls and membranes of bacteria taxonomically related to the producer strain. Bacteriocin-like inhibitory substances or BLIS are unidentified compounds that might be small bacteriocins or possibly low-molecular-mass compounds with surfactant properties (e.g. small peptides linked to fatty acids, small glycopeptides or possibly glycolipids or even oligosaccharides). They might also be large molecules that mask pathogen-binding sites on the intestinal epithelium without having actual toxic effects against pathogens (Fujiwara *et al.*, 1997), although this type of mechanism would be very difficult to observe *in vitro*. This masking effect would seem to be a more efficient mechanism, since it is difficult to imagine large molecules in bulk solution not being neutralized or interfered with in such an environment. Since epithelial receptor sites are stationary and fixed in number, it is not difficult to imagine these being masked by substances produced in small molar quantities by bacteria in close proximity. The popular commercial probiotic *S. boulardii* apparently does not produce any bacteriocin and production has been confirmed for just a single bifidobacterial isolate (Yildirim *et al.*, 1999). However, there is considerable evidence that many bifidobacteria do produce BLIS.

The antibacterial activity of bifidobacteria *in vitro* and *in vivo* is well established. It has been known for some time that bifidobacteria have inhibitory effects against several intestinal pathogens, including *Escherichia coli* O157:H7, *Salmonella enterica* serovar Typhimurium (*Salmonella typhimurium*), *Shigella dysenteriae*, *Staphylococcus aureus*, *Candida albicans* and *Clostridium difficile* (Kanbe, 1991; Orrhage *et al.*, 1994; Yusof *et al.*, 2000). Some reports attribute the antibacterial activity of bifidobacteria to their acid production (Ibrahim and Bezkorovainy, 1993), while other authors believe that bifidobacteria also produce antimicrobial compounds of some type other than organic acid (Fujiwara *et al.*, 1997; Gibson and Wang, 1994; Meghrous *et al.*, 1990). More recently, Liévin *et al.* (2000) screened 14 human bifidobacterial isolates from infant stools for their antimicrobial activity and found two that appeared antagonistic to *S. typhimurium*, which was attributed to the production of lipophilic compounds with a molecular mass below 3500 Daltons. Also, Touré *et al.* (2003) examined the ability of 34 bifidobacterial isolates from newborn stools for the production of inhibitory substances and reported the ability of six isolates to produce heat-stable proteinaceous compounds active mainly against *Listeria monocytogenes*.

Today, most scientists consider bifidobacteria to be among the most effective probiotic organisms. The reasons for this status are more theoretical and empirical than factual. Their presence as members of the intestinal flora has been associated with better health status, especially in infants, in reports dating back a century (Tissier, 1900; Drasar and Roberts, 1989), while their relative numbers in healthy adults also suggest a significant role. Although mechanisms explaining how these bacteria might provide their benefits have been proposed, neither direct effects on host physiology nor inhibition of potential pathogens have been proven in controlled experiments with humans. Certain mechanisms, such as excretion of hydrogen peroxide or the production of bacteriocins or BLIS, remain highly speculative. It is also admitted that the immune 'alertness' effect would be short-term and applicable to specific situations only, such as travel or declared disease outbreak.

Production of antimicrobial substances

Organic acids

Fermentation of carbohydrates by lactic acid bacteria (LAB) results in the accumulation of organic acids and reduction of the pH of the medium. LAB and related organisms can produce large quantities of a variety of acids, with some species producing two or more acids. The type and quantity of acid(s) produced depend largely on species, culture composition and growth conditions. Although lactic acid is the main end product of carbohydrate fermentation by these bacteria, some genera produce mixtures of lactic acid and acetic acid. The antimicrobial activity of a particular organic acid is attributed to the reduction in pH as well as activities of the undissociated form of the molecule. These activities can exert deleterious effects on bacterial cell function in a synergistic manner. The drop in the pH of the medium forces cells to tolerate acidification of the cytoplasm or expend energy reversing this effect, while the un-dissociated acid, being soluble in lipids, can diffuse passively across the cell membrane and interfere with normal metabolism.

Lactic acid has been shown to possess antibacterial activity against spore-forming organisms but little or no antifungal activity. The L-isomer is more potent than the D-isomer in this respect (Benthin and Villadsen, 1985). Acetic and propionic acids are generally more effective than lactic acid as antimicrobial agents, since their higher pKa values gives them the possibility of diffusing across the cell membrane at higher pH. Both have been used widely and extensively for decades as antimicrobial additives in manufactured foodstuffs. Both are added to hay to suppress the growth of fungi. It has been shown that propionic acid alone is more active than acetic acid at inhibiting fungi and more effective at inhibiting fungi than yeasts, while bacteria are inhibited to a lesser degree. It is still not certain how acetic acid exerts its effect, especially at pH > 5, at which it is dissociated. Nevertheless, 25 g/l of sodium acetate included in the Norris medium has been credited with exerting a selective effect for bifidobacteria.

Bifidobacteria produce lactic and acetic acids in large amounts, that is, larger than the amounts secreted by lactococci or lactobacilli, even though the latter are known to be very acid-tolerant. Talwalkar and Kailasapathy (2003) provide acid production data for probiotic lactobacilli and four species of oxygen-tolerating bifidobacteria. It has long been known that bifidobacteria differ from lactobacilli and other lactic acid bacteria in using the enzymes fructose-6-phosphate phosphoketolase, transketolase and transaldolase to ferment sugars by means of a distinct pathway that is neither Embden-Meyerhoff glycolysis nor hexose monophosphate shunt. The result is the channelling of all sugars through the splitting of pentoses to lactic and acetic acids, forming two lactic acid and three acetic acid molecules from two hexose molecules, without gas production. This molar ratio is only theoretical, since strain-specific factors and culture conditions all contribute to decreasing the amount of acetate or lactate. For example, some bifidobacteria possess an enzymatic system for splitting the two-carbon fragment to formic acid and CO_2 instead of excreting it as acetic acid, while the presence of oxygen causes the diversion of NADH and thus decreases the reduction of pyruvate to lactate (see the following section). Maklas and De Vuyst (2006) obtained more acetate, even when formic acid was produced, while Talwalkar and Kailasapathy (2003) obtained more lactate under anaerobic conditions and more acetate under atmospheric conditions, to degrees that depended on the bifidobacterial strain.

The effects of the acids, in particular acetic acid, have long been considered one of the pillars of the antagonistic, and hence probiotic, effect of bifidobacteria. Of course, some effect owing to lowered pH is expected. However, in an environment that also contains propionic acid, formic acid, butyric acid and hydrogen, attributing significant inhibitory effects to acetate, one of two fermentation products produced by just one-hundredth of the total microbial flora, is perplexing to say the least. At pH 4.5, acetic acid might be expected to penetrate cells more easily than lactic acid, since a significant proportion of it would be in an uncharged state, in addition to being a smaller molecule. However, no clear relationship has been found between the ratio of acetic to lactic acids produced and the antagonistic power of bifidobacterial strains. Furthermore, it must not be overlooked that as short-chain fatty acids, acetic, propionic and butyric acids are all absorbed, along with salts and water, by intestinal epithelial cells. It should also be remembered that there are various ways for the intestinal flora to utilize acetate, including methane formation.

By the 1990s, most articles were attributing the effects of organic acids in general to lowering of the pH. Currently exploited probiotic lactobacilli appear to produce as much acid as probiotic bifidobacteria and the strongest acid producers in the latter genus (e.g. *Bifidobacterium animalis* subsp. *lactis* Bb12) produce mostly lactic acid (Hütt *et al.*, 2006). One study that examined the question of whether the pH effect alone could explain the inhibitory nature of bifidobacteria (Gibson and Wang, 1994) reached the conclusion that some other secreted factor or factors must be involved, prompting the search for bifidobacterial bacteriocins and BLIS (see below). A later study (Makras and De Vuyst, 2006) showed clearly that adjusting the pH of culture medium to 4.5 with lactic and acetic acids inhibited the growth of *E. coli* C1845 and *S. typhimurium* more than the same adjustment using hydrochloric acid, providing support for the idea that organic acids inhibit by some mechanism that is independent of pH. However, the inhibition of these pathogens and others examined in this study depended very much on the type of test. Culture supernatants from 37 bifidobacterial strains had no effect on any of 17 test organisms including various lactobacilli, innocuous intestinal bacteria and a few pathogens, while colonies on agar inhibited at least the *E. coli* and *S. typhimurium* test strains, although there was no correlation between inhibition and the acid-producing capability of the bifidobacterial strain. This is surprising in view of the results obtained with diluted culture supernatant in a study of *Clostridium perfringens* and *E. coli* inhibition by *Bifidobacterium infantis* (Gibson and Wang, 1994). This strong inhibitor was isolated from infants found resistant to diarrheal infections and demonstrates (in comparison with the results obtained by Makras and De Vuyst) how much inhibitory power varies among bifidobacteria.

Finally, the intestine is not a very acidic environment, owing to the rapid absorption of the organic acids produced. At least one interesting study has pointed out that the most useful effect of lactic acid in the large intestine might not be pH reduction but rather as a precursor for butyric acid production by other helpful bacteria (Duncan *et al.*, 2004). The value of butyric acid as a nutrient or anti-inflammatory factor for intestinal epithelial cells has long been widely acknowledged (Cummings, 1995).

Hydrogen peroxide

One antagonistic mechanism cited repeatedly in the literature is the production, and presumably release, of hydrogen peroxide in quantities sufficient to inhibit other bacteria. There have been several studies over the years devoted to understanding how the supposedly anaerobic bifidobacteria tolerate oxygen (Poupard *et al.*, 1973; Shimamura *et al.*, 1992; Talwalkar and Kailasapathy, 2003), as several species do. Being a highly reactive molecule, oxygen will inevitably inflict irreversible damage on essential enzymes in any organism not equipped to deal with it as quickly as it penetrates the cell. Of course in aerobic organisms, oxygen is used as the terminal electron acceptor in the complete oxidation of sugars or fatty acids for energy production. In organisms lacking this ability, the highly reactive species known as the superoxide radical (O_2^-) and hydrogen peroxide (H_2O_2) are produced as oxygen is reduced with respectively one or two electrons in reactions with unsaturated fatty acids and other reduced compounds. In order to tolerate oxygen, the organism must deal with these reactive species as they form, which they do to some extent even in aerobic organisms.

In both aerobic and air-tolerating non-aerobic organisms, the enzyme superoxide dismutase catalyses the reduction of superoxide to hydrogen peroxide, while catalase reduces the latter to water and oxygen. Organisms not equipped with these two enzymatic activities have only one other known option: spending reduced nicotinamde adenine dinucleotide or NADH. In this case, the reduction of oxygen to hydrogen peroxide is catalysed by NADH oxidase, followed by the reduction of the hydrogen peroxide to water, catalysed by NADH-dependent peroxidase. Other reduced substances in smaller supply (e.g. flavoproteins) might contribute to reducing oxygen and other enzymes such as amino acid oxidases or xanthine oxidase could also participate. One consequence (not necessarily bad) of this diversion of NADH is making pyruvic acid (the penultimate product of sugar catabolism) available for use other than excretion as lactic acid (e.g. slime or capsule formation).

It has been shown that in at least four species of bifidobacteria, the activities of NADH oxidase and peroxidase are so close to balancing each other that no accumulation or excretion of hydrogen peroxide is likely and that these organisms also have some superoxide dismutase activity (Shimamura *et al.*, 1992). It has also been shown that some lactobacilli have the ability to produce hydrogen peroxide when exposed to ambient air, even after anaerobic growth (Eschenbach *et al.*, 1989). However, no study published so far indicates the amount of oxygen needed by lactobacilli or any other intestinal bacteria in order to produce hydrogen peroxide in amounts inhibitory to specific organisms. Reports of actual antagonism with hydrogen peroxide go back decades (Thompson and Johnston, 1950; Dahiya and Speck, 1968), but do not refer to the intestinal environment, in which oxygen should normally be rare. The most likely target would of course be *Clostridium*, since just about any other pathogen would be equipped at least as well as *Bifidobacterium* to deal with hydrogen peroxide. However, the oxygen required to make hydrogen peroxide would by itself be sufficient to prevent the growth of clostridia or any other strict anaerobe.

If a catalase-negative organism such as *Bifidobacterium* were to possess the ability to produce hydrogen peroxide in amounts sufficient to inhibit other organisms without killing itself, one would expect the following minimal requirements: strong NADH oxidase and possibly amino acid oxidase activities; strong superoxide dismutase activity; weak peroxidase activity; concentration of the oxidase and dismutase activities near the cell wall, ideally in the cell membrane and with directional operation in order to function as a reverse permease (or alternatively a membrane equipped with a dedicated H_2O_2 reverse permease). However, we have not found any reports of these activities.

Bacteriocins and bacteriocin-like inhibitory substances

Compared with other genera currently used as probiotic organisms, in particular the lactobacilli, very little is known about the production of bacteriocins by *Bifidobacterium*. Even quite recently, studies of the inhibitory powers of bifidobacteria were being published without a mention of the word 'bacteriocin'. One of these studies in fact appeared more concerned with the antioxidant power of the probiotic (Hütt *et al.*, 2006), a concept that seems as opportunistic as scientific. Why expect bifidobacteria to produce bacteriocins to begin with? Could they not do everything they do without this capability? The answer is probably yes, since they could establish themselves and be competitive by producing organic acids and other antimicrobial substances. In describing the role of terminal metabolic products such as acetic, lactic and possibly formic acids in the inhibition of pathogens, De Vuyst *et al.* (2006), also mention lipophilic substances that are not bacteriocins or even peptides, based on insensitivity to proteinase K. The culture supernatant of at least one bifidobacterial strain contained such an agent and decreased *E. coli* viability after 18-h exposure.

In spite of the absence of experimental proof, bifidobacteria are widely believed to produce at least six categories of inhibitory substance other than small terminal metabolites or other non-peptide compounds. These are true bacteriocins (called bifidin, bifilong and bifidocin), siderophores (iron sequestrants, see below), lectin-like bacteriocins (believed involved in binding of biopolymers, intestinal bio-film formation, co-function with cytokines, pathogen antagonism), S-layer adhesin-like and lectin-like proteins, unspecified inhibitors of pathogen adhesion and BLIS. While reports of complex substances that either inhibit pathogens or block their adhesion sites are numerous, only bifidocin has been purified and characterized in detail.

The available descriptions of BLIS produced by bifidobacteria indicate protein-like composition, and molecular masses are sometimes suggested. The term BLIS thus seems to designate bacteriocins waiting for confirmation of existence, rather than a distinct chemical class. Another criterion suggested for the definition of BLIS is broader spectrum of activity than bacteriocins, although this characteristic might not turn out to be the case once some of them have actually been purified and chemically characterized.

Among the inhibitory substances produced by LAB, bacteriocins fuel the most hope because of their potency and high specificity for targeted strains. Bacteriocins are a heterogeneous group of antibacterial peptides and proteins that vary considerably in their spectrum of activity, mode of action, molecular mass, genetic origin and biochemical properties (Abee *et al.*, 1995). The mode of action of bacteriocins usually involves creating holes in the membrane of the target cell and causing leakage of ions or decreasing the proton-motive force across the membrane, and thus interfering with ATP generation. This is done at specific sites, adjacent to specific membrane-embedded molecules that act as unwitting receptors. Organisms not affected by the bacteriocin thus either lack the receptor or are equipped with some other factor that interferes with the bacteriocin–receptor association. This implies that the bacteriocin must be a small molecule capable of slithering through the cell envelope or perhaps slipping through during cell wall synthesis.

Bacteriocins produced by LAB are inhibitory only to closely related Gram-positive bacterial species, while Gram-negative bacteria are usually less or not at all susceptible to them. The spectrum of activity of bacteriocins does not adhere to any known principle, thus raising the question of how many species of any given genus of bacteriocin-producing intestinal bacteria make up the flora in a given host. The expression 'closely related' in the characterization bacteriocins as, 'substances that target bacteria closely related to the bacteriocin-producing strain' means in some cases strains of the same species. For example, the colicin produced by *E. coli* V identified by Gratia (1925) attacks *E. coli* S, and lactococcins A, B and M kill only species of *Lactococcus*. In contrast, nisin A, produced by *Lactococcus lactis*, kills a wide range of Gram-positive bacteria, such as *Actinomyces, Bacillus, Clostridium, Corynebacterium, Enterococcus, Gardnerella, Lactococcus, Listeria, Micrococcus, Mycobacterium, Propionibacterium, Streptococcus* and *Staphylococcus* (Mota-Meira *et al.*, 2000, 2005). As biologists, our training makes us assume that the spectrum of activity must be determined by the nature of the competitors of the producing strain. However, what would be gained in the biological sense, from the elimination of a nearly identical competitor?

It has been speculated that almost all bacterial species produce some kind of bacteriocin or BLIS waiting to be discovered. Indeed, bacteriocin-like operons have been identified in most sequenced genomes. Our understanding of the biology underlying bacteriocin production by bacterial inhabitants of animal intestines has been gained mostly by studying the production of colicins and microcins by Gram-negative bacteria. When we try to understand a biological phenomenon with manifestations as diverse as for bacteriocin production, we must remember that we are looking at only a snapshot of a long-term evolutionary process. Although they have been called virulence factors, the purpose of colicins is not to make the host more ill. The opposite might be true. From the perspective of the bacterial community in an intestine, it is not advantageous to end up as sewage or inside a rotting corpse. In order to expand its existence and therefore become 'biologically successful', a microbe must not prevent its host from interacting with other hosts. Of course, some pathogens do just that by producing potent toxins that immobilize the host with serious illness. Biologists regard these organisms as primitive and inept 'newcomers' to the ecosystem. Using anthropocentric reasoning, a healthy intestinal flora could experience such pathogens as invaders that ruin conditions for the entire community and respond by elaborating bacteriocins to deal with them. In some cases, the response amounts to a 'you're either with us or against us' approach. For example, it has been demonstrated that microcin-producing *E. coli* F-18 kills a strain differing only by the inability to produce this bacteriocin (McCormick, 1989). This would make the bacteriocins of the intestinal flora a record of past and occasionally current intestinal turf wars. In any case, it remains a fact that bacteriocins produced by bifidobacteria remain elusive, even though in the colon these bacteria outnumber known producers such as lactobacilli by at least one hundred fold.

The bacteriocin bifidin, produced by *Bifidobacterium bifidum* NCDO 1452, was identified in 1984 but not characterized further (Anand *et al.*, 1984). A second bacteriocin, identified by Yildirim and Johnson (1998) and called bifidocin B, is produced by *B. bifidum* NCFB 1454. This bacteriocin was purified and partially characterized by Yildirim *et al.* (1999). The lack of knowledge about the bacteriocin production capability of this important probiotic bacterial genus is being addressed through research, but results are emerging very slowly, especially in the area of purification and characterization.

Learning to control the production of bacteriocins by bifidobacteria would provide another tool for inhibiting pathogens and contribute to better understanding of bacterial competition in the intestinal ecosystem. Screening new isolates for bacteriocin production, purifying and characterizing bacteriocins and identifying their mode of action should lead to the formulation of new probiotic products with improved pathogen-inhibiting capabilities. We have isolated an infant-derived strain with specific activity against *L. monocytogenes* (Touré *et al.*, 2003; Gagnon *et al.*, 2004; Von Ah *et al.*, 2008), a pathogen responsible for serious food-borne illness worldwide, especially menacing to persons with compromised immune systems, children, pregnant women and the elderly. Strains such as this or its bacteriocin could be useful for inhibiting this pathogen and possibly provide an adjunct or alternative to antibiotic therapy, especially in cases of antibiotic resistance.

Bacteriocins produced by lactic acid bacteria and related organisms in general potentially represent significant economic value and public health interest. Since the organisms usually have GRAS (generally recognized as safe) status, the bacteriocins should be acceptable as preservatives and regarded as natural, particularly in foods to which such organisms have been added to bring about fermentation. Nisin, produced by *Lactococcus lactis* subsp. *lactis* and having a broad range of inhibitory action, is the only bacteriocin recognized by the World Health Organization as a preservative in food (Bennik *et al.*, 1997) and is currently used as such in more than 50 countries.

The need for novel bacteriocins and antimicrobial compounds

The production of bacteriocins seems quite separate from probiotic status. Many bacteriocins with potent activities of potential commercial value have been identified in association with GRAS organisms, but so far no regulatory agency has put the approval of new bacteriocins for use as food additives on the fast track. Since reports of bacteriocin production by bifidobacteria are very few in number and no bifidobacterial bacteriocin with astonishing activity has been discovered, it seems reasonable to ask why this genus should be screened for this characteristic. Characterizing the bacteriocin production of the intestinal flora should shed light on how the flora performs its beneficial functions in human and animal hosts, but it makes more sense to focus first on known producers of bacteriocins, such as the lactobacilli, of which the intestinal flora in humans contains close to 60 known species.

Antibiotic therapy is often unsuccessful, perturbs the protective intestinal flora and increases the prevalence of antibiotic-resistant microbial strains (Bengmark, 1998). It is feared that the slow rate at which new effective antibiotics are being developed will not keep pace with the acquisition of resistance by pathogens.

The increased prevalence of multi-resistant bacterial pathogens has stimulated the search for alternative chemotherapeutic agents. Bacteriocins show promise for meeting the challenge of multi-resistant Gram-positive bacteria, since resistance to these is developing more slowly than anticipated (Severina *et al.*, 1998). In a recent study by Mota-Meira *et al.* (2000), nisin A and mutacin B-NY 266 showed stronger inhibitory activity than vancomycin and oxacillin against various pathogenic bacteria *in vitro*. Mutacin remained active against strains that were resistant to nisin, vancomycin and oxacillin. Giacometti *et al.*, (1999a) studied *in vitro* activities of polycationic peptides, nisin, magainin and ranalexin alone or in combination with other pharmceutical antibacterial agents (ampicillin, ceftriaxone, rifabutin, rifampin, azithromycin, clarithromycin and vancomycin) against six clinical isolates of

Rhodococcus equi. All isolates were more susceptible to nisin. A positive (synergistic) interaction was observed when cationic peptides were combined with any of the antibiotics. In a similar study, Giacometti *et al.* (1999b) reached a similar conclusion regarding nisin and its synergy with antibiotics against 40 clinical isolates of *Pseudomonas aeruginosa*.

In spite of the slowness with which bacteriocins in general are being approved for use as food additives, they are viewed as adding value to probiotics. Scientists would argue that most bacteria, including those used in food processing and fermentations for centuries, naturally produce antimicrobial peptides or other complex molecules fitting the definition of either bacteriocins or BLIS and that these substances are therefore consumed regularly, making it unnecessary for regulatory authorities to be especially cautious. Nevertheless, it is a fundamental principle of all regulatory agencies to allow the use of agents that have withstood the test of time, while withholding approval for novel agents, no matter how similar to existing ones, until sufficient proof of safety has accumulated. This means that although some bacteriocins or BLIS may have snuck through the system into our foods along with the bifidobacteria or lactobacilli that produce them, it would be difficult to market for human consumption any novel probiotic *selected* for its bacteriocin-producing ability, particularly if this ability were acquired or enhanced through the intervention of scientists.

If bacteriocins affect primarily bacteria that are closely related to the organism that produces them, how exactly are bacteriocins of probiotic organisms such as bifidobacteria supposed to be beneficial? To ask this question is to emphasize that the effect of bacteriocins produced by proposed probiotics on other desirable flora in the host must be taken into consideration right from the beginning of the strain development project. Furthermore, it is clear that only some types of bacteriocins have the physicochemical stability making them suitable as potential antimicrobial agents in foods. Bacteriocins as a general category include large heat-labile proteins and complex molecules of mixed chemistry with unpredictable behaviour in the digestive system. Who wants to be on the hook for a bacteriocin that gets digested to a peptide that causes a reaction analogous to the coeliac illness? It is generally agreed that the bacteriocins with potential commercial value should be small, stable peptides (< 10,000 Da). It is also admitted that bifidobacteria are of no special interest as a source of bacteriocins to be used alone as food additives. Interest in the possible bacteriocins or BLIS of this genus has more to do with understanding how these bacteria function as probiotics and the marketing opportunities this might represent.

The veterinary and medical use of bacteriocins and/or the bacteria that produce them as adjuncts to antimicrobial chemotherapy and as bio-preservatives in feeds and foods represents a promising strategy for reducing antibiotic use and the spread of antibiotic resistance. The continued discovery and characterization of bacteriocins produced by lactic acid bacteria in general is thus a valid and worthwhile research endeavour. In this context we should mention the work of Corr *et al.* (2007) who have clearly shown that *Lactobacillus salivarius* UCC118, a probiotic strain of human origin, produces a bacteriocin *in vivo* that can provide significant protection of mice against infection with the invasive food borne pathogen *Listeria monocytogenes*. This result was recently confirmed by Darbour *et al.* (2009) who showed the effectiveness of pediocin PA-1 and *Pediococcus acidilactici* UL5 at inhibiting *Listeria monocytogenes in vivo*.

Siderophores

The term siderophore in the probiotic context refers to any substance secreted by microbial cells for the purpose of sequestering iron. Iron is essential for the growth of most organisms including bacteria and in many situations is a growth-limiting nutrient. It is poorly soluble at neutral pH and often biologically unavailable in intestinal contents, tied up with phytic or oxalic acids. Nevertheless, iron deficiency is rare in meat eaters and animals have proteins in their blood and tissue fluid for the purpose of sequestering iron and thereby preventing bacterial invaders from using it. It is therefore not surprising that some bacteria have responded by elaborating compounds that enable them to grab the iron back. Bacterial siderophores may be peptides, carbohydrates or combinations thereof and need only include any structure that places at least two anionic groups (e.g. carboxyl or phosphate) in spatial proximity suitable for capturing the iron cation. Colonies of siderophore-producing strains can be selected by culture on iron-poor agar-based media. In the case of bifidobacteria, this was achieved fairly recently and is described in US patent number 6746672. It is hypothesized that siderophore production would be a considerable asset to a probiotic microbe attempting to compete as a member of the intestinal flora, as long as iron was in short supply or in less readily available forms. This condition of course would be more easily met in individuals that are on a diet that is poor or deficient in red meat but rich in leafy vegetables and legumes (and thus containing high levels of oxalic and phytic acids).

Competition for attachment sites

Competition with pathogens for attachment to and colonization of mucosal adhesion sites is a major criterion for the selection of bifidobacterial strains for probiotic applications. This factor is considered significant for the ability of bifidobacteria to implant at least temporarily among the intestinal flora and exert their effects rather than be removed quickly by normal gut contraction. Following adhesion and colonization, bifidobacteria may exert immune system modulating effects (Schiffrin *et al.*, 1997) and/or prevent the attachment of pathogens to the intestinal surface (Resta-Lenert and Barrett, 2003).

The capacity of bifidobacterial strains to adhere to intestinal cells and to compete with enteric pathogens for adhesion sites is usually evaluated *in vitro* using models consisting of cultured mono-layers of human intestinal cells. The human colon adenocarcinoma-based cell lines Caco-2 and HT-29 are the models most widely used to evaluate the ability of bifidobacteria to adhere to the intestinal wall. Although both show structural and functional features of mature human enterocytes, HT-29 cells provide the additional advantage of maintaining a constant proliferation rate with practically no differentiation when cultured in the presence of glucose (Pinto *et al.*, 1983; Zweibaum *et al.*, 1985; Servin, 2004).

Bifidobacteria adhere very well to intestinal cells, enabling them to colonize the large intestine while increasing their competitiveness against pathogenic organisms. Numerous *in vitro* studies have reported the effectiveness of bifidobacteria at inhibiting the adhesion of many enteric pathogens to Caco-2 and HT-29 cells, including *E. coli* O157:H7, *S. typhimurium*, *L. monocytogenes*, *Enterobacter sakazakii* and *C. difficile* (Moroni *et al.*, 2006; Collado *et al.*, 2005; Gagnon *et al.*, 2004; Gopal *et al.*, 2001; Liévin *et al.*, 2000). The study of Bernet *et al.*, (1993) was the first to show the superior ability of human isolates of bifidobacteria to adhere to Caco-2 cells compared to pathogenic organisms such as enteropathogenic *E. coli* and *S. typhimurium*. They suggested that the ability of bifidobacteria to inhibit the adhesion of pathogens to Caco-2 cells was due to bio-film formation and to competition for specific adhesion sites. Several non-biofilm mechanisms have been proposed to explain the ability of bifidobacteria to inhibit the adhesion of pathogens to intestinal cells and reduce the incidence of infection, including steric obstruction, specific blocking of cellular receptor sites and competition for adhesion sites. Bifidobacteria could inhibit pathogens by one or more of these mechanisms, although none has been fully characterized yet.

Finally, the ability of bifidobacteria to reduce the incidence of infectious diseases might also be due to interference with the invasion of intestinal cells, a critical step in bacterial pathogenesis (Henriksson and Conway, 2001; Silva *et al.*, 2004; Gagnon *et al.*, 2006; Mahoney and Henriksson, 2003; Podoprigora *et al.*, 1999).

Bifidobacteria have also been shown effective in reducing, preventing and treating intestinal infections in humans. In children, *B. bifidum* reduced the incidence of diarrhoea associated with rotavirus infection (Saavedra *et al.*, 1994). Bifidobacteria are also effective in preventing antibiotic-associated diarrhoea (Fooks and Gibson, 2002) and travellers' diarrhoea (Black *et al.*, 1989).

Since intestinal infection is initiated by the adhesion of pathogen to intestinal epithelial cells, blocking or occupying adhesion sites is generally regarded as the most practical first line of defence (Gill, 2003; Servin and Coconnier, 2003; Servin, 2004). Consequently, the identification of the adhesion sites on intestinal cells, production of adhesive elements by probiotics and characterization of probiotic strain cell wall composition have been the subjects of many studies in recent years. It is believed that once the mechanisms underlying the attachment of bacterial cells to intestinal cells are understood, there will be a rational basis for selecting and designing better probiotics.

Adhesive elements of macromolecules of various natures have been identified and found to be involved in bacterial adhesion to intestinal cells. Proteinaceous adhesive factors, carbohydrates, and lipoteichoic acids found on the cell wall of various lactic acid bacteria have been identified (Granato *et al.*, 1999; Greene and Klaenhammer, 1994; Morata de Ambrosini *et al.*, 1999; Ventura *et al.*, 2002). Similar adhesive elements involved in the adherence of bifidobacteria to intestinal cells have also been identified (Op den Camp *et al.*, 1985; Bernet *et al.*, 1993; Bibilonni *et al.*, 1999; Guglielmetti *et al.*, 2008). The production of a proteinaceous component found in *Bifidobacterium breve* 4 cultures has been proposed as a factor of adhesion to human intestinal cells (Bernet *et al.*, 1993). These authors observed that *B. breve* 4 adhesion was decreased by only 50% when the supernatant was washed away, suggesting that the component was strongly attached to the bacterial cell surface. This suggests that labile surface-associated adhesin-like protein(s), which appear to be species specific, play a role in the adhesion of bifidobacteria to intestinal cells. In another study, Bibilonni *et al.* (1999) suggested that cell-wall-attached glycoproteins or carbohydrate chains might be responsible for the adhesion of bifidobacteria to Caco-2 cells. More recently, Guglielmetti *et al.* (2008) identified a cell surface protein called BopA as an element responsible for *B. bifidum* MIMBb75 adhesion to Caco-2 cells. In silico analysis of its gene and globomycin experiments indicated that BopA is a cysteine-anchored lipoprotein, which is expressed as a precursor polypeptide. On the other hand, some studies have shown the importance of carbohydrate receptors on intestinal cells for the attachment of bifidobacteria (Fujiwara *et al.*, 2001; Servin, 2004). These studies suggested that reduced attachment of enteric pathogens to intestinal cells in the presence of bifidobacteria could be attributed to the latter being more competitive for carbohydrate sites on intestinal cells.

The ability of probiotics to induce the secretion of mucus (or mucin, a glycoprotein gel present on the surface of the intestinal epithelium) by intestinal cells has been suggested as another mechanism by which probiotics could inhibit the adhesion of pathogens to intestinal cells. As a layer covering the inside surface of the intestine, the main function of mucin is to protect the epithelium from direct contact with the intestinal contents (Juntunen *et al.*, 2001; Mack *et al.*, 1999; Lee *et al.*, 2003; Mattar *et al.*, 2002; Collado *et al.*, 2005). In addition, mucin could act as a site for colonization by intestinal commensals and possibly as an adhesion site for some pathogens (Gill, 2003).

Competitive exclusion of enteric pathogens from the human intestine

It has been shown that acquisition of resistance to bile salts can confer cross-resistance to low pH and induce stable changes in carbohydrate fermentation profiles and some glycosidase activities in bifidobacteria (Margolles *et al.*, 2003; Noriega *et al.*, 2004). Bile-resistant variants are usually characterized by increased ability to adhere to human intestinal mucus and thus may have an enhanced ability to inhibit the adhesion of enteric pathogens by competitive exclusion (Gueimonde *et al.*, 2005; 2007).

In the case of *B. bifidum*, both bile-resistant variants and their parent strains have shown the ability to displace attached cells of pathogens such as *E. coli*, *C. difficile*, *Ent. sakazakii*, *S. typhimurium* and *L. monocytogenes* from human intestinal mucus (Gueimonde *et al.*, 2007). These authors observed displacement efficiencies that varied from 15 to 70% depending on the particular *B. bifidum* strain and the pathogen. The mechanism by which bifidobacteria displaces other cells is not yet known, but having a higher affinity for the attachment surface is a plausible explanation. This ability also appears to be due to a very specific relationship that depends on both the bifidobacterial and pathogenic strains. Furthermore, there is no evidence that the increased ability of a given strain to attach automatically improves its ability to displace other cells (Bibiloni *et al.*, 1999; Collado *et al.*, 2005). In order to gain better understanding of the ability of bifidobacteria to displace pathogens already attached to human intestinal mucus, Gueimonde *et al.* (2007) treated bacterial cells with proteinase K, lipase, metaperiodate or heating at 100°C for 10 min prior to the adhesion assay. Their results indicate the carbohydrate nature of the adhesive elements of tested bifidobacterial strains and also support the hypothesis proposed by Fujiwara *et al.* (2001) and Servin (2004) that bifibidobacteria could exclude enteric pathogens by competing with them for the same carbohydrate receptors in the gut.

The abilities of two *Bifidobacterium* strains with acquired resistance to bile to inhibit adhesion and to displace selected pathogens from human intestinal mucus were assessed and compared with those of their bile-sensitive parent strains. A preliminary characterization of the macromolecules involved in the adhesion was also carried out. Inhibition of adhesion and displacement of pathogens were found to be strain-specific. The cholate-resistant strain *B. bifidum* M6dCo inhibited adhesion and displaced pathogens from mucus significantly more efficiently than its cholate-sensitive parent strain *B. bifidum* M6. Meanwhile, two strains showing similar adhesion to mucus, *B. bifidum* A1 and its bile-resistant derivative *B. bifidum* A1dOx, did not display any differences in such assays. Different cell surface molecules appeared to be involved in the adhesion of strains *B. bifidum* M6 and *B. bifidum* M6dCo. These may explain the differences in competitive exclusion observed between these strains (Gueimonde *et al.*, 2007).

Antimicrobial effects of bifidobacteria in animal studies

The activity of bifidobacteria against a variety of pathogens has been investigated. Examples of effects against *E. coli*, *Salmonella*, *C. albicans*, *Clostridium* and *Helicobacter pylori* in animal studies are described below.

Bifidobacterium metabolites

Bifidobacterium longum MB metabolites have been shown to reduce the number of *E. coli* translocating to mesenteric lymph nodes, but without improving survival (Suzuki *et al.*, 1997). A study of the effects of whey on the intestinal flora of conventional mice showed faecal *Clostridium* suppression to degrees that depended on the strain of *B. longum* used to ferment the milk. One of these, *B. longum* 6, was further tested and found to reduce *Bacillus* spp. and *Bacteroides fragilis* as well and to increase faecal *Bifidobacterium* levels, while levels of other measured microbial groups did not change (Romond *et al.*, 1997a).

Anti-*E. coli* activity

An unidentified *B. longum* strain was found to reduce the translocation of *E. coli* to various organs in mono-associated mice. Nevertheless, this provided only a limited reduction in mortality (Yamazaki *et al.*, 1991). It is not known whether this strain is able to produce antimicrobial components other than primary metabolites.

Another strain of *B. longum* (FPLC 117) was also found to reduce the translocation of *E. coli* to visceral organs in mono-associated mice, but not to affect faecal *E. coli* counts (Romond *et al.*, 1997a). It is not known what specific antimicrobial components this strain might produce either.

Bifidobacterium thermacidophilum RBL71 has been shown to reduce adhesion of *E. coli* O157:H7 to Caco-2 cells and to exert antimicrobial activity against this pathogen *in vitro*. This activity was retained even after adjusting the pH to neutrality in an overlay assay (Gagnon *et al.*, 2004). The weight loss caused by infection of conventional mice with *E. coli* O157:H7 could be reduced by prior colonization with *B. thermacidophilum* RBL71, while post-infection colonization did not have this effect. Faecal shedding of *E. coli* O157:H7 was reduced by prior colonization and also to some extent by feeding strain RBL71 after infection. These changes in *E. coli* colonization were accompanied by an improvement in immune response, as exhibited by higher specific faecal IgA and specific serum IgM titres (Gagnon *et al.*, 2006).

Similarly, antimicrobial activity *in vitro* and reduction of the severity of *E. coli* O157:H7 infection in mice has been shown for *B. animalis* subsp. *lactis* HN019 (Gopal *et al.*, 2001). The latter was accompanied by increased phagocytic activity of blood leucocytes and peritoneal macrophages and increased specific mucosal IgA titres (Shu *et al.*, 2001). The same strain also reduced *E. coli* weanling diarrhoea in piglets, again accompanied by increased blood leucocyte phagocytic capacity and higher specific faecal IgA, IgG and IgM titres (Shu *et al.*, 2001). It has been found to adhere well to tissue culture cell lines such as Caco-2, HT-29 and HT-29MTX and its cell-free culture supernatant has been shown to reduce *E. coli* O157:H7 binding to and invasion of these cell lines.

Anti-salmonella activity

Colonization of germfree or conventional mice with *B. longum* Bb46 and subsequent challenge with *S. typhimurium* significantly improved survival, but did not influence faecal levels of the pathogen. It has been suggested that this strain produces a bacteriocin (Bifilong Bb-46) with activity against pathogens such as *S. aureus*, *S. typhimurium*, *Bacillus cereus* and *E. coli*, although no other details are available (Cheikyoussef *et al.*, 2008). The results suggest that the protective effect observed in mice was not mediated by direct antagonism, but rather by modulation of the immune system (Silva *et al.*, 2004).

In contrast, administration of *B. breve* Yakult to conventional mice reduced faecal shedding of *S. typhimurium* and translocation to the spleen and mesenteric lymph nodes. *Bifidobacterium pseudocatenulatum* DSM 20439 reduced translocation similarly, while *B. bifidum* ATCC 15696 and *Bifidobacterium catenulatum* ATCC 27539 did not (Asahara *et al.*, 2001a). It was suggested that the effect is mediated through production of organic acids and reduction of pH. Total short chain fatty acids and in particular acetic acid were found to be significantly increased in the caecum (Asahara *et al.*, 2001b).

B. animalis AHC7 was found to reduce translocation to the liver and spleen and to exhibit antimicrobial activity against pathogens such as *S. typhimurium*, *E. coli* O157:H45, *L. monocytogenes* and *Listeria inocua*. Although the strain adhered to HT-29 cells, its interference with pathogen adhesion was not tested. Furthermore, results with mice did not reveal effects on faecal shedding or changes in immune function. The mechanism underlying the reduction in symptoms of infection therefore remains unclear (O'Mahony *et al.*, 2009).

Colonization of conventional mice with *B. animalis* subsp. *lactis* HN019 reduced mortality after challenge with *S. typhimurium*. This was associated with reduced translocation to visceral organs and improved immune responses such as increased phagocytic activity and greater serum and intestinal antibody titres (Shu *et al.*, 2000).

Mono-association with either *Bifidobacterium* CA1 or *Bifidobacterium* F9 resulted in significantly lower mortality in mice infected with *S. typhimurium*, compared to the control group. However, it is not clear whether the improved survival was associated with reduced translocation or colonization by *Salmonella* (Liévin *et al.*, 2000). Although the strain produces antimicrobial substances and reduces adhesion (Gopal *et al.*, 2001), it is not known whether these are active against *Salmonella*. The underlying mechanism for the observed protection thus remains unclear.

Anti-*Candida* activity

The protective effect of *B. animalis* subsp. *lactis* Bi-07 against *C. albicans* infection has been investigated in mono-associated mice. The Bi-07 strain was shown to reduce colonization by the yeast in most parts of the gastrointestinal tract. Translocation was also found lower than in control mice, while mortality was reduced in both adult and newborn immune-deficient (bg/bg-nu/nu) mice. These changes were associated with higher levels of serum IgA, IgG and IgM (Wagner *et al.*, 1997). It is not known if Bi-07 produces antimicrobial components. However, it binds well to Caco-2 tissue culture cells and inhibits the adhesion of *S. typhimurium*, *Yersinia enterocolitica* and *E. coli* to these cells. Inhibition of the binding of *Candida* was not tested (Candela *et al.*, 2005).

Anti-*H. pylori* activity

Treatment of mice infected with *H. pylori* and subsequently exposed to *B. animalis* subsp. *lactis* B94 did not result in reduced colonization by the pathogen. Reductions in gastric inflammation, gastric neutrophil infiltration and gastric interleukin (IL)-1b as well as increased gastric IL-10 were nevertheless noted. Serum levels of *H. pylori*-specific IgG were also reduced compared to control mice (Zhang *et al.*, 2008). It is not known whether strain B94 produces substances that are inhibitory to *Helicobacter*.

Anti-*Clostridium* activity

As mentioned above, metabolites of *B. longum* 6 have been found to reduce faecal *Clostridium* levels in mice (Romond *et al.*, 1997b). In healthy dogs, *B. animalis* AHC7 was found to reduce faecal levels of total clostridia and of *C. difficile*. In contrast, *Clostridium perfringens* remained unchanged as did the other microbial groups tested (O'Mahony *et al.*, 2009). Strain AHC7 has not been tested for specific antimicrobial activities against *Clostridium* spp.

From the findings of the research reviewed above one can draw the conclusion that numerous strains of *Bifidobacterium* are able to suppress infections by pathogens in animal models. However, it appears that this anti-infectious activity is in most cases due to host immune modulation rather than to antimicrobial effects, although the latter cannot be ruled out. It should be noted that the bifidobacteria in these studies were tested as single strains and that antimicrobial activities of bifidobacteria in combination with other probiotic bacteria have been suggested in other animal studies.

Inhibition of pathogens by strains of *Bifidobacterium* in clinical studies

Bifidobacteria are believed to bring numerous benefits to human health (Rastall *et al.*, 2005). This has been well documented for specific probiotic strains of *Bifidobacterium*, in particular the species *B. animalis* subsp. *lactis* (Leahy *et al.*, 2005). The beneficial role of endogenous bifidobacteria is less clear. In breast-fed infants, the faecal levels of bifidobacteria are usually higher than that in formula-fed, age-matched infants (Harmsen *et al.*, 2000). This correlates with a lower incidence of certain common illnesses. However, the correlation between health and faecal *Bifidobacterium* level is less clear in adults. A possible mechanism for the proposed beneficial effects of both probiotic and endogenous bifidobacteria is competition with pathogens, as mentioned in the context of animal studies. As in the case of animal studies, it is not feasible to distinguish between antimicrobial activity and other possible mechanisms in the human case. The results presented below should therefore be interpreted as the result of competition and host resistance rather than antimicrobial activity alone.

The indirect way to determine whether bifidobacteria or any microbe exert an antagonistic effect on other microbes in the human host is to measure changes in the composition of the endogenous microbial flora. Such changes may be induced by the consumption of bifidobacteria as probiotics. Alternatively, the levels of endogenous bifidobacteria can be changed in response to either antibiotics or prebiotics. This change may result in further changes in the intestinal floral composition. Effects of bifidobacteria on selected groups of intestinal bacteria, in particular endogenous bifidobacteria and lactobacilli, have been evaluated in many studies. However, studies focused on reducing the numbers of potentially harmful microbes in the human gastrointestinal tract are scarce, owing primarily to ethical constraints. *In vitro* studies of direct inhibitory effects of pure cultures are relatively straightforward to conduct. Animal models are based on experimental infection with a specific dose of the targeted pathogen and allow postmortem assessment of pathogen infectivity and translocation. In human studies, evaluation of anti-pathogen activity is constrained by the initial levels of the targeted pathogen in the subjects. These levels are of course highly variable and usually very low.

Anti-*C. difficile* effects

C. difficile is a major pathogen and has been the object of clinical evaluations of the efficacy of probiotics. In the context of antibiotic-associated diarrhoea, *Lactobacillus* and *S. boulardii* have been evaluated as probiotics in numerous studies, while studies with bifidobacteria are rare. In most cases, bifidobacteria have been used in combination with other probiotics (McFarland *et al.*, 2006). The use of probiotics, including strains of *Bifidobacterium*, to treat or prevent *C. difficile*-associated diarrhoea has proven very promising, although effects specific to certain strains or strain combinations require further study. A few studies have examined a variety of strains superficially, with focus on antibiotic-associated diarrhoea in general rather than that caused by *C. difficile*. Bifidobacteria, typically in combination with other probiotics, have also been used to prevent or treat diarrhoea caused by pathogens such as rotavirus (Dubey *et al.*, 2008; Saavedra *et al.*, 1994) and diarrhoea of unspecified or variable cause, such as acute diarrhoea (Weizman *et al.*, 2005; Jirapinyo *et al.*, 2002) and traveller's diarrhoea (McFarland *et al.*, 2007).

Anti-*Helicobacter* effects

The potential of probiotics against *H. pylori* in humans has been evaluated in several clinical studies, which were usually based on biomarkers of *H. pylori* colonization, such as positive ^{13}C-urea breath test (UBT) and serum pepsinogen (PG), which respectively indicate inflammation and atrophy associated with *H. pylori* infection. Bifidobacteria have not been studied extensively in this respect, possibly because they inhabit primarily the colon and are obligate anaerobes, while probiotics such as lactobacilli are facultative anaerobes and tolerate oxygen present in the upper gastrointestinal tract and stomach, which is the main site of *H. pylori* colonization. Nevertheless, some clinical studies with bifidobacteria have been reported.

Miki *et al.*, 2007 administered an acid-tolerant variant of *B. bifidum* YIT 4007 to volunteers with positive UBT or PG test and observed modest reductions correlated with reductions in PG values at 8 weeks of administration, but not at 4, 12 or 20 weeks.

Similar results were obtained in a study by Wang *et al.* (2004) who administered yoghurt containing a combination of strains including *B. animalis* subsp. *lactis* Bb-12 and other lactic acid bacteria. A modest decline in UBT values was noted after six weeks of treatment.

However, none of these studies show convincingly the complete eradication of *H. pylori* by bifidobacteria or other probiotics in humans. Combinatorial antibiotic therapies such as the triple therapy (combination of amoxicillin, clarithromycin and a proton pump inhibitor) and other antibiotic combinations therefore remain the gold standard for *H. pylori* treatment. Probiotics including *Bifidobacterium* as adjuvants to standard antibiotic therapy may nevertheless aid the eradication of *H. pylori*. In the published clinical studies, strains of *Bifidobacterium* have been used in combination with other strains, typically lactic acid bacteria.

Improved levels of *H. pylori* eradication have been reported using yoghurts containing *B. animalis* subsp. *lactis* Bb-12 (Sheu *et al.*, 2002, 2006) or *Bifidobacterium longum* HY8001 (Kim *et al.*, 2008). In studies with a probiotic fruit drink containing *B. bifidum* Bb99 (Myllyluoma *et al.*, 2005) and a probiotic yoghurt containing an unspecified strain of *B. animalis* (Goldman *et al.*, 2006), eradication was not significantly higher than in the placebo group, although there were non-significant tendencies towards improvement. All of the above-mentioned products also contained other probiotic strains and it is therefore difficult to estimate what effects were due to *Bifidobacterium*. The observed effects are not necessarily due to direct anti-microbial activities of bifidobacteria. Probiotics may aid *H. pylori* eradication by increasing patient tolerance of antibiotic therapy, as suggested in a study by Myllyluoma *et al.* (2005). However, this explanation was not supported in a further study by Kim *et al.* (2008).

Anti-pathogenic activity outside the gastrointestinal tract

Anti-pathogenic activity of bifidobacteria in humans has also been observed outside the gastrointestinal tract. Ingesting a combination of *B. animalis* subsp. *lactis* 420 and three lactic acid bacteria was found to reduce nasal colonization by *S. aureus*, *Streptococcus pneumoniae* and β-haemolytic streptococci. This effect is almost certainly not due to antimicrobial activity since direct interaction between the probiotics and the target strains is unlikley. The authors therefore speculated that activation of the mucosal lymphoid tissue was involved (Glück *et al.*, 2003).

Consumption by healthy adults of milk fermented with *B. animalis* subsp. *lactis* DN-173010 was found to result in reduced oral colonization by *Streptococcus mutans*, while not affecting salivary levels of lactobacilli (Cildir *et al.*, 2009). However, it is not known whether the strain possesses antimicrobial activity directed against *S. mutans* and thus the mechanism of this reduction remains unidentified.

The limited use of bifidobacteria for problems outside the gastrointestinal tract is no doubt due to the fact that the colon appears to be their principal and perhaps only habitat in the body.

When a probiotic therapy leads to reduced levels of a pathogen, of biomarkers of pathogenic activity or of symptoms such as diarrhoea, it does not necessarily imply that probiotics have a direct inhibitory effect on pathogens *in vivo.* Improved resistance to pathogens or reduction of symptoms may also be mediated by way of mechanisms other than antimicrobial activity. Possible mechanisms of probiotic action include enhancement of innate and acquired immune responses and improvement of epithelial barrier integrity.

General conclusion

Although compelling experimental evidence has accumulated in support of the direct inhibition of pathogens as a major contributor to the probiotic efficacy of bifidobacteria, much uncertainty continues to surround the relative significance of the various mechanisms proposed. The ability of these bacteria to acidify the surrounding medium by fermenting a wide range of carbohydrates is very likely a factor involved in inhibiting pathogens such as *E. coli* and *C. difficile,* but the nature of the acid secreted also appears to matter, suggesting that the effect is due to more specific metabolic mechanisms than just the burden of tolerating acidity. Secreted hydrogen peroxide as an inhibitor of pathogens raises questions about how much oxygen is present in the large intestine and how organisms other than *Clostridium* might be affected. The subject of bacteriocin production by bifidobacteria has inspired more speculation than research and although investigation in this area is of significant scientific interest, there is no reason to expect these bacteria to be sources of amazing new bacteriocins. It appears more likely that bifidobacteria owe their biological success in the large intestine and their inhibitory effects against pathogens to a combination of effects, including in particular the ability to compete for attachment sites on intestinal epithelial cells. Furthermore, if they do secrete large molecules that contribute to their value as probiotics, these too could act by interfering with pathogen attachment to epithelial cells. Common sense suggests that mechanisms able to act through highly localized effects such as adhesion to specific cell surface receptors or other binding sites would be more efficient than those acting by modifying the bulk chemical characteristics of the surrounding liquid medium. The latter is more difficult to imagine, since the large intestine of an adult human being is a microbial jungle and bifidobacteria make up at the very most only 2% of the 10^{13}–10^{14} cells jostling for a niche. In any case, pathogen-inhibiting effects of bifidobacteria all appear to be strongly dependent on bacterial strain and this will fuel much future research, both for scientific purposes and for the quest to add value to functional foods and nutraceutical products.

References

Abee, T., Krockel, L., and Hill, C. (1995). Bacteriocins: modes of action and potentials in food preservation and control of food poisoning. Int. J. Food Microbiol. *28,* 169–185.

Anand, S.K., Srinivasan, R.A., and Rao, L.K. (1984). Antimicrobial activity associated with *Bifidobacterium bifidum*-I. Cult. Dairy Products J. *2,* 6–7.

Asahara, T., Nomoto, K., Shimizu, K., Watanuki, M., and Tanaka, R. (2001). Increased resistance of mice to *Salmonella enterica* serovar Typhimurium infection by synbiotic administration of bifidobacteria and transgalactosylated oligosaccharides. J. Appl. Microbiol. *91,* 985–996.

Asahara, T., Shimizu, K., Nomoto, K., Watanuki, M., and Tanaka, R. (2001). Antibacterial effect of fermented milk containing *Bifidobacterium breve, Bifidobacterium bifidum* and *Lactobacillus acidophilus* against indigenous *Escherichia coli* infection in mice. Microb. Ecol. Health Dis. *13,* 16–24.

Bengmark, S. (1998). Ecological control of the gastrointestinal tract. The role of probiotic flora. Gut *42,* 2–7.

Bennik, M.H., Verheul, A., Abee, T., Naaktgeboren-Stoffels, G., Gorris, L.G., and Smid, E.J. (1997). Interactions of nisin and pediocin PA-1 with closely related lactic acid bacteria that manifest over 100-fold differences in bacteriocin sensitivity. Appl. Environ. Microbiol. *63,* 3628–3636.

Bernet, M.F., Brassart, D., Neeser, J.R., and Servin, A.L. (1993). Adhesion of human bifidobacterial strains to cultured human intestinal epithelial cells and inhibition of enteropathogen–cell interactions. Appl. Environ Microbiol. *59,* 4121–4128.

Black, F.T., Andersen, P.L., Orskov, J., Orskov, F., Gaarslev, K., and Laulund, S. (1989). Prophylactic efficacy of lactobacilli on travelers diarrhoea. Travel Med. *7,* 333–335.

Bibilonni, R., Pérez, P.F., and de Antoni, G.L. (1999). Factors involved in adhesion of bifidobacterial strains to epithelial cells in culture. Anaerobe *5,* 483–485.

Bouvier, M., Meance, S., Bouley, C., Berta, J.L., and Grimaud, J.C. (2001). Effects of consumption of a milk fermented by the probiotic strain *Bifidobacterium animalis* DN-173, 010 on colonic transit times in healthy humans. Bioscience Microflora *20,* 43–48.

Candela, M., Seibold, G., Vitali, B., Lachenmaier, S., Eikmanns, B.J., and Brigidi, P. (2005). Real-time PCR quantification of bacterial adhesion to Caco-2 cells: competition between bifidobacteria and enteropathogens. Res Microbiol. *156,* 887–895.

Cheikhyoussef, A., Pogori, N., Chen, W., and Zhang, H. (2008). Antimicrobial proteinaceous compounds obtained from bifidobacteria: From production to their application. Int. J. Food Microbiol, *125*, 215–222.

Cildir, S.K., Germec, D., Sandalli, N., Ozdemir, F.I., Arun, T., Twetman, S., *et al.* (2009). Reduction of salivary mutans streptococci in orthodontic patients during daily consumption of yoghurt containing probiotic bacteria. Eur. J. Orthod. *31*, 407–411.

Collado, M.C., Gueimonde, M., Hernández, M., Sanz, Y., and Salminen, S. (2005). Adhesion of selected *Bifidobacterium* strains to human intestinal mucus and the role of adhesion in enteropathogen exclusion. J. Food Protect. *68*, 2672–2678.

Corr, S., Li, U., Riedel, C.U., O'Toole, P.W., Hill, C., and Gahan, C.G.M. (2007). Bacteriocin production as a mechanism for the antiinfective activity of *Lactobacillus salivarius* UCC118. Proc. Natl. Acad. Sci. U.S.A *104*, 7617–7621.

Cummings, J.H. (1995). Short-chain fatty acids. In: Human Colonic Bacteria: Role in Nutrition, Physiology and Pathology, Gibson, G.R., and Macfarlane, G.T., eds. (Boca Raton: CRC Press), pp. 101–130.

Dahiya, R.S., and Speck, M.L. (1968). Hydrogen peroxide formation by lactobacilli and its effect on *Staphylococcus aureus*. J. Dairy Sci. *51*, 1568–1572.

Dabour, N., Zihler, A., Kheadr, E., Lacroix, C., and Fliss, I. (2009). *In vivo* study on the effectiveness of pediocin PA-1 and *Pediococcus acidilactici* UL5 at inhibiting *Listeria monocytogenes*. Int. J. Food Microbiol. *133*, 225–233.

De Melo, A., Cook, G., Miles, R., and Poole, R. (1996). Nisin stimulates oxygen consumption by *Staphylococcus aureus* and *Escherichia coli*. Appl. Environ. Microbiol. *62*, 1831–1834.

de Vuyst, L., and Leroy, F. (2007), Bacteriocins from lactic acid bacteria: production, purification, and food applications. J. Mol. Microbiol. Biotechnol.*13*, 194–199.

Drasar, B.S., and Roberts, A.K. (1989). Control of the large bowel microflora. In Human Microbial Ecology, Hill, M.J. , and Marsh, P.D. , P.D., eds. (Boca Raton: CRC Press), pp. 87–106.

Dubey, A.P., Rajeshwari, K., Chakravarty, A., and Famularo, G. (2008). Use of VSL#3 in the treatment of rotavirus diarrhea in children: preliminary results. J. Clin. Gastroenterol. *42*, S126–S129.

Duncan, S.H., Louis, P., and Flint, H.J. (2004). Lactate-utilising bacteria, isolated from human faeces, that produce butyrate as a major fermentation product. Appl. Environ. Microbiol. *70*, 5810.

Eschenbach, D.A., Pamella, R., Davick, I., Williams, B.L., Klebanoff, S.J., Young-Smith, K., Critchlow, C.M., and Holmes, K.K. (1989). Prevalence of hydrogen peroxide-producing *Lactobacillus* species in normal women and women with bacterial vaginosis. J. Clin. Microbiol. *27*, 251–256.

Fooks, L.J., and Gibson, G.R. (2002). Probiotics as modulators of the gut flora. Brit. J. Nutr. *88*, S39–S49.

Fujiwara, S., Hashiba, H., Hirota, T., and Forstner, J.F. (1997). Proteinaceous factor(s) in culture supernatant fluids of bifidobacteria which prevents the binding of enterotoxigenic *Escherichia coli* to gangliotetraosylceramide. Appl. Environ. Microbiol. *63*, 506–512.

Fujiwara, S., Hashiba, H., Hirota, T., and Forstner, J.F. (2001). Inhibition of the binding of enterotoxigenic *Escherichia coli* Pb176 to human intestinal epithelial cell line HCT-8 by an extracellular protein fraction containing BIF of *Bifidobacterium longum* SBT2928: suggestive evidence of blocking of the binding receptor gangliotetraosylceramide on the cell surface. Int. J. Food Microbiol. *67*, 97–106.

Gagnon, M., Kheadr, E.E., Dabour, N., Richard, D., and Fliss, I. (2006). Effect of *Bifidobacterium thermacidophilum* probiotic feeding on enterohemorrhagic *Escherichia coli* O157:H7 infection in BALB/c mice. Int. J. Food Microbiol. *111*, 26–33.

Gagnon, M., Kheadr, E.E., Le Blay, G., and Fliss, I. (2004). *In vitro* inhibition of *Escherichia coli* O157:H7 by bifidobacterial strains of human origin. Int. J. Food Microbiol. *92*, 69–78.

Gibson, G.R., and Wang, X. (1994). Regulatory effects of bifidobacteria on the growth of other colonic bacteria. J. Appl. Bacteriol. *77*, 412–420.

Gill, H.S. (2003). Probiotics to enhance anti-infective defences in the gastrointestinal tract. Best Pract. Res. Clin. Gastroenterol. *17*, 755–773.

Glück, U., and Gebbers, J.-O. (2003). Ingested probiotics reduce nasal colonization with pathogenic bacteria (*Staphylococcus aureus, Streptococcus pneumoniae*, and b-hemolytic streptococci). Am. J. Clin. Nutr. *77*, 517–520.

Gopal, P.K., Prasad, J., Smart, J., and Gill, H.S. (2001). *In vitro* adherence properties of *Lactobacillus rhamnosus* DR20 and *Bifidobacterium lactis* DR 10 strains and their antagonistic activity against an enterotoxigenic *Escherichia coli*. Int. J. Food Microbiol. *67*, 207–216.

Granato, D., Perotti, F., Masserey, I., Rouvet, M., Golliard, M., Servin, A., and Brassart, D. (1999). Cell surface-associated lipoteichoic acid acts as an adhesion factor for attachment of *Lactobacillus johnsonii* La1 to human enterocyte-like Caco-2 cells. Appl. Environ. Microbiol. *65*, 1071–1077.

Gratia, A. (1925). Sur un remarquable exemple d'antagonisme entre deux souches de colibacille. comp. Rend. Soc. Biol. *93*, 1040–1041.

Greene, J.D., and Klaenhammer, T.R. (1994). Factors involved in adherence of lactobacilli to human Caco-2 cells. Appl. Environ. Microbiol. *60*, 4487–4494.

Gueimonde, M., Margolles, A., de los Reyes-Gavilán, C.G., and Salminen, S. (2007). Competitive exclusion of enteropathogens from human intestinal mucus by *Bifidobacterium* strains with acquired resistance to bile – A preliminary study. Int. J. Food Microbiol. *113*, 228–232.

Gueimonde, M., Noriega, L., Margolles, A., de los Reyes-Gavilán, C.G., and Salminen, S. (2005). Ability of *Bifidobacterium* strains with acquired resistance to bile to adhere to human intestinal mucus. Int. J. Food Microbiol. *101*, 341–346.

Guglielmetti, S., Tamagnini, Mora, I. , D., Minuzzo, M., Scarafoni, A., Arioli, S., Hellman, J., Karp, M., and Parini, C. (2008). Implication of an outer surface lipoprotein in adhesion of *Bifidobacterium bifidum* to Caco-2 cells. Appl. Environ. Microbiol. *74*, 4695–4702.

Harmsen, H.J., Wildeboer-Veloo, A.C., Raangs, G.C., Wagendorp, A.A., Klijn, N., Bindels, J.G., *et al.* (2000). Analysis of intestinal flora development in breast-fed and formula-fed infants by using molecular identification and detection methods. J. Pediatr. Gastroenterol. Nutr. *30*, 61–67.

Henriksson, A., and Conway, P.L. (2001). Isolation of human faecal bifidobacteria which reduce signs of *Salmonella* infection when orogastrically dosed to mice. J. Appl. Microbiol. *90*, 223–228.

Hütt, P., Shchepetova, J., Loivukene, K., Kullisaar, T., and Mikelsaar, M. (2006). Antagonistic activity of probiotic lactobacilli and bifidobacteria against entero- and uropathogens J. Appl. Microbiol. *100*, 1324–1332.

Ibrahim, S.A., and Bezkorovainy, A. (1993). Inhibition of *Escherichia coli* by bifidobacteria. J. Food Prot. *56*, 713–715.

Jirapinyo, P., Densupsoontorn, N., Thamonsiri, N., and Wongarn, R. (2002). Prevention of antibiotic-associated diarrhea in infants by probiotics. J. Med. Assoc. Thai. *85*, S739–S742.

Joakim, S., Tiina, I., Michaela, R., and Sahl, G. (1996). Immunity to lantibiotics. Antonie van Leeuwenhoek *69*, 151–159.

Juntunen, M., Kirjavainen, P.V., Ouwehand, A.C., Salminen, S.J., and Isolauri, E. (2001). Adherence of probiotic bacteria to human intestinal mucus in healthy infants and during rotavirus infection. Clin. Diag. Lab. Immunol. *8*, 293–296.

Kanbe, M. (1991). Use of intestinal lactic acid bacteria and health. In Function of fermented milk, Nakazawa, Y. , and Hosono, A. , eds. (London: Elsevier Science Publishers)

Kim, M.N., Kim, N., Lee, S.H., Park, Y.S., Hwang, J.H., Kim, J.W., *et al.* (2008). The effects of probiotics on PPI-triple therapy for *Helicobacter pylori* eradication. Helicobacter *13*, 261–268.

Leahy, S.C., Higgins, D.G., Fitzgerald, G.F., and van Sinderen, D. (2005). Getting better with bifidobacteria. J. Appl. Microbiol. *98*, 1303–1315.

Liévin, V., Peiffer, I., Hudault, S., Rochat, F., Brassart, D., Neeser, J.-R., and Servin, A.L. (2000). *Bifidobacterium* strains from resident infant human gastrointestinal microflora exert antimicrobial activity. Gut *47*, 646–652.

Mack, D.R., Michail, S., Wei, S., McDougall, L., and Hollingsworth, M.A. (1999). Probiotics inhibit enteropathogenic *Escherichia coli* adherence *in vitro* by inducing intestinal mucin gene expression. Am. J. Physiol. *276*, G941–G50.

Makras, L., and de Vuyst, L. (2006). The *in vitro* inhibition of Gram-negative pathogenic bacteria by bifidobacteria is caused by the production of organic acids. Int. Dairy J. *16*, 1049–1057.

Mahoney, M., and Henriksson, A. (2003). The effect of processed meat and meat starter cultures on gastrointestinal colonisation and virulence of *Listeria monocytogenes* in mice. Int. J. Food Microbiol. *84*, 255–261.

Margolles, A., García, L., Sánchez, B., Gueimonde, M., and de los Reyes-Gavilán, C.G. (2003). Characterisation of a *Bifidobacterium* strain with acquired resistance to cholate: a preliminary study. Int. J. Food Microbiol. *82*, 191–198.

Mattar, A.F., Teitelbaum, D.H., Drongowski, R.A., Yongyi, F., Harmon, C.M., and Coran, A.G. (2002). Probiotics up-regulate MUC-2 mucin gene expression in a Caco-2 cell-culture model. Pediatrie Sur. Int. *18*, 586–590.

McFarland, L.V. (2006). Meta-analysis of probiotics for the prevention of antibiotic associated diarrhoea and the treatment of *Clostridium difficile* disease. Am. J. Gastroenterol. *101*, 812–822.

McFarland, L.V. (2007). Meta-analysis of probiotics for the prevention of traveler's diarrhea. Travel Med. Infect. Dis. *5*, 97–105.

Meghrous, J., Eulogy, P., Junelles, A., Ballongue, J., and Petitdemange, H. (1990). Screening of bifidobacteria strains for bacteriocin production. Biotechnol. Lett. *12*, 575–580.

Rajilić-Stojanović, M., Heilig, H.G.H.J., Molenaar, D., Kajander, K., Surakka, A., Smidt, H., and de Vos, W.M. (2009). Development and application of the human intestinal tract chip, a phylogenetic microarray: analysis of universally conserved phylotypes in the abundant microbiota of young and elderly adults.e Environ. Microbiol. *11*, 1736–1751mi_1900, 1736.

Miki, K., Urita, Y., Ishikawa, F., Iino, T., Shibahara-Sone, H., Akahoshi, R., *et al.* (2007). Effect of *Bifidobacterium bifidum* fermented milk on *Helicobacter pylori* and serum pepsinogen levels in humans. J. Dairy Sci. *90*, 2630–2640.

Morata de Ambrosini, V.I., Gonzalez, S.N., and Oliver, G. (1999). Study of adhesion of *Lactobacillus casei* CRL431 to ileal intestinal cells of mice. J. Food Prot. *62*, 1430–1434.

Moroni, O., Kheadr, E.E., Boutin, Y., Lacroix, C., and Fliss, I. (2006). Inactivation of adhesion and invasion of food borne *Listeria monocytogenes* by bacteriocin-producing *Bifidobacterium* strains of human origin. Appl. Environ. Microbiol. *72*, 6894–6901.

Mota-Meira, M., LaPointe, G., Lacroix, C., and Lavoie, M. (2000). MICs of mutacin BNy266, nisin A, vancomycin, and oxacillin against bacterial pathogens. Antimicrob. Agents Chemother. *44*, 24–29.

Mota-Meira, M., Morency, H., and Lavoie, M. (2005). *In vivo* activity of mutacin B-Ny266. J. Antimicrob. Chemother. *56*, 869–871.

Myllyluoma, E., Veijola, L., Ahlroos, T., Tynkkynen, S., Kankuri, E., Vapaatalo, H., *et al.* (2005). Probiotic supplementation improves tolerance to *Helicobacter pylori* eradication therapy -a placebo-controlled, double-blind randomized pilot study. Aliment. Pharmacol. Ther. *21*, 1263–1272.

Noriega, L., Gueimonde, M., Sánchez, B., Margolles, A., and de los Reyes-Gavilán, C.G. (2004). Effect of the adaptation to high bile salts concentrations on glycosidic activity, survival at low pH and cross-resistance to bile salts in *Bifidobacterium*. Int. J. Food Microbiol. *94*, 7986.

O'Mahony, D., Murphy, K.B., Macsharry, J., Boileau, T., Sunvold, G., Reinhart, G., et al. (2009). Portrait of a canine probiotic *Bifidobacterium*-From gut to gut. Vet. Microbiol. *139*, 106–112.

Op den Camp, H.J., Oosterhof, A., and Veerkamp, J.H. (1985). Interaction of bifidobacterial lipoteichoic acid with human intestinal epithelial cells. Infect. Immun. *47*, 332–334.

O'Sullivan, D.J. (2004). Isolated bifidobacteria that produce siderophores which inhibit growth of *Lactococcus lactis*. US Patent No: 6,746,672.

Orrhage, K., Brismar, B., and Nord, C.E. (1994). Effects of supplements of *Bifidobacterium longum* and *Lactobacillus acidophilus* on the intestinal microbiota during administration of clindamycin. Microbioal Ecol. Health Dis. *7*, 17–25.

Pinto, M., Robine-Leon, S., Appay, M.D., Kedinger, M., Triadou, N., Dussaulx, E., Lacroix, B., Simon-Assmann, P., Haffen, K., Fogh, J., and Zweibaum, A., (1983). Enterocyte-like differentiation and polarisation of the human colon carcinoma cell line Caco-2 in culture. Biol. Cell *47*, 323–330.

Podoprigora, G.L., Comunian, L.B., Pimentel, E.F., Moura, L.N., Cara, D.C., Nicoli, J.R., and Vieira, E.C. (1999). Stimulatory effect of bifidobacteria on the host mononuclear phagocyte System using gnotobiotic animal models. Anaerobe *5*, 509- 512.

Poupard, J.A., Husein, I., and Norris, R.F. (1973). Biology of the Bifidobacteria. Bacteriol. Rev. *37*, 136–165.

Rajilić-Stojanović, M., Heilig, H.G.H.J., Molenaar, D., Kajander, K., Surakka, A., Smidt, H., *et al.* (2009). Development and application of the human intestinal tract chip (HITChip), a phylogenetic microarray: Analysis of universally conserved phylotypes in the abundant microbiota of young and elderly adults. Environ. Microbiol. doi:10.1111/j.1462–2920.2009.01900.x.

Rastall, R.A., Gibson, G.R., Gill, H.S., Guarner, F., Klaenhammer, T.R., Pot, B., *et al.* (2005). Modulation of the microbial ecology of the human colon by probiotics, prebiotics and synbiotics to enhance human health: an overview of enabling science and potential applications. FEMS Microbiol. Ecol. *52*, 145–152.

Resta-Lenert, S., and Barrett, K.E. (2003). Live probiotics protect intestinal epithelial cells from the effects of infection with enteroinvasive *Escherichia coli* (EIEC). Gut *52*, 988–997.

Romond, M.B., Haddou, Z., Mielcareck, C., and Romond, C. (1997a). Bifidobacteria and human health: regulatory effects of indigenous bifidobacteria on *Escherichia coli* intestinal colonisation. Anaerobe *3*, 131–136.

Romond, M.B., Ais, A., Yazourh, A., and Romond, C. (1997b). Cell-free wheys from bifidobacteria fermented milks exert a regulatory effect on the intestinal microflora of mice and humans. Anaerobe *3*, 137–143.

Saavedra, J.M., Bauman, N.A., Oung, I., Perman, J.A., and Yolken, R.H. (1994). Feeding of *Bifidobacterium bifidum* and *Streptococcus thermophilum* to infants in hospital for prevention of diarrhoea and shedding of rotavirus. Lancet *344*, 1046–1049.

Severina, I.S. (1998)Role of soluble guanylate cyclase in the molecular mechanism underlying the physiological effects of nitric oxide. Biochemistry (Mosc) *63*, 794–801.

Schiffrin, E.J., Brassart, D., Servin, A.L., Rochat, F., and Donnet-Hughes, A. (1997). Immunemodulation of blood leucocytes in human by lactic acid bacteria: criteria for strain selection. Am. J. Clin. Nutr. *66*, 515S-529S.

Servin, A.L. (2004). Antagonistic activities of lactobacilli and bifidobacteria against microbial pathogens. FEMS Microbiol. Rev. *28*, 405–440.

Servin, A.L., and Coconnier, M.-H. (2003). Adhesion of probiotic to the intestinal mucosa and interaction with pathogens. Best Practice & Research Clinical Gastroenterol. *17*, 741–753.

Shimamura, S., Abe, F., Ishibashi, N., Miyakawa, H., Yaeshima, T., Araya, T., and Tomita, M. (1992). Relationship between oxygen sensitivity and metabolism of *Bifidobacterium* species. J. Dairy Sci. *75*, 3296–3306.

Silva, A.M., Barbosa, F.H.F., Duarte, R., Vieira, L.Q., Arantes, R.M.E., and Nicoli, J.R. (2004). Effect of *Bifidobacterium longum* ingestion on experimental salmonellosis in mice. J. Appl. Microbiol. *97*, 29–37.

Sheu, B.S., Wu, J.J., Lo, C.Y., Wu, H.W., Chen, J.H., Lin, Y.S., *et al.* (2002). Impact of supplement with *Lactobacillus*- and *Bifidobacterium*-containing yoghurt on triple therapy for *Helicobacter pylori* eradication. Aliment. Pharmacol. Ther. *16*, 1669–1675.

Sheu, B.S., Cheng, H.C., Kao, A.W., Wang, S.T., Yang, Y.J., Yang, H.B., *et al.* (2006). Pretreatment with *Lactobacillus*- and *Bifidobacterium*-containing yoghurt can improve the efficacy of quadruple therapy in eradicating residual *Helicobacter pylori* infection after failed triple therapy. Am. J. Clin. Nutr. *83*, 864–869.

Shu, Q., Lin, H., Rutherfurd, K.J., Fenwick, S.G., Prasad, J., Gopal, P.K., *et al.* (2000). Dietary *Bifidobacterium lactis* (HN019) Enhances resistance to oral *Salmonella typhimurium* infection in mice. Microbiol. Immunol. *44*, 213–222.

Shu, Q., and Gill, H.S. (2001). A dietary probiotic (*Bifidobacterium lactis* 019) reduces the severity of *Escherichia coli* 0157:H7 infection in mice. Med. Microbiol. Immunol. *189*, 147–152.

Shu, Q., Qu, F., and Gill, H.S. (2001). Probiotic treatment using *Bifidobacterium lactis* HN019 reduces weanling diarrhoea associated with rotavirus and *Escherichia coli* infection in a piglet model. J. Ped. Gastroenterol. Nutr. *33*, 171–177.

Suzuki, T., Itoh, K., Kaneko, T., and Suzuki, H. (1997). Inhibition of bacterial translocation from the gastrointestinal tract of mice by oral administration of a culture condensate of *Bifidobacterium longum*. J. Vet. Med. Sci. *59*, 665–669.

Talwalkar, A., and Kailasapathy, K. (2003). Metabolic and biochemical responses of probiotic bacteria to oxygen. J. Dairy Sci. *86*, 2537–2546.

Tissier, H. (1900). Recherchers sur la flora intestinale normale et pathologique du nourisson. Thesis, University of Paris, Paris, France.

Thompkinson, D.K., and Singh, A.K. (2000). Natural food preservation systems. Indian Food Industry *19*, 330–339.

Thompson, R., and Johnston, A. (1950). The inhibitory action of saliva on the diphtheria bacillus: hydrogen peroxide, the inhibitory agent produced by salivary streptococci. J. Infect. Dis. *88*, 81–85.

Touré, R., Kheadr, E.E., Lacroix, C., Moroni, O., and Fliss, I. (2003). Production of antibacterial substances by bifidobacterial isolates from infant stool active against *Listeria monocytogenes*. J. Appl. Microbiol. *95*, 1058–1069.

Ventura, M., Jankovic, I., Walker, D.C., Pridmore, R.D., and Zink, R. (2002). Identification and characterisation of novel surface proteins in *Lactobacillus johnsonii* and *Lactobacillus gasseri*. Appl. Environ. Microbiol. *68*, 6172–6181.

von Ah, U., Mozzetti, V., Lacroix, C., Kheadr, E.E., Fliss, I., and Meile, L. (2007). Classification of a moderately oxygen-tolerant isolate from baby faeces as *Bifidobacterium thermophilum*. BMC Microbiol. 7, 79.

Wagner, R.D., Pierson, C., Warner, T., Dohnalek, M., Farmer, J., Roberts, L., *et al.* (1997). Biotheraputic effects of probiotic bacteria on candidiasis in immunodeficient mice. Infect. Immun. *65*, 4165–4172.

Wang, K.Y., Li, S.N., Liu, C.S., Perng, D.S., Su, Y.C., Wu, D.C., *et al.* (2004). Effects of ingesting *Lactobacillus*- and *Bifidobacterium*-containing yoghurt in subjects with colonized *Helicobacter pylori*. Am. J. Clin. Nutr. *80*, 737–741.

Weizman, Z., Asli, G., and Alsheikh, A. (2005). Effect of a probiotic infant formula on infections in child care centers: comparison of two probiotic agents. Pediatrics *115*, 5–9.

Yamazaki, S., Tsuyuki, S., Akashiba, H., Kamimura, H., Kimura, M., Kawashima, T., *et al.* (1991). Immune response of *Bifidobacterium*-monoassociated mice. Bifidobacterial Microflora *10*, 19–31.

Yildirim, Z., and Johnson, M. (1998). Characterisation and antimicrobial spectrum of Bifidocin B produced by *Bifidobacterium bifidum* NCFB 1454. J. Food Prot. *61*, 47–51.

Yildirim, Z., Winters, D., and Johnson, M. (1999). Purification, amino acid sequence and mode of action of bifidocin B produced by *Bifidobacterium bifidum* NCFB 1454. Appl. Microbiol. *86*, 45–54.

Yusof, R.M., Haque, F., Ismail, M., and Hassan, Z. (2000). Isolation of *Bifidobacteria infantis* and its antagonistic activity against ETEC O157 and *Salmonella typhimurium* S-285 in weaning foods. Asia Pasific J. Clin. Nutr. *9*, 130–135.

Zhang, L., Su, P., Henriksson, A., O'Rourke, J., and Mitchell, H. (2008). Investigation of the immunomodulatory effects of *Lactobacillus casei* and *Bifidobacterium lactis* on *Helicobacter pylori* infection. Helicobacter *13*, 183–190.

Zweibaum, A., Pinto, M., Chevalier, G., Dussaulx, E., Triadou, N., Lacroix, B., Haffen, K., Brun, J.L., and Rousset, M. (1985). Enterocytic differentiation of a subpopulation of the human colon tumour cell line HT-29 selected for growth in sugar-free medium and its inhibition by glucose. J. Cell. Physiol. *122*, 21–29.

Interaction of Bifidobacteria with the Intestinal Mucosa with a Focus on Immunomodulating Effects

8

Eiji Miyauchi, Jin-zhong Xiao and Soichi Tanabe

Abstract

Bifidobacteria predominate in the gastrointestinal tract (GIT) of mammals. They are thought to benefit the host in various ways through interaction with intestinal mucosal cells. The interaction is the initiating event in immuno-modulation and merits particular attention. Intestinal epithelial cells (IECs) help to maintain gut homeostasis physically and immunologically. Bifidobacteria communicate with IECs and can ameliorate their inflammation. *In vitro* studies show that bifidobacteria can induce maturation of dendritic cells as well as differentiation and maturation of naïve T cells. These bifidobacteria-host cell interactions are, at least in part, triggered by pattern-recognition receptors such as toll-like receptors. Bacterial cell wall constituents, such as peptidoglycans and lipoteichoic acids, and bacterial DNA are involved in modulation of immune responses. Many clinical trials have been conducted to clarify the effectiveness of bifidobacteria with regards to their ability to prevent and/or alleviate allergic disorders, inflammatory bowel diseases and irritable bowel syndrome.

Introduction: intestinal barrier and indigenous (commensal) bacteria

Humans live in a bacterial world, and the body is in permanent contact with microbes in the environment. The GIT is the largest source of microbes that can be both harmful and beneficial to host health. The number of intestinal bacteria is estimated to be approximately 10^{14}, one logarithm higher than the total estimated number of human cells (Schiffrin and Blum, 2002; Macpherson and Harris, 2004). Recent research efforts have demonstrated that the microbiota in the GIT have a major impact on GIT and mucosal immune functions, and consequently impact the health of their host significantly. Since the huge intestinal surface area is in contact with microbiota, barrier function is essential to protect the host from intestinal bacteria, including and especially pathogenic microbes (bacteria, viruses and parasites). In general, both physical (tight junction) and biological barriers (mucin, immunoglobulin A (IgA), digestive enzymes, bactericidal substances such as defensins and lysozyme, and microbiological barriers) constitute a first line of defence against microbial invaders in the intestine.

Immune cells in the GIT have to respond quickly to invasive pathogens, but have to tolerate commensal bacteria at the same time. Various organs participate in this immune response. For example, the central lymphoid organs (bone marrow and thymus) contribute to the ontogenesis of different immune cells, while the peripheral lymphoid organs (spleen, lymph nodes and mucosal lymphoid tissue) co-ordinate the immune responses. Antigens from invading microbes are often the main players in inducing an immune response. As most antigens penetrate the body through the mucosa, the mucosal immune system of the host plays a key role in the defence response to pathogens.

Gut-associated lymphoid tissue (GALT) is the largest lymphoid tissue in the human body. Its earliest and greatest exposure to microbial antigens occurs during initial intestinal colonization, which starts at birth. The types of colonizing organisms depend on the mode of birth and feeding, hygiene levels and use of medication. Stimuli from the colonizing commensal bacteria are essential for the development of a fully functional and balanced immune system. However, it remains a puzzle as to how intestinal immune cells can differentiate between pathogens and commensals.

Among the bacteria of the GIT microbiota, lactobacilli and bifidobacteria have received the most intense scientific attention. Lactobacilli are predominantly present in the ileum, while bifidobacteria are present in the colon. Many studies have demonstrated that these commensal bacteria interact with intestinal mucosa in a mutually beneficial way. This chapter focuses on the immuno-modulating effects of bifidobacteria with some cross references to probiotic lactobacilli, provides an overview of the current understanding of the interaction between microbes and the host's intestinal immune system, and presents evidence of clinical effects of bifidobacteria in immuno-modulation.

For a full understanding of the bifidobacterial effects and those of other commensal and probiotic organisms, we have included a brief description of the function of the host immune system in relation to its interaction with intestinal microbiota.

Interaction with host intestinal immune systems

GALT can be divided into areas where lymphocytes are scattered throughout the epithelium, lamina propria and organized lymphoid tissues, including Peyer's patches (PPs) and mesenteric lymph nodes (MLNs). Interaction with intestinal microbes affects differentiation and homing of lymphocytes (B and T cells), differentiation of T cells and expansion to lamina propria, and the maturation of IgA plasmocytes and IgA production, as well as tolerance induction towards innocuous food and microbial antigens.

Dendritic cells

Dendritic cells (DCs) are generally found in the vicinity of the epithelial barrier (Fig. 8.1A). They play pivotal roles in early bacterial recognition in both innate and adaptive immunity, since they are the most potent antigen-presenting cells (APCs) and have the ability to activate naïve T cells. Physical and biological barriers of the intestinal mucosa restrict bacterial invasion as described above. However, lamina propria DCs can directly sample antigens, including enteropathogenic and commensal bacteria, by extruding dendrites between epithelial cells (Fig. 8.1A) (Rescigno *et al.*, 2001; Boesten and de Vos, 2008).

In the intestine, PPs and associated M cells essentially represent the primary site for uptake and presentation of ingested antigens (Fig. 8.1B) (Neutra, 1998). DCs are also present in the interfollicular regions of PPs. PPs have IgA-inducing functions and play a central role in controlling the interaction between the host immune system and commensal bacteria, which we will describe later.

Using fluorescent-tagged bacteria in mice, the interaction of probiotics and PP-M cells has been established, and the importance of these cells in the transport of antigens across the intestinal epithelium has received considerable attention (Muscettola *et al.*, 1994; Claassen *et al.*, 1995; Galdeano and Perdigon, 2004). Furthermore, it has been shown using bone marrow-derived murine DCs that lactobacilli and bifidobacteria can induce DC maturation (Iwabuchi *et al.*, 2007).

DCs interact with bacteria using pattern-recognition receptors (PRRs) (Zeuthen *et al.*, 2008). PRRs recognize multiple components termed microbe- or pathogen-associated molecular patterns (PAMPs) and trigger intracellular signal transduction. The signal induces up-regulation of various inflammatory-related gene products, including tumour necrosis factor (TNF)-α, interleukin (IL)–6 and IL-8.

The most studied PRRs are the toll-like receptor (TLR) family comprising at least 12 receptors, which trigger immune responses through the nuclear factor-κB (NF-κB) dependent pathway. TLR2 and TLR9 recognize bacterial cell surface compounds such as lipoteichoic acid (LTA) and oligodeoxynucleotides containing unmethylated CpG motifs (CpG DNA), respectively (Table 8.1). TLR expression is tightly regulated and differs between DC populations at different anatomical sites. For example, intestinal lamina propria $CD11c^+$ DCs express TLR5 but not TLR4, which might prevent induction of immune responses

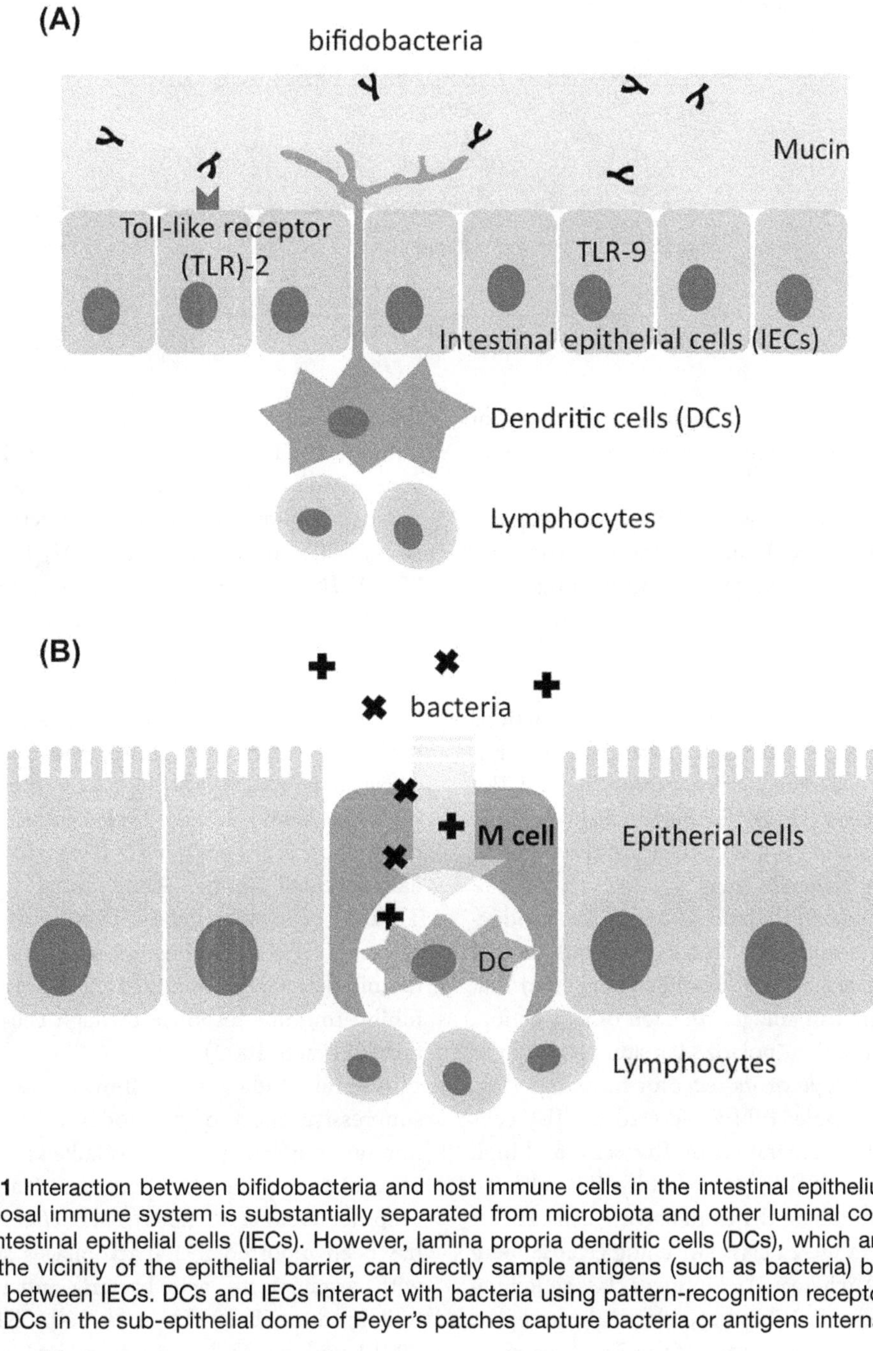

Figure 8.1 Interaction between bifidobacteria and host immune cells in the intestinal epithelium. (A) The host mucosal immune system is substantially separated from microbiota and other luminal contents by a layer of intestinal epithelial cells (IECs). However, lamina propria dendritic cells (DCs), which are generally found in the vicinity of the epithelial barrier, can directly sample antigens (such as bacteria) by extruding dendrites between IECs. DCs and IECs interact with bacteria using pattern-recognition receptors such as TLRs. (B) DCs in the sub-epithelial dome of Peyer's patches capture bacteria or antigens internalized by M cells.

towards commensal Gram-negative bacteria (Uematsu *et al.*, 2006).

Another family of PRRs is the nucleotide-binding oligomerization domain (NOD) -like receptor family, including NOD1 and NOD2, which are expressed intracellularly.

T cell differentiation

Lymphocytes (B and T cells) are the essential players in the adaptive immune response. However, T cells do not respond to free-antigens but only to antigens presented by APCs. APCs such as DCs regulate the development of effector T cells from naïve cells. T cells primarily act in

Table 8.1 Toll-like receptors (TLRs) involved in the recognition of bacterial components

Receptor	Ligands (examples)
TLR 2	Lipoproteins Lipopeptides Lipoteichoic acid
TLR 4	Lipopolysaccharide
TLR 5	Flagelin
TLR 9	CpG DNA

cell-mediated immunity. There are two types of T-cell-mediated immune responses, T helper cells ($CD4^+$, also called Th) and T cytotoxic cells ($CD8^+$). $CD4^+$ and $CD8^+$ cells recognize class II or class I major histocompatibility complex (MHC)-associated peptide antigens of phagocytosed microbes, respectively. $CD4^+$ T cells differentiate into several subsets of effector cells that produce distinct sets of cytokines, and therefore, perform distinct effector functions (Fig. 8.2). The best defined subsets of effector T cells of the $CD4^+$ helper lineage are Th1 and Th2 cells; interferon (IFN)-γ is the signature cytokine of Th1 cells and IL-4 and IL-5 are the defining cytokines of Th2 cells.

The balance of Th1 and Th2 cytokine production can determine the direction and outcome of an immune response. It has been suggested that Th1 and Th2 can antagonize each other's action, by blocking either polarized maturation of the opposite cell type or its receptor functions (Fig. 8.2). For example, IFN-γ secreted by Th1 cells can suppress proliferation of Th2 cells, and high concentration of IL-4 produced by Th2 cells can block generation of Th1 from naïve T cells. IFN-γ is involved in IgG2a class switching (Kapsenberg *et al.*, 1992), whereas IL-4 induces class switching to IgG1 and IgE. It has been classically accepted that an abnormal response of the Th2 type characterized allergic reactions.

Some probiotic strains, particularly lactobacilli, have been shown to promote production of Th1-skewing cytokines such as IL-12 and IFN-γ, and suppress the production of IgE and Th2 cytokines (Shida *et al.*, 1998; Fujiwara *et al.*, 2004; Mohamadzadeh *et al.*, 2005; Pochard *et al.*, 2005). On the other hand, several studies have demonstrated that bifidobacteria suppress Th2 cytokine and IgE production, without significant induction of Th1 cytokines. These data imply that suppression of Th2 response is, at least partially, independent of Th1 cytokine induction (Ohno *et al.*, 2005; Iwabuchi *et al.*, 2007; Inoue *et al.*, 2009). They also indicate that probiotics induce IL-12 production by APCs, such as macrophages and DCs, and maturation of naïve T cells.

Besides Th1 and Th2 cells, there are regulatory T cells controlling Th1 and Th2 responses, largely through production of IL-10 and transforming growth factor (TGF)-β (Fig. 8.2). These include the antigen-induced type 1 regulatory T (Tr) cells that secrete high levels of IL-10 and low to moderate levels of TGF-β (Groux *et al.*, 1997), type 3 T (Th3) cells (Chen *et al.*, 1994) that primarily secrete TGF-β as well as the naturally-occurring thymic-derived $CD4^+$ $CD25^+$ T cells (Treg) that inhibit immune-response through cell–cell contact (Shevach, 2000).

Several studies have shown that the Th2 suppressive effect of probiotics is mediated by immune-regulatory functions linked to induction of regulatory cytokines such as IL-10 and TGF-β (von der Weid *et al.*, 2001; Niers *et al.*, 2005). It has been reported that macrophages stimulated with certain lactic acid bacteria and bifidobacteria produce IL-10 (He *et al.*, 2002; Niers *et al.*, 2005), and that TGF-β is induced by *Lactobacillus paracasei* in mixed lymphocyte reactions (von der Weid *et al.*, 2001). Furthermore, it has been reported that neonatal application of probiotics inhibited allergic sensitization and airway disease in a murine model of asthma by induction of regulatory T cells associated with increased TGF-β production (Feleszko *et al.*, 2007).

In addition, several recent studies have shown that most inflammatory disorders are

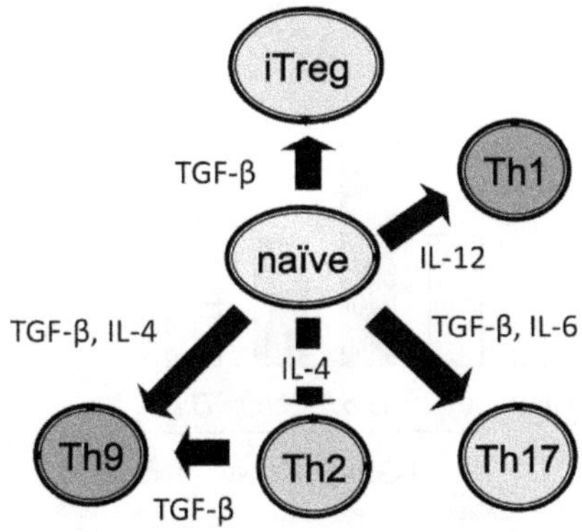

Figure 8.2 Differentiation of naive CD4⁺ T cells into effector T cells and iTreg. Cytokines that promote lineage differentiation of naive CD4⁺ T cells into effector T cells and iTreg (induced regulatory T cell) are shown by the arrows. More than 20 years ago, effector T cells were divided into two subsets, Th1 and Th2. According to the conventional classification, Th1 drives cell-mediated responses, leading to tissue damage, and IgG2a-mediated responses. Th2 drives certain antibody-mediated responses, particularly those that are involved in allergies dominated by IgE. Several years ago, a distinct subset Th17, which has an important role for inflammatory responses, was identified. TGF-β acts at the centre of the differentiation of Treg and Th17. In addition, Th9 has recently been proposed, and the plasticity of CD4⁺ differentiation is now well recognized.

also characterized by a sustained production of cytokines by a distinct lineage of Th cells, termed Th17 cells (Steinman, 2007) (Fig. 8.2). IL-17 acts as a potent inflammatory cytokine *in vitro* and *in vivo*; it is produced by Th17 and other non-Th cells, and has pleiotropic activities, one of which is to co-ordinate tissue inflammation by inducing expression of pro-inflammatory cytokines such as IL-6 and TNF-α, chemokines such as keratinocyte chemoattractant (KC), monocyte chemoattractant protein (MCP)–1, and macrophage inflammatory protein (MIP)–2 and matrix metalloproteases, that mediate tissue infiltration and destruction. IL-17 is also involved in proliferation, maturation and chemotaxis of neutrophils.

Tanabe *et al.* (2008) examined the effect of bifidobacteria on IL-17 production *in vitro* and *ex vivo*. *Bifidobacterium infantis* JCM 1222^T inhibited IL-17 production in TGF-β plus IL-6-stimulated murine splenocytes (Fig. 8.3A), and IL-17 and eotaxin production from a dextran sodium sulphate (DSS)-treated colon organ culture (Fig. 8.3-B). IL-10 induction by *B. infantis* was observed in the splenocytes and colon culture (Table 8.2), and, to a certain extent, was assumed to be important in suppressing IL-17 production.

Immunoglobulin A production

In the adaptive immune response, defence against invading microbes is provided by antibodies, particularly IgA, the most abundant antibody isotype produced in our bodies. IgA provides mucosal immune protection as a form of secretory IgA (sIgA) complexes. sIgA promotes immune exclusion by entrapping dietary antigens and microorganisms in the mucus. PPs are the main induction sites of IgA-producing plasma B cells by DCs loaded with commensal bacteria. PPs are rich in cytokines with IgA-inducing functions, including TGF-β.

IgA can participate in intestinal immune homeostasis by interacting with commensal bacteria. Mice kept under germ-free conditions and lacking an intestinal microbiota have highly reduced intestinal IgA, suggesting a direct correlation between IgA production and commensals. IgA usually limits the penetration of intestinal bacteria to MLNs. In addition, IgA antibodies to commensal bacteria can also limit the inflammatory response of intestinal immune cells (Cerutti and Rescigno, 2008; Macpherson *et al.*, 2008).

The effects of orally administered probiotics, including some strains of bifidobacteria, on IgA production have recently been revealed. Prescott

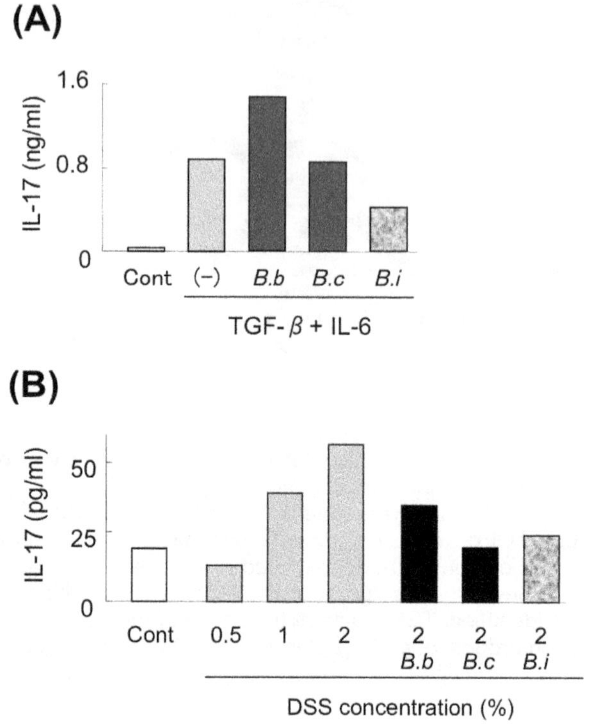

Figure 8.3 Effects of bifidobacteria on IL-17 production in murine splenocytes (A) and colon organ culture (B). (A) The spleen was removed and splenocytes were incubated with TGF-β plus IL-6. *B. bifidum* (*B.b.*), *B. catenulatum* (*B.c.*) or *B. infantis* (*B.i.*) were added to the cultures. The culture alone (cont.) and the culture without addition of bacteria (-) were included as controls. Culture supernatants were harvested and assayed for IL-17 concentration. (B) Segments of the colon were removed from three mice, cut into strips, and pooled. The strips were placed in a culture plate and incubated with the indicated concentrations of DSS.B. *bifidum* (*B.b.*), *B. catenulatum* (*B.c.*) or *B. infantis* (*B.i.*) were added to the culture under stimulation with DSS. The culture alone (cont.) and the culture without the addition of bacteria (-) were included as controls. The culture supernatants of the four strips were harvested and assayed for IL-17 concentration.

Table 8.2 Effects of *B. infantis* on cytokine profile released from colon organ culture

	Control	2% DSS	
		(–)	*B. infantis*
IL-2	nd	nd	nd
IL-4	0.45±0.08	nd	nd
IL-5	14.1±5.8	9.8±2.6	7.0±1.0
IL-10	1.2±0.0	0.45±0.12	1.0±0.4
IL-12 (p70)	0.29±0.12	0.17±0.03	0.08±0.04
IL-17	8.7±1.4	19.3±2.1	10.9±0.8

Data represent the means±SE (pg/ml) of three cultures.

et al. (2008) studied the effects of maternal probiotic supplementation on immune markers in umbilical cord blood and breast milk, and reported that supplementation with *L. rhamnosus* or *B. lactis* during pregnancy increases the IgA level in breast milk. In addition, Mohan *et al.* (2008) examined whether oral application of *B. lactis* Bb12 (probiotic) may improve indicators of health status in pre-term infants and found that faecal IgA was higher in the probiotic group compared to the placebo group.

With regard to prebiotics, Scholtens *et al.* (2008) investigated the effect of an infant milk formula with short-chain galacto- and long-chain fructooligosaccharides (ratio 9:1) on the development of the faecal sIgA response and the composition of intestinal microbiota in healthy infants during the first 26 weeks of life. The obtained results showed that the concentration of sIgA and percentages of bifidobacteria were both higher in the prebiotic group than in the control group, suggesting a positive effect on mucosal immunity.

Interaction with host epithelial cells

The host mucosal immune system is substantially separated from the microbiota and other luminal contents by a layer of intestinal IECs. The integrity of IECs is maintained by intercellular junctional complexes, for example, tight junctions (TJs). TJs comprise several proteins, among which zonula occludens (ZO)–1 was the first to be identified. It binds to transmembrane proteins, such as claudins and occludin, and links them to cytoskeletal actin (Fig. 8.4). TJs provide continuous belt-like cell–cell junctions and act as a barrier to regulate paracellular permeability (Turner, 2006).

IECs are capable of expressing a number of cytokines and chemokines that could regulate local immune responses (Eckmann *et al.*, 1993; Jung *et al.*, 1995). For example, IL-8 (also known

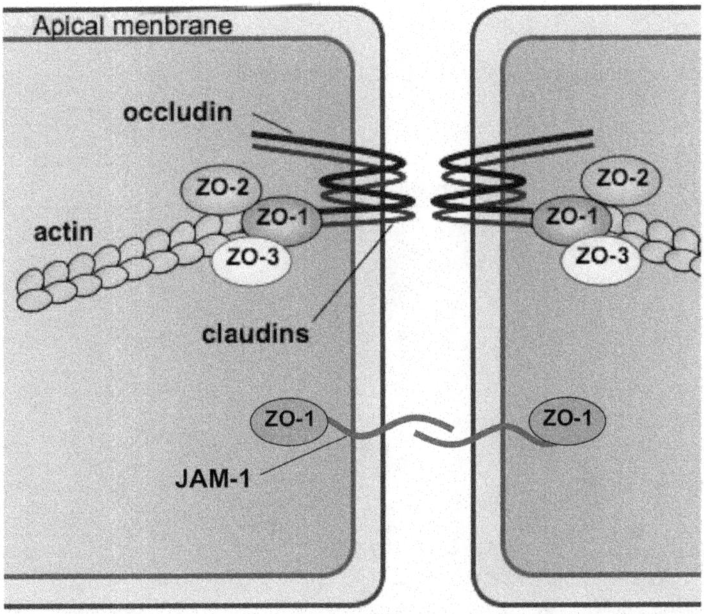

Figure 8.4 Schematic drawing of the structure of tight junction. TJs are composed of many components. Occludin has been recognized as a major component of TJs with four transmembrane domains and two extracellular loops. Occludin associates with ZO-1 through the C-terminal intracellular domain of occludin. Three proteins, ZO-1, ZO-2 and ZO-3, have been identified as members of the ZO protein family. ZO-1 binds to transmembrane proteins, such as claudins and occludin, and links them to cytoskeletal actin. There are more than 20 types of claudin. They have four transmembrane domains similar to occludin, although little homology is apparent between occludin and claudins.

as CXCL-8), one of the pro-inflammatory chemokines, attracts neutrophils into sites of infection as a host defence against pathogens (Jaye and Parkos, 2000). Under normal conditions, IECs are exposed to numerous commensal and pathogenic bacteria, and these have multiple effects on IECs. In particular, pathogens induce a series of pro-inflammatory cytokines and chemokines such as IL-1, IL-6, IL-8 and TNF-α (Harrington *et al.*, 2005). Although these events are necessary for the host to mount a defence system, chronic overproduction of pro-inflammatory molecules could lead to local tissue damage and tight junction barrier dysfunction (Murata *et al.*, 1995; Ma *et al.*, 2004; Al-Sadi and Ma, 2007). In contrast, epithelial cells are generally tolerant to commensal bacteria, including bifidobacteria and lactobacilli, which contribute to epithelial homeostasis. In addition, increasing evidence suggests that commensal bacteria directly exert homeostatic effects on epithelial cells, indicating that commensal bacteria-host epithelium cross-talk is a key event in the intestine.

Pathway of commensal/probiotic bacteria and host epithelial cell cross-talk

PRR-dependent pathway

Commensal and probiotic bacteria interact with IECs through PRRs such as TLRs, as described above. It has been shown that TLRs are predominantly expressed on macrophages and DCs and play key roles in innate immune responses. In addition, several types of TLRs are also expressed on IECs which bacteria contact most often (Gribar *et al.*, 2008; Table 8.3). IECs cross-talk with commensal/probiotic bacteria and pathogens, at least partly, through TLRs.

Among TLR ligands, the most studied is lipopolysaccharide (LPS), a TLR4 ligand. It has been reported that TLR4 signalling induces an inflammatory response with IL-8 production in an intestinal epithelial cell line and murine epithelial cells on which TLR4 and its adaptor molecules, MD-2 and MyD88, were expressed (Funda *et al.*, 2001). It is thought that the

Table 8.3 The expression pattern of human TLRs in the intestinal epithelial cells

	Small Intestine		Large Intestine	
	mRNA	Protein	mRNA	Protein
TLR1	●		●	
TLR2	●	●	●	●
TLR3		●	●	●
TLR4	●	●	●	●
TLR5		●	●	●
TLR6	Not determined		●	
TLR7	●		●	
TLR8	–	–	●	
TLR9		●	●	
TLR10	Not determined		–	–
TLR11	Not determined		Not determined	

●Expression was observed at mRNA or protein level.

–Absent expression was confirmed.

Taken from Gribar *et al.* (2008a).

overwhelming activation of TLR signalling may lead to intestinal inflammation including inflammatory bowel disease (IBD) (Gribar *et al.*, 2008).

Concurrently, it has recently been revealed that TLR signalling in IECs has a crucial role in the maintenance of intestinal homeostasis (Rakoff-Nahoum *et al.*, 2004). For instance, TLR2 signalling in IECs enforces the ZO-1-associated epithelial barrier and contributes to suppression of intestinal inflammation (Cario *et al.*, 2007). Apical TLR9 stimulation in IECs delivers negative signals that curtail inflammatory responses induced by basolateral stimulation by other TLRs ligands from entero-invasive pathogens (Lee *et al.*, 2006).

We have shown that commensal LTA, which is the TLR2 ligand, can ameliorate TNF-α-induced intestinal epithelial barrier dysfunction (Miyauchi *et al.*, 2008). It has also been reported that bacterial DNA from bifidobacteria, the TLR9 ligand, inhibits IL-8 secretion from a human epithelial cell line (Jijon *et al.*, 2004; Imaoka *et al.*, 2008). These studies imply that TLR ligand occurring in commensal or probiotic bacteria may be a factor contributing to epithelial homeostasis, although further *in vivo* studies are required to determine the clinical significance. Nevertheless, epithelial TLR signalling contributes to both homeostasis and the inflammatory response, and this bilateral function of TLR is strictly regulated by several inhibitory molecules, such as toll-interacting protein (Tollip) (Zhang and Ghosh, 2002) and single immunoglobulin-domain-containing interleukin 1 receptor-related protein (SIGIRR) (Wald *et al.*, 2003).

There is an important question that has not been fully answered yet. Both commensal/probiotic bacteria, including bifidobacteria, and pathogenic bacteria have PAMPs, even though the amounts and/or types of PAMPs may, to some extent, be different between them. It is currently not known as to why commensal/probiotic and pathogenic bacteria elicit opposite effects on the host. The mechanisms for the differential epithelial immune responses to commensal/probiotic vs.. pathogenic bacteria are an area of intensive research (Sansonetti, 2004; Artis, 2008).

Traditionally, the main explanation for the above paradox is based on the distribution of PRRs and the pathogen's ability to invade the epithelial barrier. For example, IECs express TLR4 (Hornef *et al.*, 2003) and NOD proteins (Girardin *et al.*, 2003) intracellularly, while TLR5 is expressed predominantly on the basolateral surface of polarized epithelial cells. Furthermore, it is reported that basolateral TLR3, TLR5 and TLR9 stimulation induces IL-8 secretion, but apical stimulation does not (Lee *et al.*, 2006). These observations suggest that epithelial PRRs could recognize pathogens that actively invade the epithelial barrier and induce pro-inflammatory responses, while tolerating luminal commensal bacteria.

However, it has been recently reported that different bacterial DNA's exert very different IL-8-inducing patterns from IECs. That is, DNA of pathogenic bacteria induce IL-8 from IEL, while those of commensal/probiotic DNA do not (Jijon *et al.*, 2004). Furthermore, bifidobacterial DNA inhibit IL-8 production induced by TNF-α and pathogenic bacterial DNA. Thus, TLR9 expressed on IECs may distinguish commensal/probiotic and pathogenic bacteria by their DNA fingerprint. Alternatively, as this author suggests, there is another possibility that different types of receptors recognize these bacterial DNA.

The other PAMPs, LTA and flagellin, also exhibit structural variations (Lee *et al.*, 2003; Draing *et al.*, 2006). However, it remains unclear whether these differences, as with bacterial DNA, facilitate discrimination between commensal and pathogenic bacteria in the intestine.

PRRs-independent pathway

In addition to PAMPs, other factors are also involved in the cross-talk between bifidobacteria and IECs.

Acetic acid, propionic acid and butyric acid are short chain fatty acids (SCFAs) that are produced mainly in the colon by bacterial fermentation of undigested carbohydrates. For example, polyfructans such as inulin are fermented to SCFAs specifically by bifidobacteria. SCFAs reduce the luminal pH of the colon, reducing the growth of pathogens. In addition, SCFAs directly exert a variety of effects on IECs.

Butyrate is not only utilized as an energy source for IECs, but also plays an important role in epithelial homeostasis. Indeed, butyrate deficiency has been associated with colon inflammation, such as inflammatory bowel diseases (IBDs) (Roediger, 1980). Thus, it has been proposed as a potential therapeutic agent for these diseases (Wächtershäuser and Stein, 2000).

Butyrate has been proven to exert a broad range of effects on IECs, such as suppressing pro-inflammatory chemokine secretion (Andoh *et al.*, 1999; Malago *et al.*, 2005), enhancing tight junction barrier function (Ohata *et al.*, 2005) and acting as an anti-carcinogen (D'Argenio *et al.*, 1996), through several mechanisms. It is well known that butyrate regulates NF-κB and activator protein (AP)–1 transcription factors through histone hyperacetylation (Andoh *et al.*, 1999). In addition, butyrate has the potential to inhibit IFN-g signalling (Klampfer *et al.*, 2003) and increase peroxisome proliferator-activated receptor-γ (Wächtershäuser *et al.*, 2000).

Besides SCFAs, bifidobacteria and lactobacilli secrete several bioactive substances that serve to maintain epithelial homeostasis. For example, *L. rhamnosus* GG has been shown to produce soluble proteins with molecular sizes of 40 and 75 kDa (named p40 and p75) promote IEC growth and survival (Yan *et al.*, 2007) and ameliorate tight junction dysfunction (Seth *et al.*, 2008). However, p40 and p75 secreted from lactobacilli are strain-specific, and the expression of similar proteins by bifidobacteria has not been reported.

Secretion of bioactive factors produced by *Bifidobacterium* species has been reported. For example, soluble factor(s) produced by *B. longum* C50 were shown to inhibit AP-1 and NF-κB activation as well as IL-8 secretion (Heuvelin *et al.*, 2009). The authors demonstrated that inhibition of LPS-induced TNF-α release in immune cells by these factor(s) is resistant to heat- and pepsin/trypsin- or proteinase K- treatment. Furthermore, Roselli *et al.* (2006) reported that the supernatant fraction of *B. animalis* MB5 is capable of suppressing the inflammatory-associated response in an intestinal epithelial cell line induced by *Escherichia coli* K88. However, the precise nature of these soluble factors is unclear.

In summary, some strains of bifidobacteria exert several beneficial effects on host IECs, including enhancing tight junction barrier function (Ewaschuk *et al.*, 2008), suppressing pro-inflammatory cytokines and chemokines (Jijon *et al.*, 2004; Roselli *et al.*, 2006; Imaoka *et al.*, 2008; Heuvelin *et al.*, 2009) and maintaining the homeostasis of IECs. However, the precise mechanisms and bioactive substances involved are unclear and seem to be strain dependent.

Clinical application of bifidobacteria

Today, the increased use of antibiotics is a significant factor that affects preservation of a healthy intestinal microbiota. The idea of probiotics is to restore and maintain microflora advantageous to the human body. Bifidobacteria are major components of intestinal microbiota in humans. As probiotic agents, bifidobacteria have been studied for their efficacy in the prevention and treatment of a broad spectrum of animal and/or human disorders, such as constipation, colonic transit disorders, intestinal infections, hypocholesterolaemia, colon cancer and allergic diseases. As described above, many studies have shown that bifidobacteria can induce the maturation of DCs and differentiation and maturation of naïve T cells. Bacterial cell constituents are involved in the modulation of immune responses. These lines of evidence suggest that probiotics are indeed effective in the treatment or prevention of immune-associated disorders.

Recent clinical studies have proven the prevention and curative features of probiotics in some immune diseases. In this section, we present some clinical results of the immunomodulating effects elicited by bifidobacteria.

Effects on allergic disease

Prevalence of atopic disorders such as eczema, allergic rhinitis and asthma has rapidly increased worldwide over the past few decades, particularly in industrialized countries. An association between the composition of intestinal microflora and development of allergies has been suggested by numerous studies. Studies on the composition of intestinal microflora in both allergic and non-allergic 2-year-old children indicated that prevalence of bifidobacteria is lower in allergic infants, whereas counts of *Staphylococcus aureus* and enterobacteria are higher (Bjorksten, 1999). In comparison with healthy infants, babies with allergies harbour fewer enterococci during the first month of life and fewer bifidobacteria during the first year of life (Bjorksten, 2001).

Many reports have demonstrated the effects of oral feeding of probiotic strains on the suppression of IgE and IL-4 production in animal models. Among the studies on bifidobacteria, Ohno *et al.* (2005) demonstrated that oral administration of *B. bifidum* G9–1 significantly lowered the serum levels of ovalbumin (OVA)-specific IgE but did not affect the levels of OVA-specific IgG1 and IgG2a; however, this strain could not induce IL-12p70 and IFN-γ production in splenocytes *in vitro*. Similar results were observed in a probiotic strain of *B. breve*, M-16V (Inoue *et al.*, 2009). Recently, Schouten *et al.* (2009) reported that oral administration of M-16V reduced the symptoms of allergy to cow's milk in a murine model of IgE-mediated hypersensitivity that mimics the human route of sensitization. Furthermore, administration of M-16V to infants with hypersensitivity to cow's milk and atopic dermatitis resulted in a significant reversal of allergic symptoms, a significant increase in the proportion of bifidobacteria and a decrease in the proportion of aerobic bacteria found in faecal microbiota, although no correlation has been observed between changes in faecal microbiota and decrease in the cutaneous symptom score (Hattori, 2003; Taniuchi, 2005).

Several clinical trials have been performed to assess the effects of a probiotic strain, *B. longum* BB536, on prevention and treatment of the seasonal allergic disease, Japanese cedar pollinosis (JCPsis), which the most common seasonal allergic rhinitis in Japan, and which is caused by exposure to Japanese cedar (*Cryptomeria japonica*) pollen (JCP). The JCP season typically begins in early February and ends in the latter part of April. Characteristic symptoms of JCPsis are sneezing, rhinorrhoea, nasal blockage and itching of the eyes. During recent decades, prevalence of this allergy has been increasing and now affects > 20% of the Japanese population (Baba and Nakae, 2008).

In a double-blind clinical trial performed in the JCP season of 2004, JCPsis subjects were given yoghurt supplemented with BB536 daily for 14 weeks. Subjective symptoms were alleviated, the increase in the blood eosinophil ratio was diminished and the decrease of IFN-g levels was suppressed in the BB536 group compared to the placebo group (Xiao *et al.*, 2006a). The same research group performed another trial using lyophilized BB536 powder during the JCP season of 2005, to confirm the efficacy of BB536 in the treatment of JCPsis (Xiao *et al.*, 2006b). BB536 intake was associated with a significant reduction in the number of subjects leaving the trial prematurely because of severe symptoms and pollinosis medication. Significant decrease in the subjective symptom scores of rhinorrhoea and nasal blockage was observed in the BB536 group compared to the placebo group. Comparison of medical symptom scores showed marked improvements of composite and individual symptom scores following BB536 intake (Fig. 8.5A and B). A Th2-skewed immune response, as indicated by a significant decrease in plasma IFN-γ and significant increase in thymus- and activation-regulated chemokine (TARC) and JCP-specific IgE levels, was observed in the JCPsis subjects during the periods of pollen dispersal. BB536 tended to suppress elevated JCP-specific IgE and significantly suppressed TARC levels (Fig. 8.5C and D).

In addition, a trial was performed to assess the preventative effect of BB536 intake on symptoms under controlled exposure to JCP in an environmental exposure unit outside the normal JCP season in a double-blind, two–way crossover design (Xiao *et al.*, 2007). In comparison to the placebo, a 4-week intake of BB536 powder significantly reduced ocular symptom scores during JCP exposure. Evaluating the delayed symptoms after exposure indicated that scores for the disruption of normal activities were significantly lower in the BB536 group compared to the placebo group. Use of medication was markedly reduced by BB536 intake.

These studies suggest the potential of probiotics in prevention and alleviation of JCPsis. Studies on the mechanisms of the anti-allergic effects of BB536 showed the following:

1 Cells of BB536 suppressed Th2 immune responses mediated by APCs such as DCs, partially independent of Th1-skewing cytokines (Iwabuchi *et al.*, 2007).
2 Cells of BB536 suppressed the production of Th2-attracting chemokines which were produced by APCs by stimulation of T cell–APC interactions (Iwabuchi *et al.*, 2009).

3 Constituents of the BB536 cell wall and genomic immunostimulatory sequence were involved in the suppressive effects of BB536 on the Th2 immune responses (Iwabuchi *et al.*, 2007; Takahashi *et al.*, 2006a,b).
4 There was a significant fluctuation of microbiota, e.g. increased prevalence of the *Bacteroides fragilis* group, during the pollen seasons, indicating a possible association of fluctuations of microbiota with allergen sensitization and symptom development (Odamaki *et al.*, 2007a,b, 2008).
5 BB536 intake suppressed the fluctuations of microbiota, suggesting a possible mechanism for the anti-allergic effects due to the stabilization of the intestinal microbiota (Odamaki *et al.*, 2007a,b, 2008).

Effects on inflammatory bowel disease

IBDs consist of a group of disorders, including Crohn's disease and ulcerative colitis, in which an immunological aetiology has been suspected for many years. Crohn's disease is an inflammatory disease caused by T cells that presumably react

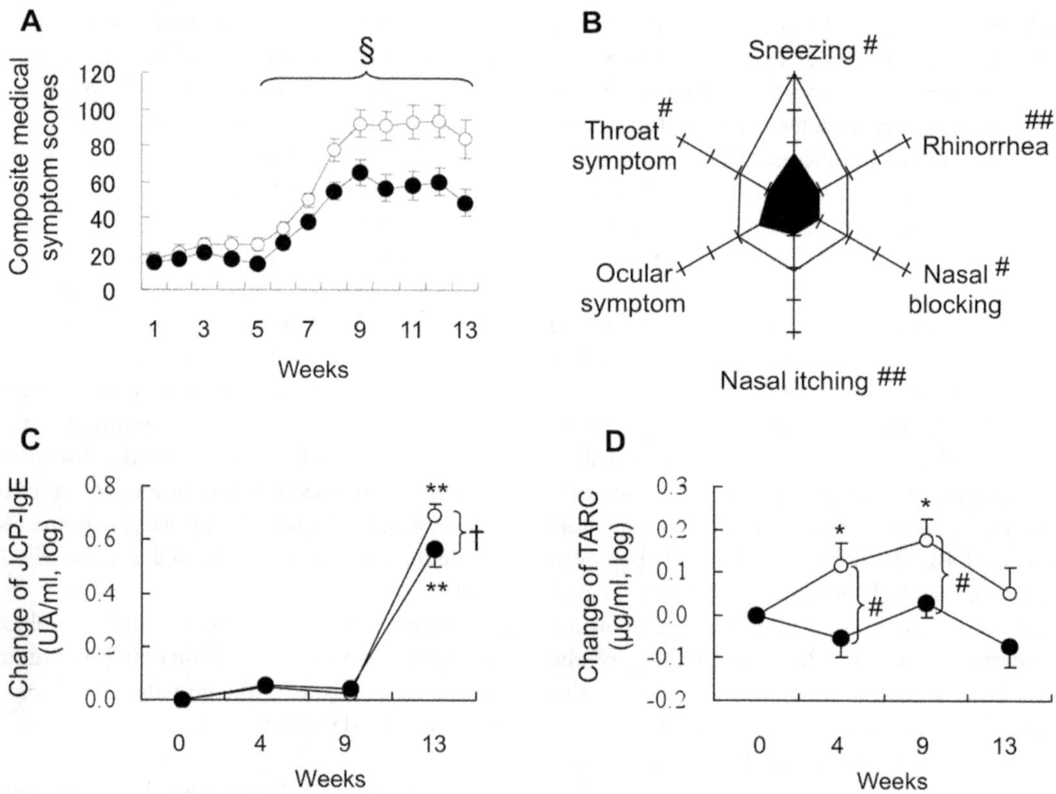

Figure 8.5 Symptom scores and blood parameters during the pollen season for BB536 (filled circles or fields) and placebo (open circles or fields) groups. (A) Means of composite medical symptom scores (sum of each of the medical symptom score as indicated in B). (B) Mean values of area-under-the-curves (AUCs) for each of the medical symptom scores. (C) Changes in blood Japanese cedar pollen (JCP)-specific IgE titres. (D) Changes in plasma thymus- and activation-regulated chemokine (TARC) levels. Error bars show standard error of the mean. § $p < 0.05$, compared between groups on the scores of 5–13 weeks with a two-way repeated ANOVA; †$P < 0.1$. *$P < 0.05$. **$P < 0.01$, compared with baseline by the Wilcoxon signed-rank test. #$P < 0.05$. ##$P < 0.01$, compared between groups with the Mann–Whitney U-test.

against intestinal microbes as well as self-antigens. Patients with IBD express increased levels of TLR2 and TLR4 in the intestinal mucosa (Szebeni *et al.*, 2008). Furthermore, polymorphisms of several TLRs and CARD15/NOD2 genes were observed in IBD patients (Török *et al.*, 2004; Pierik *et al.*, 2005, 2006). Under these conditions, IECs may possibly have abnormal susceptibility to even commensal bacteria. Several mouse models of IBD suggest that deficiency of regulatory T cells may be involved in excessive reactions to intestinal bacteria (Xavier and Podolsky, 2007). Moreover, recent studies suggest that mutation in the gene encoding the IL-23 receptor is strongly associated with IBD (Duerr *et al.*, 2006). IL-23 is the cytokine that enhances Th17 responses, and Th17 cells may be involved in inflammatory lesions (Aggarwal *et al.*, 2003; Abraham and Cho, 2009).For the treatment of IBD, recent investigations have focused on development of new therapeutics, including probiotics, which can reduce intestinal inflammation and restore balance to GIT microbiota (Prisciandaro *et al.*, 2009). Evidence from animal models of IBD indicates that probiotics can alter intestinal microbiota and ameliorate disease (Sartor, 2004). The clinical efficacy of probiotics in IBD has been investigated in a number of clinical studies. Key findings from clinical trials have recently been reviewed comprehensively by Hedin *et al.* (2007). It was demonstrated that administration of BB536 was effective in inducing remission of patients with ulcerative colitis (Takeda *et al.*, 2009). Kato *et al.* (2004) reported the effect of taking fermented milk supplemented with bifidobacteria for management of active ulcerative colitis. VSL#3, *E. coli* Nissle 1917 and *L. rhamnosus* GG have also been trialed extensively. VSL#3 is a mixture containing eight species of lactobacilli (*L. plantarum*, *L. acidophilus*, *L. casei* and *L. delbrueckii* subsp. *bulgaricus*), bifidobacteria (*B. longum*, *B. infantis* and *B. breve*) and streptococcus. In general, clinical efficiency was achieved in treatment of ulcerative colitis, but the outcome for Crohn's disease was not as positive.

Multiple mechanisms of action have been suggested to explain the protective effects of probiotics in intestinal inflammation. These can be broadly classified as follows: (1) suppression of growth or epithelial binding/invasion by pathogenic bacteria, (2) improved epithelial barrier function, and (3) immunoregulatory activities (Sartor, 2004). *B. infantis* JCM1222^{T} suppressed growth of *Bact. vulgatus*, an organism associated with IBD and experimental colitis (Shiba *et al.*, 2003). Similarly, *Bifidobacterium*-fermented milk decreased faecal *Bact. vulgatus* concentrations in ulcerative colitis (Ishikawa *et al.*, 2003). Suppression of growth and activity of intestinal pathogenic bacteria has been attributed to decreased luminal pH through production of short-chain fatty acids, secretion of bactericidal proteins, and prevention of epithelial adherence. We found that mucosal permeability was decreased by administration of *L. rhamnosus* OLL2838 to mice with colitis induced by DSS (Miyauchi *et al.*, 2009). TLR2 ligand plays a crucial role in maintaining the barrier function of TJs (Cario *et al.*, 2007; Miyauchi *et al.*, 2008; Takeda *et al.*, 2009). Several probiotics can induce protective cytokines, including IL-10 and TGF-β, and suppress pro-inflammatory cytokines, such as TNF-α, in the mucosa of patients with pouchitis and Crohn's disease (Ulisse *et al.*, 2001).

Effects on irritable bowel syndrome

Irritable bowel syndrome (IBS) follows an acute, presumably infectious diarrhoeal illness in about 15% of patients. These patients may have a persistent, mild inflammatory state, which may result in changes in mucosal function, such as increased mucosal permeability, change in mucosal structure, increases in serotonin-containing enteroendocrine cells or infiltration of inflammatory cells in bowel mucosa biopsies (Camilleri, 2006). Frequent symptoms in IBS are abdominal pain, bloating, flatulence and diarrhoea. Abdominal pain and bloating are difficult to treat and have a significant impact on the quality of life (Broekaert and Walker, 2006).

It was reported that VSL#3 significantly reduced abdominal bloating compared to the use of a placebo (Kim *et al.*, 2003). O'Mahony *et al.* (2005) showed that patients with IBS have an abnormal baseline IL-10/IL-12 ratio (the ratio of anti- to pro-inflammatory cytokines), and that administration of *B. infantis* 35624 alleviated IBS symptoms. The alleviating effects of *B. infantis* 35624 were suggested to be associated with the normalization of this ratio. It is thought that the mechanisms supporting these beneficial effects are the combination of such immune effects, changes in the intraluminal milieu and secondary effects on colonic transit (Camilleri, 2006). However, as irritable bowel disease is a heterogeneous disorder, further clinical studies are needed.

The future

Several probiotic preparations seem to show promise in preventing allergic diseases and relapse of ulcerative colitis. They may also be useful in the treatment of mild to moderately active ulcerative colitis and low-grade inflammation associated with IBS. However, results for Crohn's disease are not yet clear owing to conflicting results and a paucity of trials. In any case, a majority of the demonstrated curative and palliative effects seems to be mediated by a modulation of the intestinal immune system by microorganisms. Clinical approaches using probiotics are quite appealing because of a lack of toxicity and the desire of patients to use natural physiological approaches to treating such disease. The optimal composition, dose and length of probiotic treatment in various clinical settings, however, need to be determined by large, well-designed, placebo-controlled, prospective trials. In addition, it is clear that all lactobacilli and bifidobacteria species are not equally beneficial, each may have an individual mechanism of action, and host characteristics may determine which probiotic strain is optimal (Sartor, 2004).

References

Abraham, C., and Cho, J. (2009). Interleukin-23/Th17 pathways and inflammatory bowel disease. Inflamm. Bowel. Dis. *15*, 1090–1100.

Aggarwal, S., Ghilardi, N., Xie, M.H., de Sauvage, F.J., and Gurney, A.L. (2003). Interleukin-23 promotes a distinct CD4 T cell activation state characterized by the production of interleukin-17. J. Biol. Chem. *278*, 1910–1914.

Al-Sadi, R.M., and Ma, T.Y. (2007). IL-1beta causes an increase in intestinal epithelial tight junction permeability. J. Immunol. *178*, 4641–4669.

Andoh, A., Fujiyama, Y., Hata, K., Araki, Y., Takaya, H., Shimada, M., and Bamba, T. (1999). Counter-regulatory effect of sodium butyrate on tumour necrosis factor-alpha (TNF-α)-induced complement C3 and factor B biosynthesis in human intestinal epithelial cells. Clin. Exp. Immunol. *118*, 23–29.

Artis, D. (2008). Epithelial-cell recognition of commensal bacteria and maintenance of immune homeostasis in the gut. Nat. Rev. Immunol. *8*, 411–420.

Baba, K., and Nakae, K. (2008). National-wide epidemiological survey on the prevalence of allergic rhinitis in 2008 (in comparison with 1998): a survey on otolaryngologists and their families. Prog. Med. *28*, 2001–2012.

Björkstén, B., Naaber. P., Sepp, E., and Mikelsaar, M. (1999). The intestinal microflora in allergic Estonian and Swedish 2-year-old children. Clin. Exp. Allergy *29*, 342–346.

Björkstén, B., Sepp, E., Julge, K., Voor, T., and Mikelsaar, M. (2001). Allergy development and the intestinal microflora during the first year of life. J. Allergy Clin. Immunol. *108*, 516–520.

Boesten, R.J., and de Vos, W.M. (2008). Interactomics in the human intestine: Lactobacilli and Bifidobacteria make a difference. Clin, J. . Gastroenterol. *42* (Suppl. 3 Pt 2), S163–S167.

Broekaert, I.J., and Walker, W.A. (2006). Probiotics and chronic disease. J. Clin. Gastroenterol. *40*, 270–274.

Camilleri, M. (2006). Probiotics and irritable bowel syndrome: rationale, putative mechanisms, and evidence of clinical efficacy. J. Clin. Gastroenterol. *40*, 264–269.

Cario, E., Gerken, G., and Podolsky, D.K. (2007). Toll-like receptor 2 controls mucosal inflammation by regulating epithelial barrier function. Gastroenterology *132*, 1359–1374.

Cerutti, A., and Rescigno, M. (2008). The biology of intestinal immunoglobulin A responses. Immunity *28*, 740–750.

Chen, Y., Kuchroo, V.K., Inobe, J., Hafler, D.A., and Weiner, H.L. (1994). Regulatory T cell clones induced by oral tolerance: suppression of autoimmune encephalomyelitis. Science *265*, 1237–1240.

Claassen, E., van Winsen, R., Posno, M., and Boersma, W.J. (1995). New and safe 'oral' live vaccines based on *Lactobacillus*. Adv. Exp. Med. Biol. *371B*, 1553–1558.

D'Argenio, G., Cosenza, V., Delle, Cave, M., Iovino, P., Delle, Valle, N., Lombardi, G., and Mazzacca, G. (1996). Butyrate enemas in experimental colitis and protection against large bowel cancer in a rat model. Gastroenterology *110*, 1727–1734.

Draing, C., Pfitzenmaier, M., Zummo, S., Mancuso, G., Geyer, A., Hartung, T., and von Aulock, S. (2006). Comparison of lipoteichoic acid from different serotypes of Streptococcus pneumoniae. J. Biol. Chem. *281*, 33849–33859.

Duerr, R.H., Taylor, K.D., Brant, S.R., Rioux, J.D., Silverberg, M.S., Daly, M.J., Steinhart, A.H., Abraham, C., Regueiro, M., Griffiths, A., *et al.* (2006). A genome-wide association study identifies IL23R as an inflammatory bowel disease gene. Science *314*, 1461–1463.

Eckmann, L., Jung, H.C., Schürer-Maly, C., Panja, A., Morzycka-Wroblewska, E., and Kagnoff, M.F. (1993). Differential cytokine expression by human intestinal epithelial cell lines: regulated expression of interleukin 8. Gastroenterology *105*, 1689–1697.

Ewaschuk, J.B., Diaz, H., Meddings, L., Diederichs, B., Dmytrash, A., Backer, J., Looijer-van Langen, M., and

Madsen, K.L. (2008). Secreted bioactive factors from *Bifidobacterium infantis* enhance epithelial cell barrier function. Am. J. Physiol. Gastrointest. Liver Physiol. *295*, G1025–G1034.

Feleszko, W., Jaworska, J., Rha, R.D., Steinhausen, S., Avagyan, A., Jaudszus, A., Ahrens, B., Groneberg, D.A., Wahn, U., and Hamelmann, E. (2007). Probiotic-induced suppression of allergic sensitization and airway inflammation is associated with an increase of T regulatory-dependent mechanisms in a murine model of asthma. Clin. Exp. Allergy *37*, 498–505.

Fujiwara, D., Wakabayashi, H., Watanabe, H., Nishida, S., and Iino, H. (2005). A double-blind trial of *Lactobacillus paracasei* strain KW3110 administration for immunomodulation in patients with pollen allergy. Allergol. Int. *54*, 143–149.

Funda, D.P., Tucková, L., Farré, M.A., Iwase, T., Moro, I., and Tlaskalová-Hogenová, H. (2001). CD14 is expressed and released as soluble CD14 by human intestinal epithelial cells *in vitro*: lipopolysaccharide activation of epithelial cells revisited. Infect. Immun. *69*, 3772–3781.

Galdeano, C.M., and Perdigón, G. (2004). Role of viability of probiotic strains in their persistence in the gut and in mucosal immune stimulation. J. Appl. Microbiol. *97*, 673–681.

Girardin, S.E., Boneca, I.G., Carneiro, L.A., Antignac, A., Jéhanno, M., Viala, J., Tedin, K., Taha, M.K., Labigne, A., Zähringer, U., *et al.* (2003). Nod1 detects a unique muropeptide from gram-negative bacterial peptidoglycan. Science *300*, 1584–1587.

Gribar, S.C., Anand, R.J., Sodhi, C.P., and Hackam, D.J. (2008a). The role of epithelial Toll-like receptor signaling in the pathogenesis of intestinal inflammation. J. Leukoc. Biol. *83*, 493–498.

Gribar, S.C., Richardson, W.M., Sodhi, C.P., and Hackam, D.J. (2008b). No longer an innocent bystander: epithelial toll-like receptor signaling in the development of mucosal inflammation. Mol. Med. *14*, 645–659.

Groux, H., O'Garra, A., Bigler, M., Rouleau, M., Antonenko, S., de Vries, J.E., and Roncarolo, M.G. (1997). A CD4+ T-cell subset inhibits antigen-specific T-cell responses and prevents colitis. Nature *389*, 737–742.

Harrington, S.M., Strauman, M.C., Abe, C.M., and Nataro, J.P. (2005). Aggregative adherence fimbriae contribute to the inflammatory response of epithelial cells infected with enteroaggregative *Escherichia coli*. Cell Microbiol. 7, 1565–1578.

Hattori, K., Yamamoto, A., Sasai, M., Taniuchi, S., Kojima, T., Kobayashi, Y., Iwamoto, H., Namba, K., and Yaeshima, T. (2003). Effects of administration of bifidobacteria on fecal microflora and clinical symptoms in infants with atopic dermatitis. Arerugi *52*, 20–30.

He, F., Morita, H., Ouwehand, A.C., Hosoda, M., Hiramatsu, M., Kurisaki, J., Isorauri, E., Benno, Y., and Salminen, S. (2002). Stimulation of selection of pro-inflammatory cytokines by *Bifidobacterium* strains. Microbiol. Immunol. *46*, 781–785.

Hedin, C., Whelan, K., and Lindsay, J.O. (2007). Evidence for the use of probiotics and prebiotics in inflammatory bowel disease: a review of clinical trials. Proc. Nutr. Soc. *66*, 307–315.

Heuvelin, E., Lebreton, C., Grangette, C., Pot, B., Cerf-Bensussan, N., and Heyman, M. (2009). Mechanisms involved in alleviation of intestinal inflammation by *Bifidobacterium breve* soluble factors. PLoS One *4*, e5184.

Hornef, M.W., Normark, B.H., Vandewalle, A., and Normark, S. (2003). Intracellular recognition of lipopolysaccharide by toll-like receptor 4 in intestinal epithelial cells. J. Exp. Med. *198*, 1225–1235.

Imaoka, A., Shima, T., Kato, K., Mizuno, S., Uehara, T., Matsumoto, S., Setoyama, H., Hara, T., and Umesaki, Y. (2008). Anti-inflammatory activity of probiotic *Bifidobacterium*: enhancement of IL-10 production in peripheral blood mononuclear cells from ulcerative colitis patients and inhibition of IL-8 secretion in HT-29 cells. World J. Gastroenterol. *14*, 2511–2516.

Inoue, Y., Iwabuchi, N., Xiao, J.Z., Yaeshima, T., and Iwatsuki, K. (2009). Suppressive effects of *Bifidobacterium breve* strain M-16V on T-helper type 2 immune responses in a murine model. Biol. Pharm. Bull. 32, 760–763.

Ishikawa, H., Akedo, I., Umesaki, Y., Tanaka, R., Imaoka, A., and Otani, T. (2003). Randomized controlled trial of the effect of bifidobacteria-fermented milk on ulcerative colitis. J. Am. Coll. Nutr. 22, 56–63.

Iwabuchi, N., Takahashi, N., Xiao, J.Z., Miyaji, K., and Iwatsuki, K. (2007). *In vitro* Th1 cytokine-independent Th2 suppressive effects of bifidobacteria. Microbiol. Immunol. *51*, 649–660.

Iwabuchi, N., Takahashi, N., Xiao, J.Z., Yonezawa, S., Yaeshima, T., Iwatsuki, K., and Hachimura, S. (2009). Suppressive effects of *Bifidobacterium longum* on production of Th2-attracting chemokines induced with T cell-antigen-presenting cell interactions. FEMS Immunol. Med. Microbiol. *55*, 324–334.

Kapsenberg, M.L., Jansen, H.M., Bos, J.D., and Wierenga, E.A. (1992). Role of type 1 and type 2 T helper cells in allergic diseases. Curr. Opin. Immunol. *4*, 788–793.

Kato, K., Mizuno, S., Umesaki, Y., Ishii, Y., Sugitani, M., Imaoka, A., Otsuka, M., Hasunuma, O., Kurihara, R., Iwasaki, A., and Arakawa, Y. (2004). Randomized placebo-controlled trial assessing the effect of bifidobacteria-fermented milk on active ulcerative colitis. Aliment. Pharmacol. Ther. *20*, 1133–1141.

Kim, H.J., Camilleri, M., McKinzie, S., Lempke, M.B., Burton, D.D., Thomforde, G.M., and Zinsmeister, A.R. (2003). A randomized controlled trial of a probiotic, VSL#3, on gut transit and symptoms in diarrhoea-predominant irritable bowel syndrome. Aliment. Pharmacol. Ther. *17*, 895–904.

Jaye, D.L., and Parkos, C.A. (2000). Neutrophil migration across intestinal epithelium. Ann. N. Y. Acad. Sci. *915*, 151–161.

Jijon, H., Backer, J., Diaz, H., Yeung, H., Thiel, D., McKaigney, C., De Simone, C., and Madsen, K. (2004). DNA from probiotic bacteria modulates murine and human epithelial and immune function. Gastroenterology *126*, 1358–1373.

Jung, H.C., Eckmann, L., Yang, S.K., Panja, A., Fierer, J., Morzycka-Wroblewska, E., and Kagnoff, M.F. (1995).

A distinct array of proinflammatory cytokines is expressed in human colon epithelial cells in response to bacterial invasion. J. Clin. Invest. 95, 55–65.

Klampfer, L., Huang, J., Sasazuki, T., Shirasawa, S., and Augenlicht, L. (2004). Oncogenic Ras promotes butyrate-induced apoptosis through inhibition of gelsolin expression. J. Biol. Chem. 279, 36680–36688.

Lee, J., Mo, J.H., Katakura, K., Alkalay, I., Rucker, A.N., Liu, Y.T., Lee, H.K., Shen, C., Cojocaru, G., Shenouda, S., Kagnoff, M., Eckmann, L., Ben-Neriah, Y., and Raz, E. (2006). Maintenance of colonic homeostasis by distinctive apical TLR9 signalling in intestinal epithelial cells. Nat. Cell Biol. *8*, 1327–1336.

Lee, S.K., Stack, A., Katzowitsch, E., Aizawa, S.I., Suerbaum, S., and Josenhans, C. (2003). Helicobacter pylori flagellins have very low intrinsic activity to stimulate human gastric epithelial cells via TLR5. Microbes Infect. *5*, 1345–1356.

Ma, T.Y., Iwamoto, G.K., Hoa, N.T., Akotia, V., Pedram, A., Boivin, M.A., and Said, H.M. (2004). TNF-alpha-induced increase in intestinal epithelial tight junction permeability requires NF-kappa B activation. Am. J. Physiol. Gastrointest. Liver Physiol. *286*, G367–G376.

Macpherson, A.J., and Harris, N.L. (2004). Interactions between commensal intestinal bacteria and the immune system. Nat. Rev. Immunol. *4*, 478–485.

Macpherson, A.J., McCoy, K.D., Johansen, F.E., and Brandtzaeg, P. (2008). The immune geography of IgA induction and function. Mucosal Immunol. *1*, 11–22.

Malago, J.J., Koninkx, J.F., Tooten, P.C., van Liere, E.A., and van Dijk, J.E. (2005). Anti-inflammatory properties of heat shock protein 70 and butyrate on Salmonella-induced interleukin-8 secretion in enterocyte-like Caco-2 cells. Clin. Exp. Immunol. *141*, 62–71.

Miyauchi,E., Morita, H., Okuda, J., Sashihara, T., Shimizu, M., and Tanabe, S. (2008). Cell wall fraction of *Enterococcus hirae* ameliorates TNF-alpha-induced barrier impairment in the human epithelial tight junction. Lett. Appl. Microbiol. *46*, 469–476.

Miyauchi, E., Morita, H., and Tanabe, S. (2009). *Lactobacillus rhamnosus* alleviates intestinal barrier dysfunction in part by increasing expression of zonula occludens-1 and myosin light-chain kinase *in vivo*. J. Dairy Sci. 92, 2400–2408.

Mohamadzadeh, M., Olson, S., Kalina, W.V., Ruthel, G., Demmin, G.L., Warfield, K.L., Bavari, S., and Klaenhammer, T.R. (2005). Lactobacilli activate human dendritic cells that skew T cells toward T helper 1 polarization. Proc. Natl. Acad. Sci. U.S.A *102*, 2880–2885.

Mohan, R., Koebnick, C., Schildt, J., Mueller, M., Radke, M., and Blaut, M. (2008). Effects of *Bifidobacterium lactis* Bb12 supplementation on body weight, fecal pH, acetate, lactate, calprotectin, and IgA in preterm infants. Pediatr. Res. *64*, 418–422.

Murata, Y., Ishiguro, Y., Itoh, J., Munakata, A., and Yoshida, Y. (1995). The role of proinflammatory and immunoregulatory cytokines in the pathogenesis of ulcerative colitis. Gastroenterol, J. *30* (Suppl. 8), 56–60.

Muscettola, M., Massai, L., Tanganelli, C., and Grasso, G. (1994). Effects of lactobacilli on interferon production in young and aged mice. Ann. N. Y. Acad. Sci. *717*, 226–232.

Neutra, M.R. (1998). Current concepts in mucosal immunity, V: role of M cells in transepithelial transport of antigens and pathogens to the mucosal immune system. Am. J. Physiol. *274*, G785–G791.

Niers, L.E., Timmerman, H.M., Rijkers, G.T., van Bleek, G., van Uden, N.O., Knol, E.F., Kapsenberg, M.L., Kimpen, J.L., and Hoekstra, M.O. (2005). Identification of strong interleukin-10 inducing lactic acid bacteria which down-regulate T helper type 2 cytokines. Clin. Exp. Allergy *35*, 1481–1489.

Odamaki, T., Xiao, J.Z., Iwabuchi, N., Sakamoto, M., Takahashi, N., Kondo, S., Iwatsuki, K., Kokubo, S., Togashi, H., Enomoto, T., and Benno, Y. (2007a). Fluctuation of fecal microbiota in individuals with Japanese cedar pollinosis during pollen season. J. Invest. Allergol. Clin. Immunol. *17*, 92–100.

Odamaki, T., Xiao, J.Z., Iwabuchi, N., Sakamoto, M., Takahashi, N., Kondo, S., Miyaji, K., Iwatsuki, K., Togashi, H., Enomoto, T., and Benno, Y. (2007b). Influence of *Bifidobacterium longum* BB536 intake on faecal microbiota in individuals with Japanese cedar pollinosis during the pollen season. J. Med. Microbiol. *56*, 1301–1308.

Odamaki, T., Xiao, J.Z., Sakamoto, M., Kondo, S., Yaeshima, T., Iwatsuki, K., Togashi, H., Enomoto, T., and Benno, Y. (2008). Distribution of different species of the *Bacteroides fragilis* group in individuals with Japanese cedar pollinosis. Appl. Environ. Microbiol. *74*, 6814–6817.

Ohata, A., Usami, M., and Miyoshi, M. (2005). Short-chain fatty acids alter tight junction permeability in intestinal monolayer cells via lipoxygenase activation. Nutrition *21*, 838–847.

Ohno, H., Tsunemine, S., Isa, Y., Shimakawa, M., and Yamamura, H. (2005). Oral administration of *Bifidobacterium bifidum* G9-1 suppresses total and antigen specific immunoglobulin E production in mice. Biol. Pharm. Bull. *28*, 1462–1466.

O'Mahony, L., McCarthy, J., Kelly, P., Hurley, G., Luo, F., Chen, K., O'Sullivan, G.C., Kiely, B., Collins, J.K., Shanahan, F., and Quigley, E.M. (2005). *Lactobacillus* and *Bifidobacterium* in irritable bowel syndrome: symptom responses and relationship to cytokine profiles. Gastroenterology *128*, 541–551.

Pierik, M., De Hertogh, G., Vermeire, S., van Assche, G., van Eyken, P., Joossens, S., Claessens, G., Vlietinck, R., Rutgeerts, P., and Geboes, K. (2005). Epithelioid granulomas, pattern recognition receptors, and phenotypes of Crohn's disease. Gut *54*, 223–227.

Pierik, M., Joossens, S., Van Steen, K., Van Schuerbeek, N., Vlietinck, R., Rutgeerts, P., and Vermeire, S. (2006). Toll-like receptor-1, -2, and -6 polymorphisms influence disease extension in inflammatory bowel diseases. Inflamm. Bowel Dis. *12*, 1–8.

Pochard, P., Hammad, H., Ratajczak, C., Charbonnier-Hatzfeld, A.S., Just, N., Tonnel, A.B., and Pestel, J. (2005). Direct regulatory immune activity of lactic acid bacteria on Der *p* 1-pulsed dendritic cells from allergic patients. J. Allergy Clin. Immunol. *116*, 198–204.

Prescott, S.L., Wickens, K., Westcott, L., Jung, W., Currie, H., Black, P.N., Stanley, T.V., Mitchell, E.A., Fitzharris, P., Siebers, R., Wu, L., Crane, J., and the Probiotic Study Group. (2008). Supplementation with *Lactobacillus rhamnosus* or *Bifidobacterium lactis* probiotics in pregnancy increases cord blood interferon-gamma and breast milk transforming growth factor-beta and immunoglobin A detection. Clin. Exp. Allergy *38*, 1606–1614.

Prisciandaro, L., Geier, M., Butler, R., Cummins, A., and Howarth, G. (2009). Probiotics and their derivatives as treatments for inflammatory bowel disease. Inflamm. Bowel Dis. *15*, 1906–1914.

Rakoff-Nahoum, S., Paglino, J., Eslami-Varzaneh, F., Edberg, S., and Medzhitov, R. (2004). Recognition of commensal microflora by toll-like receptors is required for intestinal homeostasis. Cell *118*, 229–241.

Rescigno, M., Urbano, M., Valzasina, B., Francolini, M., Rotta, G., Bonasio, R., Granucci, F., Kraehenbuhl, J.P., and Ricciardi-Castagnoli, P. (2001). Dendritic cells express tight junction proteins and penetrate gut epithelial monolayers to sample bacteria. Nat. Immunol. *2*, 361–367.

Roediger, W.E. (1980). The colonic epithelium in ulcerative colitis: an energy-deficiency disease? Lancet 2, 712–715.

Roselli, M., Finamore, A., Britti, M.S., and Mengheri, E. (2006). Probiotic bacteria *Bifidobacterium animalis* MB5 and *Lactobacillus rhamnosus* GG protect intestinal Caco-2 cells from the inflammation-associated response induced by enterotoxigenic *Escherichia coli* K88. Br. J. Nutr. *95*, 1177–1184.

Sansonetti, P.J. (2004). War and peace at mucosal surfaces. Nat. Rev. Immunol. *4*, 953–964.

Sartor, R.B. (2004). Therapeutic manipulation of the enteric microflora in inflammatory bowel diseases: antibiotics, probiotics, and prebiotics. Gastroenterology *126*, 1620–1633.

Schiffrin, E.J., and Blum, S. (2002). Interactions between the microbiota and the intestinal mucosa. Eur. Clin, J. Nutr. *56* (Suppl. 3), 560–564.

Scholtens, P.A., Alliet, P., Raes, M., Alles, M.S., Kroes, H., Boehm, G., Knippels, L.M., Knol, J., and Vandenplas, Y. (2008). Fecal secretory immunoglobulin A is increased in healthy infants who receive a formula with short-chain galacto-oligosaccharides and long-chain fructooligosaccharides. J. Nutr. *138*, 1141–1147.

Schouten, B., van Esch, B.C., Hofman, G.A., van Doorn, S.A., Knol, J., Nauta, A.J., Garssen, J., Willemsen, L.E., and Knippels, L.M. (2009). Cow milk allergy symptoms are reduced in mice fed dietary synbiotics during oral sensitization with whey. J. Nutr. *139*, 1398–1403.

Seth, A., Yan, F., Polk, D.B., and Rao, R.K. (2008). Probiotics ameliorate the hydrogen peroxide-induced epithelial barrier disruption by a PKC- and MAP kinase-dependent mechanism. Am. J. Physiol. Gastrointest. Liver Physiol. *294*, G1060–G1069.

Shevach, E.M. (2000). Suppressor T cells: Rebirth, function and homeostasis. Curr. Biol. *10*, R572–R575.

Shiba, T., Aiba, Y., Ishikawa, H., Ushiyama, A., Takagi, A., Mine, T., and Koga, Y. (2003). The suppressive effect of bifidobacteria on *Bacteroides vulgatus*, a putative pathogenic microbe in inflammatory bowel disease. Microbiol. Immunol. *47*, 371–378.

Shida, K., Makino, K., Morishita, A., Takamizawa, K., Hachimura, S., Ametani, A., Sato, T., Kumagai, Y., Habu, S., and Kaminogawa, S. (1998). *Lactobacillus casei* inhibits antigen-induced IgE secretion through regulation of cytokine production in murine splenocyte cultures. Int. Arch. Allergy Immunol. *115*, 278–287.

Steinman, L. (2007). A brief history of T_H17, the first major revision in the T_H1/T_H2 hypothesis of T cell-mediated tissue damage. Nat. Med. *13*, 139–145.

Szebeni, B., Veres, G., Dezsõfi, A., Rusai, K., Vannay, A., Mraz, M., Majorova, E., and Arató, A. (2008). Increased expression of Toll-like receptor (TLR) 2 and TLR4 in the colonic mucosa of children with inflammatory bowel disease. Clin. Exp. Immunol. *151*, 34–41.

Takahashi, N., Kitazawa, H., Iwabuchi, N., Xiao, J.Z., Miyaji, K., Iwatsuki, K., and Saito, T. (2006a). Oral administration of an immunostimulatory DNA sequence from *Bifidobacterium longum* improves Th1/Th2 balance in a murine model. Biosci. Biotechnol. Biochem. *70*, 2013–2017.

Takahashi, N., Kitazawa, H., Shimosato, T., Iwabuchi, N., Xiao, J.Z., Iwatsuki, K., Kokubo, S., and Saito, T. (2006b). An immunostimulatory DNA sequence from a probiotic strain of *Bifidobacterium longum* inhibits IgE production *in vitro*. FEMS Immunol. Med. Microbiol. *46*, 461–469.

Takeda, Y., Nakase, H., Namba, K., Inoue, S., Ueno, S., Uza, N., and Chiba, T. (2009). Upregulation of T-bet and tight junction molecules by *Bifidobactrium longum* improves colonic inflammation of ulcerative colitis. Inflamm. Bowel Dis. *15*, 1617–1618.

Tanabe, S., Kinuta, Y., and Saito, Y. (2008). *Bifidobacterium infantis* suppresses proinflammatory interleukin-17 production in murine splenocytes and dextran sodium sulfate-induced intestinal inflammation. Int. J. Mol. Med. *22*, 181–185.

Taniuchi, S., Hattori, K., Yamamoto, A., Sasai, M., Hatano, Y., Kojima, T., Kobayashi, Y., Iwamoto, H., and Yaeshima, T. (2005). Administration of *Bifidobacterium* to infants with atopic dermatitis: changes in fecal microflora and clinical symptoms. J. Appl. Res. *5*, 387–396.

Török, H.P., Glas, J., Tonenchi, L., Mussack, T., and Folwaczny, C. (2004). Polymorphisms of the lipopolysaccharide–signaling complex in inflammatory bowel disease: association of a mutation in the Toll-like receptor 4 gene with ulcerative colitis. Clin. Immunol. *112*, 85–91.

Turner, J.R. (2006). Molecular basis of epithelial barrier regulation: from basic mechanisms to clinical application. Am. J. Pathol. *169*, 1901–1909.

Uematsu, S., Jang, M.H., Chevrier, N., Guo, Z., Kumagai, Y., Yamamoto, M., Kato, H., Sougawa, N., Matsui, H., Kuwata, H., *et al.* (2006). Detection of pathogenic intestinal bacteria by Toll-like receptor 5 on intestinal CD11c+ lamina propria cells. Nat. Immunol. *7*, 864–874.

Ulisse, S., Gionchetti, P., D'Alo, S., Russo, F.P., Pesce, I., Ricci, G., Rizzello, F., Helwig, U., Cifone, M.G., Campieri, M., and de Simone, C. (2001). Expression of cytokines, inducible nitric oxide synthase, and matrix metalloproteinases in pouchitis: effects of probiotic treatment. Am. J. Gastroenterol. *96*, 2691–2699.

von der Weid, T., Bulliard, C., and Schiffrin, E.J. (2001). Induction by a lactic acid bacterium of a population of CD4(+) T cells with low proliferative capacity that produce transforming growth factor beta and interleukin-10. Clin. Diagn. Lab. Immunol. *8*, 695–701.

Wächtershäuser, A., Loitsch, S.M., and Stein, J. (2000). PPAR-gamma is selectively upregulated in Caco-2 cells by butyrate. Biochem. Biophys. Res. Commun. *272*, 380–385.

Wächtershäuser, A., and Stein, J. (2000). Rationale for the luminal provision of butyrate in intestinal diseases. Eur. J. Nutr. *39*, 164–171.

Wald, D., Qin, J., Zhao, Z., Qian, Y., Naramura, M., Tian, L., Towne, J., Sims, J.E., Stark, G.R., and Li, X. (2003). SIGIRR, a negative regulator of Toll-like receptor-interleukin 1 receptor signaling. Nat. Immunol. *4*, 920–927.

Xavier, R.J., and Podolsky, D.K. (2007). Unravelling the pathogenesis of inflammatory bowel disease. Nature *448*, 427–434.

Xiao, J.Z., Kondo, S., Yanagisawa, N., Takahashi, N., Odamaki, T., Iwabuchi, N., Iwatsuki, K., Kokubo, S., Togashi, H., Enomoto, K., and Enomoto, T. (2006a). Effect of Probiotic *Bifidobacterium longum* BB536 in relieving clinical symptoms and modulating plasma cytokine levels of Japanese cedar pollinosis during the pollen season, a randomized, double-blind, placebo-controlled trial. J. Invest. Allergol. Clin. Immunol. *16*, 86–93.

Xiao, J.Z., Kondo, S., Yanagisawa, N., Takahashi, N., Odamaki, T., Iwabuchi, N., Miyaji, K., Iwatsuki, K., Togashi, H., Enomoto, K., and Enomoto, T. (2006b). Probiotics in the treatment of Japanese cedar pollinosis: a double-blind placebo-controlled trial. Clin. Exp. Allergy *36*, 1425–1435.

Xiao, J.Z., Kondo, S., Yanagisawa, Miyaji, N., K., Enomoto, K., Sakota, T., Iwatsuki, K., and Enomoto, T. (2007). Clinical efficacy of probiotic *Bifidobacterium longum* for the treatment of symptoms of Japanese cedar pollen allergy in subjects evaluated in an environmental exposure unit. Allergol. Int. *56*, 67–75.

Yan, F., Cao, H., Cover, T.L., Whitehead, R., Washington, M.K., and Polk, D.B. (2007). Soluble proteins produced by probiotic bacteria regulate intestinal epithelial cell survival and growth. Gastroenterology *132*, 562–575.

Zeuthen, L.H., Fink, L.N., and Frøkiær, H. (2008). Toll-like receptor 2 and nucleotide-binding oligomerization domain-2 play divergent roles in the recognition of gut-derived lactobacilli and bifidobacteria in dendritic cells. Immunology *124*, 489–502.

Zhang, G., and Ghosh, S. (2002). Negative regulation of toll-like receptor-mediated signaling by Tollip. J. Biol. Chem. *277*, 7059–7065.

Safety Assessment of Bifidobacteria

Maria H. Saarela

Abstract

Bifidobacteria can be considered to be very safe to humans since no cases where a probiotic bifidobacterial product has caused an infection have been reported. Also bifidobacteria belonging to the resident microbiota of the oro-gastrointestinal tract very rarely cause infections. Oral bifidobacterial species should, however, be avoided in probiotic products, since these are considered to be oral pathogens and they have also occasionally caused infections. Probiotic *Bifidobacterium* strains should be chosen among the species that either have previously been commonly used in probiotic human products or represent the commonly detected species in the human gastrointestinal tract (GIT). In addition, safety assessment of a *Bifidobacterium* strain should involve at least diverse phenotypic characterization of the strain (to indicate whether it is a typical representative of the species), antibiotic resistance pattern assessment (some strains might carry acquired resistance genes and these should be avoided), and assessment of side effects during human studies. Finally, an epidemiological surveillance of adverse incidents in consumers (post-market) should be performed at a later stage. Qualified Presumption of Safety (QPS) and its four pillars of safety assessment should be the basis of the safety evaluation of probiotics, although QPS is currently applied only to bacteria added to feed and used as biocontrol agents in the EU.

Introduction

According to Frost and Sullivan (2007), the total European food and beverage probiotics market was worth approximately $61.7 million in 2006. The dairy yoghurt and yoghurt drink application sector accounted for the highest market share. Probiotic foods contain several different *Lactobacillus* species and strains, whereas the products with bifidobacteria are dominated by *Bifidobacterium animalis* subsp. *lactis* (previously *Bifidobacterium lactis*) strain(s) (Masco *et al.*, 2005). The increasing popularity of probiotic foods in Europe unavoidably results in a situation where various consumer groups, including also immunocompromised adults and children, consume probiotics. The heterogeneity of the consumer groups (regarding their health status) and the fact that probiotic consumption can involve daily ingestion of large numbers of viable bacterial cells (daily dose 10^9–10^{11} CFU) highlight the importance of safety of the probiotic consumption.

Most people harbour bifidobacteria in their oro-gastrointestinal tract as a normal part of their oral and gut microbiota (Maukonen *et al.*, 2008). Therefore, while evaluating the safety or the consumption of probiotic bifidobacteria, a clear distinction has to be made between indigenous (resident) and ingested bifidobacteria.

Human infections caused by indigenous bifidobacteria

Bifidobacterium species are common but non-dominant members of the indigenous microbiota of the human gastrointestinal tract (GIT) (Sghir *et al.*, 2000; Lay *et al.*, 2005; Mueller *et al.*, 2006) and also of the oral cavity (Maukonen *et al.*, 2008). Like other members of the indigenous microbiota of the oro-gastrointestinal tract, they occasionally cause opportunistic infections in humans. Indigenous bifidobacteria have been connected to certain dental infections, bacteraemia, and to rare cases of other infections (Gasser, 1994; Saarela *et al.*, 2002; Hammerman *et al.*, 2006).

Bacteraemia

Bloodstream infections (BSIs), a major cause of morbidity and mortality globally, are increasing in incidence. In the US alone about 250,000 cases of BSI occur annually. In addition to increased incidence, the microbiology of BSIs seems to be changing, with a trend towards a growing predominance of Gram-positive organisms. However, there can be differences in microbiology between community acquired and nosocomial BSIs and between different patient groups. Especially the increased incidence of methicillin-resistant *Staphylococcus aureus* in BSIs is of concern (Bearman and Wenzel, 2005).

Earlier studies indicate an involvement of Gram-negative organisms in 38%, coagulase-negative staphylococci in 33%, *S. aureus* in 16%, streptococci in 7%, enterococci in 5–15%, and lactobacilli in 0.1–0.2% of bacteraemia cases (for bifidobacteria there are no figures available) (Gasser, 1994; Saxelin *et al.*, 1996; Salminen *et al.*, 2002). In nosocomial BSIs coagulase-negative staphylococci were detected in 33%, *S. aureus* in 20%, enterococci in 9%, *Candida* yeasts in 9%, and *Escherichia coli*, *Klebsiella* sp. *Pseudomonas aeruginosa* and *Enterobacter* sp. each in < 3% of the cases (Bearman and Walker, 2005).

Bacteraemias caused by bifidobacteria seem to be very rare, and only sporadically reported. In an early report of Bourne *et al.* (1978) bifidobacteria were obtained from nine patients during the 6-year period of the BSI study (1972–1977). Underlying conditions of the patients included complications with pregnancy (the most common condition), bowel obstruction, cholecystectomy, and systemic lupus erythematosus. Five of the cases were caused by *Bifidobacterium eriksonii*, reclassified as *Bifidobacterium dentium*, which is regarded as a dental pathogen. There are only few reports on septicaemia due to bifidobacteria (Guillard *et al.*, 1980; Ha *et al.*, 1999). In the paper of Guillard *et al.* (1980) the isolate was identified as 'Gram-positive rod similar to *Corynebacterium* group E (aerotolerant *Bifidobacterium adolescentis*)'. However, because of the limitations of the methodologies used, this identification must be considered doubtful. In the case report of Ha *et al.* (1999) the causative agent of septic infection was identified as *Bifidobacterium longum*. A PubMed search did not reveal any other cases of bifidobacteria and BSIs. However, in the paper of Salminen *et al.* (2002) where *Lactobacillus* bacteraemia cases in Finland were investigated, it was noted that of the 66 isolates studied (that were initially identified as lactobacilli), 18 (27%) proved to be other bacteria, among these three bifidobacteria. This paper indicates that the misidentification rate for lactic acid bacteria and bifidobacteria can be quite high in clinical laboratories which have no special expertise in the identification of commensal anaerobic bacteria. Also in the paper of Petti *et al.* (2005) it was noted that phenotypic methods commonly used in clinical laboratories can lead into misidentification – in this case bifidobacteria isolated from the blood of an endocarditis case turned out to be *Streptococcus mutans* when identification was based on partial 16S rRNA gene sequencing. These papers highlight the issue that identification of untypical bacteria (in this case bifidobacteria) in clinical samples should always be verified with appropriate molecular techniques.

Endocarditis and pericarditis

Infective endocarditis (IE) is defined as a microbial infection of the endocardial surface (lining of the heart and its valves). It may occur as a primary disorder or as a complication of or in association with another disease. Dental operations causing transient bacteraemias seem to be important starting points for IEs. When a portal of entry is created by dental operation/surgery (e.g. periodontal probing, scaling and root planing, extractions, gingival surgery), oral bacteria can reach alveolar blood and blood circulation. Host defences usually destroy these bacteria but in patients with predisposing factors that have damaged heart tissue or heart valve the bacteria can colonize the tissue/valve and cause IE. Sometimes, bacterial aggregates and emboli break away from the heart tissue and lodge in other organs leading to other systemic infections (Nisengard *et al.*, 1998). The microbiology of IE depends on whether it occurs on a native valve or on a prosthetic valve, or whether the disease is hospital or community acquired. *S. aureus*, *Streptococcus* spp. and *Enterococcus* spp. account for majority (over 80%) of all IE cases (Hill *et al.*, 2006). Similarly to BSI *S. aureus* incidence seems to be increasing also in IE. A PubMed search did not reveal any cases of bifidobacteria and IE.

Pericarditis is an inflammation of the pericardium, which is the sac-like covering around the heart. The aetiological agents of infectious pericarditis are diverse including bacteria, viruses, fungi and protozoa. The bacteria include *S. aureus*, *Neisseria meningitidis*, *Streptococcus* spp., *Haemophilus influenzae*, *E. coli*, *Klebsiella* spp., *Salmonella* spp., *P. aeruginosa*, *Staphylococcus epidermidis* and anaerobic bacteria (Brook, 2009). However, pericarditis owing to anaerobic bacteria is considered to be rare (Skiest *et al.*, 1994). There is one early report (Epstein *et al.*, 1992) about detecting bifidobacteria in a polymicrobial bacterial pericarditis (other bacteria detected included *Streptococcus milleri*, alpha-haemolytic streptococci, *Bacteroides ureolyticus*, and two unidentified Gram-positive rods). However, the paper does not give any information about the microbiological methods used and therefore the identification of *Bifidobacterium* in this case remains doubtful.

Other infections

Three *Bifidobacterium* species, *B. dentium*, *B. inopinatum*, and *B. denticolens*, have earlier been associated with human dental caries (Crociani *et al.*, 1996) and are therefore considered to be oral pathogens. Of these species two have since been transferred to other genera: *B. inopinatum* is currently *Scardovia inopinata* while *B. denticolens* is currently *Parascardovia denticolens* (http://www.dsmz.de/microorganisms/bacterial_nomenclature.php?select=L). Early studies report the isolation of bifidobacteria from otitis media, abscesses, peritonitis, and from miscellaneous other infections in children (Brook, 1996). Only in two cases (4%) were bifidobacteria isolated in pure cultures (*B. adolescentis* from otitis media and *Bifidobacterium* sp. from cervical adenitis). Isolated cases of *B. dentium* (previously *B. eriksonii*) infections have been reported in adults (Thomas *et al.* 1974; Green *et al.*, 1978; Civen *et al.*, 1993). Furthermore, *Bifidobacterium* sp. has been isolated from abdominal and reproductive tract abscesses (Brook and Frazier, 1993). However, in these early studies bifidobacteria were identified with phenotypic methods only, and since bifidobacterial identification with phenotypic tests can be difficult, there is a considerable chance that many of these were actually not bifidobacteria at all. Naturally also the opposite scenario is possible: because of the fastidious nature of bifidobacteria laboratories may fail to detect these bacteria in clinical samples.

Furthermore, it has to be kept in mind that the isolation of bifidobacteria in infection sites is not a solid evidence for any cause–effect relationship. Especially in situations when the infection is polymicrobial and antibiotic treatment had preceded the isolation, drawing any conclusions on the role of the isolated bacteria in infections is very difficult.

Safety evaluation of probiotic bifidobacteria aimed at human consumption

Owing to the importance of safe consumption of probiotic foods and supplements, several approaches have been taken to elucidate this issue. The research area of probiotic safety has been fairly fragmented and several reviews about probiotic safety have been published with somewhat variable approaches (see e.g. Borriello *et al.*, 2003; Bernardeau *et al.*, 2006). Attempts to obtain a consensus or guidelines for probiotic safety evaluation have been made in FAO/WHO working group (Anonymous, 2002) and EU funded EU-PROSAFE project carried out in 2002–2006 (Vankerckhoven *et al.*, 2008). FAO/WHO working group report (Anonymous, 2002) recommends that probiotic strains are characterized at a minimum with following: determination of antibiotic resistance patterns, assessment of certain metabolic activities (e.g. D-lactate production, bile salt deconjugation), assessment of side effects during human studies, epidemiological surveillance of adverse incidents in consumers (post-market), if the strain belongs to a species that is known to produce mammalian toxin, the toxin production needs to be tested, and if the strain belongs to a species with known haemolytic potential, this need to be tested (Fig. 9.1).

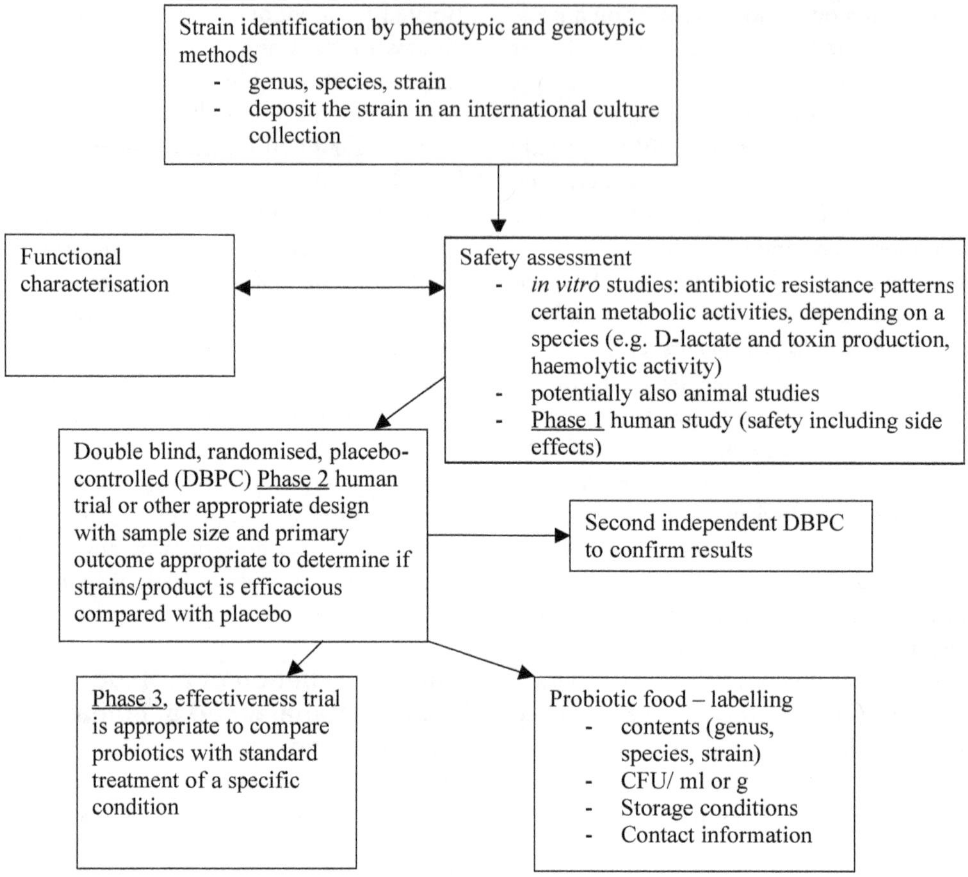

Figure 9.1 Guidelines for the evaluation of probiotics for food applications (adopted from Anonymous, 2002).

Scientific Committee on Animal Nutrition (SCAN) in EU launched an initiative towards a 'Qualified Presumption of Safety' (QPS) concept which allows probiotic strains with established safety status to enter the animal feed market without extensive testing requirements (von Wright, 2005). Since there is a need for a tool for setting priorities within the risk assessment of microorganisms used both in food and feed production referred to European Food Safety Authority (EFSA), it has been proposed that the safety assessment of a defined taxonomic group (e.g. genus or group of related species) could be made based on four pillars (establishing identity, body of knowledge, possible pathogenicity, and end use). If the taxonomic group does not raise safety concerns or, if safety concerns exist, but could be defined and excluded, the grouping could be granted QPS status. Any microorganism that could be assigned to a QPS group would then be freed from the need for further safety assessment other than satisfying any qualifications specified. Microorganisms not considered suitable for QPS would remain subject to a full safety assessment. (http://www.efsa.europa.eu/EFSA/efsa_locale1178620753812_1178667590178.htm). The QPS approach is a system similar in concept and purpose to the GRAS (generally recognized as safe) definition used in the USA, but modified to take account of the different regulatory practices in Europe. In the current list of taxonomic units proposed for QPS status following *Bifidobacterium* species are listed: *B. adolescentis, B. bifidum, B. longum, B. animalis,* and *B. breve* (Anonymous, 2007). QPS system is currently used only for microbes whose safety EFSA has to evaluate. These include bacteria added into feed and used as biocontrol agents. However, the same approach, although not compulsory, would be useful for the evaluation of human probiotics also.

Up to date the complete genome sequences of three *Bifidobacterium* strains representing two species (*B. animalis* subsp. *lactis* and *B. longum*) have been published in scientific literature (Schell *et al.*, 2002; Barrangou *et al.*, 2009). No special virulence genes that would be of concern to human health were identified in these studies.

Importance of the correct species-level identification

The first step in probiotic safety assessment is a proper identification of the strain in question to the right genus, species, and if necessary also subspecies. This may seem trivial, but considering that a fair number of probiotic products on the market are mislabelled regarding either the identity or numbers (or both) of the probiotic(s) in the product (Masco *et al.*, 2005; Theunissen *et al.*, 2005), the relevance of this issue becomes evident. For the identification of lactic acid bacteria and bifidobacteria suitable molecular tools (partial or, if necessary, complete 16S rRNA gene sequencing, or various fingerprinting methods such pulsed field gel electrophoresis, ribotyping, amplified fragment length electrophoresis) should be applied – phenotypic tools alone are not sufficient for this task (Ben Amor *et al.*, 2007; Vankerckhoven *et al.*, 2008). Not all methods suit for all (sub)species, so sometimes a combination of various techniques needs to applied to obtain the correct identification. Proper identification of a probiotic strain enables performing an initial assessment of the safety of its consumption. Information about the genus and species facilitates the evaluation of e.g. potential risk of human infections (bacteraemias, endocarditises, and other infections caused by the same genus/species earlier, antibiotic resistance carriage (does the species have any mobile/untypical antibiotic resistance genes?), carriage of potential harmful traits). This background information gives valuable guidance to the safety studies that should then be performed with the strain in question. *In vitro*, animal, and human studies can be considered for the safety assessment and all these areas are discussed below.

In vitro studies

Some microbial enzymatic activities (which can be studied *in vitro*) including deconjugation (or hydrolysis) of bile salts in the small intestine or degradation of mucus, can potentially be detrimental to the host (Marteau *et al.*, 1995). Bile salt hydrolase (BSH) activity has been detected in various GIT bacteria such as *Lactobacillus, Enterococcus, Clostridium, Bacteroides,* and *Bifidobacterium* (Begley *et al.*, 2006), including *B. animalis* and *B. longum* (Kumar *et al.*, 2006; Kim and Lee, 2008). BSH can be considered to be a virulence factor in pathogenic bacteria, since it contributes to bacterial survival and persistence in the host. However, survival and persistence in the GIT is typically a desired feature in probiotics (Begley *et al.*, 2006). In the viewpoint written by EU-PROSAFE project partners (Vankerckhoven *et al.*, 2008) it was concluded, owing to the inconclusive nature of potential harmful effects of BSH activity that 'bile salt deconjugation activity is irrelevant for safety assessment and is not recommended for *in vitro* safety testing of probiotics'.

O'Brien *et al.* (1999) and Bernardeau *et al.* (2006) listed several metabolic activities that can be considered in probiotic safety evaluation. These included biogenic amine production, bile salt hydroxylase, D- vs. L-lactate production, mucin degradation, cholyl glycine hydrolase, azoreducatse, nitroreductase, β-glucuronidase and β-glucosidase. All these properties can be studied *in vitro*. However, their *in vivo* relevance often remains speculative.

Platelet aggregation properties (Korpela *et al.*, 1997), adhesion to human intestinal mucosa and mucus (Kirjavainen *et al.*, 1999; Apostolou *et al.*, 2001) and formation of unwanted metabolites can also be studied *in vitro*. Platelet aggregation contributes to the pathogenesis of IE, and aggregation induced by bacteria can therefore be an important contributory factor in the development and progress of IE. Platelet aggregation ability of bifidobacteria has been little studied and so far it has not been detected (Zhou *et al.*, 2005). In lactobacilli it has been more extensively studied but the finding that *in vitro* platelet aggregation ability as well as adhesion to mucus, ileostomy glycoproteins or to Caco-2 cell line are variable among bacteraemia-associated lactobacilli (Kirjavainen *et al.*, 1999), undermines the value of these analyses in the safety assessment, at least for lactobacilli. Similarly to BSH activity adhesion to mucosa and mucus can be considered to be either harmful (for pathogens) or beneficial (for probiotics) (Ouwehand *et al.*, 1999; Apostolou *et al.*, 2001). Owing to the inconclusive nature of the current adhesion assays regarding probiotic safety, Vankerckhoven *et al.* (2008), in their EU-PROSAFE project viewpoint, concluded that 'adhesion assays currently are not recommended for safety assessment'.

Animal models

The general bifidobacterial safety has been studied in BALB/c mice by investigating the following parameters: feed and water intake, body weight, general health status (during the bifidobacterial intervention); haematological parameters, differential leucocyte counts, blood biochemistry, mucosal histology, and bacterial translocation to extra-gut tissues (blood, liver, spleen, kidney, and mesenteric lymph nodes), internal organ indices, and liver and kidney marker enzymes profiles (at the end of the intervention) (Zhou *et al.*, 2000; Kabeir *et al.*, 2007). In these studies bifidobacteria (*B. pseudocatenumatum* and *B. animalis* subsp. *lactis* strains) were considered to be non-pathogenic to mice and therefore likely to be safe for human use.

Animal models are usually designed to mimic the pathogenesis of infectious or inflammatory diseases. Since lactobacilli and bifidobacteria have low pathogenicity, only a few models can be used to investigate their potential virulence. The suitability of rat endocarditis model and immunocompromised mouse model has been tested with probiotics. Of these two, rat endocarditis model detected differences between probiotic lactobacilli and those recovered from IE, and thus rat model of endocarditis might be relevant for the safety assessment of probiotics. Immunocompromised mouse model, on the other hand, was shown to be unsuitable for bifidobacteria, since no translocation of bifidobacteria could be detected in this model (Vankerckhoven *et al.*, 2008). There is also a study where a mouse model of experimental autoimmune thyroiditis has been used to investigate the immunostimulatory properties of a *B. animalis* subsp. *lactis* strain (Zhou and Gill, 2005). Whether this model could be suitable for studying bifidobacterial safety cannot, however, be evaluated based on this single study.

Human studies

According to Marteau (2004) probiotics may theoretically be responsible for four types of side effects in susceptible individuals including infection, deleterious metabolic activity, excessive immune stimulation, and gene transfer. So far only infection has been observed in humans.

In the assessment of the probiotic safety several approaches, including studies on the intrinsic properties and pharmacokinetics of the probiotic strain and on interactions between the probiotic strain and the host can be taken. Information about survival of the probiotics within the GIT, their translocation and colonization properties, and the fate of probiotic-derived active components, is important for the evaluation of the effects (both positive and negative) of probiotic ingestion. The survival of different probiotic strains in different parts of the GIT varies: Some strains are rapidly killed in the stomach while others can pass through the GIT tract in fairly high numbers (Marteau *et al.*, 1993). Pharmacokinetics of probiotics in humans has been studied using aspiration, perfusion and biopsy techniques (Pochart *et al.*, 1992; Johansson *et al.*, 1993; Nielsen *et al.*, 1994; Alander *et al.*, 1997, 1999; McNaught *et al.*, 2005).

General population

Since infants have a higher risk of infection than adults, several infant intervention studies have been performed where the primary focus has been on the safety and tolerance of probiotic (including bifidobacteria) consumption (Saavedra *et al.*, 2004; Weizman *et al.*, 2006; Chouraqui *et al.*, 2008; Dekker *et al.*, 2009). The parameters followed included e.g. feeding, behaviour (e.g. crying, irritability) and stooling characteristics, growth, general health status, and gastrointestinal and related symptoms. The tested probiotics (including *B. animalis* and *B. longum*) were considered safe and well tolerated in infants.

Safety-related bifidobacterial intervention studies have also been performed in healthy adults. In the study of Mäkeläinen *et al.* (2003), well being and intestinal symptoms as well as consistency and frequency of stools were monitored.

In addition immunological parameters were studied during *B. longum* intervention. In the study of Christensen *et al.* (2006), a dose–response study was performed with *B. animalis* and *Lactobacillus paracasei* strains and various blood parameters measured. Neither study indicated any safety concerns related to probiotic consumption.

Numerous other bifidobacterial intervention studies on various volunteer groups have been performed. PubMed search resulted in 51 references with the index words *Bifidobacterium* and intervention and with the limits set to human trials. In many of these the volunteers were asked to report potential adverse effects (GIT symptoms i.e. tolerance) of the probiotic consumption. Since these studied were not focusing on the safety and typically adverse effects were only self-reported these studies were not considered for this review. However, no infections due to the consumption of probiotic bifidobacteria have been reported.

Severely ill patients

Regarding the safety of probiotic consumption, critically ill patients have been of special interest mainly due to the results of Besselink *et al.* (2008) indicating that in patients with severe acute pancreatitis probiotic consumption (mixture of *Lactobacillus acidophilus, Lactobacillus casei, Lactobacillus salivarius, Lactococcus lactis, B. bifidum* and *B. animalis* subsp. *lactis*) was associated with an increased risk of mortality. This result was surprising since earlier studies on acute pancreatitis had not indicated the same outcome. According to Madsen (2008), Morrow (2009), and Koretz (2009), data supporting the use of probiotics in the intensive care unit (ICU) are variable in scope and quality and therefore large, well designed trials are needed to evaluate the safety of probiotics in critically ill patients.

One practical issue related to safe probiotic administration in ICU is the type of formula used. Powdery preparations of the probiotic yeast *Saccharomyces boulardii* have caused several infections in ICU patients (Marteau, 2004). Thus Marteau concludes that probiotics should not be used in powder form in the ICU. Whether the same risk associated with powdery probiotic yeast preparations also applies to other probiotic organisms is not know, but this precaution is nevertheless useful with very ill patients.

Epidemiological surveillance of adverse incidents

One of the recommendations of FAO/WHO working group report (Anonymous, 2002) is epidemiological surveillance of adverse incidents in consumers. To my knowledge this has so far only been performed in Finland with a special focus on *Lactobacillus* GG. The systematic epidemiological research on lactobacillaemias in Finland where lactobacilli isolates from bacteraemias have been systematically checked for identity with commercial starter and probiotic strains (Salminen *et al.*, 2004) sets a very good example for other countries where probiotics are commonly used in foods. Only with these kind of post-marketing surveillance studies can we properly assess the risk related to probiotic (or fermented food) consumption.

Antibiotic resistance in probiotic bifidobacteria – safety implications

An additional safety consideration is the possible transferability of antibiotic resistance genes between probiotics and the members of the GIT microbiota. Bifidobacteria display a wide range of antibiotic resistances naturally, but in most cases antibiotic resistance is not of the transmissible type, and therefore it does not usually form a safety concern (Saarela *et al.*, 2002). Bifidobacteria can be intrinsically resistant to nalidixic acid, neomycin, polymyxin B, kanamycin, gentamycin, streptomycin and metronidazole. Most of the strains are susceptible to chloramphenicol, erythromycin, clindamycin and vancomycin, but some erythromycin and clindamycin resistant strains have been detected (Masco *et al.*, 2006; Ammor *et al.*, 2007; Saarela *et al.*, 2008). Susceptibility to tetracycline varies. Different tetracycline resistance genes have been characterized in resistant strains and tet(W) is the most commonly detected gene (Aires *et al.*, 2007). Further details about the antibiotic resistance in bifidobacteria can be found elsewhere in this book (Chapter 11).

From the safety point of view the presence of potentially mobile erythromycin and tetracycline resistance in some bifidobacterial strains is of concern (Ammor *et al.*, 2008; van Hoek *et al.*, 2008). In a study by Saarela *et al.* (2007), effects of oral therapy with doxycycline, a tetracycline group antibiotic, on the GIT survival and tetracycline susceptibility of probiotic strains *L. acidophilus* LaCH-5 (Tc^S) and *B. animalis* subsp. *lactis* Bb-12 (Tc^I) were studied. Several subjects who had consumed antibiotics had faecal *B. animalis* subsp. *lactis* Bb-12-like isolates with reduced tetracycline susceptibility. This was unlikely because of the acquisition of novel tetracycline resistance determinants since only *tet*(W), which is also present in the ingested *B. animalis* subsp. *lactis* Bb-12, was found in the resistant isolates. *L. acidophilus,* on the other hand, remained susceptible during antibiotic therapy. Thus, the concomitant ingestion of probiotic *L. acidophilus* and *B. animalis* subsp. *lactis* with the antibiotic did not generate a safety risk regarding the possible GIT transfer of tetracycline resistance genes to the ingested strains.

Since GIT bacteria are able to exchange genetic material, also the possibility that a probiotic strain with a mobile antibiotic resistance gene might transfer it to potentially harmful bacteria in the gut has to be kept in mind. During the past few years a lot of data has accumulated about bifidobacterial antibiotic resistance and genetic elements causing the trait. This knowledge enables choosing strains with no potential transferable antibiotic resistance genes for human and animal consumption.

Conclusions

When considering the safety risk of probiotic consumption we have to keep in mind that lactic acid bacteria have been globally consumed in a multitude of fermented food varieties (milk, meat, vegetable, and cereal products) for a very long time without an indication that they could be generally harmful to the consumers' health. Thus, the long history of safe use of lactobacilli and bifidobacteria remains the best proof of their safety. However, since the number of elderly people (often with serious underlying diseases, clinical or subclinical) is increasing, also the predisposing factors for serious bacterial infections are becoming more common in the population. This sets great demands for the safety of probiotic products as well as for any other product containing living microbes. In seriously ill patient groups one possibility to minimize any safety risks could be the use of non-viable probiotic preparations. Non-viable cells do not have all the same properties as their living counterparts but they have proved promising in the treatment of certain disorders (e.g. lactose maldigestion and treatment of gastroenteritis) (Ouwehand and Salminen, 1998; Ouwehand *et al.*, 2000).

Assessing the risks of probiotic consumption can be very expensive and time-consuming. A low risk may have to be accepted when recommended to immunocompromised individuals, but the risk to benefit ratio needs to be clearly established in such cases. This requires relevant information on the efficacy and safety of the products. However, probiotics have been administered to various patient groups (Crohn's, diarrhoeal, AIDS, cancer), infants (allergic and healthy) and elderly without any reported cases of infection. Basis for the safety evaluation comes from the following facts: probiotic foods should be as safe as other foods, host factors are essential determinants of safety evaluation, and zero risk can never be applied to the ingestion of live microbes (whether probiotics, starters, or components of natural microbiota of fresh foods, e.g. fruits and vegetables). Suggesting direct cause–effect relationship between a probiotic and infectious disease should be based on critically assessed facts to rule out misjudgements. According to Hammerman *et al.* (2006), 'safety can not be defined as an absolute risk-free condition but should be evaluated in context'.

Bifidobacteria are very safe since no cases where a probiotic bifidobacterial product has caused an infection have been reported. Also bifidobacteria belonging to the resident microbiota of the oro-gastrointestinal tract very rarely cause infections. Even among the few cases reported some are doubtful regarding the bifidobacterial identification. It is well known that bifidobacterial identification without special expertise in the genus may be difficult and lead into misidentifications. Thus valid microbiological methods combined to discriminative molecular techniques needs to be used in the identification. Since human oro-gastrointestinal tract is inhabited by several species of bifidobacteria including the ones used in probiotic products, any suggestion that the source of *Bifidobacterium* infection is exogenous (e.g. a probiotic product) should be very carefully considered. The review paper of Hammerman *et al.* (2006) states that 'current evidence suggest that the risk of infection with probiotic lactobacilli or bifidobacteria is similar to that of infection with commensal strains, and that consumption of such products presents a negligible risk to consumers, including immunocompromised hosts'.

References

Aires, J., Doucet-Populaire, F., and Butel, M.J. (2007). Tetracycline resistance mediated by *tet*(W), *tet*(M), and *tet*(O) genes of *Bifidobacterium* isolates from humans. Appl. Environ. Microbiol. *73*, 2751–2754.

Anonymous. (2002). Guidelines for the evaluation of probiotics in food. Report of a joint FAO/WHO working group on drafting guidelines for the evaluation of probiotics in foods. ftp://ftp.fao.org/es/esn/food/wgreport2.pdf. London, Ontario, Canada, April 30 and May *1*, 2002.

Anonymous. (2007). Opinion Opinion of the Scientific Committee on the Introduction of a qualified presumption of safety (QPS) approach for assessment of selected microorganisms referred to EFSA. Adopted 19/11/2007. The EFSA Journal *587*, 1–16.

Alander, M., Korpela, R., Saxelin, M., Vilpponen-Salmela, T., Mattila-Sandholm, T., and von Wright, A. (1997). Recovery of *Lactobacillus rhamnosus* GG from human colonic biopsies. Lett. Appl. Microbiol. *24*, 361–364.

Alander, M., Satokari, R., Korpela, R., Saxelin, M., Vilpponen-Salmela, T., Mattila-Sandholm, T., and von Wright, A. (1999). Persistence of colonization of human colonic mucosa by a probiotic strain, *Lactobacillus rhamnosus* GG, after oral consumption. Appl. Environ. Microbiol. *65*, 351–354.

Ammor, M.S., Florez, A.B., and Mayo, B. (2007). Antibiotic resistance in non-enterococccal lactic acid bacteria and bifidobacteria. Food Microbiol. *24*, 559–570.

Ammor, M.S., Florez, A.B., van Hoek, A.H.A.M., de los Reyes-Gavilan, C.G., Aarts, H.J.M., Margolles, A., and Mayo, B. (2008). Molecular characterisation of intrinsic and acquired antibiotic resistance in lactic acid bacteria and bifidobacteria. J. Mol. Microbiol. Biotechnol. *14*, 6–15.

Apostolou, E.,. Kirjavainen, Saxelin, P.V. , M., Rautelin, H., Valtonen, V., Salminen, S.J., and Ouwehand, A.C. (2001). Good adhesion properties of probiotics: a potential risk for bacteremia? FEMS Immunol. Med. Microbiol. *31*, 35–39.

Barrangou, R., Briczinski, E.P., Traeger, L.L., Loquasto, J.R., Richards, M., Horvarth, P., Coute-Monvoisin, A.-C., Leyer, G., Rendulic, S., Steele, J.L., *et al.* (2009). Comparison of the complete genome sequences of *Bifidobacterium animalis* subsp. *lactis* DSM 10140 and B1-04. J. Bact. *191*, 4144–4151.

Bearman, G.M.L., and Wenzel, R.P. (2005). Bacteremias: a leading cause of death. Arch. Med. Res. *36*, 646–659.

Begley, M., Hill, C., and Gahan, C.G.M. (2006). Bile salt hydrolase activity in probiotics. Appl. Environ. Microbiol. *72*, 1729–1738.

Ben Amor, K., Vaughan, E.E., and de Vos, W.M. (2007). Advanced molecular tolls for the identification of lactic acid bacteria. J. Nutr. *137*, 741S-747S.

Bernardeau, M., Guguen, M., and Vernoux, J.P. (2006). Beneficial lactobacilli in food and feed: long-term use, biodiversity and proposals for specific and realistic safety assessments. FEMS Microbiol. Rev. *30*, 487–513.

Besselink M.G.H., van Santvoort, H.C., Buskens, E., Boermeester, M.A., van Goor, H., Timerman, H.M., Nieuwenhuijs, V.B., Bollen, T.L., van Ramshorts, B., Witteman, B.J.M., *et al.* (2008). Probiotic prophylaxis in predicted severe acute pancreatitis: a randomised, double-blind, placebo-controlled trial. Lancet *371*, 651–659.

Borriello, S.P., Hammes, W.P., Holzapfel, W., Marteau, P.M., Schrezenmeier, J., Vaara, M., and Valtonen, V. (2003). Safety of probiotics that contain lactobacilli or bifidobacteria. Clin. Infect. Dis. *36*, 775–780.

Bourne, K.A., Beebe, J.L., Lue, Y.A., and Ellner, P.D. (1978). Bacteremia due to *Bifidobacterium, Eubacterium* or *Lactobacillus*; twenty-one cases and review of literatute. Yale J. Biol. Med. *51*, 505–512.

Brook I. (1996). Isolation of non-sporing anaerobic rods from infection in children. J. Med. Microbiol. *45*, 21–26.

Brook I. (2009). Pericarditis caused by anaerobic bacteria. Int. J. Antimicrob. Agents *33*, 297–300.

Brook, I., and Frazier, E.H. (1993). Significant recovery of non-sporulating anaerobic rods from clinical specimens. Clin. Infect. Dis. *16*, 476–480.

Chouraqui. J.P., Grathwohl, D., Labaune, J.M., Hascoet, J.M., de Montgolfier, I., Leclaire, M., Giarre, M., and Steenhout, P. (2008). Assessment of the safety, tolerance, and protective effect against diarrhea of infant formulas containing mixtures of probiotics or probiotics and prebiotics in a randomised controlled trial. Am. J. Clin. Nutr. *87*, 1365–1373.

Christensen, H.R., Nexmann Larsen, C., Kætel, P., Buus Rosholm, L., Sternberg, C., Fleischer Michaelsen, K., and Frøkiær, H. (2006). Immunomodulating potential of supplementation with probiotics: a dose–response study in healthy adults. FEMS Immunol. Med. Microbiol. *47*, 380–390.

Civen, R., Väisänen, M.L., and Finegold, S.M. (1993). Peritonsillar abscess, retropharyngeal abscess, mediastinitis, and nonclostridial anaerobic myonecrosis: a case report. Clin. Infect. Dis. *Suppl.* 4, S299–303.

Crociani, F., Biavati, B., Alessandrini, A., Chiarini, C., and Scardovi, V. (1996). *Bifidobacterium inopinatum* sp. nov. and *Bifidobacterium denticolens* sp. nov., two new species isolated from human dental caries. Int. J. Syst. Bacteriol. *46*, 564.571.

Dekker, J.W., Wickens, K., Black, P.N., Stanley, T.V., Mitchell, E.A., Fitzharris, P., Tannock, Purdie, G.W. , G., and Crane, J. (2009). Safety aspects of probiotic bacterial strains *Lactobacillus rhmanosus* HN001 and *Bifidobacterium animalis* subsp. *lactis* HN019 in human infants aged 0–2 years. Int. Dairy J. *19*, 149–154.

Epstein, S.K., Winslow, C.J., Brecher, S.M., and Faling, L.J. (1992). Polymicrobial bacterial pericarditis after transbrochial needle aspiration. Am. Rev. Respir. Dis. *148*, 523–525.

Frost and Sullivan. (2007). Strategic Analysis of the European Food and Beverage Probiotics Markets. [volume number?] B956–88.

Gasser, F. (1994). Safety of lactic acid bacteria and their occurence in human clinical infections. Bull. Inst. Pasteur *92*, 45–67.

Green, S.L. (1978). Case report: fatal anaerobic pulmonary infection due to *Bifidobacterium eriksonii*. Postgrad. Med. *63*, 187–192.

Guillard, F., Appelbaum, P.C., and Sparrow, F.B. (1980). Pyelonephritis and septicemia due to gram-positive rods similar to *Corynebacterium* group E (aerotolerant *Bifidobacterium adolescentis*). Ann. Int. Med. *92*, 635–636.

Ha, G.Y., Yang, C.H., Kim, H., and Chong, Y. (1999). Case of sepsis caused by *Bifidobacterium longum*. J. Clin. Microbiol. *37*, 1227–1228.

Hammerman, C., Bin-Nun, A., and Kaplan, M. (2006). Safety of probiotics: comparison of two popular strains. BMJ *333*, 1006–1008.

Hill, E.E., Herijgers, P., Herregods, M.-C., and Peetermans, W.E. (2006). Evolving trends in infective endocarditis. Clin. Microbiol. Infect. *12*, 5–12.

Johansson, M.-L., Molin, G., Jeppson, B., Nobaek, S., Ahrné, S., and Bengmark, S. (1993). Administration of different *Lactobacillus* strains in fermented oatmeal soup. *In vivo* colonization of human intestinal mucosa and effect on the indigenous flora. Appl. Environ. Microbiol. *59*, 15–20.

Kabeir, B.M., Yazid, A.M., Stephenie, W., Nazrul Hakim, M., Muhammad Anas, O., and Shuhaimi, M. (2007). Safety evaluation of *Bifidobacterium pseudocatenulatum* G4 as assessed in BALB/c mice. Lett. Appl. Microbiol. *46*, 32–37.

Kim, G.-B., and Lee, B.H. (2008). Genetic analysis of a bile salt hydrolase in *Bifidobacterium animalis* subsp. *lactis* KL612. J. Appl. Microbiol. *105*, 778–790.

Kirjavainen, P.V., Tuomola, E.M., Crittenden, R.G., Ouwehand, A.C., Harty, D.W.S., Morris, L.F., Rautelin, H., Playne, M.J., Donohue, D.C., and Salminen, S.J. (1999). *In vitro* adhesion and platelet aggregation properties of bacteremia-associated lactobacilli. Infect. Immun. *67*, 2653–2655.

Koretz, R.L. (2009). Probiotics, critical illness and methodological bias. Nutr. Clin. Pract. *24*, 45–49.

Korpela, R., Moilanen, E., Saxelin, M., and Vapaasalo, H. (1997). *Lactobacillus rhamnosus* GG (ATCC 53013) and platelet aggregation *in vitro*. Int. J. Food Microbiol. *37*, 83–86.

Kumar, R.S., Brannigan, J.A., Prabhune, A.A., Pundle, A.V., Dodson, G.G., Dodson, E.J., and Suresh, C.G. (2006). Structural and functional analysis of a conjugated bile salt hydrolase from *Bifidobacterium longum* reveals an evolutionary relationship with penicillin V acylase. J. Biol. Chem. *281*, 32516–32525.

Lay, C., Rigottier-Gois L., Holmstrom, K., Rajilic, M., Vaughan, E.E., de Vos, W.M., Collins, M.D., Thiel, R., Namsolleck, P., Blaut, M., and Dore, J. (2005). Colonic microbiota signatures across five Northern European countries. Appl. Environ. Microbiol. *71*, 4153–4155.

Madsen, K. (2008). Probiotics in critcally ill patients. J. Clin. Gastroenterol. *42*, S116–S118.

Mäkeläinen, H., Tahvonen, R., Salminen, S., and Ouwehand, A.C. (2003). *In vivo* safety assessment of two *Bidifobacterium longum* strains. Microbiol. Immunol. *12*, 911–914.

Marteau P. (2004). Tolerance of probiotics and prebiotics. J. Clin. Gastroenterol. *38*, S67–S69.

Marteau, P., Gerhardt, M.F., Myara, A., Bouvier, E., Trivin, F., and Rambaud, J.C. (1995). Metabolism of bile salts by alimentary bacteria during transit in the human small intestine. Microb. Ecol. Health Dis. *8*, 151–157.

Marteau, P., Pochart, P., Bouhnik, Y., and Rambaud, J.C. (1993). Fate and effects of some transiting microorganisms in the human gastrointestinal tract. World Rev. Nutr. Diet *74*, 1–21.

Masco, L., Huys, G., de Brandt, E., Temmerman, R., and Swings, J. (2005). Culture-dependent and culture-independent qualitative analysis of probiotic products claimed to contain bifidobacteria. Int. J. Food Microbiol. *105*, 221–230.

Masco, L., Van Hoorde, K., De Brandt, E., Swings, J., and Huys, G. (2006). Antimicrobial susceptibility of *Bifidobacterium* strains from humans, animals and probiotic products. J. Antimicr. Chemother. *58*, 85–94.

Maukonen, J., Mättö, J., Suihko, M.-L., and Saarela, M. (2008). Intraindividual diversity and similarity of salivary and faecal microbiota. J. Med. Microbiol. *57*, 1560–1568.

McNaught C.E., Woodcock, N.P., Anderson, A.D., and MacFie, J. (2005). A prospective randomised trial of probiotics in critically ill patients. Clin. Nutr. *24*, 211–219.

Mueller, S., Saunier, K., Hanisch, C., Norin, E., Alm, L., Midtvedt, T., Cresci, A., Silvi, S., Orpianesi, C., Verdenelli, M.C., *et al.* (2006). Differences in fecal microbiota in different European study populations in relation to age, gender and country: a cross-sectional study. Appl. Environ. Microbiol. 72, 1027–1033.

Morrow, L.E. (2009). Probiotics in the intensive care unit. Curr. Op. Crit. Care *15*, 144–148.

Nielsen, O.H., Jorgensen, S., Pedersen, K., and Justesen, T. (1994). Microbiological evaluation of jejunal aspirates and faecal samples after oral administration of bifidobacteria and lactic acid bacteria. J. Appl. Bacteriol. 76, 469–474.

Nisengard, R.J., Zambon, J.J., and Newman, M.G. (1998). Medical infections of interest. In Oral microbiology and Immunology, Nisengard, R.J., and Newman, M.G. , eds. (Philadelphia, PA, USA: Saunders, W.B. Company), pp. 398–401.

O'Brien, J., Crittenden, R., Ouwehand, A.C., and Salminen, S. (1999). Safety evaluation of probiotics. Trends Food Sci. Technol. *10*, 418–424.

Ouwehand, A.C., Isolauri, E., Kirjavainen, P.V., and Salminen, S.J. (1999). Adhesion of four *Bifidobacterium* strains to human intestinal mucus from subjects in different age groups. FEMS Microbiol. Lett. *172*, 61–64.

Ouwehand, A.C., and Salminen, S.J. (1998). The health effects of cultured milk products with viable and non-viable bacteria. Int. Dairy J. *8*, 749–758.

Ouwehand, A.C., Tölkkö, S., Kulmala, J., Salminen, S.J., and Salminen, E. (2000). Adhesion of inactivated probiotic strains to intestinal mucus. Lett. Appl. Microbiol. *31*, 82–86.

Petti, C.A., Polage, C.R., and Schreckenberger, P. (2005). The role of 16S rRNA gene sequencing in identification of microorganisms misidentifies by conventional methods. J. Clin. Microbiol. *43*, 6123–6125,

Pochart, P., Marteau, P., Bouhnik, Y., Goderel, I., Bourlioux, P., and Rambaud, J.-C. (1992). Survival of bifidobacteria ingested via fermented milk during their passage through the human small intestine: an *in vivo* study using intestinal perfusion. Am. J. Clin. Nutr. *55*, 78–80.

Saarela, M., Mättö, J., and Mattila-Sandholm, T. (2002). Safety aspects of *Lactobacillus* and *Bifidobacterium* species originating from human oro-gastrointestinal tract or from probiotic products. Microbial Ecol. Health Dis. *14*, 233–240.

Saarela, M., Maukonen, J., von Wright, A., Vilpponen-Salmela, T., Patterson, A.J., Scott, K.P., Hämynen, H., and Mättö, J. (2007). Tetracycline susceptibility of the ingested *Lactobacillus acidophilus* LaCH-5 and *Bifidobacterium animalis* subsp. *lactis* Bb12 strains during antibiotic/probiotic intervention. Int. J. Antimicrob. Agents *29*, 271–280.

Saarela, M., Mayrhofer, S., Domig, K.J., Danielsen, M., Mayo, B., and Mättö, J. (2008). Susceptibility of bifidobacteria originating from different sources to tetracycline, erythromycin, streptomycin and vancomycin. Int. J. Prob. Preb. *3*, 267–270.

Saavedra, J.M., Abi-Hanna, A., Moore, N., and Yolken, R.H. (2004). Long-term consumption of infant formulas containing live probiotic bacteria: tolerance and safety. Am. J. Clin. Microbiol. *79*, 261–267.

Salminen, M.K., Rautelin, H., Tynkkynen, S., Poussa, T., Saxelin, M., Valtonen, V., and Järvinen, A. (2004). *Lactobacillus* bacteremia, clinical significance, and patient outcome, with special focus on probiotic *L. rhamnosus* GG. Clin. Infect. Dis. *38*, 62–69.

Salminen, M.K., Tynkkynen, S., Rautelin, H., Saxelin, M., Vaara, M., Ruutu, P., Sarna, S., Valtonen, V., and Järvinen, A. (2002). *Lactobacillus* bacteremia during a rapid increase in probiotic use of *Lactobacillus rhmanosus* GG in Finland. Clin. Infect. Dis. *35*, 1155–1160.

Saxelin, M., Rautelin, H., Salminen, S., and Mäkelä, P.H. (1996). Safety of commercial products with viable *Lactobacillus* strains. Inf. Dis. Clin. Pract. *5*, 331–335.

Schell, M.A., Karmirantzou, M., Snel, B., Vilanova, D., Berger, B., Pessi, G., Zwahlen, M.-C., Desiere, F., Bork, P., Delley, M., *et al.* (2002). The genome sequence of *Bifidobacterium longum* reflects its adaptation to the human gastrointestinal tract. Proc. Natl. Acad. Sci USA *99*, 14422–14427.

Sghir, A., Gramet, G., Suau, A., Rochet, V., Pochart, P., and Dore, J. (2000). Quantification of bacterial groups within human fecal flora by oligonucleotide probe hybridization. Appl. Environ. Microbiol. *66*, 2263–2266.

Skiest, D.J., Steiner, D., Werner, M., and Garner, J.G. (1994). Anaerobic pericarditis: case report and review. Clin. Infect. Dis. *19*, 435–440.

Theunissen, J., Britz, T.J., Torriani, S., and Witthuhn, R.C. (2005). Identification of probiotic microorganisms in South African products using PCR-based DGGE analysis. Int. J. Food Microbiol. *98*, 11–21.

Thomas, A.V., Sodeman, T.H., and Bentz, R.R. (1974). *Bifidobacterium (Actinomyces) eriksonii* infection. Am. Rev. Respirat. Dis. *110*, 663–668.

van Hoek A.H.A.M., Mayrhofer, S., Domig, K.J.,. and Aarts H.J.M. (2008). Resistance determinant *erm*(X) is borne by transposon Tn*5432* in *Bifidobacterium thermophilum* and *Bifidobacterium animalis* subps. *lactis*. Int. J. Antimicrob. Agents *31*, 544–548.

Vankerckhoven, V., Huys, G., Vancanneyt, M., Vael, C., Klare, I., Romond, M.-B., Entenza, J.M., Oreillin, M., P., Wind, R.D., Knol, J., *et al.* (2008). Biosafety assessment of probiotics used for human consumption: recommendations from the EU-PROSAFE project. Trends Food Sci. Tech. *19*, 102–114.

von Wright, A. (2005). Regulating the safety of probiotics – the European approach. Curr. Pharmac. Design *11*, 17–23.

Weizman, Z., and Alsheikh, A. (2006). Safety and tolerance of a probiotic formula in early infancy comparing two probiotic agents: a pilot study. J. Am. Coll. Nutr. *25*, 415–419.

Zhou, J.S., and Gill, H.S. (2005). Immunostimulatory probiotic *Lactobacillus rhamnosus* HN001 and *Bifidobacterium lactis* HN019 do not induce pathological inflammation in mouse model of experimental autoimmune thyroiditis. Int. J. Food Microbiol. *103*, 97–104.

Zhou, J.S., Rutherfurd, K.J., and Gill, H.S. (2005). Inability of probiotic bacterial strains *Lactobacillus rhamnosus* HN001 and *Bifidobacterium lactis* HN019 to induce human platelet aggregation *in vitro*. J. Food Prot. *68*, 2459–2464.

Zhou, J.S., Shu, Q., Rutherfurd, K.J., Prasad, J., Birtles, M.J., Gopal, P.K., and Gill, H.S. (2000). Safety assessment of potential probiotic lactic acid bacterial strains *Lactobacillus rhamnosus* HN001, *Lb. acidophilus* HN017, and *Bifidobacterium lactis* HN019 in BALB/c mice. Int. J. Food Microbiol. *56*, 87–96.

Synthesis and Utilization of Exopolysaccharides and Prebiotics

10

Sandra Macfarlane, Katie E. Blackett and George T. Macfarlane

Abstract

Many bacteria and fungi secrete exopolysaccharides (EPS), which have been demonstrated to have significant ecological, physiological and pathological functions in the human body. These substances have also been harnessed for industrial purposes, and are widely used in the food industry. EPS formation has been studied extensively in various species of lactic bacteria, but relatively few investigations have been performed on these processes in bifidobacteria. However, although yields are generally low in comparison to other microorganisms, studies have shown that various bifidobacterial species secrete rhamnose-, glucose- and galactose-based EPS that have immunopotentiating activities, anti-mutagenic properties and the ability to promote or inhibit adhesion of other organisms to mucosal surfaces. Bifidobacteria have also been found to utilise the polymers they produce, as well as those formed by other microbial species, suggesting that these molecules may have prebiotic properties. Bifidobacterial communities in the large intestine are not only targets for prebiotics, but β-galactosidases produced by different species have been used to synthesize galacto-oligosaccharides with enhanced selectivity for the producing organisms.

Introduction

Many bacteria and fungi produce extracellular polymeric substances known as exopolysaccharides (EPS), which can be intimately associated with the cell wall (capsular exopolysaccharides), loosely attached, or completely excreted from the cell (Welman and Maddox, 2003; Chen and Seviour, 2007; Deutsch *et al.*, 2008; Vu *et al.*, 2009). Studies have shown that physical removal of EPS does not affect cell growth, and these substances are not believed to be used as energy sources or food reserves by the organisms that form them. Bacterial EPS have variously been associated with health and disease processes in humans (see later). The reason for EPS production has not been fully established, but in pathogens they may provide mechanisms that enable the producer to evade host immune responses, whereas in commensal bacteria, EPS have been postulated to be formed as a response to starvation and other stress conditions, or to allow them to establish in the host. The formation of mutans EPS by *Streptococcus mutans* enhances bacterial colonization and the formation of dental plaque, while the production of O-acetylated alginate by *Pseudomonas aeruginosa* increases its resistance to opsonization (Pier *et al.*, 2001). EPS are believed to play a protective role in sequestration of cations (Looijesteijn *et al.*, 2001), in microbial colonization of surfaces by biofilm formation (Roberts, 1996), cell/cell recognition, evasion of phagocytosis, protection from antibiotics and antimicrobial peptides, bacteriophages, lysozyme and toxic compounds (Looijesteijn *et al.*, 2001), while they may also possess prebiotic activity (Korakli *et al.*, 2002). The importance of EPS in biofilm production is shown by observations indicating that the ability to form biofilms is lost in mutants unable to synthesize EPS (Sutherland, 2001). The production of EPS may also aid in bacterial adhesion to the intestinal mucosa (Ruas-Madiedo *et al.*, 2006a). Several lactic acid bacteria (LAB) including streptococci, lactobacilli, leuconostocs and lactococci synthesize a large number of EPS, which differ in compositional and structural characteristics, with a size range between 4×10^4 to 6×10^6 Da (de Vuyst and Degeest, 1999). The majority of EPS-forming LAB have been isolated from fermented products, including yoghurts, kefir, cereal fermented products, meat, olives, and cider (Korakli and Vogel, 2006). Many strains of bifidobacteria and propionibacteria are also able to produce EPS. Some of these polymers might be important in the food industry, since they can improve the rheological, textural and organoleptic properties of dietary constituents, and can be used as viscosifiers, gelling agents or thickeners (Hess *et al.*, 1997). They may also have other benefits. Since LAB, bifidobacteria and some propionibacteria have a generally recognized as safe (GRAS) status, and are commonly used as probiotics, their EPS may provide safer and healthier options for the replacement of chemically modified additives, thickeners and stabilizers produced by non-food grade bacteria (Becker, 1998; Sutherland, 1998).

Exopolysaccharides

Synthesis of EPS

EPS are secreted from bacterial cells predominantly, but not exclusively, during active growth, and differences in EPS classification depend on the method of production and the monomeric precursors required for their synthesis (Sutherland, 1993). The two main categories of EPS are homo- and heteropolysaccharides. Homopolysaccharides consist of one monosaccharide (usually glucose or fructose), and can be divided into two major groups – glucans and fructans – containing both α-D-glucose and β-D-fructose, which are synthesized by extracellular or cell-associated glycansucrases (Korakli *et al.*, 2003; van Hijum *et al.*, 2006). Several genera of LAB including *Streptococcus, Leuconostoc, Lactobacillus* and *Weissella* have been shown to form glucans and β-fructans (Korakli and Vogel, 2006). Dextran is produced commercially by *Leuconostoc mesenteroides* subsp. *dextranicum* and *Leuconostoc mesenteroides* subsp. *mesenteroides*. Only a few species of LAB have been found to produce β-glucans; these include *Pediococcus damnosus, Pediococcus parvulus, Oenococcus oeni,* and *Lactobacillus* sp. G-77 (Korakli and Vogel, 2006). Recently, a polysaccharide synthase gene was detected in the genome of *Propionibacterium freundenreichii* subsp. *shermanii,* which resembles the polysaccharide synthase gene of *Streptococcus pneumoniae* type 37, and which produces a β-glucan capsular polysaccharide (Deutsch *et al.*, 2008).

Heteropolysaccharides have much greater structural variety, and product yields are lower (ca. 2 g/l) than homopolysaccharides (up to 40 g/l) (Korakli and Vogel, 2006). The difference in EPS yield is probably owing to the fact that the synthesis of heteropolysaccharides is a more energy consuming process, while in homopolysaccharide production, energy resulting from cleavage of the osidic bond in sucrose may be utilized for the glycosyltransferase reactions. Heteropolysacharides are principally composed of glucose, galactose and rhamnose; however, *N*-acetylglucosamine and N-acetylgalactosamine can also be present, while L-fucose has also been identified. In addition, a variety of non-carbohydrate moieties can exist in the side chains, which further contribute towards their heterogeneity. Individual bacterial species are frequently able to use several different systems to transport sugar monomers and dimers into the cells, whence they are converted to sugar nucleotides by several dedicated enzymes. These sugar nucleotides are then used by glycosyltransferases to form an EPS-repeating unit, which is assembled on a C_{55}-isoprenoid-lipid carrier molecule attached to the cell's cytoplasmic membrane. The first step in the assembly of the repeating unit is carried out by a priming glycosyltransferase (priming GT), which catalyses the transfer of a sugar −1-phosphate to the phosphorylated undecaprenol anchored in the membrane. Unlike glycosyltransferases, priming GTs do not form glycosidic linkages and have been shown to be essential in EPS formation as inactivation of genes encoding priming GT's results in disruption of EPS synthesis (Provencher *et al.*, 2003; Videira *et al.*, 2005). The completed repeating unit is then translocated across the membrane, polymerized and released. Glucose-6-phosphate is a key intermediate linking the pathways of sugar catabolism and the anabolic pathway of EPS production. In the anabolic pathway, the enzyme phosphoglucomutase has a key role in the generation of glucose-1-phospate and formation of the sugar nucleotides UDP-glucose and dTDP-glucose (de Vuyst and Degeest, 1999; Boels *et al.*, 2001; Jolly *et al.*, 2002). In some LAB, galactose is also converted to glucose-1-phosphate by way of the Leloir pathway (Groissiord *et al.*, 1998; de Vuyst and Degeest, 1999) (Fig. 10.1).

Structure and formation of EPS

Studies have shown that EPS structures are diverse (Harding *et al.*, 2005). They can vary in their molecular mass, degree of branching, sugar composition, glycosidic and other linkages, and charged groups (Tuinier, 1999, 2001; Jolly *et al.*, 2002). The viscosity of EPS depends on the molecular mass and size of the polymer. The biological functionality of the EPS is determined by the structure of the molecule and the composition of its monosaccharide constituents, and this depends in large part on the number and type of glycosyltransferase genes present (Delcenserie *et al.*, 2008). Chain stiffness plays a role in the viscosity of EPS, and relatively minor changes in sugar linkages have been demonstrated to result in stiffer or more flexible chains (Ruas-Madiedo *et al.*, 2002). *Streptococcus thermophilus* strain ST11 was shown to produce a 5000-kDa EPS composed of galactose and rhamnose, while other streptococcal strains can produce low molecular mass EPS or a combination of both high and low molecular mass EPS. Ropy cultures of lactobacilli have been reported to be able to synthesize high molecular mass EPS, and several studies have found that these strains increase the viscosity of yoghurts and fermented milk. However, no direct correlation between viscosity and EPS production has been found (Vaningelgem *et al.*, 2004; Lin and Chang Chien, 2007).

While many investigations have been carried out to elucidate the structure and function of LAB EPS, there are comparatively few studies on bifidobacterial exopolysaccharides in the literature. However, some work has been done, and structural differences have been observed in various bifidobacterial EPS (Roberts *et al.*, 1995; Ricciardi and Clementi, 2000; Ruas-Madiedo *et al.*, 2002; Kohno *et al.*, 2009). *Bifidobacterium*

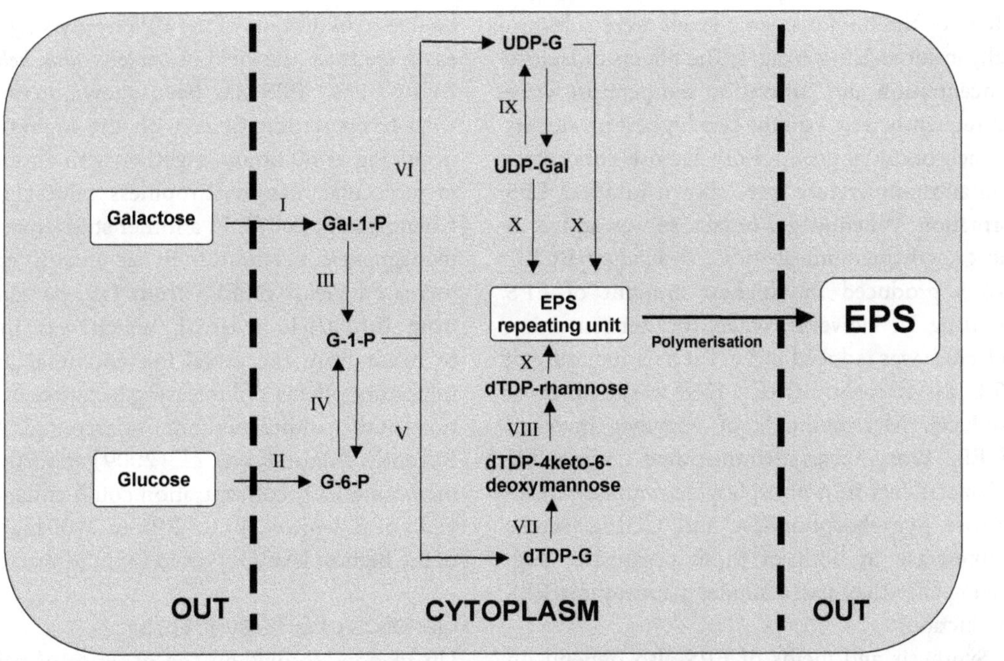

Figure 10.1 Conversion of galactose and glucose to exopolysaccharide (EPS) in the formation of heteropolysaccharides in lactic acid bacteria. G, glucose; Gal, galactose; I, galactokinase; II, glucokinase; III, galactose-1-phosphate uridylyltransferase; IV, phosphoglucomutase; V, dTDP-glucose pyrophosporylase; VI, UDP-glucose pyrophosphorylase; VII, dTDP-glucose 4,6-dehydratase; VIII, dTDP-rhamnose synthetic enzyme system; IX, UDP-galactose-4-epimerase; X, glycosyltransferases.

breve and *Bifidobacterium bifidum* have been shown to produce EPS heteropolymers that are rich in rhamnose (Nagoaka *et al.*, 1994) whereas EPS from a *Bifidobacterium longum* EPS was found to have glucose and galactose as its main constituents, accounting for 30% of the organism's total polysaccharide production (Abbad Andalousi *et al.*, 1995). Recently, *B. longum* JBL05 was shown to produce EPS containing glucose, galactose and rhamnose in a molar ratio of 2:4:1.

Yields of bacterial EPS vary markedly, depending on whether they are homo-EPS or hetero-EPS, but large variations are observed between different species or strains. Thus, *Strep. thermophilus* can produce up to 100 mg/l of hetero-EPS (Vaningelem *et al.*, 2004) in fermented milk, but *Lactobacillus rhamnosus* RW 9595 forms up to 1275 mg/l of a homo-EPS when grown on lactose or glucose (Van Calsteren *et al.*, 2002), while early studies reported that *B. longum* BB-79 could produce 470 mg/l (Roberts *et al.*, 1995). In one investigation, EPS production was compared in six bifidobacterial strains and 13 lactobacilli (Liu *et al.*, 2009). EPS produced by the lactobacilli ranged from 4.1 to 29.2 mg/l, while for the bifidobacteria, levels were substantially lower at 2.5–8.6 mg/l. The effects of lactose concentration and cultivation temperature were subsequently tested on the two highest producers in each bacterial group. Both lactose concentration and temperature were shown to affect EPS formation. When 40 g/l of lactose was added to the growth medium at 40°C, *B. bifidum* BCRC 14615 produced the highest amount of EPS (17.0 mg/l). However, when the concentration of lactose was reduced to 5 g/l at a temperature of 35°C, *Lb. salivarius* BCRC 14759 was the highest producer. Measurements of enzymes involved in EPS biosynthesis demonstrated comparable enhancements in α-phosphoglucomutase, UDP-glucose pyrophosphorylase and UDP-galactose 4-epimerase in both of these organisms, suggesting that they use a similar pathway for EPS production.

Synthesis and yields of EPS also depend on the constitution of the culture medium, bacterial growth phase, temperature, fermentation time and environmental conditions (Degest *et al.*, 2001; Ruas-Madiedo *et al.*, 2002; Dueñas *et al.*, 2003; Lin and Chang Chien, 2007; Ruas-Madiedo *et al.*, 2009; Velasco *et al.*, 2009). While the composition and length of EPS branches can affect its rheological properties, Grobben *et al.* (1996) demonstrated that EPS production by *Lactobacillus delbreuckii* subsp. *bulgaricus* NCFB 2772 depends to a large degree on the carbon source used for growth. When this bacterium was grown on fructose, the EPS contained glucose and galactose in a ratio of 1:2.4; however, when the bacterium was cultured in the presence of fructose and glucose, the exo-polysaccharide was shown to consist of glucose, galactose and rhamnose (ratio 1:7:0.8). A subsequent study by the same authors found that when grown on glucose and fructose, the organism formed two EPS fractions with molecular masses of 1700 and 40 kDa (Grobben *et al.*, 1997). The ability of fermentation time to affect the molecular mass and composition of EPS has been demonstrated in several investigations. Pham *et al.* (2000) observed that the amount of EPS produced by *Lb. rhamnosus* declined concomitantly with fermentation time, and similarly, Cerning *et al.* (1988) demonstrated a decrease in *Strep. thermophilus* EPS after 48 hours growth, which was thought to be due to hydrolysis of the EPS by glycohydrolases. In contrast, the yield of *Lactobacillus. helveticus* BCRC14030 EPS has been shown to increase with fermentation time, with the highest yield occurring at 60 hours, together with an increase in molecular mass and ropiness value (Lin and Chang Chen, 2007). In another study, increasing manganese concentration in the growth medium increased yields of EPS from *Ped. parvulus* 2.6 from 0.12 g/l to 4.10 g/l, which was thought to result from the metal ion counteracting the inhibitory effects of increasing lactate concentrations in the culture medium (Velasco *et al.*, 2006). Recently, Ninomiya *et al.* (2009) reported that increasing CO_2 concentration could enhance the yield of *B. longum* JBL05 EPS to 2900 mg/l, one of the highest levels reported for bifidobacteria.

Genetics of EPS synthesis

Up to now, studies on the genetics of bacterial EPS production have been fairly limited. Genes that direct hetero-EPS synthesis (*eps*) are located in highly conserved clusters (Boels *et al.*, 2001). These assemblages were first identified in *Strep. thermophilus* Sfi6 (Stingele *et al.*, 1996) and

Lactococcus lactis NIZO B40 (van Kranenburg *et al.*, 1997). The ability of lactococci to produce EPS can be plasmid-associated (Vedamuthu and Nevill, 1986; van Kranenburg *et al.*, 1997; van Kranenburg *et al.*, 1999) or be encoded by the chromosome (Stingele *et al.*, 1996; Lamothe, 2000). A common operon structure has been found in EPS-producing LAB. The *eps* genes are transcribed in one direction, as a single transcriptional mRNA. In the study by Stingele *et al.* (1996), it was found that the *eps* genetic locus was located on a plasmid, and contained a 14.5 kb region which encoded 13 genes involved in EPS synthesis. In *L. lactis* NIZO B40, a 12 kb gene cluster was detected on a 40 kb plasmid (van Kranenburg *et al.*, 1997). Interestingly, in this investigation, the authors were able to ascribe functionality to several of the genes involved in EPS manufacture, while such genes were found to have sequence homology to those involved in the production of other cellular polysaccharide, lipopolysaccharide and teichoic acid biosynthesis. Similarity in gene clusters for EPS production has also been shown to occur between different LAB (Jolly and Stingele, 2001).

The genome of *B. longum* NCC2705 was the first of any intestinal commensal organism to be sequenced (Schell *et al.*, 2002), and several genes responsible for EPS synthesis were detected. Efforts to elucidate polysaccharide encoding gene sequences in bifidobacteria have shown that in *B. longum* NCC2705, polysaccharide synthesis is located in two regions, differing in G + C content, which are believed to be acquired by horizontal gene transfer (Klijn *et al.*, 2005). In the *B. breve* UCC2003 genome, at least 17 putative determinants of EPS biosynthesis have been detected on a 21.3-kb-long DNA region (Ventura *et al.*, 2007). Similar to the *eps* loci of other bacteria, the *eps* loci of *B. breve* UCC2003 and *B. longum* biotype *longum* NCC2705 have both been shown to be flanked by IS elements, and diverge in G + C content from the rest of the genome (Ventura *et al.*, 2007).

Improvement of EPS production

Levels of EPS produced by LAB and bifidobacteria are lower than in Gram-negative non-food-grade bacteria, which may be due to their relatively inefficient conversion of carbohydrates to energy that limits the amount of EPS that can be formed (between 40 and 800 mg/l). This compares poorly with *Xanthomonas campestris*, which is used commercially, and which can convert 60% of sugars in the growth medium to xanthan, and produces up to 25 g/l EPS. Nevertheless, there is increasing interest in enhancing EPS production in LAB and bifidobacteria to obtain EPS that has optimum commercial/industrial properties, and improved functional attributes (Welman and Maddox, 2003). Several mechanisms have been postulated that could enhance the production of EPS from LAB, including the introduction of new glycosyltransferases (Boels, 2001), increasing glycosyltransferase activity (van Kranenburg *et al.*, 1999; Jolly *et al.*, 2002), manipulation of EPS gene function (van Kranenburg *et al.*, 1999), and transfer of *eps* operons to plasmids. Several of these methods have been used to modify the physical structure of the EPS, or to increase yields; these include the use of bacterial polysaccharide hydrolases and lyases, mild chemical degradation (Nakajima *et al.*, 1992a; Robijn *et al.*, 1995), a combination of both (van Casteren *et al.*, 1998, 2000), tightly controlling culture conditions (Ruas-Madiedo *et al.*, 2002), metabolic engineering (de Vos, 1996), modifying sugar catabolism (Boels *et al.*, 2001) and genetic modification (Breedveld *et al.*, 1998; Boels *et al.*, 2001). Recently, Werning *et al.* (2008) were able to express plasmidic gycosyltransferase from *Ped. parvulus* into *Lc. lactis*, which enabled the lactococci to synthesize and secrete a β-D-glucan which on analysis was found to be the same as that of *Ped. parvulus* EPS. Production of high levels of *Lb. rhamnosus* RW-9595M EPS have been observed by immobilization of the cells on solid porous supports in a two-stage continuous culture system (Bergmaier *et al.*, 2005).

Production of EPS by other bacteria

Dextran produced by *Leuc. mesenteroides* has been shown to have applications in medicine and biotechnology, while xanthan formed by *X. campestris* and gellan from *Sphingomonas paucimobilis* have been used in the food industry. Levan is formed by *Zymomonas mobilis*, *Gluconacetobacter diazotrophicus*, and several *Bacillus* species. Acetan, which is made by *Acetobacter xylinum*, and *Sphingomonas* capsular EPS are used as gelling agents and stabilizers. Several strains of *Burkholderia* can produce the EPS cepacian which is thought to play a key role in cystic fibrosis infections (Chung *et al.*, 2003), and *B. cenecepacia* EPS has been shown to scavenge reactive oxygen species and inhibit neutrophil chemotaxis (Bylund *et al.*, 2006). *Alcaligenes faecalis* var. *myxogenes* produces curdlan, a low molecular weight soluble polysaccharide, which when heated above 55° C, can form an elastic gel that improves the stability and texture of a number of food products (Rehm, 2009; Vu *et al.*, 2009). Cellulose is an exopolysaccharide, which can be a major product of bacterial biofilms involving certain *Salmonella*, *Citrobacter*, and *Enterobacter* species, as well as several strains of *Escherichia coli*. Cellulose is produced by acetobacters, pseudomonads, and rhizobia, and is used as a food matrix, thickening agent or dietary fibre. Recently, *Bacillus coagulans* was shown during both the stationary and exponential phases of growth to produce an EPS, with a unique composition of sugars containing fucose (Kodali *et al.*, 2009).

Interactions of EPS with human physiological processes

A number of putative health benefits and disease associations (summarized in Table 10.1) linked to bacterial EPS have been reported (Ruas-Madiedo *et al.*, 2008). The probiotic abilities of many bacteria have also been linked to their production of EPS, and the variety of polymeric structures that can be formed may relate to the differences seen in biological properties of individual species and strains (Welman and Maddox, 2003). The production of EPS may enhance a probiotic's ability to survive in the presence of bile salts (Alp and Aslim, 2010), modulate the immune system, inhibit the growth and establishment of pathogens, and increase its adherence to the intestinal mucosa. LAB EPS have been shown to have B-cell mitogen activity, modify macrophage and

Table 10.1 Putative health effects associated with bacterial EPS

Respiratory system
Inhibition of bacterial attachment to epithelial cells
Stomach
Anti-ulcer effects
Gut
Act as prebiotics
Promote probiotic function
Increase bifidobacteria and lactobacillus numbers
Increase bacterial fermentation product formation
Inhibit the adhesion of pathogens
Promote colonisation of pathogens and commensal species
Anti-bacterial activity
Anti-tumour effects
Free radical activity
Antioxidant activity
Blood
Lower cholesterol
Reduce blood pressure
Immune system
Macrophage activation
Stimulate T-lymphocyte proliferation
Increase cytokine production
Some EPS inhibit phagocytosis
Anti-inflammatory effects
Increase IgA production

splenocyte functions, and to have antimicrobial properties and to lower cholesterol (Nakijima *et al.*, 1992b; Vinderola *et al.*, 2006). EPS from *Lb. rhamnosus* RW-5995M was shown to be able to stimulate TNF-α, IL-6 and IL-12 in human and mouse immunocompetent cells (Chabot *et al.*, 2001). Intraperitoneal injection of slime-forming *Lc. lactis* subsp. *cremoris* isolated from Scandinavian ropy sour milk has been shown to have anti-tumour activity, and was able to inhibit the growth of sarcoma-180 in mice (Kitazawa *et al.*, 1991). Using an *in vitro* model system, de Palencia *et al.* (2009) demonstrated that EPS-producing *Ped. parvulus* 2.6 could resist gastrointestinal stress, induce polarized macrophage formation of pro-inflammatory cytokines, and adhere to Caco-2 cells. It had previously been shown that a ropy, oat-based product fermented by this strain was able to reduce serum cholesterol levels, and stimulate bifidobacterial growth in the gut when added to oat-based food (Martensson *et al.*, 2005).

The EPS content of *Lb. casei strain* Shirota has been linked to its ability to suppress macrophage activation (Yasuda *et al.*, 2008). Vinderola *et al.* (2006) investigated the ability of an EPS producing strain of *Lactobacillus kefiranofaciens* to modulate murine immune system reactivity, and found that administration of the EPS was able to regulate cytokine formation and maintain intestinal homeostasis, influence systemic immunity and enhance IgA production. Overall molecular charge due to the presence of phosphate groups in the EPS may play a role in lymphocyte and macrophage activation, since studies have shown that the mitogenic activity of *Lb. delbreuckii* subsp. *bulgaricus* 1073R-1 EPS in lymphocytes is lost by dephosphorylation (Kitazawa *et al.*, 1998; Nishimura-Uemura *et al.*, 2003), while neutralization of charged groups interferes with the ability of capsular EPS to interact with T cells (Amrouche *et al.*, 2006). EPS-producing probiotic strains have been reported to be beneficial in attenuating experimental colitis in rats (Sengul *et al.*, 2006), and released EPS from the probiotic *Lb. acidophilus* was able to reduce enterhaemorrhagic *E. coli* 0157:H7 biofilm formation (Kim *et al.*, 2009), possibly reducing its persistence. Dextran has been observed to inhibit the attachment of *Legionella pneumophila* to human respiratory epithelial cell *in vitro* (Thomas and Brooks, 2004), and, recently, an exopolysaccharide from a probiotic *B. coagulans* was shown to have free radical scavenging and antioxidant activities (Kodali and Sen, 2008).

There have been much fewer studies on the health benefits of bifidobacterial EPS, although Hosono *et al.* (1997) demonstrated that *B. adolescentis* M101–4 produced a water-soluble polysaccharide fraction with immunopotentiating activity in a mouse model. *B. longum* EPS has been reported to promote mucosal adhesion (Abbad Andaloussi *et al.*, 1995), have anti-mutagenic properties (Sreekumar and Hosono, 1998), promote self-colonization (Roberts *et al.*, 1995), and be able to modulate the immune system (Kohno *et al.*, 2009), which may benefit its employment as a probiotic. In another study, EPS from *B. longum* NB667 and *Bifidobacterium animalis* IPLA-R1 were found to inhibit the adherence of probiotics, and increase colonization by enteropathogens in an *in vitro* human intestinal mucus model (Ruas-Madiedo *et al.*, 2006b). Both EPS were shown to inhibit the adhesion of probiotic strains of *B. longum* and *Lb. rhamnosus*, and to enhance adherence of *Enterobacter sakazakii* ATCC 29544 and *E. coli* NCTC 8603, while NB667 EPS facilitated greater adherence by *Salmonella enterica* serovar Typhimurium ATCC 29631 and *Clostridium difficile* ATCC 9689. In a similar study by the same authors, EPS from the Scandinavian fermented milk viili was also reported to inhibit adherence of probiotics to human intestinal mucus, but no inhibition of the enteropathogens was observed (Ruas-Madiedo *et al.*, 2006a), suggesting that EPS's are involved in adhesion and may benefit probiotic-producing strains *in vivo*. In a comparison of the immuno-stimulatory abilities of three bifidobacteria isolated from infant faeces, the bifidobacterial cell wall and cytoplasmic components effected cytokine production, and increased INF-γ, IL-10 and cell proliferation, but no effect was seen with EPS recovered from the culture broths (Amrouche *et al.*, 2006).

Bacterial utilization of EPS

Several attempts have been made to determine the abilities of intestinal bacteria to degrade EPS. *In vitro* experiments have shown that differential EPS degradation occurs. Using anaerobic incubations in faecal slurries Ruijssenaars *et al.* (2000)

demonstrated that, while EPS from *Strep. thermophilus* Sfi39 and Sfi12 were broken down by colonic microorganisms, EPS produced by several other LAB were not. In one *in vivo* study in rats fed *Lc. lactis* subsp. *cremoris* B40 for two weeks, no EPS was digested and EPS recovery from faecal material was 100% (Looijesteijn *et al.*, 2001).

Studies on EPS as prebiotics are limited; however, recently there has been an upsurge of interest in their nutritional benefits, and whether they can increase microbial populations perceived as being beneficial to the host, such as bifidobacteria. Martensson *et al.* (2005) reported that the addition of a β-glucan-producing *Ped. damnosus* to an oat base product *in vitro* increased levels of bifidobacteria. Levan-type and fructan-type EPS produced by one strain of *Lb. sanfranciscensis* have been shown to be utilized by a number of bifidobacterial strains, including *B. breve, B. adolescentis, B. bifidum, B. infantis and B. minimum* (Dal Bello *et al.*, 2001; Korakli *et al.*, 2002). Interestingly, while the study by Korakli *et al.* (2002) found that the bifidobacteria were also able to degrade water-soluble polysaccharides such as polyfructan and starch that were present in rye and wheat, these were actually degraded during the baking process, while bacterial EPS was not. Consequently, fructan formation may enhance the nutritional benefits of sourdough fermented products, by exerting a prebiotic effect and increasing levels of intestinal bifidobacteria. Low molecular mass β-(2- > 6)-fructooligosaccharides (FOS) from levan have also been found to support growth of bifidobacteria (Marx *et al.*, 2000). Unsurprisingly, the FOS was found to be resistant to amylolytic and proteolytic digestion in *in vitro* incubations, with *B. adolescentis* strains being the most efficient utilizers of the substrate.

Ruas-Madiedo *et al.* (2007) demonstrated that several bifidobacteria isolated from Spanish volunteers harboured genes for heteropolysaccharide synthesis, and were able to produce low levels of EPS, which was shown to contain galactose, glucose and rhamnose. In a subsequent study by the same group (Salazar *et al.*, 2008), the abilities of intestinal bacteria to ferment bifidobacterial EPS were investigated. The authors compared the fermentation of eleven different EPS from a range of intestinal bifidobacteria (*B. animalis, B. pseudocatenulatum, B. longum*) with inulin and glucose, using *in vitro* batch cultures inoculated with faecal material. Higher levels of short-chain fatty acids compared to the control without carbohydrate were found in the EPS, inulin and glucose cultures, again demonstrating that EPS can be fermented by colonic microorganisms. Shifts in fermentation profiles were observed, with inulin and EPS, with a reduction in the acetate:propionate ratio being observed, compared to glucose. The greatest changes occurred with EPS from four isolates of *B. longum*, and one strain of *B. pseudocatenulatum*. It has been suggested that a reduction in the acetate:propionate ratio could be useful as an indicator of the potential hypolipidaemic properties of prebiotics (Delzenne *et al.*, 2001). Interestingly, acetate:propionate ratios with EPS were found to be lower than with the established prebiotic inulin. Real-time PCR analysis indicated that the EPS was able to increase levels of bifidobacteria, and denaturing gradient gel electrophoresis (DGGE) demonstrated different banding patterns associated with various bacteria in the gut microbiota, suggesting differential bacterial fermentation of the EPS. However, since the DGGE was not quantitative, actual changes in bacterial numbers could not be ascertained (Salazar *et al.*, 2008).

Recently, EPS formed by two bifidobacterial species, one of which was the intestinal isolate *B. longum* IPLA E44, and the other, *B. animalis* subsp. *lactis* IPLA R1 of dairy origin, were tested for their abilities to induce changes in the intestinal microbiota, using pH-controlled batch cultures inoculated with human faecal communities (Salazar *et al.*, 2009). Both EPS were shown to be resistant to breakdown by colonic bacteria under conditions assumed to simulate those extant in the large bowel. FISH was used to analyse bacterial populations, and the results were compared with batch cultures inoculated with glucose, or the prebiotic inulin. Increased levels of bifidobacteria were detected in both the inulin and EPS fermentations, while increases were also detected with the *Bacteroides/Prevotella* and *Lactobacillus/Enterococcus* probes for both substrates. EPS R1 was reported to increase the *Atopobium* group in the initial stage of fermentation, however, no increases in responses to probes designed to detect *Clostridium* clusters XIVa+b and IV were found for any of the substrates.

Bifidobacteria and prebiotics

A prebiotic is a selectively fermented food ingredient that induces specific changes, both in the composition and/or activity of the gastrointestinal microbiota in a way that confers benefits on host health and wellbeing. For the most part, the main targets for prebiotics in the human large bowel have been bifidobacteria and lactobacilli. In contradistinction to probiotics, where invading species are being introduced to the digestive tract, and have to compete against existing colonic communities, a significant advantage of prebiotics is that the target organisms are already established commensals in the large intestine. It can therefore be argued that the use of prebiotics could be a more practical and efficient way to manipulate the composition and metabolic activities of the colonic microbiota than probiotics. Conversely, however, if for any reason, such as changes in diet or environment, disease, ageing or antibiotic therapy, the desired health-promoting species are not present in the bowel, the prebiotic is unlikely to be effective, at least in the short term.

Although there is no hard and fast consensus, at the moment, regarding exactly what constitutes a prebiotic, the majority of contenders are non-digestible oligosaccharides (NDO). Prebiotic oligosaccharides are distinguished on the basis of their chemical constituents and degree of polymerization, and include lactulose (Tuohy *et al.*, 2002), soybean-oligosaccharides (Saito *et al.*, 1992), pectic-oligosaccharides (Olano-Martin *et al.*, 2003), manno-oligosaccharides (Zentek *et al.*, 2002), isomalto-oligosaccharides (Morgan *et al.*, 1992) and xylo-oligosaccharides (Campbell *et al.*, 1997). However, most of the research that has been carried out on prebiotics has centred on inulin, FOS, galacto-oligosaccharides (GOS) and transgalacto-oligosaccharides (TOS), reflecting the ready availability of their progenitor substrates and their perceived commercial application.

So far, the majority of research on prebiotics has focused on fermentability, and their abilities to increase the relative numbers of beneficial species in faecal microbiotas, while suppressing putatively pathogenic organisms. Nevertheless, many other types of studies have been undertaken, and it has been shown that some oligosaccharides are able to imitate eukaryotic cell surface structures that serve as receptors for pathogenic bacteria. Thus, TOS was reported to enhance the protective abilities of *B. breve* in a mouse model of salmonella infection (Asahara *et al.*, 2001), while inulin, FOS and GOS were found to inhibit the attachment of enteropathogenic *E. coli* (EPEC) to HEp-2 and Caco-2 cells (Shoaf *et al.*, 2006). Furthermore, in addition to the anti-adherent effects manifested by these substances, short-chain chito-oligosaccharides were shown to have anti-bacterial characteristics (Sekiguchi *et al.*, 1994). If these properties are expressed in the same way *in vivo*, prebiotic oligosaccharides could be protective in the small bowel, as well as the large intestine, but it is unclear whether sufficient amounts could be delivered to mucosal surfaces in the gut for them to have significant protective value.

Bifidobacteria in combination with FOS have also been observed to reduce 1,2-dimethyhydrazine-induced carcinogenesis in mice (Koo *et al.*, 1991) and can also have anti-tumour activities (Challa *et al.*, 1997; Singth *et al.*, 1997). Consumption of prebiotic oligosaccharides may also have protective effects in relation to other genotoxic processes in the gut. In rats, TOS has been reported to reduce caecal pH, as well as β-glucuronidase and nitrate reductase activities, although β-glucosidase was increased in that study (Rowland and Tanaka, 1993). Similarly, Kikuchi *et al.* (1996) found that GOS feeding in rats induced β-galactosidase synthesis and inhibited β-glucuronidase formation. Human studies (Buddington *et al.* 1996) have shown that daily consumption of four grams of FOS reduced faecal β-glucosidase and glycholate hydrolase, but had no effect on nitroreductase, while van Dokkum *et al.* (1999) observed that daily feeding of either 15 grams of inulin or GOS per day to healthy young volunteers significantly reduced β-glucuronidase, though FOS had little effect on the activities of this enzyme. However, there are some inconsistencies in data obtained from different studies. For example, Kleessen *et al.* (1997) found no changes in β-glucosidase or β-glucuronidase in stools from elderly people who were fed inulin, while Bouhnik *et al.* (1996) reported that feeding 12.5 grams of FOS daily did not affect faecal azoreductase, nitroreductase or β-glucuronidase.

In vitro modelling experiments using a three-stage gut model (McBain and Macfarlane, 2001) suggested that 5% GOS was only weakly bifidogenic in the fermentation system. However, GOS strongly inhibited arylsulphatase, β-glucosidase and β-glucuronidase synthesis in all of the fermentation vessels. The study also showed that nitroreductase and azoreductase activities were increased by addition of the oligosaccharide, making it difficult to assess its anti-mutagenic properties. In the same investigation, the effects of FOS on bacterial community structure were found to be generally similar to GOS, but there were no significant inhibitory effects on either β-glucosidase or β-glucuronidase.

Utilization of prebiotics by bifidobacteria

The growth dynamics of several bifidobacterial strains on a range of oligosaccharides were investigated by Hopkins *et al.* (1998), and it was shown that considerable interspecies differences existed in the abilities of the test organisms to utilize these substances. GOS and FOS supported the best growth, while pyrodextrins and xylo-oligosaccharides were found to be relatively poor substrates. Growth was often found to be better on oligosaccharides in comparison with their component monosaccharides, supporting previous suggestions that some bifidobacterial species may be particularly adapted to competing for oligosaccharides in the large gut (Degnan and Macfarlane, 1995). Bifidobacteria seem to grow better than lactobacilli on GOS, as indicated by the work of Ohtsuka and co-workers (1989). These authors reported that several pure cultures of bifidobacteria (*B. bifidum, B. adolescentis, B. longum, B. infantis, B. breve*) grew well on GOS, but only one out of four *Lactobacillus* species tested (*Lb. acidophilus*) utilized the substrate. No growth was seen with *Lb. gasseri, Lb. salivarius* or *Lb. casei*.

A great many *in vivo* and *in vitro* studies have now shown that prebiotics such as FOS and GOS increase numbers of bifidobacteria in a generally selective way (Tuohy *et al.*, 2001; Gibson *et al.*, 1995; Macfarlane *et al.*, 2008). In a double-blind randomized controlled trial, Bouhnik *et al.* (2004) administered several candidate prebiotics for seven days to healthy humans, and demonstrated that a greater bifidogenic effect was seen with GOS compared to short-chain FOS. Although bifidobacteria can also utilize some polysaccharides, at least to a certain extent, studies have shown that they prefer oligosaccharides (Degnan and Macfarlane, 1995; Palframen *et al.*, 2003; Amaretti *et al.*, 2006). Genome sequencing has indicated that these organisms contain several oligosaccharide uptake systems, and carbohydrate modifying enzymes, that vary between species. The genome sequence of *B. longum* NCC2705 contains over 30 glycosyl hydrolases that are likely to be involved in oligosaccharide degradation (Schell *et al.*, 2002). This may be one of the reasons for differences in oligosaccharide substrate preferences reported in bifidobacteria. For instance, arabinoxylan-derived oligosaccharides can be fermented by *B. adolescentis* and *B. longum*, but not by *B. breve*. Study of the genome sequence of *B. breve* UCC2003 suggests that it may contain more than 40 glycosyl hydrolases, and while the organism contains a *fos* operon which encodes a β-fructofuranosidase, enabling it to utilize FOS, it does not appear to harbour enzymes that degrade arabinoxylan (Ryan *et al.*, 2005; Van den Broek and Voragen, 2008). Hernot *et al.* (2009) used *in vitro* fermentation models to compare the metabolic profiles and bifidogenic properties of several short-chain, medium-chain and long-chain FOS, inulin, GOS and polydextrose. Results indicated that the short-chain oligosaccharides were more rapidly fermented, and promoted the greatest increases in bifidobacterial growth and fermentation acid production.

Synthesis and utilization of novel prebiotics

GOS can be produced from lactose in transgalactosylation reactions by bacterial or fungal β-galactosidases, which are usually immobilized on microparticle carriers such as ion exchange resins, chitosan, cellulose and agarose beads, or fibrous supports such as cotton cloths. This often results in the formation of a variety of different GOS products (Crittenden and Playne, 1996). The final product and its degree of polymerization depend on conditions used for the reaction, as well as the microbial source of the enzyme (Macfarlane *et al.*, 2008). Therefore, it may be possible to produce new prebiotics and oligosaccharides using bifidobacterial enzymes involved in modifying galactose and fructose-based carbohydrates with transglycosylation activity (Gibson *et al.*, 2000; van den Broek and Voragen, 2008) such as glycoside hydrolases. These processes have an added advantage in that the starting materials for the reactions are inexpensive, and that their use can result in the formation of more elongated and branched oligosaccharides. They usually take longer for bacteria to digest in the large intestine, with consequent health benefits for the host resulting from increased fermentation product formation lower down in the colon. Moreover, enhanced selectivity for particular bifidobacterial strains can be designed using glycotechnology to produce GOS with specific glycosidic linkages containing monosaccharides that will enhance selectivity for specific types of bifidobacteria, because they are more efficiently utilized by species that produce the enzyme. This was shown in the study by Rabiu *et al.* (2001), in which the authors synthesized oligosaccharides, which had distinct thin layer chromatography spectra, using β-galactosidases from five bifidobacterial species (*B. pseudolongum* DSM-20099, *B. angulatum*, *B. bifidum* BB-12, *B. adolescentis* ANB-7, *B. infantis* DSM-20088). The spectra were observed to differ from the commercial oligosaccharide Oligimate 55. The *B. angulatum* oligosaccharides had a prebiotic activity comparable to that of the commercial GOS preparation, and all of the bifidobacteria, except *B. adolescentis*, exhibited enhanced growth (when compared to Oligimate 55) on oligosaccharides synthesized by their own enzymes. The concept was further tested using a β-galactosidase produced using whole cell extracts of the faecal *B. bifidum* strain NCIMB 41171 (Tzortis *et al.*, 2005). The GOS obtained from 50% (w/w) lactose contained 25% disaccharides, 35% trisaccharides, 25% tetrasaccharides and 15% pentasaccharides. Maximum conversion of lactose (35%) was obtained with 55% (w/w) lactose as the starter material. In batch culture experiments, the ability of *B. bifidum* to grow on the novel prebiotic was compared to a range of other nominal prebiotics including FOS and a commercial GOS. The *B. bifidum* strain was shown to have the highest specific growth rate on the novel GOS (0.89 ± 0.06) compared to FOS (0.51 ± 0.04) and the commercial GOS (0.63 ± 0.06). Fermentation of the novel GOS fraction also produced the highest levels of acetate and lactate.

Subsequently, the authors used this bifidobacterial-enzyme produced GOS preparation (designated B-GOS) in a three-stage continuous culture fermentation system, designed to simulate environmental conditions extant in the human proximal, transverse and distal colon, as well as in *in vivo* studies in pigs (Tzortis *et al.*, 2005a). The GOS was shown to increase numbers of bifidobacteria in the first two vessels of the fermentation system, while its administration (4% w/w) to the pigs' diet increased faecal and colonic levels of bifidobacteria. Bifidobacterial numbers in the proximal colon were higher than those obtained using inulin. Prebiotics may also act as decoy molecules by mimicking bacterial receptors, thereby inhibiting the binding of pathogens to gut epithelial surfaces. *In vitro* analyses demonstrated that the B-GOS was able to inhibit the attachment of an enteropathogenic *E. coli* and *Salmonella enterica* serotype Typhimurium to HT29 cells (Tzortis *et al.*, 2005b). Further evaluation of the B-GOS was determined in a two-part randomized, crossover, placebo-controlled intervention trial using healthy volunteers (Depeint *et al.*, 2008). The subjects were either fed B-GOS, a placebo or a commercial GOS (designated V-GOS) produced by a β-galactosidase isolated from *Bacillus circulans* ATCC 4516. In the feeding study, a dose-related increase in bifidobacteria occurred with B-GOS, levels of which were found to be significantly higher than either the placebo or V-GOS. In the same investigation, the B-GOS was also shown to increase growth of a number of intestinal bifidobacterial isolates *in vitro*, resulting in significantly higher numbers of *B. bifidum* and *B. longum* than the commercial V-GOS.

Different bifidobacterial and lactobacillus species have been found to exhibit substrate preferences during growth on galactose-containing oligosaccharides. This was demonstrated when *B. adolescentis* was shown to be able to degrade oligosaccharides with a degree of polymerization (d.p.) greater than 3, whereas *B. infantis* and *Lb. acidophilus* could only utilize molecules with a d.p. of less than 3. The ability of *B. adolescentis* to use high concentrations of *trans*-galactosylated oligosaccharides (TOS) more effectively was believed to be due to a novel cell envelope-associated β-galactosidase (β-galactosidase II), with activity towards TOS, but not lactose (van

Laere *et al.*, 2000). Similarly, in another study, substrate preferences were seen with respect to GOS utilization, where *Bifidobacterium animalis* subsp. *lactis*, unlike *Lb. rhamnosus* was able to metabolize trisaccharide and tetrasaccharide fractions (Gopal *et al.*, 2001). As mentioned earlier, intestinal organisms other than bifidobacteria and lactobacilli have been shown to be able to grow on GOS, including bacteroides and clostridia, which can utilize the trisaccharides 4′ galactosyllactose and 6′ galactosyllactose (Ohtsuka *et al.*, 1989; Sako *et al.*, 1999), and it seems that the pattern of glycosidic linkages in different GOS products influences the abilities of different bacteria to grow on them.

A number of investigations have shown that the selectivity of polysaccharides for bifidobacteria and lactobacilli can be increased by hydrolysing them to oligosaccharides, before they are added to the diet (Mountzouris *et al.*, 1999; van Laere *et al.*, 2000). The use of pectic oligosaccharides as potential prebiotic sources has been evaluated in a number of studies. These substances can be obtained from a range of citrus products, and bifidobacteria have been reported to grow well on them (Olano-Martin *et al.*, 2002; Manderson *et al.*, 2005). They can also enhance the growth of eubacterial species (Duncan *et al.*, 2003; Manderson *et al.*, 2005), which may be beneficial to health, since these organisms are able to utilize lactate produced in bifidobacterial fermentations to produce butyrate, which is the main fuel for the colonic epithelium, and may be protective in colon cancer and ulcerative colitis (Macfarlane *et al.*, 2009). The prebiotic properties of pectic oligosaccharides prepared from bergamot peel were compared to FOS using pure cultures, and *in vitro* pH controlled fermenters inoculated with faecal bacteria (Mandalari *et al.*, 2006). Bifidobacteria were shown to have the highest specific growth rates on the pectic product, while clostridial and bacteroides' growth rates were low. Changes in intestinal bacterial profiles were detected by FISH in the batch culture fermenters, and addition of bergamot was shown to increase numbers of lactobacilli, bifidobacteria and eubacteria, while clostridial populations decreased. The effects of bergamot on microbial communities were greater than those observed with FOS at 24 hours, which demonstrates the potential prebiotic effectiveness of these pectic oligosaccharides.

Similarly, arabino-oligosaccharides can be obtained from arabinan, a pectic-associated oligomer present in sugar beet. The fermentation profiles of arabinan and several hydrolysis fractions were tested after pectinase digestion in batch cultures with a faecal inoculum (Al-Tamimi *et al.*, 2006). Bifidobacteria were able to grow on all of the oligosaccharide fractions, with the lower molecular weight components being more selective for bifidobacteria, and promoting the greatest increases in their populations. Numbers of clostridia decreased during fermentation of all of the fractions.

Utilization of potential prebiotics such as honey polysaccharides and their effects on intestinal bacteria has been reported (Kajiwara *et al.*, 2002; Sanz *et al.*, 2005). Honey contains approximately 55–75% monosacharides, mainly consisting of glucose and fructose. It has been shown to inhibit pathogens (Molan *et al.*, 1992; Cooper *et al.*, 1999) and may have antioxidant properties (Gheldof *et al.*, 2002; Schram *et al.*, 2003). Kajiwara *et al.* (2002) demonstrated that the growth of several strains of bifidobacteria could be enhanced when cultured with honey. Other workers have investigated putative prebiotic effects of honey in *in vitro* incubations, after removal of its monosaccharide constituents. Three honey fractions, together with untreated honey, were inoculated into batch culture fermentation systems, and the effects on various microbial populations were compared to FOS using FISH. Increased numbers of lactobacilli and bifidobacteria were detected with all of the honey samples; however, growth stimulation of these organisms was found to be lower than with FOS (Sanz *et al.*, 2005), indicating that honey has a lower prebiotic potential as compared to this latter prebiotic.

Effects of prebiotics on bifidobacteria populations in the elderly

Changes in gastrointestinal function such as reductions in colonic transit times and gastric acid secretion, as well as reduced immune system functionality, are associated with the ageing process. Together with reduced exercise and food intake, this can also lead to profound changes in intestinal bacterial communities, a decline in populations of protective bifidobacteria (Hopkins *et al.*, 2001), increased numbers of putrefactive organisms, and other more detrimental species such as clostridia and enterobacteria. These phenomena are also associated with a general reduction in bacterial species diversity in the colon (Woodmansey *et al.*, 2004; Bartosch *et al.*, 2005). Such changes in microbial community structure and gut functionality contribute towards making older people more susceptible to intestinal infections and other gut diseases. There have been several investigations on the benefits of prebiotics in the elderly, and their effects on intestinal bifidobacteria in particular. In a crossover study, 12 healthy elderly volunteers were fed 8 grams of short-chain FOS per day, for 4 weeks (Bhounik *et al.*, 2007). During the feeding period, numbers of bifidobacteria in stools increased significantly, along with a reduction in faecal pH. No differences were observed in gut transit times or stool weights, although significantly more flatus was experienced. In another study involving healthy elderly volunteers, feeding a synbiotic combination of the inulin-based prebiotic Synergy 1, with *B. lactis* BL-01 and *B. bifidum* BB-02, was also shown to increase levels of bifidobacteria, in the gut, and this persisted for up to 4 weeks after synbiotic consumption was stopped (Bartosch *et al.*, 2005). Guigoz *et al.* (2002) showed that the administration of FOS for 3 weeks to frail elderly subjects was able to increase faecal bifidobacteria, and in addition, reduce the expression of IL-6 in peripheral blood mononuclear cells and phagocytic activities. In another study, Vulevic *et al.* (2008) administered B-GOS, the novel prebiotic produced by *B. bifidum* (see earlier), or a placebo in a double-blinded trial to 44 elderly subjects for 10 weeks. Significantly increased levels of faecal bifidobacteria, lactobacilli/enterococci, and the *Clostridium coccoides-Eubacterium rectale* group were observed in the study, with concomitant reductions in bacteroides, *E. coli*, desulfovibrios and the *Clostridium histolyticum* group. Additionally, in the prebiotic group, there was a beneficial effect on the immune system, with significant reductions in pro-inflammatory cytokines and increased phagocytosis, NK cell activity and immunomodulatory IL-10 production.

In conclusion, over the last 30 years, many studies have demonstrated that bifidobacteria growing in the large intestine have important health-promoting properties, while more recent work has indicated that these organisms also occupy an important ecological niche in the colonic ecosystem. As a consequence, bifidobacteria have been the focus of considerable research interest, with one of the main aims being to find ways of increasing their numbers in the gut through dietary means. In practice, this has focused on carbohydrate catabolism, and the fermentation of non-digestible oligosaccharides. These studies, involving the use of rapidly fermentable substrates, have had a fair degree of success. However, the challenge now is to identify carbohydrates that can be targeted to the distal bowel, which have a higher level of selectivity for particular bifidobacterial species that have well-characterized health benefits.

References

Abbad Andalloussi, S., Talbaoui, H., Maeczak, R., and Bonaly, R. (1995). Isolation and characterization of exocellular polysaccharides produced by *Bifidobacterium longum*. Appl. Microbiol. Biotechnol. *43*, 995–1000.

Alp, G., and Aslim, B. (2010). Relationship between the resistance to bile salts and low pH with exopolysaccharide (EPS) production of *Bifidobacterium* spp. isolated from infants faeces and breast milk. Anaerobe *16*, 101–105.

Al-Tamimi, M.A.H.M., Palframan, R.J., Gibson, G.R., and Rastall, R.A. (2006). *In vitro* fermentation of sugar beet arabinan and arabinooligosaccharides by the human gut microflora. J. Appl. Microbiol. *100*, 407–414.

Amaretti, A., Tamburini, E., Benardi, T., Pompei, A., Zanoni, S., Vaccari, G., Matteuzzi, D., and Rossi, M. (2006). Substrate preference of *Bifidobacterium adolescentis* MB 239: compared growth on single and mixed carbohydrates. Appl. Microbiol. Biotech. *73*, 654–662.

Amrouche, T., Boutin, Y., Prioult, G., and Fliss, I. (2006). Effects of bifidobacterial cytoplasm, cell wall and exopolysaccharide on mouse lymphocyte proliferation and cytokine production. Int. Dairy J. *16*, 70–80.

Asahara, T., Nomoto, K., Shimizu, K., Watanuki, M., and Tanaka, R. (2001). Increased resistance of mice to *Salmonella enterica* serovar Typhimurium infection by synbiotic administration of bifidobacteria and transgalactosylated oligosaccharides. J. Appl. Microbiol. *91*, 985–996.

Bartosch, S., Woodmansey, E.J., Paterson, J.C.M., McMurdo, M.E., and Macfarlane, G.T. (2005). Microbiological effects of consuming a synbiotic containing *Bifidobacterium bifidum*, *Bifidobacterium lactis*, and oligofructose in elderly persons, determined by real-time polymerase chain reaction and counting of viable bacteria. Clin. Infect. Dis. *40,* 28–37.

Becker, A., Katzen, F., Puhler, A., and Lelpie, L. (1998). Xanthan gum biosynthesis and application: A biochemical/genetic perspective. Appl. Microb. Biotech. *50,* 145–152.

Bergmaier, D., Champagne, C.P., and Lacroix, C. (2005). Growth and exopolysaccharide production during free and immobilized cell chemostat culture of *Lactobacillus rhamnosus* RW-9595M. J. Appl. Microbiol. *98,* 272–284.

Boels, I.C., van Kranenburg, R., Hugenholtz, J., Kleerebezem, M., and de Vos, W.M. (2001). Sugar catabolism and its impact on the biosynthesis and engineering of exopolysaccharide production in lactic acid bacteria. Int. Dairy J. *11,* 723–732.

Bouhnik, Y., Achour, L., Paineau, D., Riottot, M., Attar, A., and Bornet, F. (2007). Four-week short chain fructooligosaccharides ingestion leads to increasing fecal bifidobacteria and cholesterol excretion in healthy elderly volunteers. Nutr. J. *6,* 42.

Bouhnik, Y., Flourie, B., Riottot, M., Bisetti, N., Gailing, M.-F., Guibert, S., Bornet, F., and Rambaud, J.-C. (1996). Effects of fructooligosaccharides ingestion on fecal bifidobacteria and selected metabolic indexes of colon carcinogenesis in healthy humans. Nutr. Can. *26,* 21–29.

Bouhnik, Y., Raskine, L., Simoneau, G., Vicaut, E., Neut, C., Flourie, B., Brouns, F., and Bornet, F.R. (2004). The capacity of nondigestable carbohydrates to stimulate fecal bifidobacteria in healthy humans: a double-blind, randomized, placebo-controlled, parallel group, dose response relation study. Am. J. Clin. Nutr. *80,* 1658–16664.

Breedveld, M., Bonting, K., and Dijkhuizen, L. (1998). Mutational analysis of exopolysaccharide biosynthesis by *Lactobacillus sakei* 0–1. Microb. Lett. *169,* 241–249.

Buddington, R.K., Williams, C.H., Chen, S.-C., and Witherly, S.A. (1996). Dietary supplement of neosugar alters the fecal flora and decreases activities of some reductive enzymes in human subjects. Am. J. Clin. Nutr. *63,* 709–716.

Bylund, J., Burgess, L.A., Cescutti, P., Ernst, R.K., and Speert, D.P. (2006). Exopolysaccharides from *Burkholderia cenocepacia* inhibit neutrophil chemotaxis and scavenge reactive oxygen species. Biol, J. . Chem. 281, 2526–2532.

Campbell, J.M., Fahy, G.C., and Wolf, B.W. (1997). Selected indigestible oligosaccharides affect large bowel mass and fecal short-chain fatty acids, pH and microflora in rats. J. Nutr. *127,* 130–136.

Cerning, J., Bouillane, C., Desmazeaud, C., and Landon, M. (1988). Exocellular polysaccharide production by *Streptococcus thermophilus.* Biotech. Letts. *10,* 255–260.

Chabot, S., Yu, H.L., De Leseleuc, L., Cloutier, D., Van Calsteren, M.R., and Lessard, M. (2001). Exopolysaccharides from *Lactobacillus rhamnosus* RW-9595M stimulates TNFα, IL-6 and IL-12 in humans and mouse cultured immunocompetent cells, and INF-g in mouse splenocytes. Lait *81,* 683–697.

Challa, A., Rao, D.R., Chawan, C.B., and Shackelford, L. (1997). *Bifidobacterium longum* and lactulose suppress azoxymethane-induced colonic aberrant crypt foci in rats. Carcinogenesis *18,* 517–521.

Chen, J., and Seviour, R. (2007). Medicinal importance of fungal β-(1®3), (1®6)-glucans. Mycol. Res. *111,* 635–652.

Chung, J.W., Altman, E., Beveridge, T.J., and Speert, D.P. (2003). Colonial morphology of *Burkholderia cepacia* complex genomovar III: implications in exopolysaccharide production, pilus expression, and persistence in the mouse. Infect. Immun. *71,* 904–909.

Cooper, R.A., Molan, P.C., and Harding, K.G. (1999). Antibacterial activity of honey against strains of *Staphylococcus aureus* from infected wounds. J. R. Soc. Med. *92,* 283–285.

Crittenden, R.G., and Playne, M.J. (1996). Production properties and applications of food grade oligosaccharides. Trends Food Sci. Technol. *7,* 353–361.

Dal Bello, F., Walter, J., Hertel, C., and Hammes, W.P. (2001). *In vitro* study of prebiotic properties of levan-type exopolysaccharides from lactobacilli and non-digestable carbohydrates using denaturing gradient gel electrophoresis. Syst. Appl. Microbiol. *24,* 1–6.

Degeest, B., Janssens, B., and de Vuyst, L. (2001). Exopolysaccharide (EPS) biosynthesis by *Lactobacillus sakei* 0–1: production kinetics, enzyme activities and EPS yields. J. Appl. Microbiol. *91,* 470–477.

Degnan, B.A., and Macfarlane, G.T. (1995). Arabinogalactan utilization in continuous cultures of *Bifidobacterium longum*: Effect of co-culture with *Bacteroides thetaiotaomicron.* Anaerobe *1,* 103–112.

Delcenserie, V., Lessard, M.H., LaPointe, G., and Roy, D. (2008). Genome comparison of *Bifidobacterium longum* strains NCC2705 and CRC-002 using suppression subtractive hybridization. FEMS Microbiol. Lett. *280,* 50–56.

Delzenne, N.M., and Kok, N. (2001). Effects of fructan-type prebiotics on lipid metabolism. Am. J. Clin. Nutr. *73,* 456S-458S.

de Palencia, P.F., Werning, M.L., Sierra-Filardi, E., Dueñas, M.T., Irastorza, A., Corbi, A.L., and Lopez, P. (2009). Probiotic properties of the 2-substituted (1,3)-D-glucan-producing bacterium *Pediococcus parvulus* 2.6. Appl. Environ. Microbiol. 75, 4887–4891.

Depeint, F., Tzortis, G., Vulevic, J., I'Anson, K., and Gibson, G. (2008). Prebiotic evaluation of a novel galactooligosaccharide mixture produced by the enzymatic activity of *Bifidobacterium bifidum* NCIMB 41171, in healthy humans: a randomized, double-blind, crossover, placebo-controlled intervention study. Am. J. Clin. Nutr. *87,* 785–791.

Deutsch, S.-M., Falentin, H., Dols-Lafargue, M., LaPointe, G., and Roy, D. (2008). Capsular exopolysaccharide biosynthesis gene of *Propionibacterium freudenreichii* subsp. *shermanii.* Int. J. Food Microbiol. *125,* 252–258.

de Vos, W.M. (1996). Metabolic engineering of sugar catabolism in lactic acid bacteria. Antonie von Leeuwenhoek *70,* 223–242.

de Vuyst, L and Degeest, B. (1999). Heteropolysaccharides from lactic acid bacteria. FEMS Microbiol. Rev. *21*, 153–177.

Dueñas, M., Munduate, A., Perea, A., and Irastorza, A. (2003). Exopolysaccharide production by *Pediococcus damnosus* 2.6 in a semidefined medium under different growth conditions. Int. J. Food Microbiol. *87*, 113–120.

Deutsch, S.-M., Falentin, H., Dols-Lafargue, M., LaPointe, G., and Roy, D. (2008). Capsular exopolysaccharide biosynthesis gene of *Propionibacterium freudenreichii* subsp. *shermanii*. Int. J. Food. Microbiol. *125*, 252–258.

Duncan, S.H., Scott, K.P., Ramsay, A.G., Harmsen, H.J.M., Welling, G.W., Stewart, C.S., and Flint, J.H. (2003). Effects of alternative dietary substrates on competition between human colonic bacteria in an anaerobic fermentor system. Appl. Environ. Microbiol. *69*, 1136–1142.

Gheldof, N., Wang, X.H., and Engeseth, N.J. (2002). Identification and quantification of antioxidant components of honeys from various floral sources. J. Agric. Food Chem. *50*, 5870–5877.

Gibson, G.R., Beatty, E.R., and Cummings, J.H. (1995). Selective stimulation of bifidobacteria in the human colon by oligofructose and inulin. Gastroenterology *108*, 975–982.

Gibson, G.R., Ottaway, P.B., and Rastall, R.A. (2000). Prebiotics: new developments in functional foods. London: Chadwick House Group Ltd.

Gopal, P.K., Sullivan, P.A., and Smart, J.B. (2001). Utilisation of galacto-oligosaccharides as selective substrates for growth by lactic acid bacteria including *Bifidobacterium lactis* DR10 and *Lactobacillus rhamnosus* DR20. Int. Dairy J. *11*, 19–25.

Grobben, G.J., Smith, M.R., Sikkema, J., and de Bont, J.A.M. (1996). Influence of fructose and glucose on the production of exopolysaccharides and the activities of enzymes involved in the sugar metabolism and the synthesis of sugar nucleotides in *Lactobacillus delbreuckii* subsp. *bulgaricus* NCFB 2772. Appl. Microbiol. Biotech. *46*, 279–284.

Grobben, G.J., van Casteren, W.H.M., Schols, H.A., Oosterveld, A., Sala, G., Smith, M.R., Sikkema, J., and De Bont, J.A.M. (1997). Analysis of the exopolysaccharides produced by *Lactobacillus delbreuckii* subsp. *bulgaricus* NCFB 2772 grown in continuous culture on glucose and fructose. Appl. Microbiol. Biotech. *48*, 516–521.

Guigoz, Y., Rochat, F., Peruisseau-Carrier, G., Rochat, I., and Schiffrin, E.S. (2002). Effects of oligosaccharide on the faecal flora and non-specific immune system in elderly people. Nutr. Res. *22*, 13–25.

Harding, L.P., Marshall, V.M., Hernandez, Y., Gu, Y., Maqsood, M., McLay, N., and Laws, A.P. (2005). Structural characterisation of a highly branched exopolysaccharide produced by *Lactobacillus delbrueckii* subsp. *bulgaricus* NCFB2074. Carbohydr. Res. *340*, 1107–1111.

Hernot D., Boileau, T.W., Bauer, L.L., Middelbos, I.S., Murphy, M.R., Swanson, K.S., and Fahey, G.C. (2009). *In vitro* fermentation profiles, gas production rates, and microbiota modulation as affected by certain fructans, galactooligosaccharides, and polydextrose. J. Agric. Food Chem. *57*, 1354–1361.

Hess, S.J., Roberts, R.F., and Ziegler, G.R. (1997). Rheological properties of non-fat yoghurt stabilized using *Lactobacillus delbreuckiis sp. bulgaricus* producing exopolysaccharide or using commercial stabilizer systems. J. Dairy Sci. *80*, 252–263.

Hopkins, M.J., Cummings, J.H., and Macfarlane, G.T. (1998). Inter-species differences in maximum specific growth rates and cell yields of bifidobacteria cultured on oligosaccharides and other simple carbon sources. J. Appl. Microbiol. *85*, 381–386.

Hopkins, M.J., Sharp, R., and Macfarlane, G.T. (2001). Age and disease-related changes in intestinal bacterial populations assessed by cell culture, 16S rRNA abundance and community cellular fatty acid profiles. Gut *48*, 198–205.

Hosono, A., Lee, J., Ametani, A., Natsume, M., Hirayama, M., Adachi, T., and Kaminogawa, S. (1997). Characterization of a water soluble polysaccharide fraction with immunopotentiating activity from *Bifidobacterium adolescentis* M101-4. Biosci. Biotech. Biochem. *61*, 312–316.

Jolly, I., and Stingele, F. (2001). Molecular organisation and functionality of exopolysaccharide gene clusters in lactic acid bacteria. Int. Dairy J. *11*, 733–745.

Jolly, L., Vincent, S.J.F., Duboc, P., and Neeser, J.-R. (2002). Exploiting exopolysaccharides from lactic acid bacteria. Antonie van Leeuwenhock. *82*, 367–374.

Kajiwara, S., Gandhi, H., and Ustonol, Z. (2002). Effect of honey on the growth of and acid production by human intestinal *Bifidobacterium* spp.: an *in vitro* comparison with commercial oligosaccharides and inulin. J. Food Prot. *65*, 214–218.

Kikuchi, H., Andrieux, C., Riottot, M., Bensaada, M., Popot, F., Beaumatin, P., and Szylit, O. (1996). Effect of two levels of transgalactosylated oligosaccharide intake in rats associated with human fecal microflora on bacterial glycolytic activity, end-products of fermentation and bacterial steroid transformation. J. Appl. Bacteriol. *80*, 439–446.

Kim, Y., Oh, S., and Kim, S.-H. (2009). Released exopolysaccharide (r-EPS) produced from probiotic bacteria reduce biofilm formation of enterohemorrhagic *Escherichia coli* O157:H7. Biochem. Biophys. Res. Comm. *379*, 324–329.

Kitazawa, H., Harata, T., Uemura, J., Saito, T., Kaneko, T., and Itoh, T. (1998). Phosphate group requirement for mitogenic activation of lymphocytes by extracellular phosphopolysaccharide from *Lactobacillus delbrueckii* subsp. *bulgaricus*. Food Microbiol. *17*, 109–118.

Kitazawa, H., Toba, T., Itoh, T., Kumano, N., Adachi, S., and Yamaguchi, T. (1991). Antitumoral activity of slime-forming encapsulated *Lactobacillus lactis* subsp. *cremoris* isolated from Scandinavian ropy sour milk, 'viili'. Anim. Sci. Tech. *62*, 277–283.

Kleessen, B., Sykura, B., Zunft, H.J., and Blaut, M. (1997). Effects of inulin and lactose on fecal microflora, microbial activity and bowel habit in elderly constipated persons. Am. J. Clin. Nutr. *65*, 1397–1402.

Klijn, A., Mercenier, A., and Arigoni, F. (2005). Lessons from the genomes of bifidobacteria. FEMS Microbiol. Rev. *29*, 491–509.

Kodali, V.P., and Sen, R. (2008). Antioxidant and free radical scavenging activities of an exopolysaccharide from a probiotic bacterium. Biotech. J. *3*, 245–251.

Kodali, V.P., Das, S., and Sen, R. (2009). An exopolysaccharide from a probiotic: Biosynthesis dynamics, composition and emulsifying activity. Food Res. Int. *42*, 695–699.

Kohno, M., Suzuki, S., Kanaya, T., Yoshino, T., Matsuura, Y., Asada, M., and Kitamura, S. (2009). Structural characterization of the extracellular polysaccharide produced by *Bifidobacterium longum* JBL05. Carbohyd. Poly. *77*, 351–357.

Koo, M., and Rao, A.V. (1991). Long term effect of bifidobacteria and neosugar on precursor lesions of colonic cancer in mice. Nutr. Cancer *16*, 249–257.

Korakli, M., and Vogel, R.F. (2006). Structure/function relationship of homopolysaccharide producing glycansucrases and therapeutic potential of their synthesised glycans. Appl. Microbiol. Biotechnol. *71*, 790–803.

Korakli, M., Ganzle, M.G., and Vogel, R.F. (2002). Metabolism by bifidobacteria and lactic acid bacteria of polysaccharides from wheat and rye, and exopolysaccharides produced by *Lactobacillus sanfranciscensis*. J. Appl. Microbiol. *92*, 958–965.

Korakli, M., Pavlovic, M., Ganzle, M.G., and Vogel, R.F. (2003). Exopolysaccharide and ketose production by *Lactobacillus sanfranciscensis* LTH2590. Appl. Environ. Microbiol. *69*, 2073–2079.

Lamothe, G.T. (2000). Molecular characterization of exopolysaccharide biosynthesis by *Lactobacillus delbreuckii* subsp. *bulgaricus*. Thesis, L'Universitie' de Lausanne, Switzerland.

Lin, T.Y., and Chang Chien, M.-F. (2007). Exopolysaccharides production as affected by lactic acid bacteria and fermentation time. Food Chem. *100*, 1419–1423.

Liu, C.-T., Hsu, I.-T., Chou, C.-C., Lo, P.-R., and Yu, R.-C. (2009). Exopolysaccharide production of *Lactobacillus salivarius* BCRC 14759 and *Bifidobacterium bifidum* BCRC 14615. World J. Microbiol. Biotechnol. *25*, 883–890.

Looijesteijn, P.J., Trapet, L., de Vries, E., Abee, T., and Hugenholtz, J. (2001). Physiological function of polysaccharides produced by *Lactococcus lactis*. Int. J. Food. Microbiol. *64*, 71–80.

Macfarlane, G.T., Steed, H., and Macfarlane, S. (2008). Bacterial metabolism and health-related effects of galacto-oligosaccharides and other prebiotics. J. Appl. Microbiol. *104*, 305–344.

Macfarlane, S., Steed, H., and Macfarlane, G.T. (2009). Intestinal bacteria and inflammatory bowel disease. Crit. Rev. Clin. Lab. Sci. *46*, 25–54.

Mandalari, G., Neuno Palop, C., Touhy, K., Gibson, G.R., Bennett, R.N., Waldron, K.W., Bisignano, G., Narbad, A., and Faulds, C.B. (2007). *In vitro* evaluation of the prebiotic activity of a pectic oligosaccharide-rich extract enzymatically derived from bergamotpeel. Appl. Microbiol. Biotechnol. 73, 1173–1179.

Manderson, K., Pinart, M., Tuohy, K.M., Grace, W.E., Hotchkiss, A.T., Widmer, W., Yadhav, M.P., Gibson, G.R., and Rastall, R.A. (2005). *In vitro* determination of prebiotic properties of oligosaccharides derived from orange juice manufacturing by-product stream. Appl. Environ. Microbiol. *71*, 8383–8389.

Martensson, O., Biorklund, Lambo, M. , M.A., Dueñas-Chasco, M.T., Irastorza, A., Holst, O., Norin, E., Walling, G., Oste, R., and Onning, G. (2005). Fermented ropy, oat-based products reduce cholesterol levels and stimulate the bibidobacteria flora in humans. Nutr. Res. *25*, 429–442.

Marx, S.P., Winkler, S., and Hartmeier, W. (2000). Metabolism of β-(2,6)-linked fructose-oligosaccharides by different bifidobacteria. FEMS Microbiol. Lett. *182*, 163–169.

McBain, A.J., and Macfarlane, G.T. (2001). Modulation of genotoxic enzyme activities by non- digestible oligosaccharide metabolism in *in-vitro* human gut bacterial ecosystems. J. Med. Microbiol. *50*, 833–842.

Molan, P.C. (1992). The antibacterial activity of honey. 1. The nature of the antibacterial activity. Bee World *73*, 5–28.

Morgan, A.J., Mul, A.J., Beldman, G., and Voragen, A.G.J. (1992). Dietary oligosaccharides-new insights. AGRO Fd. Indust. High Technol. Nov/Dec, 35–38.

Mountzouris, K., Gilmour, S., Grandison, A., and Rastall, R. (1999). Modeling of oligodextran production in an ultrafiltration stirred cell membrane reactor. Enzyme Microb. Technol. *24*, 75–85.

Nagaoka, M., Hashimoto, S., Watanabe, T., Yokokura, T., and Mori, Y. (1994). Anti-ulcer effects of lactic acid bacteria and their cell wall polysaccharides. Biol. Pharm. Bull. *17*, 1012–1017.

Nakajima, H., Hirota, T., Toba, T., Itoh, T., and Adachi, S. (1992a). Structure of the exopolysaccharide from slime-forming *Lactococcus lactis* subsp. *cremoris* SBT0495. Carb. Res. *224*, 245–253.

Nakijima, H., Suzuki, Y., Kaizu, H., and Hirota, T. (1992b). Cholesterol lowering activity of ropy fermented milk. J. Food Sci. *57*, 1327–1329.

Ninomiya, K., Matsuda, K., Kawahata, T., Kanaya, T., Kohno, M., Katakura, Y., Asada, M., and Shioya, S. (2009). Effect of CO_2 concentration on the growth and exopolysaccharide production of *Bifidobacterium longum* cultivated under anaerobic conditions. J. Biosci. Bioeng. *107*, 535–537.

Nishimura-Uemura, J., Kitazawa, H., Kawai, Y., Itoh, T., Oda, M., and Saito, T. (2003). Functional alteration of murine macrophages stimulated with extracellular polysaccharides from *Lactobacillus delbreuckii* subsp. *bulgaricus* OLL1073R-1. Food Microbiol. *20*, 267–273.

Ohtsuka, K., Benno, Y., Endo, A., Ueda, H., Ozawa, O., Uchida, T., and Mitsuoka, T. (1989). Effects of 4' -galactosyl-lactose on human intestinal microflora. Bifidus *2*, 143–149.

Olano-Martin, E., Gibson, G.R., and Rastall, R.A. (2002). Comparison of the *in vitro* bifidogenic properties of pectins and pectic-oligosaccharides. J. Appl. Microb. *93*, 505–511.

Olano-Martin, E., Williams, M.R., Gibson, G.R., and Rastall, R.A. (2003). Pectins and pectic-oligosaccharides inhibit *Escherichia coli* O157 H7 shiga toxin as directed towards the human colonic cell line HT27. FEMS Microbiol. Lett. *218*, 101–105.

Palframen, R.J., Gibson, G.R., and Rastall, R.A. (2003). Carbohydrate preferences of *Bifidobacterium* species isolated from the human gut. Curr. Issues Intest. Microbiol. *4*, 71–75.

Pham, P.L., Dupont, I., Roy, G., Lapointe, G., and Cerning, J. (2000). Production of exopolysaccharide by *Lactobacillus rhamnosus* R and analysis of its enzymatic degradation during prolonged fermentation. Appl. Environ. Microbiol. *66*, 2302–2310.

Pier, G.B., Coleman, F., Grout, M., Frankling, M., and Ohman, D.E. (2001). Role of alginate *O* acetylation in resitance of mucoid *Pseudomonas aeruginosa* to opsonic phagocytosis. Infect. Immun. *69*, 1895–1901.

Provencher, C., LaPointe, G., Sirois, S., Van Calsteren, M.R., and Roy, D. (2003). Consensus–degenerate hybrid oligonucleotide primers for amplification of priming glycosyltransferase genes of the exopolysaccharide locus in strains of the *Lactobacillus casei* group. Appl. Environ. Microbiol. *69*, 3299–3307.

Rabiu, B.A., Jay, A.J., Gibson, G.R., and Rastall, R.A. (2001). Synthesis and fermentation properties of novel galactoligosaccharides by β-galactosidases from *Bifidobacterium* species. App. Environ. Microbiol. *67*, 2526–2530.

Rehm, B. (2009). Microbial exopolysaccharides: Variety and potential applications. In: Microbial production of biopolymers and polymer precursors: applications and perspectives. Rehm, B., ed. (Norfolk, UK: Caister Academic), pp. 229–254.

Ricciardi, A., and Clementi, F. (2000). Exopolysaccharides from lactic acid bacteria: structure, production and technological applications. Ital. J. Food Sci. *12*, 23–45.

Roberts, C.M., Fett, W.F., Osman, S.F., Wijey, C., O'Connor, J.V., and Hoover, D.G. (1995). Exopolysaccharide production by *Bifidobacterium longum* BB-79. J. Appl. Bacteriol. *78*, 463–468.

Roberts, I.S. (1996). The biochemistry and genetics of capsular polysaccharide production in bacteria. Ann. Rev. Microb. *50*, 285–315.

Robijn, G.W., van den Berg, D.J.C., Haas, H., Kamerling, J.P., and Vliegenhart, J.F.G. (1995). Determination of the structure of the exopolysaccharide produced by *Lactobacillus sake* 0–1. Carb. Res. *276*, 117–136.

Rowland, I.R., and Tanaka, R. (1993). The effects of transgalactosylated oligosaccharides on gut flora metabolism in rats associated with a human faecal microflora. J. Appl. Bacteriol. *74*, 667–674.

Ruas-Madiedo, P., Abraham, A., Mozzi, F., and de los Reyes-Madiedo, C.G. (2008). Functionality of exopolysaccharides produced by lactic acid bacteria. In Molecular aspects of lactic acid bacteria for traditional new applications. Mayo, B. , Lopez, P. , and Perez-Martin, G. , eds. (Kerala, India: Research Signpost), pp. 137–166.

Ruas-Madiedo, P., Gueimonde, M., Arigoni, F., de los Reyes-Gavilan, C.G., and Margolles, A. (2009). Bile affects the synthesis of exopolysaccharides by *Bifidobacterium animalis*. Appl. Environ. Microbiol. *75*, 1204–1207.

Ruas-Madiedo, P., Gueimonde, M., de los Reyes-Gavilan, C.G., and Salminen, S. (2006a). Effect of exopolysaccharide isolated from 'villi' on the adhesion of probiotics and pathogens to intestinal mucus. J. Dairy Sci. *89*, 2355–2358.

Ruas-Madiedo, P., Gueimonde, M., Margolles, A., de los Reyes-Gavilan, C.G., and Salminen, S. (2006b). Exopolysaccharides produced by probiotic strains modify the adhesion of probiotics and enteropathogens to human intestinal mucus. J. Food Prot. *69*, 2011–2015.

Ruas-Madiedo, P., Hugenholtz, J., and Zoon, P. (2002). An overview of the functionality of exopolysaccharides produced by lactic acid bacteria. Int. Dairy J. *12*, 163–171.

Ruas-Madiedo, P., Moreno, J.A., Salazar, N., Delgado, S., Mayo, B., Margolles, A., and de los Reyes-Gavilan, C.G. (2007). Screening of exopolysaccharide-producing *Lactobacillus* and *Bifidobacterium* strains isolated from the human intestinal microbiota. Appl. Environ. Microbiol. 73, 4385–4388.

Ruijssenaars, H.J., Stingele, F., and Hartmans, S. (2000). Biodegradibility of food-associated extracellular polysaccharide. Current Microbiol. *40*, 194–199.

Ryan, S.M., Fitzgerald, G.F., and van Sinderen, D. (2005). Transcriptional regulation and characterization of a novel beta-fructofuranosidase-encoding gene from *Bifidobacterium breve* UCC2003. Appl. Environ. Microbiol. *71*, 3475–3482.

Saito, Y., Tanaka, T., and Rowland, I.R. (1992). Effects of soybean oligosaccharides on the human gut microflora *in vitro* culture. Microb. Ecol. Hlth. Dis. *5*, 105–110.

Sako, T., Matsumoto, K., and Tanaka, R. (1999). Recent progress on research and applications of non-digestable galactooligosaccharides. Int. Dairy J. *9*, 69–80.

Salazar, N., Gueimonde, M., Hernandez-Barranco, A.M., Ruas-Madiedo, P., and de los Reyes-Gavilan, C.G. (2008). Exopolysaccharides produced by intestinal *Bifidobacterium* strains act as fermentable substrates for human intestinal bacteria. Appl. Environ. Microbiol. *74*, 4737–4745.

Salazar, N., Ruas-Madiedo, P., Kolida, S., Collins, M., Rastall, R.A., Gibson, G.R., and de los Reyes-Gavilan, C.G. (2009). Exopolysaccharides produced by *Bifidobacterium longum* IPLA E44 and *Bifidobacterium animalis* subsp. *lactis* IPLA R1 modify the composition and metabolic activity of human faecal microbiota in pH-controlled batch cultures. Int. J. Food Microbiol. *135*, 260–267.

Sanz, M.L., Polemis, N., Morales, V., Corzo, N., Drakoularakou, A., Gibson, G.R., and Rastall, R.A. (2005). *In vitro* investigation into the potential prebiotic activity of honey oligosaccharides. J. Agric. Food Chem. 53, 2914–2921.

Schell, M.A., Karmirantzou, M., Snel, B., Vilanova, D., Berger, Pessi, B. , G., Zwahleen, M.C., Desiere, F., Bork, P., Delley, M., Pridmore, R.D., and Arigoni, F. (2002). The genome sequence of *Bifidobacterium longum* reflects its adaptation to the human gastrointestinal tract. Proc. Natl. Acad. Sci. U.S.A *99*, 14422–14427.

Schramm, D.D., Karim, M., Schrader, H.R., Holt, R.R., Cardetti, M., and Keen, C.L. (2003). Honey with high levels of antioxidants can provide protection to healthy subjects. J. Agric. Food Chem. *51*, 1732–1735.

Sekiguchi, S., Miura, Y., Kaneko, H., Nishimura, S.-I., Nishi, N., Iwase, M., and Tokura, S. (1994). Molecular weight dependency of antimicrobial activity by chitosan oligomers. In: Food Hydrocolloids: Structures, Properties and Functions. Nishinari, K., and Doi, E., eds. (New York, USA: Plenum Press), pp. 71–76.

Sengul, N., Aslim, B., Ucar, G., Yucel, N., Bozkurt, H., Sakaogullari, Z., and Atalay, F. (2006). Effects of exopolysaccharide-producing probiotic strains on experimental colitis in rats. Dis. Colon Rectum. *49*, 250–258.

Singh, J., Rivenson, A., Tomita, A., Shimamura, S., Ishibashi, N., and Reddy, B.S. (1997). *Bifidobacterium longum*, a lactic acid-producing intestinal bacterium inhibits colon cancer and modulates the intermediate biomarkers of colon carcinogenesis. Carcinogenesis *18*, 833–841.

Shoaf, K., Mulvey, G.L., Armstrong, G.D., and Hutkins, R.W. (2006). Prebiotic galactooligosaccharides reduce adherence of enteropathogenic *Escherichia coli* to tissue culture cells. Infect. Immun. *74*, 6920–6928.

Sreekumar, O., and Hosono, A. (1998). The antimutagenic properties of a polysaccharide produced by *Bifidobacterium longum* and its cultured milk against some heterocyclic amines. Can J. Microbiol. *44*, 1029–1036.

Stingele, F., Neeser, J.R., and Mollet, B. (1996). Identification and characterisation of the *eps* (exopolysaccharide) gene cluster from *Streptococcus thermophilus* Sfi6. J. Bacteriol. *178*, 1680–1690.

Sutherland, I.W. (1993). Microbial polysaccharides. In: Industrial gums: Polysaccharides and their derivates. Whistler, R.L., and Miller, J.N., eds. (San Diego, USA: Academic Press), pp. 69–85.

Sutherland, I.W. (1998). Novel and established applications of microbial polysaccharides. Trends Biotech. *16*, 41–46.

Sutherland, I.W. (2001). Biofilm exopolysaccharides: a strong and sticky framework. Microbiol. *147*, 3–9.

Thomas, R.J., and Brooks, T.J. (2004). Oligosaccharide receptors mimics inhibit *Legionella pneumophilia* attachment to human respiratory epithelial cells. Microb. Pathog. *36*, 83–92.

Tuinier, R., van Casteren, W.H.M., Looijesteijn, P.J., Scols, H.A., Voragen, A.G.J., and Zoon, P. (2001). Effects of structural modifications on some physical characteristics of exopolysaccharides from *Lactococus lactis*. Biopolymers *50*, 641–646.

Tuinier, R., Zoon, P., Cohen-Stuart, M.A., Fleer, G.J., and Kruijf, C.G. (1999). Concentration and shear-rate dependance of the viscosity of an exocellular polysaccharide. Biopolymers *50*, 641–646.

Tuohy, K.M., Kolida, S., Lustenberger, A.M., and Gibson, G.R. (2001). The prebiotic effects of biscuits containing partially hydrolysed gum guar and fructooligosaccharides- a human volunteer study. Br. J. Nutr. *86*, 341–348.

Tuohy, K.M., Ziemer, C.J., Klinder, A., Knobel, Y., Pool-Zobel, B.L., and Gibson, G.R. (2002). A human volunteer study to determine the prebiotic effects of lactulose powder on human colonic microbiota. Microb. Ecol. Hlth. Dis. *14*, 165–173.

Turroni, F., Ribbera, A., Foroni, E., van Sinderen, D., and Ventura, M. (2008). Human gut microbiota and bifidobacteria: from composition to functionality. Antonie van Leeuwenhock. *94*, 35–50.

Tzortis, G., Goulas, A.K., and Gibson, G.R. (2005a). Synthesis of prebiotic galactooligosaccharides using whole cells of a novel strain, *Bifidobacterium bifidum* NCIMB 41171. Appl. Microbiol. Biotechnol. *68*, 412–416.

Tzortis, G., Goulas, A.K., Gee, J.M., and Gibson, G.R. (2005b). A novel galactoligosaccharide mixture increases the bifidobacterial population numbers in a continuous *in vitro* fermentation system and in the proximal colonic contents of pigs *in vivo*. J. Nutr. *135*, 1726–1731.

van Calsteren, M.R., Pau-Roblet, C., Begin, A., and Roy, D. (2002). Structure determinations of the exopolysaccharide produced by *Lactobacillus rhamnosus* strains RW-9595 M and R. Biochem. J. *363*, 7–17.

van Casteren, W.H.M., Dijkema, C., Schols, H.A., Beldman, G., and Voragen, A.G.J. (1998). Characterization and modification of the exopolysaccharides produced by *Lactococcus lactis* subsp. *cremoris* B40. Carbohyd. Polymers *37*, 123–130.

van Casteren, W.H.M., Eimermann, Van, M. den Broek, L.A.M., Vincken, J.P., Schols, H.A., and Voragen, A.G.J. (2000). Purification and characterization of a β-galactosidase from *Aspergillus aculeatus* with activity towards (modified) exopolysaccharides from *Lactococcus lactis* susp. *cremoris* B39 and B891. Carb. Res. *327*, 411–422.

van den Broek, L.A.M., and Voragen, A.G.J. (2008). *Bifidobacterium* glycoside hydrolases and (potential) prebiotics. Innov. Food Sci. Emerg. Technol. *9*, 401–407.

van Dokkum, W., Wexendonk, B., Srikumar, T.S., and van den Heuvel E.G. (1999). Effect of nondigestible oligosaccharides on large bowel functions, blood lipid concentrations and glucose absorption in young healthy male subjects. Eur. J. Clin. Nutr. *53*, 1–7.

van Hijum, S.A., Kralj, S., Ozimek, L.K., Dijkhuizen, L., van Geel-Schutten, I.G. (2006). Structure–function relationships of glucansucrase and fructansucrase enzymes from lactic acid bacteria. Microbiol. Mol. Biol. Rev. *70*, 157–176.

van Kranenburg, R., Boels, I.C., Kleerebezem, M., and de Vos, W.M. (1999). Genetics and engineering of microbial exopolysaccharides for food: approaches for the production of existing and novel polysaccharides. Curr. Opin. Biotechnol. *10*, 498–504.

van Kranenburg, R., Marugg, J.D., van Swam, I.I., Willem, N.J., and de Vos, W.M. (1997). Molecular characterization of the plasmid-encoded eps gene cluster essential for polysaccharide biosynthesis in *Lactococcus lactis*. Mol. Microbiol. *24*, 387–397.

Vaningelgem, F., Zamfir, M., Mozzi, F., Adriany, T., Vancanneyt, M., Swings, J., and de Vuyst, L. (2004). Biodiversity of exopolysaccharides produced by *Streptococcus thermophilus* strains is reflected in their production and their molecular and functional characteristics. Appl. Environ. Microbiol. *70*, 2900–2912.

van Laere, K.M.J., Abee, T., Schols, H.A., Beldman, G., and Voragen, A.G.J. (2000). Characterization of a novel β-galactosidase from *Bifidobacterium adolescentis* DSM 20083 active towards transgalactooligosaccharides. Appl. Environ. Microbiol. *66*, 1379–1384.

Vedamuthu, E.R., and Neville, J.M. (1986). Involvement of a plasmid in production of ropiness (mucoidness) in milk cultures by *Streptococcus cremoris* MS. Appl. Environ. Microbiol. *51*, 1385–1386.

Velasco, S., Arskold, E., Paese, M., Grage, H., Irastorza, A., Radstrom, P., and van Niel, E.W.J. (2006). Environmental factors influencing growth of and exopolysaccharide formation by *Pediococcus parvulus* 2.6. Int. J. Food Microbiol. *111*, 252–258.

Velasco, S.E., Areizaga, J., Irastorza, A., Dueñas, M.T., Santamaria, A., and Muñoz, M.E. (2009). Chemical and rheological properties of the β-glucan produced by *Pediococcus parvulus* 2.6. J. Agric. Food Chem. *57*, 1827–1834.

Ventura, M., Connell-Motherway, O., M., Leahy, S., Moreno-Muñoz, J.A., Fitzgerald, G.F., and van Sinderen, D. (2007). From bacterial genome to functionality; case bifidobacteria. Int. J. Food Microbiol. *120*, 2–12.

Videira, P.A., Garcia, A.P., and Sa-Correia, I. (2005). Functional and topological analysis of the Burkholderia cenocepacia priming glucosyltransferase BceB, involved in the biosynthesis of the cepacian exopolysaccharide. J. Bacteriol. *187*, 5013–5018.

Vinderola, G., Perdigon, G., Duarte, J., Farnworth, E., and Matar, C. (2006). Effects of the oral admınıstratıon of the exopolysaccharide produced by *Lactobacillus kefiranofaciens* on the gut mucosal immunity. Cytokine. *36*, 254–260.

Vu, B., Chen, M., Crawford, R.J., and Ivanova, E.P. (2009). Bacterial extracellular polysaccharides involved in biofilm formation. Molecules. *14*, 2535–2554.

Vulevic, J., Drakoularakou, A., Yaqoob, P., Tzortzis, G., and Gibson, G.R. (2008). Modulation of the fecal microflora profile and immune function by a novel trans–galactooligosaccharide mixture (B-GOS) in healthy elderly volunteers. Am. J. Clin. Nutr. *88*, 1438–1446.

Welman, A.D., and Maddox, I.S. (2003). Exopolysaccharides from lactic acid bacteria: perspectives and challenges. Trends Biotechnol. *21*, 269–274.

Werning, M.L., Corrales, M.A., Prieto, A., Fernandez de Palencia, P., Navas, J., and Lopez, P. (2008). Herterologous expression of a position 2-substituted (1®3)-β-D-Glucan in *Lactococcus lactis*. Appl. Environ. Microbiol. *74*, 5259–5262.

Woodmansey, E.J., McMurdo, M.E., Macfarlane, G.T., and Macfarlane, S. (2004). Comparison of compositions and metabolic activities of fecal microbiotas in young adults and in antibiotic-treated and non-antibiotic-treated elderly subjects. Appl. Environ. Microbiol. *70*, 6113–6122.

Yasuda, E., Serata, M., and Sako, T. (2008). Suppresive effects on activation of macrophages by *Lactobacillus casei* strain Shirota genes determining the synthesis of cell wall-associated polysaccharides. Appl. Environ. Microbiol. *74*, 4746–4755.

Zentek, J., Marquart, B., and Pietrzak, T. (2002). Intestinal effects of mannooligosaccharides, transoligosaccharides, lactose and lactulose in dogs. J. Nutr. *132*, S1682–S1684.

Antibiotic Resistance/Susceptibility Profiles of Bifidobacteria and Molecular Analysis of Atypical Resistances

11

Baltasar Mayo, Sigrid Mayrhofer, Susana Delgado and Konrad J. Domig

Abstract

Antibiotics have greatly improved human life expectancy and quality, being undisputed winners in the fight against infectious diseases over the past 70 years. However, their continued success is severely threatened by the global spread of antibiotic resistance (AR). There is significant and wide-spread concern that commensal and beneficial bacterial populations, such as food-borne and gastrointestinal tract (GIT)-associated bacteria, will become a reservoir for AR-conferring genes, from which these could readily be transferred to opportunistic and pathogenic organisms.

Bifidobacteria represent a ubiquitous commensal bacterial group in the GIT of humans and animals, where they are supposed to exert protective and beneficial effects on their host. This has led to the use of selected strains as probiotics. High numbers of bifidobacteria are deliberately brought into contact with a vast array of intestinal microorganisms during probiotic consumption, which means that AR spread is possible if probiotic strains carry transferable AR-conferring genes. Strains carrying such genes should be excluded from use as probiotics. This chapter addresses key topics on the methodology of surveying AR in bifidobacteria, their intrinsic resistance/susceptibility profiles, and summarizes the molecular basis and genetic organization of atypical resistances from bifidobacteria. Among these, AR encoded by acquired genes, particularly those harboured in mobile genetic elements, pose the greatest risk of transfer. Knowledge on AR-conferring genes and their transfer potential is critical in preventing dissemination of AR among intestinal bacteria through the use of probiotics.

Introduction

Antibiotics and antibiotic resistance

The clinical use of antibiotics has achieved a significant reduction in morbidity and mortality from infectious diseases, and has a profound influence on the success of surgery, transplantation, and cancer therapy (Levy and Marshall, 2004). Antibiotics have been extensively employed in veterinary medicine as therapeutic and prophylactic agents and as animal growth promoters (Wegener, 2003). Furthermore, they have been used in aquaculture and agriculture practices (McManus *et al.*, 2002). It has been estimated that about 10 million tons of antibiotics have already been released into the biosphere (European Commission, 2005). Unsurprisingly, a clear correlation between antibiotic use and the level of resistant bacteria has been repeatedly reported (Bronzwaer *et al.*, 2002). AR complicates the treatment of infections (Levy and Marshall, 2004), while increasing economic costs by doubling the average stay at hospitals and doubling mortality and morbidity rates (Cosgrove and Carmelli, 2003). The presence of antibiotics in an ecosystem promotes the appearance and spread of resistant bacteria by killing susceptible bacteria and allowing resistant organisms to grow and dominate the niche (Baquero *et al.*, 2002). AR has been a cause for concern since antibiotics were introduced, but it is the increase in number, diversity, and range of resistant organisms that has become a huge clinical problem worldwide (Tenover, 2006). The phenomenon has recently been defined by the Alliance for the Prudent Use of Antibiotics as a shadow epidemic (http://www.apua.org).

AR-conferring genes are usually located on mobile genetic elements (such as plasmids and transposons) that have a high genetic transfer capacity (Normark and Normark, 2002). These elements pass from one cell to another, in some cases even crossing species and genus barriers. Furthermore, some AR genes are found in integrons, gene capture systems containing dedicated places for the insertion of AR and other adaptive genes and gene cassettes that are all expressed from a unique promoter (Nandi *et al.*, 2004). Integrons and their associated genes spread by a site-specific recombination mechanism (Nandi *et al.*, 2004). The collection of all genetic determinants with the potential to function as resistance genes is termed the antibiotic resistome (D'Acosta *et al.*, 2006). The resistome consists of an impressive list of genes, including AR-specifying genes of pathogenic and non-pathogenic microorganisms, resistance genes from antibiotic producers, cryptic and silent AR genes, and precursor genes that, under appropriate selection pressure, might evolve into genuine AR genes. The spread of AR is linked with the discontinued research on new antimicrobials, which the pharmaceutical industry attributes to an ever-increasing stringent criterion of regulatory agencies for new drugs and a short time protection of the discoveries (Shales *et al.*, 2004).

Intrinsic vs. acquired AR

Resistance can be inherent to a bacterial species or genus (referred to as intrinsic or natural resistance) or acquired (European Commission, 2005). Intrinsic or natural resistance involves an absence of the target for the antibiotic activity or presence of low affinity targets, low cell permeability, enzymatic inactivation of the antibiotic, and activity of efflux mechanisms (Tenover, 2006). In contrast, acquired AR occurs via mutation of pre-existing genes or, more frequently, by horizontal transmission. With some exceptions, intrinsic resistance and resistance by mutation are unlikely to be disseminated; horizontally transferred genes, particularly those carried on mobile genetic elements, are those most likely to be transmitted (Normak and Normak, 2002).

Acquired AR is associated with the volume of antimicrobials utilized (Bronzwaer *et al.*, 2002), the frequency of mutation of the microorganisms, and the biological cost of the resistance (Bjorkman and Andersson, 2000). However, the biological cost of most resistances is not strong, because the rapid appearance of resistances contrasts with the slow decline in the absence of the selective agent (Andersson, 2003). Resistances can be maintained by continued exposure to a residual concentration of antibiotic in the environment (Donoghue, 2003). Nevertheless, resistance genes are frequently associated with resistance determinants against other toxic compounds, meaning that resistances can be maintained by co-selection (Aarestrup, 2000; Borgen *et al.*, 2002).

AR in commensal and beneficial bacteria

Commensal bacteria can act as a reservoir of resistant elements (plasmids, transposons) able to be transferred to pathogen microorganisms, a fact that has been firmly stated in some bacterial groups (Levy and Marshall, 2004). For example, the commensal bacterium *Haemophilus parainfluenzae* supplies plasmids encoding β-lactamases to *Haemophilus influenzae* strains. Similarly, strains of *Staphylococcus epidermidis* serve as a reservoir of resistance genes and plasmids for the pathogenic *Staphylococcus aureus* strains. Although not encoded on mobile elements, the origin of penicillin resistance in *Streptococcus pneumoniae* was found to reside in the commensal species *Streptococcus viridans,* which is naturally resistant to this antibiotic (Balsalobre *et al.*, 2003). Thus, the public health concern of commensal populations acting as reservoirs for AR that can transfer across species boundaries to pathogenic organisms is well supported by an ever-growing body of scientific evidence (Levy and Marshall, 2004; Salyers *et al.*, 2004; http://www.roarproject.org). Little is known, however, about the abundance, diversity, and distribution of resistance genes in commensal and beneficial bacterial groups because until recently bacterial pathogens have been the primary focus of studies on AR (Sommer *et al.*, 2009). Moreover, the mechanisms involved in horizontal gene transfer from and to commensals are equally poorly understood (Netherwood *et al.*, 1999; Salyers *et al.*, 2004).

The first data regarding antibiotic susceptibility levels of bifidobacteria were published for human isolates by de Lavergne *et al.* (1959), Miller and Finegold (1967), Matteuzzi *et al.* (1983), and Lim *et al.* (1993), and those regarding animal sources by Biavatti *et al.* (1992) and Herreman *et al.* (1994). Over the past 10 years, the vast array of antibiotic surveys since undertaken has expanded the database of resistance/susceptibility profiles of bifidobacteria species. Besides susceptibility data, the genetic basis of some atypical resistances has also been determined.

Ecology and applications of bifidobacteria

Microbial ecology of bifidobacteria

Bifidobacteria were first isolated from the faeces of breast-fed infants by Tissier (1900), who called them *Bacillus bifidus*. They are high G+C Gram-positive, non-gas-producing, non-motile, non-sporeforming, pleomorphic (bifid) rod-shaped anaerobes belonging to the phylum *Actinobacteria*. The genus *Bifidobacterium* is part of the Bifidobacteriales family, which includes six other genera: *Aeriscardovia, Alloscardovia, Gardnerella, Metascardovia, Parascardovia,* and *Scardovia. Bifidobacterium* species are common inhabitants of the GIT and genitourinal tract of mammals (Scardovi, 1986; Biavatti and Mattarelli, 2006). At present, the genus contains about 37 species, of which 12 have been detected in human samples (Euzeby *et al.*, 2009; Turroni *et al.*, 2009).

Applications of bifidobacteria

Bifidobacteria are among the first colonizers of the human GIT, where they form dominant populations contributing to the maturation of this organ and the development of a competent immune system (Leahy *et al.*, 2005; Turroni *et al.*, 2009). These facts have lead to their intensive use as probiotics. Consumption of bifidobacteria can both contribute to rising numbers in the GIT and increase their alleged beneficial activities (FAO/WHO, 2006). In addition, bifidobacteria have recently been used for *in situ* delivery of therapeutic agents such as antigens and tumour-suppressing substances (Fujimori, 2006; Xu *et al.*, 2007), as well as a means of increasing beneficial detoxifying activities into the GIT (Park *et al.*, 2005). Traditional and new applications require strains with assured safety properties.

Safety aspects of bifidobacteria

Based on their naturally high numbers in the human GIT soon after birth, and a growing history of safe consumption in fermented products and dietary supplements (FAO/WHO, 2006), bifidobacteria enjoy a 'generally regarded as safe' (GRAS) status (see Chapter 9 for a full coverage of safety aspects of bifidobacteria). Except for the dental caries-causing species *B. dentium*, they are non-pathogenic and have rarely been implicated in human infections. The greatest current safety concern of bifidobacterial probiotics is the possibility of spreading AR genes through the GIT (Saarela *et al.*, 2002; Vankerckhoven *et al.*, 2008; Wassenaar and Klein, 2008). Consequently, distinction between natural (intrinsic) and acquired resistances, and distinguishing between those caused by mutation from those encoded on transferable elements, is extremely important (FAO/WHO, 2006; European Commission, 2008). A validated methodology and access to appropriate control strains with known susceptibility profiles are essential for the detection of potentially harmful strains (Flórez *et al.*, 2008).

Antimicrobial susceptibility testing of bifidobacteria

In contrast with clinically-relevant bacteria, for which resistance monitoring is indispensable, standard procedures and interpretive breakpoints for categorizing susceptible and resistant strains of non-pathogenic bacteria are non-existent (Masco *et al.*, 2006; Mättö *et al.*, 2006). Thus, various techniques have been used to address antibiotic susceptibility levels for bifidobacteria, including disc diffusion, broth dilution, microdilution, agar dilution and the Etest (Fig. 11.1 and Table 11.1). Standard media for antimicrobial susceptibility testing, such as Mueller-Hinton and Iso-Sensitest medium, which provide minimal interaction between antimicrobial agents and components of the medium, do not always support sufficient growth of bifidobacteria (Mättö *et al.*, 2006). As

Table 11.1. Representative AR surveys of bifidobacteria strains from human and animal origin by different methods.

Method	Medium	Antibiotics	Number of species	Number of strains	Atypical resistances	Reference
Agar dilution	Eugonagar	Bac, Ceph, Cm, Em, Km, Lin, Nal, Neo, Nitro, PG, Pol B, Sm, Tc, Vm	7	34	n.r.	Miller and Finegold (1967)
Broth dilution	TPY	Bac, Clin, Cm, Em, Gm, Km, Lin, Metro, Nal, Neo, Nitro, PG, Pol B, Sm, Tc, Vm	15	459	n.r.	Matteuzzi *et al.* (1983)
Broth macrodilution	Tryptic Soy Broth + yeast extract + cysteine	Am, Bac, Ceph, Cm, Em, Gm, Km, Lin, Met, Nal, Neo, Nitro, Paro, PG, Pol B, Sm, Tc, Vm	6	37	n.r.	Lim *et al.* (1993)
Soft agar overlay disc diffusion	TPY	Am, Amk, Amo, Azt, Bac, Cefa, Cefac, Cefi, Cefop, Cefot, Cefox, Cefta, Cefti, Ceftri, Cefur, Ceph, Cephra, Cipro, Clin, Cm, Col, Cotri, Em, Fura, Fus, Gm, Km, Metro, Nal, Netil, Nitro, Norflo, Oflo, PG, Pip, Pol B, Rif, Sm, Sulmet, Tc, Tmp, Vm	6	16	n.r.	Charteris *et al.* (1998)
Disc diffusion	TPY	Am, Amk, Amo, Bac, Car, Cefa, Cefam, Cefop, Cefox, Clox, Cm, Col, Em, Fura, Fus, Gm, Imi, Km, Met, Moxa, Nal, Neo, Netil, Nitro, Nitrof, Novo, Ole, Oxa, PG, Pol B, Sm, Sulfa, Sulfi, Tc, Tm	10	18	n.r.	Yazid *et al.* (2000)
Broth microdilution	MRS + cysteine	Am, Clox, Cm, Em, Km, Nal, Neo, Novo, PG, Pir, Rif, Sm, Tc, Vm	9	18	n.r.	Kheadr *et al.* (2004)
Broth microdilution	Brucella Standard Broth + haemin + vitamin K_1	Amo, Cefox, Clin, Cm, Em, Imi, Metro, Moxi, PG, Pip, Tc, Vm	4	76	32% Tc 24% Em 18% Clin	Delgado *et al.* (2005)
Disc diffusion	Brucella Agar + laked sheep blood+ vitamin K_1	Amo, Cefot, Cefott, Cefox, Ceph, Clin, Cm, Em, Fus, Gati, Gm, Imi, Km, Line, Metro, Mino, Nal, Oflo, Oxa, Pip, PG, Pris, Rif, Spi, Tc, Teic, Teli, Tic, Vm	8	50	14% Tc 14% Mino 4% Rif	Moubareck *et al.* (2005)
Agar dilution	Brucella Agar + laked sheep blood+ vitamin K_1	Amo, Cefox, Em, Gati, Line, PG, Tc, Teli, Vm	8	50		Moubareck *et al.* (2005)
Agar overlay disc diffusion	LSM + cysteine	Amo, Cipro, Clin, Cm, Em, Gm, Pol B, Q/D, Rif, Sulmet, Tc, Tmp	11	100	n.r.	Masco *et al.* (2006)

Broth microdilution	LSM + cysteine	Cipro, Clin, Cotri, Mino, Pol B, Sulmet, Tc, Tmp, Vm	11	26	n.r.	Masco *et al.* (2006)
Disc diffusion	TPY	Am, Amk, Amo, Apra, Bac, Ceftio, Ceph, Cepha, Cm, Col, Doxy, Em, En, Flor, Flu, Gm, Lin, Nal, Neo, PG, Sm, Spec, Spi, Sulmet, Tc, Tmp,	5	57	n.r.	Vlková *et al.* (2006)
Etest	LSM + cysteine	Am, Clin, Em, Gm, Sm, Tc, Vm	10	203	13% Tc 2% Em 2% Clin	Mätto *et al.* (2007)
Broth microdilution	LSM + cysteine	Am, Clin, Em, Gm, Sm, Tc, Vm	2	162	48% Tc 4% Em 4% Clin	Mayrhofer *et al.* (2007)
Broth microdilution	TPY	Am, Azt, Bac, Ceph, Clin, Cm, Cy, Di, Em, Gm, Km, Lin, Metro, Nal, Neo, Novo, Paro, PG, Pol B, Rif, Sm, Spec, Tc, Vm	2.	22	n.r.	D'Aimmo *et al.* (2007)
Etest	LSM + cysteine	Clin, Cm, Em, Sm, Tc, Vm	8	34	53% Tc 24% Em 24% Clin	Ammor *et al.* (2008a)
Broth microdilution	Brucella Standard Broth + haemin + vitamin K_1	Amo, Cefox, Clin, Cm, Em, Imi, Metro, Moxi, PG, Pip, Tc, Vm	4	11	n.r.	Delgado *et al.* (2008)

Key of antibiotics: Ampicillin (Am), Amikacin (Amk), Amoxicillin (Amo), Apramycin (Apra), Aztreonam (Azt), Bacitracin (Bac), Carbenicillin (Car), Cefaclor (Cefac), Cefamandole (Cefam), Cefazolin (Cefa), Cefixime (Cefi), Cefoperazone (Cefop), Cefotaxime (Cefot), Cefotetan (Cefott), Cefoxitin (Cefox), Ceftazidime (Cefta), Ceftizoxime (Cefti), Ceftiofur (Ceftio), Ceftriaxone (Ceftri), Cefuroxime (Cefur), Cephalothin (Ceph), Cephalexin (Cepha), Cephradine (Cephra), Chloramphenicol (Cm), Ciprofloxacin (Cipro), Clindamycin (Clin), Cloxacillin (Clox), Colistin (Col), Cotrimoxazole (Cotri), Cycloserin (Cy), Dicloxacillin (Di), Doxycyclin (Doxy), Enrofloxacin (En), Erythromycin (Em), Florfenicol (Flor), Flumequine (Flu), Furazolidone (Fura), Fusidic acid (Fus), Gatifloxacin (Gati), Gentamicin (Gm). Imipenem (Imi), Kanamycin (Km), Lincomycin (Lin), Linezolid (Line), Methicillin (Met), Metronidazole (Metro), Minocycline (Mino), Moxalactam (Moxa), Nalidix acid (Nal), Neomycin (Neo), Netilmicin (Netil), Nitrofurantoin (Nitro), Nitrofurazone (Nitrof), Norfloxacin (Norflo), Novobiocin (Novo), Ofloxacin (Oflo), Oleandomycin (Ole), Oxacillin (Oxa), Paromomycin (Paro), Penicillin (PG), Piperacillin (Pip), Pirlimycin (Pir), Polymyxin B (Pol B), Pristinamycin (Pris), Quinupristin/Dalfopristin (Q/D), Rifampicin (Rif), Spectinomycin (Spec), Spiramycin (Spi), Streptomycin (Sm), Sulfadiazine (Sulfa), Sulfamethoxazole (Sulmet), Sulfisoxazole (Sulfi), Teicoplanin (Teic), Telithromycin (Teli), Tetracycline (Tc), Ticarcillin (Tic), Tobramycin (Tm), Trimethoprim (Tmp), and Vancomycin (Vm).

n.r. = not reported: number of strains displaying higher MIC values not indicated or increased MIC value not further characterized.

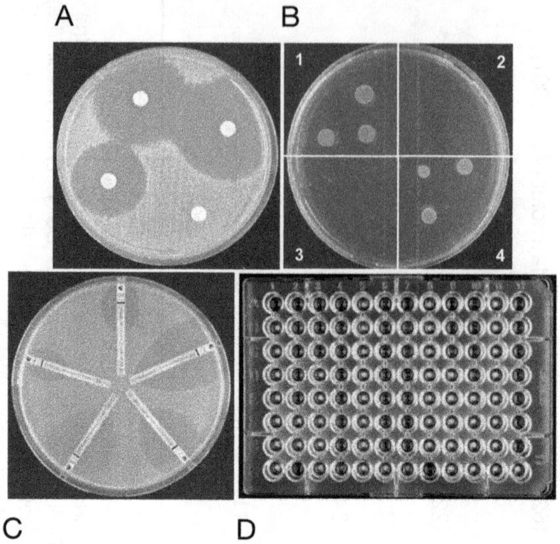

Figure 11.1 Composite photography showing common methods to assay AR in bifidobacteria. (A) Disc diffusion in which antibiotics are fixed at defined concentrations on filter paper disks. Diameters of the halo of inhibition correlate inversely with resistance. (B) Agar dilution which is performed by incorporating two-fold antibiotic concentrations in an agar medium dispensed into Petri plates. Strains are then inoculated in an ordered manner (in triplicate on the picture), and their MICs correspond to the minimal antibiotic concentration inhibiting appearance of growth buttons (strains 1 and 4 on the picture are still resistant to the antibiotic concentration in this plate). (C) The Etest, a commercial product, consists of thin plastic strips impregnated with a dried antibiotic gradient on the underside and are marked with a concentration scale for visual interpretation of results on the upper surface. MICs are determined by the intersection of the bacterial growth ellipse with the strip. Panel D. Broth dilution involves preparing two-fold dilutions of antibiotics in a liquid medium. Antibiotic-containing tubes (macrodilution) or wells (microdilution) are then inoculated with a standardized bacterial suspension. MICs coincide with the lowest antibiotic concentration preventing cell growth (observed by formation of pellets on the bottom of the wells).

a consequence, various growth media have been used to meet their complex nutritional requirements (Table 11.1). However, particular components of these media can influence the resistance levels to certain antibiotics (Kheadr *et al.*, 2007). The use of different techniques and media has inevitably led to discrepancies in the reported resistance/susceptibility profiles of bifidobacteria (Mättö *et al.*, 2007).

Recently, the Lactic acid bacteria Susceptibility test Medium with cysteine (LSM-C) has been developed (Klare *et al.*, 2005) and extensively used in a majority of subsequent surveys (Table 11.1). Essentially, LSM-C consists of a mixture of Iso-Sensitest broth (90%, v/v) and MRS broth (10%, v/v), supplemented with 0.3 g/l L-cysteine hydrochloride. Compared with trypticase phytone yeast extract medium (TPY), the most commonly used test medium for antimicrobial susceptibility testing of bifidobacteria (Mättö *et al.*, 2006), type and reference strains grow better in LSM-C (Klare *et al.*, 2005). Furthermore, the minimum inhibitory concentration (MIC) of most antibiotics for a set of international control strains obtained in LSM-C were in good agreement with those determined in cation-adjusted Mueller-Hinton broth with or without lysed horse blood by the CLSI method (CLSI, 2004).

Comparing agar disc diffusion and broth microdilution using LSM-C shows a clear relationship between the two methods. The results demonstrate that the disc diffusion zone diameters are highly reproducible for the tested conditions (Domig *et al.*, 2007). Therefore, the disc diffusion technique is a useful alternative for antimicrobial screening of bifidobacteria, but broth microdilution is still necessary for MIC determination. In general, dilution methods and the Etest are preferred as reference methods over conventional diffusion-based tests because they allow determination of MIC values, which provide a more reliable indication of the intrinsic or acquired nature of a given resistance phenotype (Vankerckhoven *et al.*, 2008). The Etest (AB-Biodisk/bioMérieux) is an exact and handy method, however, as a proprietary commercial product, it cannot be used as a reference. A similar product, M.I.C. evaluator, is also available from Oxoid. Based on broth microdilution and LSM-C, an ISO-standard for antimicrobial susceptibility testing of bifidobacteria is under development (ISO/DIS, 2008).

Antibiotic susceptibility profiles of bifidobacteria

As many different protocols have been used in surveys and different sets and concentrations of antibiotics have been assayed, comparison of the reported results is difficult (Table 11.1). Discrepancies among MICs and breakpoints to antibiotics should, therefore, be taken with some precaution until new and definitive cut-offs are determined.

Susceptibility to β-lactams

β-lactam antibiotics are a broad class of antibiotics that include penicillins, cephalosporins, monobactams, and carbapenems. They all share the common, central, four-member β-lactam ring and the principal mode of action to inhibit cell wall synthesis (CLSI, 2006). As reported in distinct surveys, bifidobacteria are usually susceptible to penicillin and its derivatives ampicillin, amoxicillin, carbenicillin, dicloxacillin, ticarcillin, and piperacillin. This susceptibility may be explained by the lack of β-lactamase activity in bifidobacteria (Moubareck *et al.*, 2005; Masco *et al.*, 2006). Interestingly, cloxacillin was described to be less effective than other penicillins (Zhou *et al.*, 2005). That finding is supported by Yazid *et al.* (2000), who observed variable effectiveness to cloxacillin as well as oxacillin. Susceptibility has also been observed for imipenem (Yazid *et al.*, 2000; Delgado *et al.*, 2005; Moubareck *et al.*, 2005), a carbapenem antibiotic, whereas bifidobacteria are highly resistant to the monobactam aztreonam (Charteris *et al.*, 1998; D'Aimmo *et al.*, 2007).

A variety of different cephalosporins have been tested, of which the most thoroughly examined is cephalothin. Variable results have been reported for this antibiotic, as well as for cefoxitin, moxalactam, and cefotetan. In contrast, most bifidobacteria are susceptible to cefotaxime, ceftiofur, cefpodoxime, cefazolin, and cephalexin. Production of β-lactamase or cephalosporinase in the supposed resistant strains was never detected (Charteris *et al.*, 1998; Moubareck *et al.*, 2005). Cell wall impermeability has been suggested as the main mechanism of cephalosporin resistance (Charteris *et al.*, 1998; Delgado *et al.*, 2005).

Susceptibility to glycopeptides

Glycopeptides, which include vancomycin and teicoplanin, share a complex chemical structure and a principal mode of action of inhibition of cell wall synthesis at a different site than that of the β-lactams (CLSI, 2006). There are some discrepancies concerning the susceptibility of bifidobacteria to vancomycin. Most studies have reported susceptibility of bifidobacteria to this antibiotic, except for that of Charteris *et al.* (1998) and Kheadr *et al.* (2004). While limited diffusion ability of vancomycin in agar media has been suggested as the main factor for the discrepancy in the results obtained (Moubareck *et al.*, 2005; Masco *et al.*, 2006), differences in the composition of the medium employed have also been proposed (Kheadr *et al.*, 2004). Susceptibility to teicoplanin has been examined only once (Moubareck *et al.*, 2005), and all strains were found to be highly susceptible.

Susceptibility to aminoglycosides

Most bifidobacteria have been reported as resistant to aminoglycosides such as gentamicin, kanamycin, neomycin and streptomycin. These antibiotics inhibit bacterial protein synthesis at the ribosomal level (CLSI, 2006). Bifidobacteria are particularly resistant to kanamycin (Matteuzzi *et al.*, 1983; D'Aimmo *et al.*, 2007), whereas gentamicin is reported to be more active than the other mentioned aminoglycosides (Lim *et al.*, 1993). In the majority of relevant scientific reports (Table 11.1), no marked difference in MIC distribution between species to the investigated aminoglycosides was observed. In contrast, strains of *B. longum* and *B. asteroides* have been reported to be more sensitive than all other species examined (Matteuzzi *et al.*, 1983; Masco *et al.*,

2006). In general, anaerobes such as bifidobacteria possess a natural resistance to aminoglycosides because of the lack of cytochrome-mediated drug transport. Indeed, phenotypic resistances to these antibiotics could not be traced back to specific genes by microarray hybridization (Kastner *et al.*, 2006; Ammor *et al.*, 2008a; see below). Furthermore, exposure of bifidobacterial strains to oxgall, a detergent-like substance that can disorganize the lipid bilayer structure of the cell membrane increasing permeability (Noriega *et al.*, 2005), enhances susceptibility to aminoglycosides (Charteris *et al.*, 2000; Kheadr *et al.*, 2007).

In addition, resistance has also been detected to amikacin (Charteris *et al.*, 1998; Yazid *et al.*, 2000; Vlková *et al.*, 2006), apramycin (Vlková *et al.*, 2006; Ouoba *et al.*, 2008), paromomycin (Lim *et al.*, 1993; D'Aimmo *et al.*, 2007) and tobramycin (Yazid *et al.*, 2000; Kastner *et al.*, 2006). Using agar disc diffusion, variable results were found for spectinomycin (Zhou *et al.*, 2005; Vlková *et al.* 2006), whereas MIC values between 16 and 128 μg/ml were observed using dilution procedures for antimicrobial susceptibility testing (D'Aimmo *et al.*, 2007; Ouoba *et al.*, 2008).

Susceptibility to macrolides and lincosamides

Macrolides are structurally-related antimicrobial agents that inhibit bacterial protein synthesis at the ribosomal level (CLSI, 2006). Erythromycin, the most frequently tested macrolide, has been found to inhibit growth of bifidobacteria at a very low concentration. The few strains reported to be resistant displayed MICs higher than 4 μg/ml. Prevalence has been found to range from 2% to 24% (Table 11.1), and includes strains of *B. animalis*, *B. catenulatum*, *B. longum*, *B. pseudocatenulatum*, and *B. thermophilum* (Delgado *et al.*, 2005; Mättö *et al.*, 2007; Mayrhofer *et al.*, 2007; Ammor *et al.*, 2008a). Erythromyin-resistant strains often display cross-resistance to clindamycin, a lincosamide antibiotic. Although structurally different, lincosamides have a similar action to macrolides and inhibit protein synthesis at a ribosomal binding site overlapping with that of macrolides (Roberts, 2002). A dedicated mechanism (*erm*(X), see below) has been identified in *B. thermophilum* strains from pigs, resistant to erythromycin and clindamycin (Mayrhofer *et al.*, 2007).

Testing of bifidobacteria to other macrolide antibiotics, such as azithromycin (de Backer *et al.*, 2006), clarithromycin (Collado *et al.*, 2005), and oleandomycin (Yazid *et al.*, 2000) has shown that these bacteria are highly susceptible to all of them, whereas susceptibility or moderate susceptibility has been reported for some strains to spiramycin (Moubareck *et al.*, 2005; Vlková *et al.*, 2006). Variable results have been referred for the lincosamides lincomycin (Matteuzzi *et al.*, 1983; Lim *et al.*, 1993; Vlková *et al.*, 2006; D'Aimmo *et al.*, 2007) and pirlimycin (Kheadr *et al.*, 2004).

Susceptibility to tetracyclines

Tetracyclines inhibit protein synthesis at the ribosomal level (CLSI, 2006). Variable levels of susceptibility to this antibiotic have been reported (Table 11.1). The study of tetracycline MICs distribution has shown that true resistance (i.e. resistance encoded by acquired genes) are present in strains of many works. Atypical resistances have been detected in strains of *B. adolescentis*, *B. animalis*, *B. asteroides*, *B. bifidum*, *B. boum*, *B. breve*, *B. catenulatum*, *B. longum*, *B. pseudocatenulatum*, *B. pseudolongum*, and *B. thermophilum* (Moubareck *et al.*, 2005; Flórez *et al.*, 2006; Masco *et al.*, 2006; Vlková *et al.*, 2006; D'Aimmo *et al.*, 2007; Ammor *et al.*, 2008a). Furthermore, the presence of tetracycline resistance genes has been established in resistant strains of many species (Moubareck *et al.*, 2005; Flórez *et al.*, 2006; Masco *et al.*, 2006; Mättö *et al.*, 2007; Mayrhofer *et al.*, 2007; Ammor *et al.*, 2008a). Resistance to minocycline is always associated with resistance to tetracycline (Moubareck *et al.*, 2005; Mättö *et al.*, 2008). In contrast, tigecycline, a novel glycylcycline derived from minocycline and structurally similar to tetracyclines but differing on binding to the bacterial ribosome, has been shown to inhibit bifidobacteria at a low concentration (Mättö *et al.*, 2008).

Susceptibility to phenicols

Bifidobacteria are usually sensitive to chloramphenicol, a phenicol antibiotic interfering with the synthesis of proteins. Normal MICs to chloramphenicol have been reported to range between 0.5 and 8 μg/ml. Higher MICs up to 64 mg/ml were reported in the survey by Kheadr *et al.* (2004), although these authors declare that strains may still be susceptible. Until now, acquired resistance and the presence of chloramphenicol resistance genes have not been detected (Mättö *et al.*, 2007; Ammor *et al.*, 2008a; Ouoba *et al.*, 2008). Bifidobacterial strains have also been reported to be susceptible to florfenicol, a fluorinated phenicol (Vlková *et al.*, 2006; Ouoba *et al.*, 2008).

Susceptibility to quinolones

Quinolones are structurally-related antimicrobial agents that function primarily by inhibiting the activity of DNA gyrases and topoisomerases (CLSI, 2006). Regarding bifidobacteria, nalidixic acid exhibits a low inhibitory activity and resistance is described as intrinsically present in this genus (Masco *et al.*, 2006). Generalized resistance has also been observed for flumequine (Vlková *et al.*, 2006), norfloxacin (Charteris *et al.*, 1998), and enrofloxacin (Vlková *et al.*, 2006). Susceptibility to ciprofloxacin (Masco *et al.*, 2006; Ouoba *et al.*, 2008) and ofloxacin (Moubareck *et al.*, 2005), however, has been found to be variable and strain-specific. A distribution of highly variable MICs among bifidobacteria has recently been reported for moxifloxacin (Delgado *et al.*, 2005, 2008). In contrast, tested bifidobacterial strains have been found to be susceptible to gatifloxacin, a new antianaerobic fluoroquinolone (Moubareck *et al.*, 2005).

Susceptibility to folate pathway inhibitors

Sulphonamides and trimethoprim are chemotherapeutic agents with similar spectrum of activity resulting from the inhibition of the bacterial folate pathway (CLSI, 2006). Bifidobacteria are usually intrinsically resistant to sulphonamides such as sulfamethoxazole and sulfadiazine (Yazid *et al.*, 2000; Masco *et al.*, 2006; Vlková *et al.*, 2006). Reduced susceptibility towards trimethoprim has also been repeatedly reported (Charteris *et al.*, 1998; Vlková *et al.*, 2006; Ouoba *et al.*, 2008). In contrast, susceptibility to this antibiotic was found to be variable and strain-specific by Masco *et al.* (2006) using an agar overlay disc diffusion test. Antibiotic antagonist compounds in the test medium may allow limited growth for some strains. For this reason, it has been recommended to correct the end point of the concentration at which growth is reduced by at least 80% (CLSI, 2006).

Susceptibility to nitrofurans

Nitrofurans act by inhibiting several protein synthesis and assembly steps at the ribosomal level (CLSI, 2006). Variable results have been described for the activity of furazolidone, nitrofurazone, and nitrofurantoin on bifidobacteria (Matteuzzi *et al.*, 1983; Lim *et al.*, 1993; Yazid *et al.*, 2000).

Susceptibility to nitroimidazoles

Metronidazole is a nitroimidazole antibiotic frequently used in cases of anaerobic infections. Resistance of bifidobacteria to this antibiotic has been described as a common feature in some works (Matteuzzi *et al.*, 1983; Charteris *et al.*, 1998; Collado *et al.*, 2005), whereas variable susceptibility has been reported in others (Moubareck *et al.*, 2005; Delgado *et al.*, 2005, 2008; D'Aimmo *et al.*, 2007). Resistance to metronidazole appears to be more frequent among animal bifidobacteria (Rada, 1997). The relative insensitivity of bifidobacteria to metronidazole (Moubareck *et al.*, 2005; Delgado *et al.*, 2008) may be explained by a low intracellular redox potential, which is necessary for reduction of metronidazole to cytotoxic radicals and other reactive species. It may be considered a desirable trait for probiotic strains during treatment of intestinal infections with this antibiotic (Saarela *et al.*, 2002; Vankerckhoven *et al.*, 2008).

Susceptibility to mupirocin

Bifidobacteria are consistently resistant to mupirocin, an inhibitor of protein synthesis. Resistance to mupirocin may serve as a simple criterion to distinguish bifidobacteria from lactobacilli (Rada, 1997), as intestinal lactobacilli are usually susceptible. A medium incorporating mupirocin has been developed for the selective isolation of bifidobacteria (Rada, 1997).

Susceptibility to other antibiotics

Bifidobacteria are generally susceptible to Gram-positive spectrum antibiotics, such as novobiocin, an aminocoumarin, and bacitracin, a polypeptide. Furthermore, they are usually inhibited at relatively low concentrations by the broad spectrum antimicrobial rifampicin, a rifamycin antibiotic. Resistance to the latter antibiotic has only rarely been detected (Moubareck *et al.*, 2005). Low level, unimodal MIC distributions have been reported for quinupristin/dalfopristin, a streptogramin antibiotic (Masco *et al.*, 2006), telithromycin, a new antibiotic of the ketolide class (Moubareck *et al.*, 2005), and linezolid, a new, synthetic oxazolidinone (Moubareck *et al.*, 2005; de Backer *et al.*, 2006; Ouoba *et al.*, 2008), suggesting bifidobacteria are all susceptible. In contrast, resistant strains were detected for cycloserine, another oxazolidinone (D'Aimmo *et al.*, 2007). Generalized resistance to polymyxin B, colistin and fusidic acid has also been reported in most works (Table 11.1).

Acquired resistance in probiotic, human and animal bifidobacteria strains

Generally, it is difficult to compare the results of different studies, as tests have been conducted with different methods. The limited reliability of the disc diffusion method to determine the occurrence of vancomycin resistance within bifidobacteria is well known. Thus, the detection of vancomycin resistance in one (Zhou *et al.*, 2005) and two (Temmerman *et al.*, 2003) out of four probiotic *B. lactis* strains tested has been questioned by the authors themselves. In contrast, the finding of *tet*(W) in *B. lactis* strains, isolated from yoghurt, confirms the presence of acquired resistance (Kastner *et al.*, 2006). The construction of antibiotic sensitive derivatives of probiotic strains by selective removal of the resistances shows that the biosafety issue of probiotics is taken very seriously (Arigoni and Delley, 2007; Vankerckhoven *et al.*, 2008).

Tetracycline and erythromycin-clindamycin resistant strains have been described in many surveys analysing human (Delgado *et al.*, 2005; Masco *et al.*, 2005; Moubareck *et al.*, 2005; Flórez *et al.*, 2006; Mättö *et al.*, 2007) and animal (Mayrhofer *et al.*, 2007) isolates. Erythromycin- and clindamycin-resistant strains were shown to be rather rare and occur at nearly the same frequency in human and animal bifidobacteria (Mättö *et al.*, 2007; Mayrhofer *et al.*, 2007; Table 11.1). In contrast, higher atypical resistances to tetracycline were reported for human and animal species. The high proportion of tetracycline-resistant strains is in accordance with this resistance being common among bacteria from all environments, and correlates with the large use of tetracyclines in human and veterinary medicine (Chopra and Roberts, 2001).

Molecular basis of atypical resistances

Phenotypic tests alone cannot confirm the presence or absence of transferable resistance genes. Atypical resistances may be caused by mutations in the molecular target, in elements involved in the internalization of antibiotics or by acquired AR genes (added genes), which have been obtained from strains inhabiting the same ecosystem by horizontal gene transfer mechanisms (such as conjugation, transduction, transformation). Therefore, phenotypic assays have to be complemented by molecular methods in which bacterial strains are directly screened for the presence of AR-conferring determinants. These methods include amplification by PCR with specific primers of single or multiplex AR genes, real-time quantitative PCR, or the use of DNA microarrays containing large collections of AR genes. Finally, determinants have to be analysed by sequencing and compared with similar sequences, if present in public databases. Acquired AR genes have already been identified and characterized in bifidobacteria species (Table 11.2). In particular, the *tet*(W) gene encoding tetracycline resistance seems to be widely distributed among human and animal species (Scott *et al.*, 2000; Moubareck *et al.*, 2005; Flórez *et al.*, 2006; Masco *et al.*, 2006; Ammor *et al.*, 2007). Analysis of AR determinants and their immediate flanking regions should help to identify their mechanism of transfer and design strategies to diminish or prevent AR spread by the use of probiotics.

The tetracycline resistance gene *tet*(W)

The *tet*(W) gene encodes a ribosomal protection protein and is one of the most widespread tetracycline resistance genes in environmental samples. In particular, it is abundant among commensal anaerobic organisms from rumen and intestinal sources (Roberts, 2005), but it has also been identified in a large variety of Gram-positive and Gram-negative aerobic and anaerobic

Table 11.2. Dedicated AR genes and genes from the normal metabolism of bifidobacteria involved in antibiotic resistance

Species/strain	Resistance gene	Database accession number(s)	Reference
Bifidobacterium adolescentis LMG11579	*tet*(W)	AM181319	Masco *et al.* (2006)
Bifidobacterium animalis R11	*tet*(W)	DQ988358	Aires *et al.* (2007)
G14	*tet*(W)	DQ988363	Aires *et al.* (2007)
Bifidobacterium animalis subsp. *lactis* LMG18314^{T} (Bb12)	*tet*(W)	AM181318	Masco *et al.* (2006)
E43	*tet*(W)	EU434754	Flórez *et al.* (2006)
LM624	*tet*(W)	AM181317	Masco *et al.* (2006)
R39, R63, R73	*tet*(W)	DQ988360, DQ988362, DQ988361	Aires *et al.* (2007)
B0456	*erm*(X) (Tn5432)	AM748803	van Hoek *et al.* (2008b)
AD11	*Tet38, tet*(W)	CP001213 (Bla_0695, Bla_1480)	Kim *et al.* (2009)
Bifidobacterium bifidum ATCC 29521	*rpoB* (mutat on)	AB198730	Vitali *et al.* (2007)
L22	*tet*(W)	EU434755	Ammor *et al.* (2008a)
131, 145, 163	*tet*(W)	DQ988355, DQ988354, DQ988356	Aires *et al.* (2007)
CCTCCM203049	*tet*(Q) (pBIF10)	DQ093580	–
Bifidobacterium breve 55	*tet*(W)	DQ988352	Aires *et al.* (2007)
UCC2003	*bmb*R	DQ115902	Margolles *et al.* (2005)
UCC2003	*bbmA, bbmB*	DQ486860	Margolles *et al.* (2006)
Bifidobacterium longum NCC2705	*tet*(Q)	NC_004307.2 (BL0072)	Schell *et al.* (2002)
H66, M21, L42	*tet*(W)	DQ060146	Flórez *et al.* (2006)
F8	*tet*(W)	DQ294299	Kazimierczak *et al.* (2006)
NCIMB702259^{T}	*CTR*	DQ017587 (BL1102 orthologue)	Price *et al.* (2006)
R29	*tet*(W)	DQ988359	Aires *et al.* (2007)
M21, L42	*tet*(W)	EU434757, EU434756	Ammor *et al.* (2008a)

Table 11.2. Continued

Species/strain	Resistance gene	Database accession number(s)	Reference
LMG13197, E111, B94	*tet*(W)	EU434752, EU434751, EU434750	Ammor *et al.* (2008b)
DJO10A	*tet*(W)	NC_010816 (BLD_0034)	Lee *et al.* (2008)
Bifidobacterium longum biovar infantis ATCC15697	*mfs*	ACJ53073.1 (Blon_2006)	Sela *et al.* (2008)
R5	*tet*(W)	DQ988357	Aires *et al.* (2007)
BI07	*rpoB* (mutation)	AB198734	Vitali *et al.* (2007)
Bifidobacterium longum biovar longum R29	*tet*(W)	DQ988359	Aires *et al.* (2007)
64	*tet*(W)	DQ988353	Aires *et al.* (2007)
Bifidobacterium pseudocatenulatum LMG10505^T	*tet*(W)	AM181316	Masco *et al.* (2006)
LMG11593	*tet*(W)	AM181315	Masco *et al.* (2006)
Bifidobacterium pseudolongum subsp. *globosum* LMG11569	*tet*(W)	EF015603	Flórez *et al.* (2008)
Bifidobacterium thermophilum LMG21813	*tet*(W)	EU434753	Ammor *et al.* (2008b)
B0219, B0241, B0242, B0253, B0256	*tet*(O/W)	AM889118 through AM889122	van Hoek *et al.* (2008a)
B0219, B0241, B0242, B0253, B0256	*tet*(W/32/O)	AM710601 through AM710605	van Hoek *et al.* (2008a)
B0219	*tet*(L)	AM710601	van Hoek *et al.* (2008a)
B0196, B0213, B0221, B0222, B0225, B0252	*erm*(X) (Tn*5432*)	AM748797 through AM748802	van Hoek *et al.* (2008b)

tet(W), *tet*(Q), *tet*(L), and *mfs* (major facilitator superfamily MFS_1) encode tetracycline resistance.

erm(X) encodes resistance to erythromycin and clindamycin.

rpo(B) encodes the beta chain of the RNA polymerase, whose mutation at different positions circumvents translocation blockage of the RNA molecule for antibiotics such as rifaximin (a structural analogue of rifamycin).

CTR encodes an efflux cholate transporter; over expression of *CTR* conferrers resistance to a number of unrelated antimicrobial compounds.

bbmR encodes an integral membrane protein whose heterologous expression in *Lactococcus lactis* conferred moderate resistance to macrolides.

bbmA and *bbmB* genes are organized in an operon-like structure and constitute an ABC multidrug transporter; expression of the two genes in *L. lactis* correlated with a resistance increase to ciprofloxacin and polymyxin B.

bacteria (Chopra and Roberts, 2001; Roberts, 2005). The gene was originally identified in the rumen bacterium *Butyrivibrio fibrisolvens* 1.230, where it was located in a large conjugative chromosomal element (Scott *et al.*, 1997). Copies of *tet*(W) have also been found in other isolates of *Buty. fibrisolvens* and in isolates of *Selenomonas* spp., *Mitsuokella* spp., *Clostridium* spp., *Roseburia* spp., and *Megasphaera elsdenii* from bovine and sheep rumen, as well as from porcine and human faeces (Roberts, 2005). The *tet*(W) gene was also reported to occur in isolates of the animal pathogen *Arcanobacterium pyogenes*, where it is also carried on a mobile genetic element (Billington *et al.*, 2002). The tetracycline-conferring genes from these different bacteria display an identical molecular organization and build-up, including nearly identical nucleotide and protein sequences. This high degree of conservation suggests a recent and rapid horizontal transference between these microorganisms from a common ancestor.

The *tet*(W) gene was recently reported to have spread among many intestinal *Bifidobacterium* species of human origin, including *B. animalis, B. bifidum, B. catenulatum, B. longum, B. pseudocatenulatum, B. pseudolongum*, and *B. thermophilum* (Table 11.2). It is also present in the type strains of *B. longum* biotype *suis, B. pseudolongum* subsp. *globosum, B. thermophilum* (Flórez *et al.*, 2008), and in all closely-related *B. animalis* subsp. *lactis* strains examined so far (Masco *et al.*, 2006; Arigoni and Delley, 2007; Saarela *et al.*, 2007; Barrangou *et al.*, 2009; Kim *et al.*, 2009), among which Bb12, the very widely used commercial probiotic strain sold by Chr. Hansen, is the paradigm. All *tet*(W) sequences share a core DNA region of 2638 nucleotides, except for genes from *B. animalis* subsp. *lactis*, which share a sequence of only 2154 nucleotides. This core region starts 658 bp upstream of *tet*(W), ending at different positions in the 3′ end of the gene. Putative conserved regulatory regions were observed in the first 250 bp upstream of the gene, including a short 14 amino acid open reading frame (ORF), thought to be involved in translation attenuation, a perfect 17 bp long inverted repeat with a loop of 3 bp (AATCCCCAGTGGTAAAA-GTA-TTTTACTGCTGGGGATT), which may act as a σ-independent terminator, and a putative ribosome binding site (AAAGGGAGG) 9 bp upstream of the presumed start codon of *tet*(W). In this region, the nucleotide identity of the sequences to one another ranges from 98.2 to 100%, with the least conserved sequences being those of *B. animalis* subsp. *lactis* strains. In all cases, the *tet*(W) gene was composed of 1912 bp and the deduced Tet(W) protein contained 639 amino acids.

While in *Buty. fibrisolvens* the gene is encoded on a highly mobile transposable chromosomal element, Tn*B1230* (Scott *et al.*, 1997), transposon-related sequences have never been detected in *Bifidobacterium* species (Flórez *et al.*, 2006; Kazimierczak *et al.*, 2006; Aires *et al.*, 2007). This means that the transfer machinery of *tet*(W) among bifidobacteria is yet to be discovered.

Though the molecular organization of the core regions are identical, the genetic context of *tet*(W) varies widely between species and strains (Fig. 11.2). However, the immediate flanking regions of the gene reveal several conserved features, which, supposedly, might contribute to its transference. Interestingly, *tet*(W) genes and flanking sequences from different *Bifidobacterium* species are more similar to genes and sequences from other species, suggesting independent and repeated transfer events (Flórez *et al.*, 2006). In most cases, *tet*(W) genes seem to be flanked on one or both sides by complete, interrupted or remnant sequences of insertion sequence (IS) elements and transposon-related sequences. Kazimierczak *et al.* (2006) propose the existence of a circular mini-element or transfer cassette of approximately 3.3 kilobase pair (kbp) within the transposon Tn*B1230* of *Buty. fibrisolvens*, which somehow can exist as a circular intermediate, based on the presence of conserved sequences and identical ORFs upstream and downstream of *tet*(W). This element was further identified by PCR and hybridization experiments in some transconjugants receiving transposon Tn*B1230*. While this may be true for some *tet*(W) carriers, there must be other ways of transmission, because the above mentioned DNA repeats flanking *tet*(W) are not present in bifidobacteria and some other species. Indeed, frequently, and particularly in bifidobacterial strains, tetracycline determinants are directly followed by chromosomal genes (Kazimierczak *et al.*, 2006; Ammor *et al.*, 2008b). Whether complete or truncated, it is conceivable that transposase- and integrase-like sequences present in the vicinity of the tetracycline resistance determinants may provide sufficient nucleotide homology (when internalized into the cell of chromosomally-encoded transposase and integrase sequences) to allow homologous recombination to occur on at least one side of the gene. This would agree with integration of the genes at distinct locations (i.e. positions with intact or incomplete IS copies) and the occasional interruption of chromosomal genes. Integration at a second point would be mediated by any of the cellular integrases, which only require short identical nucleotide sequences, and would be driven by the strong selective pressures imposed by the presence of antibiotic residues in the diet (Donoghue, 2003). Nevertheless, except for the strains whose genome has been completely sequenced, the presence of larger organizational units cannot be ruled out, as compatibility with transposon-like structures is frequently observed.

Mosaic *tet*(W) genes

Hybrid or mosaic *tet*(O)/*tet*(W) genes separated by 36% nucleotide sequence divergence were firstly detected in isolates of the propionate-forming intestinal anaerobe from swine *M. elsdenii* (Stanton and Humphrey, 2003). Mosaic genes have been shown to provide increased tetracycline resistance in this species (Stanton *et al.*, 2004). Mosaic *tet* genes comprising *tet*(O), *tet*(W), and *tet*(32) sequences have recently been reported in cloned DNA extracted from pig and human faecal samples (Patterson *et al.*, 2007). Although microorganisms were not isolated, many combinations such as *tet*(O/W/O), *tet*(O/32/O), *tet*(O/W/32/O), and *tet*(O/W/32/O/W/O) were identified, indicating a high degree of spread of such hybrid genes among intestinal bacteria. Hybrid genes containing parts of *tet*(O), *tet*(W), and *tet*(32), with the order *tet*(O/W) or *tet*(W/32/O), have recently been reported in several *B. thermophilum* strains from animal sources (van Hoek *et al.*, 2008a). Surprisingly, most strains were shown to harbour mosaic genes of the two classes, and one strain showed an additional *tet*(L) gene coding for an efflux pump (Table 11.2). Mosaic genes are usually flanked at the 5′ end by identical sequences to those found upstream of the corresponding mosaic ORF (van Hoek *et al.*, 2008a). Consequently, *tet*(O/W) genes are preceded by an around 400 bp sequence that is nearly identical to a DNA region found in front of *tet*(O) genes, while *tet*(W/32/O) genes are flanked at the 5′ end by about a 600 bp DNA segment that is identical to a DNA sequence commonly found upstream of *tet*(W) genes (van Hoek *et al.*, 2008a). The similarities between the DNA sequences of various tetracycline-resistance genes and their flanking regions suggest that these hybrid genes may have arisen from the interclass recombination within the coding regions of the ribosome protection protein genes *tet*(O) and *tet*(W) (Stanton and Humphrey, 2003; Patterson *et al.*, 2007; van Hoek *et al.*, 2008a).

Silent *tet*(W) genes

PCR and hybridization analysis of tetracycline-resistant and susceptible bifidobacteria for the presence of *tet*(W) has occasionally scored positive results in susceptible strains (Flórez *et al.*, 2006, 2008; Ammor *et al.*, 2008a; Mättö *et al.*, 2008). This suggests the existence of non-functional *tet*(W) genes (whether truncated, interrupted or silent) in some strains. Ammor *et al.* (2008b) reported on a *B. longum* strain in which *tet*(W) was shown to be interrupted by a 763 bp long IS element of the IS*30* family (Fig. 11.2). Apparently, a point mutation in the middle of the transposase sequence would impede the IS a further transposition. This agrees well with the impossibility of recovering tetracycline-resistant revertants after more than 100 generations of growth (B. Mayo, unpublished). Silent *tet*(W) genes have been found in the majority of *B. animalis* subsp. *lactis* probiotic strains. Though they harbour a well-characterized *tet*(W) gene, these strains usually present a reduced MIC to tetracycline ranging from 4 to 16 μg/ml (Saarela *et al.*, 2007). However, after intestinal passage during antibiotic and probiotic interventions a subset of isolates recovered has been shown to present enhanced MICs to this antibiotic (≥ 256 μg/ml) (Saarela *et al.*, 2007). Analysis of the isolates proved the higher MICs were not because of the acquisition of new AR determinants. Moreover, sub-culturing the isolates without antibiotics for 3 to 4 weeks correlated with a MIC decrease of 1–2 $\log_2$. Activation of *tet*(W) in Bb12 was also obtained after growth of the strain in

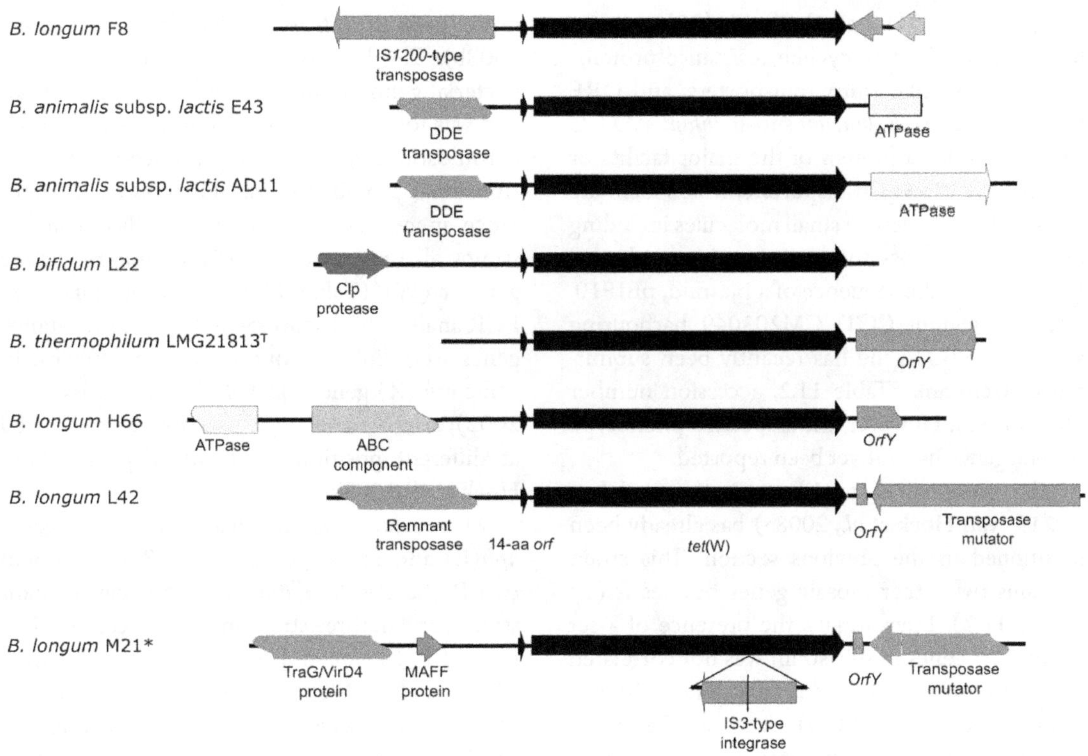

Figure 11.2 Genetic structure of the *tet*(W) loci in several bifidobacteria strains and their flanking regions. In black, conserved ORFs, which includes the *tet*(W) structural gene, a 14 amino acid peptide-encoding ORF thought to be involved in regulation, and their surrounding sequences. Proteins matching the highest homology scores of complete and truncated ORFs at the extremes of *tet*(W) are indicated. Proteins having identical or similar physiological functions are highlighted with the same colour. The asterisk marks the genetic structure of the *tet*(W) region in the tetracycline susceptible strain *B. longum* M21. Compiled from Flórez *et al.* (2006), Kazimierczak *et al.* (2006), Arigoni and Delley (2007), Ammor *et al.* (2008b), and Kim *et al.* (2009).

tetracycline-containing plates (A. Margolles, personal communication). Molecular analysis of one of the resistant isolates showed a 2 bp insertion in the promoter region of the gene, generating a −35 region closer to that of the prokaryotic consensus sequence. The presence of silent genes calls for the need to analyse such dormant resistance determinants in susceptible strains as they may become activated under certain circumstances, such as during probiotic manufacture and storage, or passage through the GIT,.

Other tetracycline resistance genes

Although less frequent than *tet*(W), the *tet*(O), and *tet*(M), resistance markers have also been reported in bifidobacteria species of different sources (Aires *et al.*, 2007; Mättö *et al.*, 2007). Aires *et al.* (2007) examined a series of 89 strains belonging to 10 species or subspecies, among which the prevalence of tetracycline resistance was found with a frequency of 33%. Twenty-three strains harboured *tet*(W), while *tet*(M) was found in just five strains. Interestingly, one *B. breve* strain was shown to contain both *tet*(W) and *tet*(M), both located on the bacterial chromosome. Transference of these genes to other bifidobacteria was not observed after conjugation (Aires *et al.*, 2007). Analysis of recently-released bifidobacterial genomes has shown the presence of several *tet*-like sequences (Table 11.2). For instance, the product of ORF BL0072 from *B. longum* NCC2705 displays homology to elongation factor G and Tet(M)-like proteins (Schell *et al.*, 2002). ORF Bla_0695 from *B. animalis* subsp. *lactis* AD11 encodes a Tet(38) protein, belonging to the tetracycline resistance proteins of the Tet(K)-like efflux transporters, and ORF Blon_2006 from *B. longum* subsp. *infantis* ATCC 15697 encodes a protein of the major facilitator superfamily MFS_1 transporters, which facilitate transport of a variety of small molecules including tetracyclines across the cytoplasmic membrane. Furthermore, the sequence of a plasmid, pBIF10, from *B. bifidum* CCTCCM203049 harbouring a further *tet*(Q) gene has recently been submitted to GenBank (Table 11.2, accession number DQ093580). However, the associated phenotype of such genes has not yet been reported.

The presence of *tet*(L) in *B. thermophilum* B0219 (van Hoek *et al.*, 2008a) has already been mentioned in the previous section. This strain contains two other mosaic genes besides *tet*(L) (Table 11.2). Interestingly, the presence of a set of three *tet* genes in this strain was not correlated with an elevated level of tetracycline resistance (van Hoek *et al.*, 2008a). Instead, the genes may function to confer tetracycline resistance under different biological and/or physiological conditions, as it has recently been found for a *Lactobacillus sakei* strain harbouring a chromosomally-encoded *tet*(M) and a plasmid located *tet*(L) (Ammor *et al.*, 2008c). In this bacterium, a basal expression of both genes was detected in the absence of tetracycline. In the presence of tetracycline, the expression of *tet*(M) was induced only at high concentrations, whereas the expression of *tet*(L) was upregulated at lower concentrations (Ammor *et al.*, 2008c).

Resistance genes to antibiotics other than tetracycline

Six *B. thermophilum* strains showing resistance to both erythromycin and clindamycin were shown to contain an *erm*(X) gene (van Hoek *et al.*, 2008b). A total of 763 bp of the resistance gene showed a nucleotide identity higher than 98% with determinants from opportunistic pathogenic bacteria such as *A. pyogenes, Propionibacterium acnes* and *Corynebacterium* spp. strains isolated from several sources. In these microorganisms, the *erm*(X) gene was found to be associated with the transposon Tn*5432*; accordingly, sequences coding a transposase similar to the IS*1249* copies found in Tn*5432* were encountered upstream of the gene in *B. thermophilum* (van Hoek *et al.*, 2008b). The location of Tn*5432* in the bifidobacterial genome was assessed for a single strain and was found to be inserted in the 3′ end of the chromosomal *rplI* gene, which encodes the 50S ribosomal protein L9 (van Hoek *et al.*, 2008b). Since in addition to *erm*(X), *B. thermophilum* strains also possessed the tetracycline resistance gene *tet*(W) (Table 11.2), these authors used PCR analysis to discover whether both resistance genes were linked. Contrary to the situation in some *erm*(X) genes (Jost *et al.*, 2003; Ross *et al.*, 2002), *erm*(X) and *tet*(W) seemed to be located at different positions in *B. thermophilum* (van Hoek *et al.*, 2008b).

Microarray analysis has further detected *aph*(E) and *sat*(3) genes in one *B. bifidum* and two *B. pseudocatenulatum* streptomycin-resistant strains, and in three streptomycin-susceptible isolates of *B. longum* and one *B. pseudocatenulatum* (Ammor *et al.*, 2008a). However, the presence of these transferases was not associated with higher MICs resistance to aminoglycosides and streptomycin (Ammor *et al.*, 2008a), suggesting that the detected enzymes play biological roles other than AR.

Transferability of acquired AR genes

The identification of AR genes with identical nucleotide and protein sequences among species and strains occupying the same ecological niche argues for a recent spread by horizontal transmission. Potential mechanisms for gene transfer between bacteria include bacteriophage-mediated transduction, natural transformation, and conjugation from a cell donor to a cell recipient. However, limited information on all these mechanisms is available in bifidobacteria. Reported phages have only been partially characterized (Sgorbati *et al.*, 1983), and natural transformation has never been observed. Plasmids, which are abundant in some intestinal species, seem to be scarce on bifidobacteria (Álvarez-Martín *et al.*, 2007), and conjugative and mobilizable modules in the characterized molecules have yet to be described (see Chapter 13). In spite of all this, transference of antibiotic resistance genes in the human GIT has long been reported (Netherwood *et al.*, 1999). Moreover, the *tet*(W) region of *B. longum*

F8 was transferred under laboratory conditions to *B. adolescentis* L2–32 (Kazimierczak *et al.*, 2006), suggesting bifidobacteria may certainly be implicated in spreading AR genes.

Transposon-associated erythromycin and clindamycin resistance has recently been identified in several *B. thermophilum* strains (van Hoek *et al.*, 2008b). The transference ability of these transposons and their encoding resistances remains to be determined. However, the presence of identical elements in different bacterial species that are resistant to these antibiotics strongly argues for an as yet unidentified genetic transmission ability (van Hoek *et al.*, 2008b).

Non-specific detoxification mechanisms

Although indirectly involved in resistance against antibiotics, the cooperation of non-specific mechanisms, such as multi-drug resistance (MDR) transporters (Putman *et al.*, 2001) and defective cell wall autolytic systems (Kim *et al.*, 1982) may contribute to the MIC differences observed between species and strains in both susceptible and resistant bacteria, of either intrinsic or acquired types. Non-specific mechanisms could also enhance resistance levels to aminoglycosides, to which anaerobic bacteria lacking an electron transport chain are thought to be naturally resistant. In that sense, heterologous expression of a hypothetical MDR-encoding gene from *Bifidobacterium breve* has been shown to confer moderate resistance to macrolides in *L. lactis,* increasing the MICs to four antibiotics by a factor of 1.7–2.7 (Margolles *et al.*, 2005). Similarly, heterologous expression in *Lactococcus lactis* of a two-gene operon encoding hypothetical membrane proteins with homology to members of the ATP-binding cassette (ABC) family of multidrug efflux transporters from *B. breve* exhibited enhanced resistance to the polypeptide antibiotic polimyxin B (Margolles *et al.*, 2006). Low resistance to some antibiotics has also been observed in *Escherichia coli* after expression of a gene encoding a cholate transporter from *B. longum* NCIMB702259^{T} (Price *et al.*, 2006).

AR by mutation

Resistances may not only be acquired by gaining exogenous DNA, but also by chromosomal mutations, by which bacteria can modify the antibiotic-binding site, increase the expression of intrinsic resistance determinants, or alter the molecular target (European Commission, 2005). By their nature, acquired resistances by mutation are considered to pose a low risk of dissemination. Rifaximin is a synthetic structural analogue of rifampicin and other rifamycin class antibiotics inhibiting the synthesis of RNA by blocking the bacterial RNA polymerase, which is used to treat gastrointestinal disorders. Recently, nucleotide mutations inside a core region of the *rpoB* gene (which encodes the β subunit of RNA polymerase) have been found in rifaximin-resistant strains of *Bifidobacterium* spp. (Vitali *et al.*, 2007).

Conclusions

Standardization of the methodology to survey AR in bifidobacteria is currently in progress, and IDF/ISO standards will probably soon be available. This methodology is necessary in order to fix the resistance/susceptibility profiles of bifidobacteria species and the resistance/susceptibility cut-offs separating susceptible from resistant strains. Monitoring AR programs for species intended to be used in food and feed systems are fully justified to avoid inclusion of strains containing AR genes in the formulation of probiotic cultures. Suspected strains displaying atypical resistances have to be carefully analysed to uncover the molecular basis of these resistances. Until now, acquired multi-resistance has never been described in bifidobacteria, but *B. thermophilum* strains resistant to erythromycin and tetracycline (with up to three different genes in one strain) have already been identified. It is worth noting that bifidobacteria species appear to share a similar profile to AR genes, which are identical to those displayed by other GIT bacteria, confirming the spread and transmission of AR genes among commensal organisms in the complex intestinal ecosystem. Analysis of the AR determinants already identified and characterized in bifidobacteria and comparison of their nucleotide sequences suggest that most AR genes emerged in microorganisms other than bifidobacteria. In fact, these bacteria might be at the end of the resistance transmission chain (although they could still become a reservoir of acquired determinants). Future work should concentrate on understanding the molecular mechanisms underlying transfer, integration, and stabilization of AR genes in bifidobacteria, which would surely help in the control of their spread.

Beyond dissemination of AR genes, the evaluation of antibiotic susceptibility patterns of bifidobacteria has several other purposes. Intrinsic AR has traditionally been used for their selective enumeration; consequently, analysis of AR susceptibility profiles of bifidobacteria might help in designing improved and new selective media for these bacteria. Antibiotic susceptibility is also of interest for anticipating the effects of antimicrobial therapy on the intestinal microbial populations, including those of bifidobacteria. In that sense, AR strains could help to inhibit excessive development of pathogenic and opportunistic bacteria during antibiotic treatment.

Acknowledgements

Work at the authors' laboratory was supported by projects from the VI EU Frame Program (ACE-ART, Ref. CT-2003-506214), the Spanish Ministry of Science and Innovation (MICINN) (Ref. AGL2000-1474-ALI and AGL2007-61869-ALI), and FICYT (Ref. IB08-005). S.D. awarded a postdoctoral contract from MICINN under Juan de la Cierva Program (Ref. JCI-2009-02391). M. Saarela, VTT Technical Research Centre of Finland, Finland, is acknowledged for the critical reading of the manuscript.

References

Aarestrup, F.M. (2000). Characterization of glycopeptide-resistant-enterococci (GRE) from boilers and pigs in Denmark: genetic evidence that persistance of GRE in pig herds is associated with coselection by resistance to macrolides. J. Clin. Microbiol. *8*, 2774–2777.

Aires, J., Doucet-Populaire, F., and Butel, M.J. (2007). Tetracycline mediated by *tet(W), tet(M), and tet*(O) genes of *Bifidobacterium* isolates from humans. Appl. Environ. Microbiol. 73, 2751–2754.

Álvarez-Martín, P., Flórez, A.B., and Mayo, B. (2007). Screening for plasmids among human bifidobacteria species: sequencing and analysis of pBC1 from *Bifidobacterium catenulatum* L48. Plasmid *57*, 165–174.

Ammor, M.S., Flórez, A.B., and Mayo, B. (2007). Antibiotic resistance in non-enterococcal lactic acid bacteria and bifidobacteria. Food Microbiol. *24*, 559–770.

Ammor, M.S., Florez, A.B., van Hoek, A.H., de los Reyes-Gavilan, C.G., Aarts, H.J., Margolles, A., and Mayo, B. (2008a). Molecular characterization of intrinsic and acquired antibiotic resistance in lactic acid bacteria and bifidobacteria. J. Mol. Microbiol. Biotechnol. *14*, 6–15.

Ammor, M.S., Flórez, A.B., Álvarez-Martín, P., Margolles, A., and Mayo, B. (2008b). Analysis of tetracycline resistance *tet*(W) genes and their flanking sequences in intestinal *Bifidobacterium* species. J. Antimicrob. Chemother. *62*, 688–693.

Ammor, M.S., Gueimonde, M., Danielsen, M., Zagorec, M., van Hoek, A.H., de los Reyes-Gavilán, C.G., Mayo, B., and Margolles, A. (2008c). Two different tetracycline resistance mechanisms, plasmid-carried *tet*(L) and chromosomally located transposon-associated *tet*(M), coexist in *Lactobacillus sakei* Rits 9. Appl. Environ. Microbiol. *74*, 1394–1401.

Andersson, D.I. (2003). Persistence of antibiotic resistant bacteria. Curr. Opinion Microbiol. *6*, 452–456.

Arigoni, F., and M. Delley. 2007. Genetic remodelling in *Bifidobacterium*. International Patent No. WO *2008019886*, 20080221.

Balsalobre, L., Ferrandiz, M.J., Linares, J., Tubau, F., and de la Campa, A.G. (2003). Viridans group streptococci are donors in horizontal transfer of topoisomerase IV genes to *Streptococcus pneumoniae*. Antimicrob. Agents Chemother. *47*, 2072–2081.

Baquero, F., Baquero-Artigao, G., Canton, R., and García-Rey, C. (2002). Antibiotic consumption and resistance selection in *Streptococcus pneumoniae*. J. Antimicrob. Chemother. *50*, 27–37.

Barrangou, R., Briczinski, E.P., Traeger, L.L., Loquasto, J.R., Richards, M., Horvath, P., Coûté-Monvoisin, A.C., Leyer, G., Rendulic, S., Steele, J.L., *et al.* (2009). Comparison of the complete genome sequences of *Bifidobacterium animalis* subsp. *lactis* DSM 10140 and Bl-04. J. Bacteriol. Doi: doi:*10*.1128/JB.00155-09.

Biavati, B., Mattarelli, P., and Crociani, P. (1992). Identification of bifidobacteria from fermented milk products. Microbiologia *15*, 7–13.

Biavati, B., and Mattarelli, P. (2006). The Familiy *Bifidobacteriaceae*. In: The Prokaryotes, 3rd Edition. Falkow, S. , Rosenberg, E. , Schleifer, K.-H. , Stackebrandt, E., and Dworkin, M., (eds.). (Heidelberg, Germany: Springer-Verlag GmbH), pp. 322–382.

Billington, S.J., Songer, J.G., and Jost, B.H. (2002). Widespread distribution of a *tet*W determinant among tetracycline-resistant isolates of the animal pathogen *Arcanobacterium pyogenes*. Antimicrob. Agents Chemother. *46*, 1281–1287.

Bjorkman, J., and Andersson, D.I. (2000). The cost of antibiotic resistance from a bacterial perspective. Drug Resistance Updates 3, 237–245.

Borgen, K., Wasteson, Y., Kruse, H., and Willems, R.J.L. (2002). Vancomycin-resistant *Enterococcus faecium* (VREF) from Norwegian poultry cluster with VREF from poultry from the United Kingdom and the Netherlands in an amplified fragment length polymorphism genogroup. Appl. Environ. Microbiol. *68*, 3133–3137.

Bronzwaer, S.L., Cars, O., Buchholz, U., Molstad, S., Goettsch, W., Veldhuijzen, I.K., Kool, J.L., Sprenger, M.J., and Degener, J.E. (2002). A European study on the relationship between antimicrobial use and antimicrobial resistance. Emerging Infect. Dis. *8*, 278–282.

Charteris, W.P., Kelly, P.M., Morelli, L., and Collins, J.K. (1998). Antibiotic susceptibility of potentially probiotic *Bifidobacterium* isolates from the human gastrointestinal tract. Lett. Appl. Microbiol. *26*, 333–337.

Charteris, W.P., Kelly, P.M., Morelli, L., and Collins, J.K. (2000). Effect of conjugated bile salts on antibiotic susceptibility of bile salt-tolerant *Lactobacillus* and *Bifidobacterium* isolates. J. Food. Prot. *63*, 1369–1376.

Chopra, I., and Roberts, M. (2001). Tetracycline antibiotics: mode of action, applications, molecular biology, and epidemiology of bacterial resistance. Microbiol. Mol. Biol. Rev. *65*, 232–260.

C.L.S.I (Clinical and Laboratory Standards Institute). (2004). Methods for antimicrobial susceptibility testing of anaerobic bacteria; approved standards- 6th ed. Vol. 24, *n*° 2. (Wayne, Pa., USA: CLSI document M11-A6).

C.L.S.I. (2006). Methods for dilution antimicrobial susceptibility tests for bacteria that grow aerobically; approved standard- 7th ed., Vol 26. (Wayne, Pa., USA: CLSI document M7-A7).

Collado, M.C., Gonzalez, A., Gonzalez, R., Hernandez, M., Ferrus, M.A., and Sanz, Y. (2005). Antimicrobial peptides are among the antagonistic metabolites produced by *Bifidobacterium* against *Helicobacter pylori*. Int. J. Antimicrob. Agents 25, 385–391.

Cosgrove, S.E., and Carmeli, Y. (2003). The impact of antimicrobial resistance on health and economic outcomes. Clin. Infect. Dis. *36*, 1433–1437.

D'Acosta, V.M., McGrann, K., Huhges, D.W., and Wright, G.D. (2006). Sampling the antibiotic resistome. Science *311*, 374–377.

D'Aimmo, M.R., Modesto, M., and Biavati, B. (2007). Antibiotic resistance of lactic acid bacteria and *Bifidobacterium* spp. isolated from dairy and pharmaceutical products. Int. J. Food Microbiol. *115*, 35–42.

de Backer, E., Verhelst, R., Verstraelen, H., Claeys, G., Verschraegen, G., Temmerman, M., and Vaneechoutte, M. (2006). Antibiotic susceptibility of *Atopobium vaginae*. BMC Infect. Dis. *6*, 51.

de Lavergne, E., Burdin, J.E., Schmitt, J., and Manciaux, M. (1959). Sensibilité de *Bifidobacterium bifidum* à onze antibiotiques. Ann. Inst. Pasteur 97, 104–107.

Delgado, S., Flórez, A.B., and Mayo, B. (2005). Antibiotic susceptibility of *Lactobacillus* and *Bifidobacterium* species from the human gastrointestinal tract. Curr. Microbiol. *50*, 202–207.

Delgado, S., O'Sullivan, E., Fitzgerald, G., and Mayo, B. (2008). *In vitro* evaluation of the probiotic properties of human intestinal *Bifidobacterium* species and selection of new probiotic candidates. J. Appl. Microbiol. *104*, 1119–1127.

Domig, K.J., Mayrhofer, S., Zitz, U., Mair, C., Petersson, A., Amtmann, E., Mayer, H.K., and Kneifel, W. (2007). Antibiotic susceptibility testing of *Bifidobacterium thermophilum* and *Bifidobacterium pseudolongum* strains: broth microdilution vs. agar disc diffusion assay. Int. J. Food Microbiol. *120*, 191–195.

Donoghue, D.J. (2003). Antibiotic residues in poultry tissues and eggs: human health concerns? Poultry Sci. *82*, 618–621.

European Commission. (2005). Opinion of the scientific panel on additives and products in substances used in animal feed on the updating of the criteria used in the assessment of bacteria for resistance to antibiotics of human and veterinary importance. EFSA J. *223*, 1–12.

European Commission. (2008). Update of the criteria used in the assessment of bacterial resistance to antibiotics of human or veterinary importance. EFSA J. *732*, 1–15.

Euzéby, J.P. (2009). List of Prokaryotic names with Standing in Nomenclature. http://www.bacterio.net; updated list of prokaryotic names.

F.A.O/W.H.O. (2006). Probiotics in food: Health and nutritional properties and guidelines for evaluation. FAO Food Nutr. Pap. *85*, 1–50.

Flórez, A.B., Ammor, M.S., Álvarez-Martín, P., Margolles. A., and Mayo, B. (2006). Molecular analysis of *tet*(W) gene-mediated tetracycline resistance in dominant intestinal *Bifidobacterium* species from healthy humans. Appl. Environ. Microbiol. 72, 7377–9737.

Florez, A.B., Ammor, M.S., Mayo, B., van Hoek, A.H.A.M., Aarts, H.J.M., and Huys, G. (2008). Antimicrobial susceptibility profiles of 32 type strains of *Lactobacillus, Bifidobacterium, Lactococcus* and *Streptococcus* spp. Int. J. Antimicrob. Agents *31*, 484–486.

Fujimori, M. (2006). Genetically engineered *Bifidobacterium* as a drug delivery system for systemic therapy of metastatic breast cancer patients. Breast Cancer *13*, 27–31.

Herreman, C., Asselin, D., and Novel, G. (1994). Identification et caracterisation de souches industrielles du genre *Bifidobacterium* par 2 methodes et taxonomie moleculaire. Lait 74, 479–484.

I.S.O/D.I.S. (2008). Milk and milk products-Determination of the minimal inhibitory concentration (MIC) of antibiotics applicable to bifidobacteria and non-enterococcal lactic acid bacteria (LAB). ISO/DIS *n*° 10932.

Jost, B.H., Field, A.C., Trinh, H.T., Songer, J.G., and Billington, S.J. 2003. Tylosin resistance in *Arcanobacterium pyogenes* is encoded by and *ermX* determinant. Antimicrob. Agents Chemother. *47*, 3519–3524.

Kastner, S., Perreten, V., Bleuler, H., Hugenschmidt, G., Lacroix, C., and Meile, L. (2006). Antibiotic susceptibility patterns and resistance genes of starter cultures and probiotic bacteria used in food. Syst. Appl. Microbiol. *29*, 145–155.

Kazimierczak, K.A., Flint, H.J., and Scott, K.P. (2006). Comparative analysis of sequences flanking *tet*(W) resistance genes in multiple species of gut bacteria. Antimicrob. Agents Chemother. *50*, 2632–2639.

Kheadr, E., Bernoussi, N., Lacroix, C., and Fliss, I. (2004). Comparison of the sensitivity of commercial strains and infant isolates of bifidobacteria to antibiotics and bacteriocins. Int. Dairy J. *14*, 1041–1053.

Kheadr, E., Dabour, N., Le Lay, C., Lacroix, C., and Fliss, I. (2007). Antibiotic susceptibility profile of bifidobacteria as affected by oxgall, acid, and hydrogen peroxide stress. Antimicrob. Agents Chemother. *51*, 169–174.

Kim, J.F., Jeong, H., Yu, D.S., Choi, S.H., Hur, C.G., Park, M.S., Yoon, S.H., Kim, D.W., Ji, G.E., Park, H.S., and Oh, T.K. (2009). Genome sequence of the probiotic bacterium *Bifidobacterium animalis* subsp. *lactis* AD011. J. Bacteriol. *191*, 678–679.

Kim, K.S., Morrison, J.O., and Bayer, A.S. (1982). Deficient autolytic enzyme activity in antibiotic-tolerant lactobacilli. Infect. Immun. *36*, 582–585.

Klare, I., Konstabel, C., Muller-Bertling, S., Reissbrodt, R., Huys, G., Vancanneyt, M., Swings, J., Goossens, H., and Witte, W. (2005). Evaluation of new broth media for microdilution antibiotic susceptibility testing of lactobacilli, pediococci, lactococci, and bifidobacteria. Appl. Environ. Microbiol. *71*, 8982–8986.

Leahy, S.C., Higgins, D.G., Fitzgerald, G.F., and van Sinderen, D. (2005). Getting better with bifidobacteria. J. Appl. Microbiol. *98*, 1303–1315.

Lee, J.H., Karamychev, V.N., Kozyavkin, S.A., Mills, D., Pavlov, A.R., Pavlova, N.V., Polouchine, N.N., Richardson, P.M., Shakhova, V.V., Slesarev, A.I., *et al.* (2008). Comparative genomic analysis of the gut bacterium *Bifidobacterium longum* reveals loci susceptible to deletion during pure culture growth. BMC Genomics. *9*, 247.

Levy, S.B., and Marshall, B. (2004). Antibacterial resistance worldwide: causes, challenges and responses. Nature Medicine *10*, 122–129.

Lim, K.S., Huh, C.S., and Baek, Y.J. (1993). Antimicrobial susceptibility of bifidobacteria. J. Dairy Sci. *76*, 2168–2174.

Margolles, A., Moreno, J.A., van Sinderen, D., and de los Reyes-Gavilán, C.G. (2005). Macrolide resistance mediated by a *Bifidobacterium breve* membrane protein. Antimicrob. Agents Chemother. *49*, 4379–4381.

Margolles, A., Flórez, A.B., Moreno, J.A., van Sinderen, D., and de los Reyes-Gavilán, C.G. (2006). Two membrane proteins from *Bifidobacterium breve* UCC2003 constitute an ABC-type multidrug transporter. Microbiology *152*, 3497–3505.

Masco, L., Van Hoorde, K., De Brandt, E., Swings, J., and Huys, G. (2006). Antimicrobial susceptibility of *Bifidobacterium* strains from humans, animals and probiotic products. J. Antimicrob. Chemother. *58*, 85–94.

Matteuzzi, D., Crociani, F., and Brigidi, P. (1983). Antimicrobial susceptibility of bifidobacteria. Ann. Microbiol (Inst. Pasteur, Paris) *134A*, 339–349.

Mättö, J., Suihko, M.L., and Saarela, M. (2006). Comparison of three test media for antimicrobial susceptibility testing of bifidobacteria using the Etest method. Int. J. Antimicrob. Agents *28*, 42–48.

Mättö, J., van Hoek, A.H.A.M., Domig, K.J., Saarela, M., Floréz, A.B., Brockmann, E., Amtmann, E., Mayo, B., Aarts, H.J.M., and Danielsen, M. (2007). Susceptibility of human and probiotic *Bifidobacterium* spp. to selcted antibiotics as determined by the Etest method. Int. Dairy J. *17*, 1123–1131.

Mättö, J., Maukonen, J., Alakomi, H.-L., Suihko, M.-L., and Saarela, M. (2008). Influence of oral doxycycline therapy on the diversity and antibiotic susceptibility of human intestinal bifidobacterial population. J. Appl. Microbiol. *105*, 279–289.

Mayrhofer, S., Domig, K.J., Amtmann, E., Van Hoek, A.H., Petersson, A., Mair, C., Mayer, H.K., and Kneifel, W. (2007). Antibiotic susceptibility of *Bifidobacterium thermophilum* and *Bifidobacterium pseudolongum* isolates from animal sources. J. Food Prot. *70*, 119–124.

McManus, P.S., Stockwell, V.O., Sundin, G.W., and Jones, A.L. (2002). Antibiotic use in plant agriculture. Ann. Rev. Phytopathol. *40*, 443–465.

Miller, L.G., and Finegold, S.M. (1967). Antibacterial sensitivity of *Bifidobacterium* (*Lactobacillus bifidus*). J. Bacteriol. *93*, 125–130.

Moubareck, C., Gavini, F., Vaugien, L., Butel, M.J., and Doucet-Populaire, F. (2005). Antimicrobial susceptibility of bifidobacteria. J. Antimicrob. Chemother. *55*, 38–44.

Nandi, S., Maurer, J.J., Hofacre, C., and Summers, A.O. (2004). Gram-positive bacteria are a major reservoir of Class 1 antibiotic resistance integrons in poultry litter. Proc. Natl. Acad. Sci. U.S.A *101*, 7118–7122.

Netherwood, T., Bowden, R., Harrosin, P., O'Donnel, A.G., Parker, D.S., and Gilbert, H.J. (1999). Gene transfer in the gastrointestinal tract. Appl. Environ. Microbiol. *65*, 5139–5141.

Noriega, L., de los Reyes-Gavilan, C.G., and Margolles, A. (2005). Acquisition of bile salt resistance promotes antibiotic susceptibility changes in *Bifidobacterium*. J. Food Prot. *68*, 1916–1919.

Normark, B.H., and Normark, S. (2002). Evolution and spread of antibiotic resistance. J. Int. Med. *252*, 91–106.

Ouoba, L.I.I., Lei, V., and Jensen, L.B. (2008). Resistance of potential probiotic lactic acid bacteria and bifidobacteria of African and European origin to antimicrobials: determination and transferability of the resistance genes to other bacteria. Int. J. Food Microbiol. *121*, 217–224.

Park, K.B., Ji, G.E., Park, M.S., and Oh, S.H. (2005). Expression of rice glutamate decarboxylase in *Bifidobacterium longum* enhances gamma-aminobutyric acid production. Biotechnol. Lett. *27*, 1681–1684.

Patterson, A.J., Rincon, M.T., Flint, H.J., and Scott, K.P. (2007). Mosaic tetracycline resistance genes are widespread in human and animal fecal samples. Antimicrob. Agents Chemother. *51*, 1115–1118.

Price, C.E., Reid, S.J., Driessen, A.J., and Abratt, V.R. (2006). The *Bifidobacterium longum* NCIMB 702259T *ctr* gene codes for a novel cholate transporter. Appl. Environ. Microbiol. *72*, 923–926.

Putman, M., van Veen, H.W., Degener, J.E., and Konings, W.N. (2001). The lactococcal secondary multidrug transporter LmrP confers resistance to lincosamides, macrolides, streptogramins and tetracyclines. Microbiology *147*, 2873–2880.

Rada, V. (1997). Detection of *Bifidobacterium* species by enzymatic methods and antimicrobial susceptibility testing. Biotechnol. Tech. *11*, 909–912.

Roberts, M.C. (2002). Resistance to tetracycline, macrolide-lincosamide-streptogramin, trimethoprim, and sulfonamide drug classes. Mol. Biotechnol. *20*, 261–283.

Roberts, M.C. (2005). Update on acquired tetracycline resistance genes. FEMS Microbiol. Lett. *245*, 195–203.

Ross, J.I., Eady, E.A., Carnegie, E., and Cove, J.H. (2002). Detection of transposon Tn*5432*-mediated macrolide-lyncosamide-streptogramin B (MLS_B) resistance in cutaneous propionibacteria from six European cities. J. Antimicrob. Chemother. *49*, 165–168.

Saarela, M., Mogensen, G., Fonden, R., Mättö, J., and Mattila-Sandholm, T. (2002). Probiotic bacteria: safety, functional and technological properties. J. Biotechnol. *84*, 197–215.

Saarela, M., Maukonen, J., von Wright, A., Vilpponen-Salmela, T., Patterson, A.J., Scott, K.P., Hamynen, H., and Mättö, J. (2007). Tetracycline susceptibility of the ingested *Lactobacillus acidophilus* LaCH-5 and *Bifidobacterium animalis* subsp. *lactis* Bb-12 strains during antibiotic/probiotic intervention. Int. J. Antimicrob. Agents *29*, 271–280.

Salyers, A.A., Gupta, A., and Wang, Y. (2004). Human intestinal bacteria as reservoirs for antibiotic resistance genes. Trends Microbiol. *12*, 412–416.

Scardovi, V. (1986). Genus *Bifidobacterium* Orla-Jensen. In: Bergey's Manual of Systematic Bacteriology, Vol. 2. Sneath, P.H.A., Mair, N.S., Sharpe, M.E., and Holt, J.G., eds. (Baltimore, USA: Williams and Wilkins, Baltimore), pp 1418- 1434.

Scott, K.P., Barbosa, T.M., Forbes, K.J., and Flint, H.J. (1997). High-frequency transfer of a naturally occurring chromosomal tetracycline resistance element in the ruminal anaerobe *Butyrivibrio fibrisolvens*. Appl. Environ. Microbiol. *63*, 3405–3411.

Scott, K.P., Melville, C.M., Barbosa, T.M., and Flint, H.J. (2000). Occurrence of the new tetracycline resistance gene *tet*(W) in bacteria from the human gut. Antimicrob. Agents Chemother. *44*, 775–777.

Sela, D.A., Chapman, J., Adeuya, A., Kim, J.H., Chen, F., Whitehead, T.R., Lapidus, A., Rokhsar, D.S., Lebrilla, C.B., German, J.B., *et al.* (2008). The genome sequence of *Bifidobacterium longum* subsp. *infantis* reveals adaptations for milk utilization within the infant microbiome. Proc. Natl. Acad. Sci. U.S.A *105*, 18964–18969.

Sgorbati, B., Smiley, M.B., and Sozzi, T. (1983). Plasmids and phages in *Bifidobacterium longum*. Microbiologica *6*, 169–173.

Shales, D.M., Projan, S.J., and Edwards, J.E. (2004). Antibiotic discovery: state of the state. ASM News *70*, 275–281.

Schell, M.A., Karmirantzou, M., Snel, B., Vilanova, D., Berger, B., Pessi, G., Zwahlen, M.C., Desiere, F., Bork, P., Delley, M., *et al.* (2002). The genome sequence of *Bifidobacterium longum* reflects its adaptation to the human gastrointestinal tract. Proc. Natl. Acad. Sci. U.S.A *99*, 14422–14427.

Sommer, M.O.A., Dantas, G., and Church, G.M. (2009). Functional characterization of the antibiotic resistance reservoir in the human microflora. Science *325*, 1128–1131.

Stanton, T.B., and Humphrey, S.B. (2003). Isolation of tetracycline-resistant *Megasphaera elsdenii* strains with novel mosaic gene combinations of *tet*(O) and *tet*(W) from swine. Appl. Environ. Microbiol. *69*, 874–882.

Stanton, T.B., McDowall, J.S., and Rasmussen, M.A. (2004). Diverse tetracycline resistance genotypes of *Megasphaera elsdenii* strains selectively cultured from swine feces. Appl. Environ. Microbiol. *70*, 3754–3757.

Temmerman, R., Pot, B., Huys, G., and Swings, J. (2003). Identification and antibiotic susceptibility of bacterial isolates from probiotic products. Int. J. Food Microbiol. *81*, 1–10.

Tenover, F.C. (2006). Mechanisms of antimicrobial resistance in bacteria. Am. J. Med. *119*, S3–S10.

Tissier, M.H. (1900). Réchérches sur la flore intestinale normale et pathologique du nourison. PhD Thesis, University of Paris, Paris, France, 1–253.

Turroni, F., van Sinderen, D., and Ventura, M. (2009). Bifidobacteria: from ecology to genomics. Front. Biosci. *14*, 4673–4684.

van Hoek, A.H.A.M., Mayrhofer, S., Domig, K.J., Flórez, A.B., Ammor, M.S., Mayo, B., and Aarts, H.J.M. (2008a). Mosaic tetracycline resistance genes and their flanking regions in *Bifidobacterium thermophylum* and *Lactobacillus johhsonii*. Antimicrob. Agents Chemother. *52*, 248–252.

van Hoek, A.H.A.M., Mayrhofer, S., Domig, K.J., and Aarts, H.J.M. (2008b). Resistance determinant *erm*(X) is borne by transposon Tn*5432* in *Bifidobacterium thermophilum* and *Bifidobacterium animalis* subsp. *lactis*. Int. J. Antimicrob. Agents *31*, 544–548.

Vankerckhoven, V., Huys, G., Vancanneyt, M., Vael, C., Klare, I., Romond, M.-B., Entenza, J.M., Moreillon, P., Wind, R.D., Knol, J., *et al.* (2008). Biosafety assessment of probiotics used for human consumption: recommendations from the EU-PROSAFE project. Trends Food Sci. Technol. *19*, 102–114.

Vitali, B., Turroni, S., Dal Piaz, F., Candela, M., Wasinger, V., and Brigidi, P. (2007). Genetic and proteomic characterization of rifaximin resistance in *Bifidobacterium infantis* BI07. Res. Microbiol. *158*, 355–362.

Vlková, E., Rada, V., Popelárová, P., Trojanová, I., and Killer, J. (2006). Antimicrobial susceptibility of bifidobacteria isolated from gastrointestinal tract of calves. Livestock Sci. *105*, 253–259.

Wassenaar T.M., and Klein, G. (2008). Safety aspects and implications of regulation of probiotic bacteria in food and food supplements. J. Food Prot. *71*, 1734–1741.

Wegener, H.C. (2003). Antibiotics in animal feed and their role in resistance development. Curr. Opinion Microbiol. *6*, 439–445.

Xu, Y.F., Zhu, L.P., Hu, B., Fu, G.F., Zhang, H.Y., Wang, J.J., and Xu, G.X. (2007). A new expression plasmid in *Bifidobacterium longum* as a delivery system of endostatin for cancer gene therapy. Cancer Gene Ther. *14*, 151–157.

Yazid, A.M., Ali, A.M., Shuhaimi, M., Kalaivaani, V., Rokiah, M.Y., and Reezal, A. (2000). Antimicrobial susceptibility of bifidobacteria. Lett. Appl. Microbiol. *31*, 57–62.

Zhou, J.S., Pillidge, C.J., Gopal, P.K., and Gill, H.S. (2005). Antibiotic susceptibility profiles of new probiotic *Lactobacillus* and *Bifidobacterium* strains. Int. J. Food Microbiol. *98*, 211–217.

Viability and Stability of Bifidobacteria in Commercial Preparations

12

Kamila Goderska and Catherine Stanton

Abstract

Probiotic bacteria, including species of *Lactobacillus* and *Bifidobacterium*, have been associated with numerous health benefits. As a result, the global market for probiotic products is continuously growing and food products enriched with probiotic bacteria are one of the most successful categories of functional foods. However, incorporating probiotic bacteria into food matrices poses significant technological challenges, and can be difficult owing to the various stresses imposed on the probiotic strains during food processing. Bifidobacteria in particular are sensitive microorganisms with low survival upon exposure to stresses, e.g. acid and temperature stress and oxygen exposure, encountered during their production, storage and consumption. Furthermore, incompatibility of the probiotic culture with the starter culture(s) used for the probiotic food fermentation may be a cause of further problems. So far, the use of probiotic bacteria is largely restricted to refrigerated fermented dairy products, where consumer health concerns, the traditional healthy image of dairy foods and the positive taste perceptions associated with these products are key catalysts driving this sector. However, non-dairy-based vehicles such as fruit juices and soy-based products are also gaining popularity for delivery of probiotic bacteria. This review focuses on the behaviour of bifidobacteria in food matrices of both dairy and non-dairy origin, and examines recent advances in strategies to enhance bifidobacterial viability and stability during the processing and storage of probiotic products.

Introduction

Probiotics are defined as 'live microorganisms which when administered in adequate numbers confer a health benefit on the host' (FAO/WHO, 2001). The most commonly used probiotic strains belong to species of either the *Lactobacillus* or *Bifidobacterium* genus. Bifidobacteria are Gram-positive bacteria, which naturally inhabit the gastrointestinal tract of humans and other warm-blooded animals (Leahy *et al.*, 2005). The presence of bifidobacteria in the gastrointestinal tract has been associated with a number of health benefits including relief of lactose intolerance, resistance to microbial infections, alleviation of constipation and diarrhoea, reduction of serum cholesterol concentrations, modulation of the immune system and indirect evidence suggests a potential role of bifidobacteria in cancer prevention (Leahy *et al.*, 2005). Interest in the incorporation of bifidobacteria into fermented products and other food commodities and commercial preparations has increased considerably during recent years, with many studies reporting human health benefits associated with the consumption of these products.

Remarkably, the global market for probiotics was valued at $14.9 billion in 2007 and is expected to reach a value of $19.6 billion by 2013 (Agheyisi, 2008). While the growth in probiotic foods containing bifidobacteria is expected to continue, food manufacturers must consider the effects of the environment of the food during processing and storage to ensure that the concentration of bacteria at the time of consumption provides a therapeutic dose to consumers. It has been suggested that up to 10^6 cfu/g of probiotic bacteria should be present in a fermented food product at the end of its shelf-life (FAO/WHO, 2003). A standard introduced by the Fermented Milks and Lactic Acid Bacteria Beverages Association in Japan stipulated that at least 10^7 viable bifidobacteria should exist per gram or millilitre of product to constitute a probiotic product (Ishibashi and Shimamura, 1993). Dried concentrated probiotic cultures are the most convenient form for their incorporation into functional foods, given the ease of storage, handling and transport, especially for shelf-stable functional products (Santivarangkna *et al.*, 2006). In addition, the foods with added bifidobacteria must maintain the characteristic sensory attributes of the traditional food (Gomes and Malcata, 1999). However, bifidobacteria are sensitive microorganisms with low survival to stresses that they endure during their production, storage and consumption. This review evaluates the various food matrices, which have been investigated to carry bifidobacteria including dairy-based and non-dairy-based vehicles, and methods that may increase survival of bifidobacteria strains in foods from encapsulation to novel strategies involving molecular genetics.

Growth and survival of bifidobacteria in dairy-based products

Growth and survival of bifidobacteria in milk

Bifidobacteria are mainly marketed through fermented dairy foods, which are well suited to promote the health image of probiotics for several reasons. Fermented foods, and dairy products in particular, already have a positive health image, and consumers are familiar with the fact that these products contain living microorganisms. Most bifidobacteria cannot be used directly in milk fermentations because of their slow growth in this matrix, because, although milk contains many essential nutrients, it lacks sufficient amino acids/peptides to support proper growth. Nevertheless, cases where bifidobacterial strains did adapt to growth in milk upon multiple transfers have been reported (Acharya and Shah, 2002). Comparison of various milk types (i.e. bovine, ovine and caprine milk) with respect to their available nitrogen fraction has also been exploited in an attempt to determine the growth abilities of this raw material with regards to selected probiotic strains (Petschow and Talbott, 1990). The growth of *Bifidobacterium longum* was studied in skimmed cows', ewes' and goats' milk, in mixtures of cows' and ewes' milk and goats' skimmed milk and whey powder products from these two types of milk (non-fat milk solids of cows' milk were replaced with non-fat milk solids ewes' and goats' milk) (Kehagias *et al.*, 2008). In the first experiment, the microbial population, the pH and the formation of D-aspartic and D-glutamic acids were studied in parallel in cows', goats' and ewes' milk with 9% NFMS. In the second experiment, the microbial population and the pH were studied in parallel in cows' milk with 9% NFMS and in cows' milk after replacing 3% of it with the same quantity of NFMS of goats' and ewes' milk. In the third experiment, the microbial population and the pH were studied in parallel in cows' milk with 9% NFMS and in cows' milk after replacing 3% of it with the same quantity of MWPP. D-aspartic acid and D-glutamic acid were determined as markers of the proteins of microbial origin. Comparing the viable counts for *B. longum* following growth on each of the three milk types as part of the fermentation process demonstrated a higher cell density when *B. longum* was grown in ewes' or goats' milk as compared to cows' milk for the first 8 h of fermentation, indicative of a superior growth ability of this organism in the two former milk types. The values of D-amino acids increased during fermentation and there was a significant correlation between D-amino acid content and microbial growth. However, it seems that this metabolic activity is also affected by other factors, mainly by the concentration of the respective L-amino acids. The formation of D-amino acids warrants further investigation, not only as growth markers, but also in relation to the formation of substances with probiotic properties (Kehagias *et al.*, 2008).

Because of bifidobacteria showing poor growth in milk, milk-derived components have been studied for their potential growth-stimulating activity. A set-type fermented milk manufactured from goat's milk was developed (Janer *et al.*, 2004). Optimal curd tension was achieved by supplementation of milk with skim milk powder and whey protein concentrate (WPC). Caseinomacropeptide (CMP) isolated from cow's or combined ewe's and goat's sweet cheese whey, and WPC enhanced growth of *B. lactis* in milk (Janer *et al.*, 2004). Supplementation of milk at 2% CMP increased counts of *B. lactis* by 1.5 log cycles after 24 h incubation at 37°C when compared with unsupplemented milk, showing no differences between the two CMPs. The highest bacterial growth was obtained when milk was supplemented with 2% WPC. The growth-promoting activity shown by WPC could be because of its CMP and whey protein content. CMP and WPC could be utilized to complement milk in order to increase counts of bifidobacteria in probiotic fermented milks (Janer at al., 2004). It was observed that growth of *B. lactis* in unsupplemented skimmed milk was poor; with mean viable cell numbers. Maximum specific growth rates for milk with CMP ranged between 75% and 80% of that obtained with WPC, not exceeding 39% for unsupplemented milk. Thus, it may be accepted that CMP and WPC both exhibit a beneficial influence on the growth of *B. lactis* in skimmed milk.

Growth and survival of bifidobacteria in cheese and yoghurt

Roy *et al.* (1995) studied the growth and acid production of bifidobacteria in milk under conditions typical of cheese-making, and evaluated the survival of the selected strains during storage in the presence of cheese starters. The bifidobacterial species differed in their ability to grow in different media, including milk, renneted milk, and acidified milk in the presence of starter cultures, and in their ability to produce acetic acid and lactic acid. *B. longum* strains demonstrated high survival rates in the presence of mesophilic starters and LAB, and *B. longum* would be acceptable for use in cheese-making also. However, *Bifidobacterium adolescentis* strains were unable to grow under conditions typical of cheese-making. The viability of different *Bifidobacterium* strains during cheese-making, ripening, and storage was shown to vary greatly. Pure cultures of *B. longum, Bifidobacterium breve, Bifidobacterium catenulatum, Bifidobacterium bifidum, Bifidobacterium angulatum,* and *Bifidobacterium infantis* were added to commercial samples of cottage cheese to compare the viability of different bifidobacteria species (Boylston *et al.*, 2004). After a 14-day refrigerated storage period, viability losses ranged from 0 to almost 3 $\log_{10}$ cycles and were strain-dependent. *B. bifidum* exhibited the best survival, showing only negligible loss of viability. In a similar experiment, various *B. infantis* and *B. breve* strains, which showed good inhibitory abilities against *Pseudomonas* spp., demonstrated poor survival rates in cottage cheese with a 3 $\log_{10}$ decrease in viable counts during storage at refrigeration temperature over 14 days (O'Riordan and Fitzgerald, 1998). Two Cheddar cheeses were manufactured at pilot scale, each inoculated with a different commercial probiotic bifidobacterial strain in an effort to develop probiotic Cheddar cheese harbouring high numbers of viable *Bifidobacterium* spp. The bifidobacterial strains were introduced during Cheddar manufacture as starter adjuncts at levels of 9.9×10^7 cfu/ml of cheese milk for *B. lactis* Bb-12 and 9.2×10^6 cfu/ml for *B. longum* BB536. *B. lactis* Bb-12 survived at high numbers ($>10^8$ cfu/g cheese), while numbers of *B. longum* BB536 were reduced to 10^5 cfu/g cheese, following six months of ripening. The addition of *B. longum* BB536 did not adversely affect cheese composition, while the addition of *B. lactis* Bb-12 had a partial negative effect, resulting in a moisture level of 40%, which is slightly above the legal limit permitted for Cheddar cheese. During early ripening, more extensive proteolysis and improved flavour were observed in the *B. lactis* Bb-12 cheese compared with the control cheese. Quantitative differences in volatile compounds were observed in the cheeses harbouring bifidobacteria, most notably acetic acid in the *B. lactis* Bb-12 cheese. This study illustrates the suitability of Cheddar cheese as a probiotic functional food harbouring certain strains of bifidobacteria (Brearty *et al.*, 2001). Three batches of Cheddar cheeses (Batch 1, with only starter *Lactococcus lactis*; Batch 2, with lactococci and *Lb. acidophilus* 4962, *Lactobacillus casei* 279, *B. longum* 1941; Batch 3, with lactococci and *Lb. acidophilus* LAFTIs L10, *Lactobacillus paracasei* LAFTIs L26, *B. lactis* LAFTIsB94) were manufactured in triplicate to study the survival and influence of probiotic bacteria on proteolytic patterns and production of organic acid during a ripening period of 6 months at 4°C. All probiotic adjuncts survived the manufacturing process and maintained their viability of 7.5 $\log_{10}$ cfu/g[1] at the end of ripening. The number of *L. lactis* cells decreased by one to two $\log_{10}$ units, but their counts were not significantly different ($P > 0.05$) in control and probiotic cheeses. No significant differences were observed in composition (fat, protein, moisture, salt content), but acetic acid concentrations were higher in probiotic cheeses. Assessment of proteolysis during ripening showed no significant difference in the levels of water-soluble nitrogen (primary proteolysis), but the levels of secondary proteolysis indicated by the concentration of free amino acids were significantly higher ($P > 0.05$) in probiotic cheeses. SDS-PAGE results on hydrolysis of α_s-CN (α_s casein fractions) after 6 months were consistently higher in probiotic cheeses (19.28%, 46.99% and 63.42% in Batch 1, Batch 2 and Batch 3, respectively). Proteolytic activity, however, remained relatively low for all cheeses due to the low ripening temperature (4°C). Also another publication has shown that Cheddar cheeses can be an effective vehicle for delivery of probiotic organisms (Ong *et al.*, 2006). Six batches of Cheddar cheese were manufactured containing different combinations of commercially available probiotic cultures from three suppliers. Duplicate cheeses contained the micro organisms of each supplier, a *Bifidobacterium* spp. (each supplier), an *Lb. acidophilus* (two suppliers), and either *Lb. casei*, *Lb. paracasei*, or *Lactobacillus rhamnosus*. Using selective media, the different strains were assessed for viability during Cheddar cheese maturation over 32 weeks. The *Bifidobacterium* spp. remained viable at high numbers with the three strains being present in cheese at 4×10^7, 1.4×10^8, and 5×10^8 CFU/g after 32 weeks. Similarly the *Lb. casei* (2×10^7 CFU/g), *Lb. paracasei* (1.6×10^7 CFU/g), and *Lb. rhamnosus* (9×10^8 CFU/g) strains survived well, whereas the *Lb. acidophilus* strains performed rather poorly with both decreasing in a similar manner to be present at 3.6×10^3 CFU/g and 4.9×10^3 CFU/g after 32 weeks. The study by Phillips *et al.* (2006) corroborates earlier finding that Cheddar cheese is a good vehicle for a variety of commercial probiotics, although survival of *Lb. acidophilus* strains is rather poor and will need to be improved. The aim of the study was to develop a new type of cheese with potential health effects. The cheese was produced using cheese starter cultures in combination with *Lb. acidophilus* and bifidobacteria. The viability and growth of *Lb. acidophilus* and bifidobacteria in the cheese were monitored during ripening for 13 weeks and subsequent storage up to 4 months. Optimization of cheese manufacture by means of temperature and ripening time variation yielded a new, low-fat cheese with a strong flavour and taste. The fat content was reduced to approximately 11% with no adverse effect on the sensory properties. Upon analyses, the ripened cheese was found to contain bioactive peptides with potential antihypertensive effects. On the basis of these results, it can be concluded that the new cheese, developed on the basis of these studies, may have a beneficial impact on human health.

The influence of inulin, oligofructose and oligosaccharides from honey, combined in different proportions, on the consumers' sensory acceptance, probiotic viable count and fructan content of novel potentially symbiotic petit-suisse cheeses was investigated (Cardarelli *et al.*, 2008). Probiotic populations varied from 7.20 $\log_{10}$ cfu/g up to 7.69 $\log_{10}$ cfu/g for *B. lactis* and *Lb. acidophilus* from 6.08 $\log_{10}$ cfu/g up to 6.99 $\log_{10}$ cfu/g. The fructan-type prebiotics inulin and oligofructose may also aid survival of probiotic organisms during processing and storage of dairy products, particularly increasing or, at least, retaining the viability of *Bifidobacterium* spp. and of *Lb. acidophilus*.

Various foods may be used to deliver probiotic bacteria into the gastrointestinal tract; one such example is Requeijão, a Portuguese whey cheese, as reported by Madureira *et al.* (2005). Survival and stability of *B. animalis* strains BLC-1, Bb-12, and Bo, *Lb. acidophilus* strains LAC-1 and Ki, *Lb. paracasei* subsp. *paracasei* strain LCS-1 and *Lactobacillus brevis* strain LMG 6906, following their inoculation into Requeijão and exposure to simulated gastrointestinal tract conditions, were assessed. Homogenates of inoculated whey cheese in 0.85% (w/v) sterile saline water were exposed to a solution of hydrochloric acid (pH 2.5–3.0) and pepsin (1000 units/ml) at 37°C, and then to 0.3% (w/v) bile salts after 60 or 120 min of acid exposure. All bacterial strains retained their initial viable cell numbers. *B. animalis* strains Bb-12 and Bo, and *Lb. brevis* strain LMG 6906 exhibited the highest viable cell counts when exposed to bile salts, whereas the other strains had variable death rates (Madureira *et al.*, 2005)

It has been possible to make a good quality frozen Zabady with *B. bifidum* (Kebary, 1996). The traditionally used Zabady (a set-type yoghurt) starter culture was partially replaced at levels of 33% and 50% with either *B. bifidum* DI or *B. bifidum* BB_{12}, respectively, and frozen Zabady was made from these new starter culture mixtures. Neither the strain nor the amount of bifidobacteria significantly affected the specific gravity. Mixes made with bifidobacteria had higher freezing point and pH values, and lower acidity. Adding bifidobacteria increased the overrun, melting resistance and improved the organoleptic properties of the resulting frozen Zabady. Although all the produced frozen Zabady products were acceptable over 5 weeks frozen storage, samples made with bifidobacteria were more acceptable. The survival of bifidobacteria increased as the amount of added bifidobacteria increased. The number of bifidobacteria decreased at a different rate throughout the storage period. The number of bifidobacteria that survived even after 5 weeks of frozen storage in frozen Zabady made with bifidobacteria ($>10^7$) was higher than the minimum level that should be present to achieve the beneficial effect of bifidobacteria (10^5 to $10^6 ml^{-1}$). Therefore, frozen Zabady could be used as an effective carrier of these probiotic bacteria.

The effect of commercial fruit preparations (mango, mixed berry, passion fruit and strawberry) on the viability of probiotic bacteria, *Lb. acidophilus* LAFTIs L10 and *B. lactis* LAFTIs B94 in stirred yoghurts during storage (35 days) at refrigerated temperature (4°C) was evaluated by Kailasapathy *et al.*, 2008. *B. lactis* LAFTIs B94 decreased from 1.55×10^8 cfu/g on day 0 to 5.45×10^7 cfu/g on day 35. Overall, all yoghurts stored contained more than 10^7 cfu/g of viable *Lb. acidophilus* at the end of the 35-day shelf life. Ninety-six per cent of yoghurts contained greater than 10^7 cfu/g viable *B. lactis* with the remaining 4% containing 10^6 cfu/g. Thus in this study, all yoghurts contained the recommended level (10^6–10^7 cfu/g) of both species of incorporated probiotic bacteria. As with all fermented dairy products containing living bacteria, products containing bifidobacteria must be cooled during storage, which is necessary both to guarantee high survival rates of bifidobacteria and to ensure product stability (Shah *et al.*, 1995). Survival of probiotic lactobacilli and bifidobacteria during refrigerated storage was evaluated in ten commercial fermented milks. Acidity and pH values of these products ranged between 0.79 and 1.16, and between 3.9 and 4.2, respectively. Streptococci were present in all products at levels ranging from 10^7 to 10^9 cfu/ml during cold storage (4°C). Counts of lactobacilli decreased between 0.75 and 1.85 log units, depending on the product tested. Nonetheless, during the 30 days of cold storage the population of these microorganisms remained above 10^5 cfu/ml, the minimum level suggested by some authors for probiotic microorganisms in fermented milks in order to produce therapeutic benefits. Bifidobacteria showed in general lower population decreases as compared to lactobacilli (between 0.17 and 1.10 log units in most fermented milks), although their counts in certain products were lower than 10^5 CFU/ml from the twentieth to the 25th day of cold storage. However, if the more restrictive criterion of Schuller-Malyoth *et al.* (1968) is considered (a viable cell number between 10^6 and 10^8 cfu/ml) only a few products fulfilled this requirement. Viability losses between 0 and 3 log units have been previously reported by other authors (Dave and Shah, 1997; Medina and Jordano, 1994) during refrigerated storage of fermented milks. However, Shah *et al.* (1995) found considerably higher population decreases for *B. bifidum* (3.5–7 log units) and *Lb. acidophilus* (1.5–6 log units). Variation in probiotic viability data among different authors may be attributed to differences among the strains tested and to the influence of factors such as acidity, pH, other starter microorganisms and oxygen dissolved in the milk (Gueimonde at al., 2004). In any case, bifidobacteria in dairy products are protected by milk proteins and fats during digestion, which allows better delivery of viable cells to the host as compared to other preparations.

Growth and survival of bifidobacteria in non-dairy-based products

Fruit-based beverages, vegetable products and desserts

Since most of the probiotic foods are dairy products, they cannot be consumed by humans who are allergic to milk proteins or have severe lactose intolerance. While looking for alternative food matrices, the suitability of carrot juice as a raw material for the production of probiotic food with *Bifidobacterium* strains (*B. lactis* Bb-12, *B. bifidum* B7.1 and B3.2) has recently been investigated (Kun *et al.*, 2008). Pasteurization of freshly prepared carrot juice at 80°C for 20 min decreased the microbial population to below the detection limit (10 cfu/ml). All tested bifidobacterial strains were capable of growing well on pure carrot juice without nutrient supplementation. Moreover, the 10^7 cfu/ml initial cell concentration increased to 10^8 cfu/ml after 6 h of incubation, and this level of viability was retained until the end of fermentation (24 h). Volumetric productivities of *B. lactis* Bb-12, *B. bifidum* B7.1 and *B. bifidum* B3.2 were 2.16×10^{10} cfu/l/h, 4.65×10^{10} cfu/l/h, and 3.85×10^{10} cfu/l/h, respectively. Owing to bacterial metabolism, carrot juice-based growth medium was acidified to a pH level of less than 4.5. During the fermentation, the amounts of glucose and sucrose decreased significantly. Meanwhile the fructose concentration did not change. Degradation of carotenoids (α-carotene and β-carotene) was between 15–45% depending on the strain used. Production of lactic and acetic acids was in the range of 14.8–16.7 mg/ml and 3.3–5.3 mg/ml, respectively.

The suitability of probiotic cultures as fruit juice supplements was examined by assessing their acid tolerance and technological robustness. Survival of *Lactobacillus* and *Bifidobacterium* strains in orange juice (OJ), pineapple juice (PJ) and cranberry juice (CJ) has been monitored recently (Sheehan *et al.*, 2007). Results revealed that extensive differences exist among probiotic strains regarding their acid resistance. All of the strains screened survived longer in OJ and PJ as compared to CJ. *Lb. casei* DN-114 001, *Lb. rhamnosus* GG and *Lb. paracasei* NFBC43338 displayed the greatest level of robustness surviving at levels above 10^7 cfu/ml in OJ and above 10^6 cfu/ml in PJ for at least 12 weeks. Probiotic tolerance to thermal and non-thermal processing was studied to determine the feasibility of their addition to OJ prior to pasteurization. OJ fortified with probiotic cultures was subjected to thermal pasteurization at 76°C for 30 s and 90°C for 1 min in addition to a high pressure treatment of 400 MPa for 5 min. Results indicated no strain was capable of withstanding treatments necessary to achieve a stable juice at levels of 10^6 cfu/ml. The outcome of the overall study points to *Lb. rhamnosus* GG, *Lb. casei* DN-114 001 and *Lb. paracasei* NFBC43338 as having potential for exploitation as functional supplements in fruit juices, due to their impressive tolerance of acidic environments; however, fortification post-processing is recommended (Sheehan *et al.*, 2007). An oat-based medium was also evaluated as a matrix for the inclusion of bifidobacteria (Laine *et al.*, 2003). All of the tested *Bifidobacterium* strains were observed to acidify the oat-based medium. However, only half of the strains reduced the pH to below 5 after a 24-h incubation, which is a desirable pH value for many fermented foods. Six strains retained stable in viable numbers during the fermentation and one strain, *B. adolescentis* 34 significantly increased in numbers. Six of the seven strains that reduced the pH to below 5 survived at similar levels as inoculated or increased in number during the fermentation. The addition of soy protein to the oat-based medium affected the acidification and the survival of the tested bifidobacteria. Only five strains reduced the pH to below 5 after 24 h fermentation. These were the same strains that retained their viable numbers or grew during fermentation (Laine *et al.*, 2003). The survival of the probiotic strains *Lactobacillus reuteri* ATCC 55730, *Lb. acidophilus* DSM 20079 and *B. bifidum* DSM 20456, all of human origin, were investigated in three different oat-based, non-dairy products (Adavena M40, MG20 and G40). The products were fermented by the three strains with and without the presence of a commercial yoghurt culture (V2). Samples were stored at 6°C up to 30 days. In general, the oat-based products were shown to be a suitable support for these intestinal bacteria. *Lb. reuteri* ATCC 55730 had the highest viability in all of the products investigated. After 30 days, the cell viability of *Lb. reuteri*

ATCC 55730 was 10^8 CFU/ml in all three oat-based products (Mårtensson *et al.*, 2002). Lower viability was found when the strains were grown in the presence of the yoghurt culture compared to when they were grown as pure cultures. These products also exhibited a lower pH value in comparison to products fermented with the pure cultures. The utilization of the main fermentable carbohydrates in the products varied across the bacterial strains. A decrease in β-glucan content was seen for the products fermented by the *B. bifidum* DSM 20456 strain. This work shows that these oat-bases with different mono- and disaccharide composition can be used to support the growth of human intestinal bacteria and also maintain high cell viability during cold storage. Yosa is a newly marketed snack food made from oat bran pudding cooked in water and fermented with LAB and bifidobacteria. After fermentation, the matter is then flavoured with sucrose or fructose and fruit jam. It is mainly consumed in Finland and other Scandinavian countries (Blandiono *et al.*, 2003).

The possible use of bifidobacterial strains from different origin (adult and infant humans, and chicken) as novel starter cultures for bread making has also been evaluated (Palacios *et al.*, 2006). Fermentative parameters of doughs (pH, volume, total titrable acidity (TTA), lactic and acetic acids production and rheofermentative parameters) and technological parameters of breads (specific volume, bread shape and crumb hardness) were all analysed. Human bifidobacterial strains could replace *Lactobacillus* strains, commercialized for bread making, as they yielded breads with similar characteristics but with the advantage of having softer crumbs. Important differences between the behaviour of chicken bifidobacterial strains and human bifidobacterial strains were found when comparing bread TTA, bread shape and bread volume. Breads made with chicken strains showed significantly lower ($P < 0.05$) specific bread volume than those made with human strains, while showing similar values of TTA. The effects observed when using bifidobacterial strains from different origin as novel starter cultures for bread making seemed to depend on the strain and its origin.

A malt hydrolysate mimicking the composition of that produced in a brewery by the infusion method has been evaluated as a potential substrate for fermentation by bifidobacteria with the aim of producing a potentially probiotic beverage (Rozada-Sánchez *et al.*, 2008). Preliminary experiments indicated the need for addition of a growth promoter to the malt hydrolysate, and yeast extract at a concentration of 10 g/l was chosen. The growth of four *Bifidobacterium* spp. in the malt hydrolysate with 10 g/l of yeast extract was followed over a 24 h period at 37°C. The results showed an increase in bacterial population between 1.5 and 2.0 $\log_{10}$ cycles with a maximum growth rate of approximately 0.2/h. Large amounts of sugars and free amino nitrogen were still present at the beginning of the stationary phase of growth. The end of the exponential phase was attributed to the accumulation of toxic metabolites combined with the low pH.

Soy-derived products

Growth of *B. infantis* CCRC 14633 and *B. longum* B6 in soymilk was investigated and it was found that while soymilk could support the growth of both organisms tested, *B. infantis* grew better than *B. longum* in this medium (Chou and Hou, 2000). Supplementation of soymilk with bifitose (isomaltooligosaccharide), glucose, lactose or galactose resulted in increased growth of both *B. infantis* and *B. longum* as determined after 48 h of fermentation. Furthermore, addition of yeast extract, peptone, tryptone, casitone or N-Z-Case plus to soymilk enabled *B. infantis* to reach its maximum population in a shorter cultivation time of 24 h. Acid production by *B. longum* and *B. infantis* in soymilk was mainly non-growth associated, while in the yeast extract-supplemented soymilk, acid produced by *B. infantis* was found to be growth-associated. Populations of *B. longum* reduced more than did *B. infantis* in the prepared fermented soymilk drink during the storage period. Viable populations of both test microorganisms reduced less severely in the fermented drink stored at 5°C as compared to storage at 25°C. After 10 days of storage at 5°C, viable *B. infantis* and *B. longum* reduced by 0.44 and 3.18 log cfu/ml, respectively, in the fermented drink. Addition of sucrose to the fermented drink resulted in an increased reduction of viable bifidobacteria during the storage period. This phenomenon was observed to be most prominent for *B. infantis* in the fermented drink held at 25°C (Chou and Hou, 2000). In another study, *B. longum* B6 and *B. infantis* CCRC 14633 were grown in soymilk for a period of 48 h, and throughout the fermentation period, changes in the contents of crude protein, sugars, B-vitamins, acetic and lactic acids in soymilk were examined (Hou *et al.*, 2000). Crude protein and titratable acidity were increased during fermentation and soymilk fermented with *B. infantis* had higher titratable acidity than that fermented with *B. longum*. The degree of protein hydrolysis, thiamin and riboflavin contents were increased while niacin content was decreased in soymilk fermented with either *B. infantis* or *B. longum*. Acetic and lactic acid contents were increased while the molar ratio of acetic and lactic acids was decreased during fermentation. Stachyose, raffinose and sucrose contents were decreased, with stachyose showing the greatest level of reduction. On the other hand, contents of fructose and glucose plus galactose contents were increased during fermentation (Hou *et al.*, 2000).

To further the goal of developing a probiotic dietary adjunct using soymilk, soymilk was fermented with LAB (*Lb. acidophilus* CCRC 14079 or *S. thermophilus* CCRC 14085) and bifidobacteria (*B. infantis* CCRC 14633 or *B. longum* B6) individually, and in conjunction (Wang *et al.*, 2006). Several antioxidative activities including the inhibition of ascorbate autoxidation, the scavenging effect of superoxide anion radicals and hydrogen peroxide, and the reducing activity exerted by different varieties of fermented soymilks, were investigated. In general, anti-oxidative activity in soymilk fermented with LAB and bifidobacteria simultaneously was significantly higher ($P < 0.05$) than that fermented with either individually. Moreover, anti-oxidative activity increased as the fermentation period was extended. However, unfermented soymilk showed an H_2O_2-scavenging effect, while there was no scavenging effect except for the accumulation of H_2O_2 in fermented soymilk. In addition, the effect of spray-drying and freeze-drying on changes in anti-oxidative activity was examined. Finally, it was found that freeze-drying caused a significantly ($P < 0.05$) lesser reduction in the anti-oxidative activity of soymilk than spray-drying. Irrespective of the drying method and the starters used for fermentation, the anti-oxidative activity of fermented soymilk reduces after drying yet remains higher than that of dried unfermented soymilk. The behaviour of *B. infantis* CCRC14633 and *B. longum* B6, and of the lactic acid bacteria (LAB) species *Lb. acidophilus* CCRC14079, *Lb. bulgaricus* CCRC14009 and *S. thermophilus* CCRC 14085 during the fermentation and storage of cultured soymilk drinks were also investigated (Wang *et al.*, 2002). Soymilk was found to support the simultaneous growth of the bifidobacteria and *Lb. acidophilus* or *S. thermophilus*, but *B. infantis* and *B. longum* both had a deleterious effect on the growth of *Lb. bulgaricus* in soymilk. Values for pH, ranging from 6·13 to 4·19, and titratable acidity ranging from 0·09% to 0·25% were noted in soymilk after 48 h of fermentation with various combinations of bifidobacteria and LAB. The viable populations of bifidobacteria and LAB decreased rapidly in the fermented soymilk drinks held at 25°C. When sucrose was added, the reduction in numbers was more pronounced, especially for bifidobacteria. However, the numbers of bifidobacteria and LAB showed no marked change in drinks with or without sucrose during storage for 10 days at 5°C (Wang *et al.*, 2002).

Soymilk fermented with various combinations of LAB (*S. thermophilus* and *Lb. acidophilus*) and bifidobacteria (*B. longum* and *B. infantis*) was subjected to freeze-drying and spray-drying. Survival of the starter organisms during the drying process, subsequent rehydration at different temperatures and during a 4 month period of storage under different storage conditions was examined. After freeze-drying, LAB and bifidobacteria exhibited a survival level of 46.2–75.1% and 43.2–51.9%, respectively, higher than that noted after spray-drying. Regardless of the drying condition, *S. thermophilus* showed a higher percentage of survival than *Lb. acidophilus*, while *B. longum* survived better than *B. infantis*. Further study with soymilk fermented with *S. thermophilus* and *B. longum* revealed that the freeze-dried and spray-dried fermented soymilk rehydrated at 35–50°C and 20°C, respectively, was optimum for the recovery of the starter organisms. Both *S. thermophilus* and *B. longum* survived better in the freeze-dried than the spray-dried fermented soymilk during storage. A higher level of survival was also noted for both the starter organisms when the dried fermented soymilk was stored at 4°C than 25°C. Holding the dried fermented soymilk in the laminated pouch enabled *S. thermophilus* and *B. longum* to exhibit a higher survival rate than in the deoxidant- and desiccant-containing glass or polyester (PET) bottle. Among all the packaging materials and storage temperatures tested, starter organisms were most stable in the dried fermented soymilk held in laminated pouch and stored at 4°C. Under such storage conditions, *S. thermophilus* and *B. longum* showed a survival percentage of 51.1% and 68.8%, respectively, in the freeze-dried fermented soymilk after 4 months of storage. Meanwhile, *S. thermophilus* and *B. infantis* in the spray-dried fermented soymilk showed a survival per cent of 29.5% and 57.7%, respectively (Wang *et al.*, 2004). It is worth noting that reconstitution conditions of dried preparations have recently been shown to influence survival rates of probiotic strains (Muller *et al.*, 2009). These conditions should have to be optimized for each particular strain.

Another study involved the fermentation of soymilk with LAB species (*S. thermophilus, Lb. acidophilus*) and bifidobacteria (*B. infantis, B. longum*) both individually and simultaneously (Hsieh and Chou, 2006). This study showed that fermentation in general, significantly ($p < 0.05$) enhanced the anti-mutagenicity of soymilk and the levels of increased anti-mutagenicity of fermented soymilk varied with the starter organism and the type of mutagen tested. Although unfermented soymilk exerted lower anti-mutagenic activity against DMAB than 4-NQO, the fermented soymilk generally, showed a higher anti-mutagenic activity against DMAB than 4-NQO. Among the various fermented soymilk tested, soymilk fermented with both *S. themophilus* and *B. infantis* simultaneously exhibited the highest anti-mutagenicity of 85.07% and 85.78%, respectively, against 4-NQO and DMAB Further investigation revealed that both the anti-mutagenic factors formed during fermentation and the cells of the starter organisms contributed to the increased anti-mutagenic activity against DMAB, while the former led to the increased activity against 4-NQO (Hsieh and Chou, 2006).

The effects of *B. breve*-fermented soymilk on probiotic function was investigated and it was reported that *B. breve* Yakult strain was capable of growing in soymilk 10^9 cfu/ml without the use of additives. During storage of the fermented soymilk at 10°C for 20 days, viable counts of the strain did not change. In a human feeding study, the administered *B. breve* was recovered at a level of over 10^9 cfu/g faeces, accompanied by an increase in the total number of bifidobacteria, but the growth inhibition of the strain in a bile containing medium (*in vitro*) was lessened by the addition of soy protein. These results indicate that fermented soymilk can be used as an efficient delivery system for the probiotic *B. breve* Yakult strain (Shimakawa *et al.*, 2003).

Probiotic microorganisms were incorporated into a non-fermented, vegetarian frozen soy dessert at initial populations greater than 10^6 cfu/g and the product was assessed for the survival of probiotic microorganisms and sensory acceptability. *Lb. acidophilus* MJLA1, *Lb. rhamnosus* 100-C, *Lb. paracasei* subsp. *paracasei* 01, *B. lactis* BBDB2, *B. lactis* Bb-12 all survived the 6 month storage period at a population density of 10^7 cfu/g or higher. *Saccharomyces boulardii* 74012 did not retain sufficient viability, decreasing below the desirable level of 10^6 cfu/g. To detect sensory differences, products containing *Lb. acidophilus* MJLA1, *S. boulardii* 74012 and a control, which did not contain either of these microorganisms, were stored for 0, 4 and 7 months and compared using triangle tests. The product containing *Lb. acidophilus* MJLA1 could not be distinguished from the control sample. In contrast, the product containing *S. boulardii* 74012 differed from the control and *Lb. acidophilus* MJLA1 by the development of undesirable flavours during storage. This study demonstrated that the frozen soy dessert was a suitable food system for the delivery of bacterial probiotic strains with excellent viability and acceptable sensory characteristics (Heenan *et al.*, 2004).

Twenty-two amylolytic bifidobacteria grown in BHI-starch media were compared for the amylase activity of the intra- and extracellular enzymes. The activity of the cells grown in the liquid medium differed considerably. Among the strains tested, *B. adolescentis* Int57 and *B. adolescentis* ZS8 exhibited higher activities than others. In rice medium containing 0.05% L-cysteine HCl and 0.2% yeast extract, the amylolytic *Bifidobacterium* strains grew considerably better than non-amylolytic *Bifidobacterium* strains. *B. adolescentis* Int57, which showed highest growth and amylase activity in the rice medium, was chosen as the best candidate to perform a rice-based fermentation. The titratable acidity (TA) reached 2.43 and pH decreased to 4.4 after 24 h fermentation. The relative ratio of acetic acid to lactic acid gradually decreased during fermentation. The concentration of reducing sugar and amylase activity gradually increased and reached 14 mg maltose equivalent ml^{-1} and 35 mU/ml/min/, respectively, in 24 h. The accumulated reducing sugar was mostly maltotriose. The layer-separation of fermented product was stabilized by the addition of 1% gelatin. It was suggested that amylolytic bifidobacteria may be used for the production of fermented rice products (Lee *et al.*, 1999).

The viability of bifidobacteria in mul-kimchi, a type of kimchi with added water, was investigated under various conditions. When a mul-kimchi preparation was inoculated with five strains of *Bifidobacterium* at a concentration of 10^7 cfu/ml, *B. longum* JK-2 showed the highest viability, maintaining a population of 10^6 cfu/ml after 1 week storage at 4°C (Lee *et al.*, 1999). The influence of NaCl concentration and initial pH on viability was further investigated in mul-kimchi inoculated with *B. longum* JK-2; NaCl concentrations greater than 3% (w/w) reduced viability considerably. In kimchi with an initial pH of 6.5, the cells showed the highest survival. When mul-kimchi containing 2% NaCl (w/w) was inoculated with 10^8 cfu/ml *B. longum* JK-2, there was a 10-fold reduction in viability during 10 days of incubation at 4°C. These results demonstrate acceptable levels of viable microorganism in the product, suggesting the possible use of selected strains in commercial kimchi production.

When applying the intestinal isolates *Lb. paracasei* L26 and *B. lactis* B94 in conjunction with a traditional meat starter culture, negative impact on the sensory properties of the product was not found (de Vuyst *et al.*, 2008). Furthermore, some probiotic strains were shown to inactivate pathogenic microorganisms. Combination of *Lb. reuteri* ATCC 55730 and *B. longum* ATCC 15708 proved to increase the inactivation of *Escherichia coli* O157:H7 during sausage manufacture.

Strategies to enhance survival of bifidobacteria

Micro-encapsulation is a method that provides living cells with a physical barrier against adverse external conditions by immobilizing cells in a specific matrix (Kailasapathy, 2002). The most popular system of immobilization reported involves the entrapment of cells in a gel matrix of alginates (Champagne *et al.*, 1994). Alginate-coated gelatin microspheres were produced to encapsulate the probiotic *B. adolescentis* 15703T with the objective of enhancing survival. Encapsulation yields (EY) for uncoated and alginate-coated (internal Ca^{2+}-source) microspheres gave similar results (Annan *et al.*, 2007). However, EY for the alginate-coated microspheres prepared by the external Ca^{2+}-source method was significantly lower. These microspheres were physically stronger as indicated by their resistance to degradation in simulated gastro-intestinal juices and required longer homogenization times for complete disintegration (Annan *et al.*, 2008).

B. longum Bb-46 and *B. lactis* Bb-12 were either added as free cells to milk (2% milk fat, Farmers Dairy, Bedford, Nova Scotia, Canada) or as entrapped cells in small alginate microspheres (Poste *et al.*, 1991). Milk samples of 10 ml each were placed into test tubes (15-ml centrifuge tubes, polypropylene, Fisher Scientific) and stored at 4°C for 15 or 16 days during which triplicate samples were withdrawn regularly. Non-inoculated control samples were also included in the experiment. Surviving bifidobacteria were enumerated by pour-plating appropriate dilutions in MRS-cysteine agar. Changes in acidity were followed by measuring the pH of this medium. Differences in sensory characteristics were assessed by a triangle test conducted with 12 panellists on day 13 for *B. longum* samples, and day 12 for *B. lactis* samples. The treated milk samples (free and microencapsulated) were evaluated in separate triangle tests against non-inoculated control milk samples stored under the same conditions. The triangle tests were performed according to Poste *et al.* (1991), and panelists were invited to comment on the sensory characteristics of the samples. In addition, the sensory properties (texture, flavour and odour) of *B. lactis* samples were followed by descriptive profiling by three panellists throughout the storage period (Truelstrup Hansen *et al.*, 2002).

In another study, bifidobacteria-loaded alginate microparticles were prepared by spraying a mixture of alginate and bifidobacteria culture using an air atomization method (Cui *et al.*, 2000). Survival and stability of bifidobacteria loaded in microparticles was then evaluated. Survival of bifidobacteria from alginate poly-L-lysine microparticles was significantly increased when MRS broth or yeast extract was added in simulated intestinal fluid (pH 6.8). The number of bifidobacteria gradually increased for 8 h (10^8 cfu/g) and then reached a viable count of about 10^9–10^{10} cfu/g when incubated over 12 h in intestinal fluid containing 0.5% yeast extract and 0.05% L-cysteine. The survival of bifidobacteria was highly dependent on the pH of the exposing media. When bifidobacteria were immobilized with alginate or even poly-L-lysine treatment, their survival was highly enhanced at low pH conditions (ca. $> 10^8$ vs. $< 10^3$ cfu/g). The stability of free flowing bifidobacteria-loaded alginate poly-L-lysine microparticles was significantly improved during storage at 4°C in a refrigerator when compared to normal planctonic bifidobacterial cultures. This study concluded that the bifidobacteria-loaded alginate poly-L-lysine microparticles can be applied to various dairy products.

The viability of *B. bifidum* BB-12 and *Lb. acidophilus* LA-5 microencapsulated by either an extrusion or an emulsion technique and used in white-brined cheese has recently been monitored (Özer *et al.*, 2009). Both microencapsulation techniques were effective in keeping the viable count of probiotic bacteria above the level of the therapeutic minimum ($>10^7$ cfu/g). While the counts of probiotic bacteria decreased approximately 3 log in the control cheese in which probiotics were used as free cells, the decrease was more limited in the cheeses containing microencapsulated cells (approximately 1 log). Medium- and long-chain free fatty acid contents of the cheeses with immobilized probiotics were much higher than in the control cheese. Similarly, cheeses made with immobilized probiotics contained higher acetaldehyde and diacetyl levels than the control. Experimental cheeses containing microencapsulated probiotics were not different from the control cheese in terms of sensory properties (Özer *et al.*, 2009). In a previous study by Stanton *et al.* (1998) it was demonstrated that bifidobacteria survived well in Cheddar cheese with either encapsulated or free bacteria.

The stability of differently produced (variables being fermentation time, pH during drying, and cryoprotectant) freeze-dried *B. lactis* E-2010 (Bb-12) cells was investigated in fruit juice and low-fat milk. In addition, the effect of the food matrix on the acid and bile tolerance of the cells was studied. Cells produced in different ways had comparable stability in milk, whereas in juice, sucrose-protected cells survived better than reconstituted skimmed milk-protected cells. The acid and bile tolerance was better in cells added to milk compared with those in phosphate buffered saline or juice. Despite good stability in milk the acid and bile tolerance of cells decreased during storage. Stability data alone do not give an accurate prediction of the probiotic functionality under adverse conditions (e.g. survival in acid and bile stress), and therefore when choosing a cryoprotectant for a probiotic strain, the stability in target food applications should also be considered (Saarela *et al.*, 2006).

A microencapsulation delivery system for *B. lactis*, a possible probiotic suited for use by the rural population of South Africa, was evaluated using two existing traditional fermented foods, amasi and mahewu (McMaster *et al.*, 2005). Gellan/xanthan microcapsules containing *B. lactis*, were tested under simulated physiological conditions, and added to pasteurized beverages. The capsules protected the microorganism under simulated low pH conditions associated with the stomach and from the biocidal activity of pancreatic and bile acids. For mahewu, microencapsulation of *B. lactis* with storage aerobically at 4°C and 22°C enhanced survival over a 21-day period as compared to free cells. In amasi, differences in viability between immobilized and free cells were less obvious. An analytically trained taste panel was unable to detect a significant difference in texture in any of the samples fortified with microcapsules. Although flavour differences were noted for mahewu containing either free or immobilized cells, after 14 days refrigerated storage, these difference were acceptable by panelists from Africa. No significant flavour differences were noted between amasi containing immobilized cells and untreated controls over the period tested.

A novel acid-stable bead made of gellan gum and xanthan gum was used to immobilize bifidobacteria. The beads (0.75% gellan and 1% xanthan gum) had an average diameter of 3 mm and did not shrink in 25% lactic acid solution, pH 1.5, or 20% acetic acid solution, pH 1.5, after storage at 4°C for 4 weeks. *B. infantis* ATCC 15697, the most acid-tolerant strain tested, was immobilized in gellan–xanthan beads and its survival in peptone water, pH 4, pasteurized yoghurt, and simulated gastric juice was monitored. In peptone water, pH 4, the reduction in cell count of immobilized cells of *B. infantis* ATCC 15697 was not significantly different from that obtained with free cells during 6 weeks of storage at 4°C. However, counts of immobilized cells of *B. infantis* ATCC 15697 remained significantly higher than free cells ($P < 0.0001$) when both were exposed to simulated gastric juices at pH 2.5, 2.0 and 1.5. At pH 2.5, the viable count of free cells dropped from 1.23x10^9 cfu/ml to an undetectable level (< 10 cfu/ml) in 30 min, while the viable count of immobilized cells decreased by only 0.67 log cycle in the same time period. Immobilized cells also survived significantly better than free cells ($P < 0.05$) in pasteurized yoghurt after refrigerated storage for 5 weeks (Sun and Griffiths, 2000).

B. breve R070 (BB R070) and *B. longum* R023 (BL R023) were encapsulated as freeze-dried or fresh cultures in water-insoluble food-grade microcapsules produced by emulsion and/or spray-drying, using milk fat and/or denatured whey proteins as immobilization material (Picot and Lacroix, 2004). The encapsulation yield differed significantly according to the method and the strain used. Dispersion of fresh cells in heat-treated whey protein suspension followed by spray-drying was the least destructive immobilization technique tested, with a survival rate of 25.77 ± 0.1% and 1.47 ± 0.2% for BB R070 and BL R023, respectively. Viable counts of BB R070 cells entrapped in whey protein microcapsules using this method were significantly higher than those of free cells after 28 days in yoghurt stored at 4°C (+2.6 log cycles), and after sequential exposure to simulated gastric and intestinal juices (+2.7 log cycles). In contrast, no protective effect of encapsulation was observed for BL R023. Immobilization of probiotic cultures in whey protein-based microcapsules can increase cell survival when subjected to extreme conditions, making this approach potentially useful for delivery of viable bacteria to the gastrointestinal tract of humans by way of dairy fermented products. However, technological properties of the strains, and particularly heat resistance, should be taken into consideration for spray-drying/encapsulation of sensitive microorganisms (Picot and Lacroix, 2004).

The survival of probiotic microorganisms including *Lb. acidophilus*, *Lb. casei*, *Lb. rhamnosus* and *Bifidobacterium* spp. was evaluated in yoghurt and freeze-dried yoghurt after processing and storage. The effectiveness of microencapsulating probiotic organisms as well as adding cryoprotectants and prebiotics in improving their viability was also investigated. The viability of *B. infantis* 17930 and *Lb. rhamnosus* GG was reduced by 0.07 log, while that of *Lb. casei* 1520 and *B. longum* 1941 was reduced by 0.28 and 0.39 log, respectively. There was a 7% improvement in the viability of *L. casei* 1520 when cryoprotectant Unipectine™ RS 150′ was added at 2.5% (w/v). The prebiotic Raftilose®P95′ when added at 1.5% w/v to yoghurt improved the viability of the combined selected probiotic microorganisms by 1.42 log during four weeks of storage at 4°C. Microencapsulation with alginate improved viability of combined selected probiotic microorganisms by 0.31 log in freeze-dried yoghurt stored at 21°C (Capela *et al.*, 2006).

Whether the viability of *B. bifidum* could be substantially enhanced during spray-drying and storage time by pre-selecting protective colloids offering resistance to oxygen diffusion (high activation energy) and adding aguamiel as a thermoprotector prebiotic was investigated (Rodriquez-Huezo *et al.*, 2007). Three different protective colloid blends exhibiting relatively high, medium and low activation energies were mixed with *B. bifidum* harvested in the late log phase, with or without aguamiel, spray-dried at 130, 140 and 155°C, and stored at 4°C at a water activity of 0.32. Viability was determined by cell total count in MRS agar, and best viability was achieved when microencapsulating the microorganism in the protective colloid blend exhibiting highest activation energy (40.7 kJ/mol), with aguamiel and dried at 140°C. Viability ranged from 1.26x10^8 cfu/g immediately after drying to 6.0x 10^6 cfu/g after 5 weeks storage time at 4°C (Rodriquez-Huezo *et al.*, 2007).

A modified method using calcium alginate for the microencapsulation of probiotic bacteria has also been reported (Sultana *et al.*, 2000). Incorporation of Hi-Maize starch (a prebiotic) improved encapsulation of viable bacteria as compared to when the bacteria were encapsulated without the starch. Inclusion of glycerol (a cryoprotectant) with alginate mix increased the survival of bacteria when frozen at −20°C. The acidification kinetics of encapsulated bacteria showed that the rate of acid produced was lower than that of free cultures. The encapsulated bacteria, however, did not demonstrate a significant increase in survival when subjected to *in vitro* high acid and bile salt conditions. A preliminary study was carried out in order to monitor the effects of encapsulation on the survival of *Lb. acidophilus* and *Bifidobacterium* spp. in yoghurt over a period of 8 weeks. This study showed that the survival of encapsulated cultures of *Lb. acidophilus* and *Bifidobacterium* spp. showed a decline in viable count of about 0.5 log over a period of 8 weeks while there was a decline of about 1 log in cultures which were incorporated as free cells in yoghurt. The encapsulation method used in this study did not result in uniform bead size, and hence additional experiments need to be designed using uniform bead size in order to assess the role of different encapsulation parameters, such as bead size and alginate concentration, in providing protection to the bacteria (Sultana *et al.*, 2000).

The probiotics *Lb. acidophilus* 547, *B. bifidum* ATCC 1994, and *Lb. casei* 01 were encapsulated into uncoated calcium alginate beads and the same beads were coated with three types of material, chitosan, sodium alginate, and poly-L-lysine in combination with alginate. The thickness of the alginate beads increased with the addition of coating materials. No differences were detectable in the bead strength by texture analysis or in the thickness of the beads with different types of coating materials by transmission electron microscopy. The survivability of three probiotics in uncoated beads, coated beads, and as free cells (unencapsulated) was conducted in 0.6% bile salt solution and simulated gastric juice (pH 1.55) followed by incubation in simulated intestinal juice with and without 0.6% bile salt. Chitosan-coated alginate beads provided the best protection for *Lb. acidophilus* and *Lb. casei* in all treatments. However, *B. bifidum* did not survive the acidic conditions of gastric juice even when encapsulated in coated beads (Krasaekoopt *et al.*, 2004).

Evaluation of synbiotic preparations for enhanced survival of bifidobacteria

The survival of *B. bifidum* BBI and *Lb. acidophilus* LAI in reduced-fat (liquid) and full-fat (set) yoghurts produced with two commercial lactic starter cultures (SID and SISD) was investigated. The viability of the probiotic bacteria was also assayed in milk acidified with lactic acid at different pH values. Samples were stored at 5°C for up to 4 weeks. There was a high level of variability in the survival ability of the probiotic cultures in the two yoghurt types. *Lb. acidophilus* LAI demonstrated, in general, a lower resistance to the yoghurt environment than *B. bifidum* BBI. The rate of this loss in cell viability depended on the yoghurt type and the lactic starter used. The loss of cell viability in yoghurt samples was different (higher in some cases and lower in others) from that owing to lactic acid only. In general, pH values of 4.5 or lower posed challenges to cell viability of the probiotic microorganisms in yoghurt stored at 5°C. This work shows the importance of selecting a suitable combination of probiotic strains and starter cultures when different yoghurt types are formulated (Vinderola *et al.*, 2000).

The growth and metabolism of probiotic strains (*Lb. acidophilus* La5 and 1748, *B. animalis* Bb12, and *Lb. rhamnosus* GG) were studied in milk- and water-based puddings containing both maize and rice flour (Helland *et al.*, 2004). All pudding formulations were produced both with and without the potential prebiotic polydextrose, Litesse^{TM}, to investigate whether this addition had any effect on the growth, metabolism and survival of the probiotic strains. Puddings were analysed for viable cell counts, pH and metabolites during fermentation (12 h, 37°C) and after refrigerated storage (21 d, 4–6°C). All strains showed good growth and survival in milk-based puddings (8–9.1 log cfu/g), but *Lb. rhamnosus*

GG was the only strain with an acceptable survival in water-based puddings (8 log cfu/g). After storage, 560–9800 mg/kg lactic acid was produced, with significantly higher concentrations in milk-based puddings ($P < 0.05$). The pH values in the products were reduced to 3.4–4.4. High concentrations of diacetyl were detected in puddings inoculated with *Lb. rhamnosus* GG, reaching 18 mg/kg in milk-based puddings. In most cases, puddings prepared with or without the addition of Litesse^{TM} were not significantly different (Helland *et al.*, 2004).

Two types of synbiotic ice cream containing 1% of resistant starch with free and encapsulated *Lb. casei* (Lc-01) and *B. lactis* (Bb-12) were manufactured. The survival of *Lb. casei* and *B. lactis* were monitored during storage for 180 days at −20°C. The viable cell number of *Lb. casei* and *B. lactis* in the free state in prepared ice cream mixture was 5.1×10^9 and 4.1×10^9 CFU/ml at day one and after 180 days storage at −20°C, these numbers were decreased to 4.2×10^6 and 1.1×10^7 CFU/ml, respectively. When the probiotic was encapsulated in calcium alginate beads, the probiotic survival was significantly increased during the same period of storage at the same temperature. In general, the results indicated that encapsulation can significantly increase the survival rate of probiotic bacteria in ice cream over an extended shelf-life. The addition of encapsulated probiotics had no significant effect on the sensory properties of non-fermented ice cream in which we used the resistant starch as a prebiotic compound (Homayouni *et al.*, 2008).

New strategies to enhance probiotic viability

The induction of a stress response in probiotic cells has been achieved by exposing the cells to a sub-lethal level of the stress before use in food processing, enabling retention of bacterial viability at high levels (Corcoran *et al.*, 2008). The resulting stress response, which in essence is a form of cellular resistance, can be divided into two classes. In the first case, a specific system induced by a sublethal dose of a chemical or physical stress (e.g. heat shock) permits survival against a challenge dose of the same agent (Pichereau *et al.* 2000; Desmond *et al.* 2001; Gouesbet *et al.* 2001). The second class of resistance is more general and enables cells to survive against very different environmental stresses, without the need for cultures to have had prior exposure to that stress (Pichereau *et al.* 2000; Desmond *et al.* 2001; Gouesbet *et al.* 2001). This mechanism is known as cross-protection (Kim *et al.* 2001). For example, *Lb. paracasei* NFBC 338, pre-adapted by exposure to 0.3 M NaCl, was significantly more resistant to heat stress associated with spray drying (outlet temperatures between 95°C and 100°C) than non-adapted control cells (33.46 ± 2.3% vs. 8.27 ± 4.42% survival, respectively) (Desmond *et al.* 2001). Using an alternative approach, namely patho-biotechnology, scientists have begun to exploit the stress-survival strategies of pathogenic microorganisms as a means of increasing the survival of probiotics through molecular genetic approaches (Sleator and Hill, 2006). Indeed, cloning the listerial betaine uptake system (BetL) into the probiotic strain *B. breve* UCC2003 improved gastric transit, gastrointestinal persistence and the therapeutic efficacy of the probiotic strain using a mouse model (Sheehan *et al*, 2007). Likewise, cloning the bile resistance mechanism (BilE) into *B. Breve* UCC2003 and *L. lactis* enhanced bile tolerance of both strains, improving their survival and persistence in the murine gastrointestinal tract (Watson *et al.* 2008).

Conclusions

The viability and stability of bifidobacteria has been both a marketing and technological challenge for industrial producers, since bifidobacteria should present high levels of viability during production and product-shelf life. Improved or new manufacturing processes and formulation technologies are required for bacteria primarily selected for their functional health properties. Immobilized cell technology and long-term continuous culture have been recently proposed to efficiently produce, in a one step process, cells with enhanced tolerance to environmental stresses. In addition, cells produced by continuous immobilized-cell cultures, which are in exponential or early stationary growth phase, exhibit both a high viability and metabolic activity compared with starving cells produced by conventional free-cell batch cultures. Innovative technological solutions using microencapsulation that can generate large quantities of food-grade materials at low cost are still needed to improve stability and viability of bifidobacteria in products and during digestion. For improved cell protection, the use of food-grade materials, such as resistant starch and milk proteins might increase the range of food types into which bifidobacteria can be incorporated. Moreover, the development of controlled-release systems in foods and pharmaceuticals will also pave the way for new applications of probiotics.

References

Acharya, M.R., and Shah, R.K. 2002. Selection of human isolates of bifidobacteria for their use as probiotics. Appl. Biochem. Biotechnol. *102–103*, 81–98.

Agheyisi, R. (2008). The Probiotics Market: Ingredients, Supplements, Foods: BCC Research FOD 035B, June F.A.O/W.H.O, (2001). Evaluation of health and nutritional properties of powder milk with live lactic acid bacteria. In: FAO/WHO Expert Consulation. Cordoba, Argentina, 1–4 October 2001, 1–34.

Annan, N.T., Borza, A., Moreau, D.L., Allan-Wojtas, P.M., and Truelstrup Hansen, L. (2007). Effect of process variables on particle size and viability of *Bifidobacterium lactis* Bb-12 in genipin–gelatin microspheres. J. Microencapsulation *24*, 152–162.

Annan, N.T., Borza, A.D., and Truelstrup Hansen, L. (2008). Encapsulation in alginate-coated gelatin microspheres improves survival of the probiotic *Bifidobacterium adolescentis* 15703T during exposure to simulated gastro-intestinal conditions. Food Res. Int. *41*, 184–193.

Blandino, A., Al-Aseeria, M.E., Pandiellaa, S.S., Canterob, D., and Webba, C. (2003). Cereal-based fermented foods and beverages. Food Res. Int. *36*, 527–543.

Boylston, T.D., Vinderola, C.G., Ghoddusi, H.B., and Reinheimer, J.A. (2004). Incorporation of bifidobacteria into cheeses: challenges and rewards. Int. Dairy J. *14*, 375–387.

Brearty, S.M., Ross, R.P., Fitzgerald, G.F., Collins, J.K., Wallace, J.M., and Stanton, C. (2001). Influence of two commercially available bifidobacteria cultures on Cheddar cheese quality. Int. Dairy J. *11*, 599–610.

Capela, P., Hay, T.K.C., and Shah, N.P. (2006). Effect of cryoprotectants, prebiotics and microencapsulation on survival of probiotic organisms in yoghurt and freeze-dried yoghurt. Food Res. Int. *39*, 203–211.

Cardarelli, H.R., Buriti, F.C.A., Castro, I.A., and Saad, S.M.I. (2008). Inulin and oligofructose improve sensory quality and increase the probiotic viable count in potentially synbiotic petit-suisse cheese. LWT-Food Sci. Technol. *41*, 1037–1046.

Champagne, C.P., Lacroix, C., and Sodini-Gallot, I. (1994). Immobilized cell technologies for the dairy industry. Crit. Rev. Biotechnol. *14*, 109–134.

Chou, C.-C., and Hou, J.-W. (2000). Growth of bifidobacteria in soymilk and their survival in the fermented soymilk drink during storage. Int. J. Food Microbiol. *56*, 113–121.

Corcoran, B.M., Stanton, C., Fitzgerald, G., and Ross, R.P. (2008). Life under stress: the probiotic stress response and how it may be manipulated. Curr. Pharmaceutical Design *14*, 1382–1399.

Cui, J.H., Goh, J.S., Kim, P.H., Choi, S.H., and Lee, B.J. (2000). Survival and stability of bifidobacteria loaded in alginate poly-*l*-lysine microparticles. Int. J. Pharmaceutics *210*, 51–59.

Dave, R.I., and Shah, N.P. (1997). Viability of yoghurt and probiotic bacteria in yoghurts made from commercial starter cultures. Int. Dairy J. *7*, 31–41.

Desmond, C., Stanton, C., Fitzgerald, G.F., Collins, K., and Ross, R.P. (2001). Environmental adaptation of probiotic lactobacilli towards improvement of performance during spray drying. Int. Dairy J. *11*, 801–808.

de Vuyst, L.D., Falony, G., and Leroy, F. (2008). Probiotics in fermented sausages. Meat Sci. *80*, 75–78.

F.A.O/W.H.O., (2001). Evaluation of health and nutritional properties of powder milk with live lactic acid bacteria. In: FAO/WHO Expert Consultation. Córdoba, Argentina.

F.A.O/W.H.O. (2003). Codex standard for fermented milks. Codex Stan 243–2003.

Gomes, A.M.P., and Malcata, F.X. (1999). *Bifidobacterium* spp. and *Lactobacillus cidophilus*: biological, biochemical, technological and therapeutical properties relevant for use as probiotics. Trends Food Sci. Technol. *10*, 139–157.

Gueimonde, M., Delgado, S., Mayo, B., Ruas-Madiedo, P., Margolles, A., and Reyes-Gavilán, C.G. (2004). Viability and diversity of probiotic *Lactobacillus* and *Bifidobacterium* populations included in commercial fermented milks. Food Res. Int. *37*, 839–850.

Heenan, C.N., Adams, M.C., Hosken, R.W., and Fleet, G.H. (2004). Survival and sensory acceptability of probiotic microorganisms in a nonfermented frozen vegetarian dessert. LWT-Food Sci. Technol. *37*, 461–466.

Helland, M.H., Wicklund, T., and Narvhus, J.A. (2004). Growth and metabolism of selected strains of probiotic bacteria in milk- and water-based cereal puddings. Int. Dairy J. *14*, 957–965.

Homayouni, A., Azizi, A., Ehsani, M.R., Yarmand, M.S., and Razavi, S.H. (2008). Effect of microencapsulation and resistant starch on the probiotic survival and sensory properties of synbiotic ice cream. Food Chem. *111*, 50–55.

Hou, J.-W., Yu, R.-C., and Chou, C.-C. (2000). Changes in some components of soymilk during fermentation with bifidobacteria. Food Res. Int. *33*, 393–397.

Hsieh, M.-L., and Chou, C.-C. (2006). Mutagenicity and antimutagenic effect of soymilk fermented with lactic acid bacteria and bifidobacteria. Int. J. Food Microbiol. *111*, 43–47.

Ishibashi, N., and Shimamura, S. (1993). Bifidobacteria: research and development in Japan. Food Technol. *46*, 126–135.

Janer, C., Peláez, C., and Requena, T. (2004). Caseinomacropeptide and whey protein concentrate enhance *Bifidobacterium lactis* growth in milk. Food Chem. *86*, 263–267.

Kailasapathy, K. (2002). Microencapsulation of probiotic bacteria: technology and potential applications. Curr. Issues Intestinal Microbiol. *3*, 39–48.

Kailasapathy, K., Harmstorf, I., and Phillips, M. (2008). Survival of Lactobacillus acidophilus and *Bifidobacterium animalis* subsp. *lactis* in stirred fruit yoghurts. LWT-Food Sci. Technol. *41*, 1317–1322.

Kailasapathy, K., Harmstorf, I., and Phillips, M. (2006). Survival of free and encapsulated probiotic bacteria and their effect on the sensory properties of yoghurt. LWT-Food Sci. Technol. *39*, 1221–1227.

Kebary, K.M.K. (1996). Viability of *Bifidobacterium bifidum* and its effect on quality of frozen Zabady. Food Res. Int. *29*, 431–437.

Kehagiasa, C., Csapo, J., Kontelesa, S., Kolokitha, E., Koulourisa, S., and Csapo-Kiss, Z. (2008). Support of growth and formation of D-amino acids by *Bifidobacterium longum* in cows', ewes', goats' milk and modified whey powder products. Int. Dairy J. *18*, 396–402.

Kim, W.-S., Perl, L., Park, J.-H., Tandianus, J.E., and Dunn, N.W. (2001). Assessment of stress response of the probiotic *Lactobacillus acidophilus*. Curr. Microbiol. *43*, 346–350.

Krasaekoopt, W., Bhandari, B., and Deeth, H. (2004). The influence of coating materials on some properties of alginate beads and survivability of microencapsulated probiotic bacteria. Int. Dairy J. *14*, 737–743.

Kun, S., Rezessy-Szabo, J.M., Nguyen, Q.D., and Hoschke, A. (2008). Changes of microbial population and some components in carrot juice during fermentation with selected *Bifidobacterium* strains. Process Biochem. *43*, 816–821.

Laine, R., Salminen, S., Benno, Y., and Ouwehand, A.C. (2003). Performance of bifidobacteria in oat-based media. Int. J. Food Microbiol. *83*, 105–109.

Lee, S.K., Ji, G.E., and Park, Y.H. (1999). The viability of bifidobacteria introduced into kimchi. Lett. Appl. Microbiol. *28*, 153–156.

Lee, J.H., Lee, S.K., Park, K.H., Hwang, I.K., and Ji, G.E. (1999). Fermentation of rice using amylolytic *Bifidobacterium*. Int. J. Food Microbiol. *50*, 155–161.

Madureira, A.R., Pereira, C.I., Truszkowska, K., Gomes, A.M., Pintado, M.E., and Malcata, F.X. (2005). Survival of probiotic bacteria in a whey cheese vector submitted to environmental conditions prevailing in the gastrointestinal tract. Int. Dairy J. *15*, 921–927.

Mårtensson, A., Öste, R., and Holst, O. (2002). The effect of yoghurt culture on the survival of probiotic bacteria in oat-based, non-dairy products. Food Res. Int. *35*, 775–784.

Martín-Diana, A.B., Janer, C., Peláez, C., and Requena, T. (2003). Development of a fermented goat's milk containing probiotic bacteria. Int. Dairy J. *13*, 827–833.

McMaster, L.D., Kokott, S.A., Reid, S.J., and Abratt, V.R. (2005). Use of traditional African fermented beverages as delivery vehicles for *Bifidobacterium lactis* DSM 10140. Int. J. Food Microbiol. *102*, 231–237.

Medina, L.M., and Jordano, R. (1994). Survival of constitutive microflora in commercially fermented milk containing bifidobacteria during refrigerated storage. J. Food Protection *56*, 731–733.

Muller, J.A., Stanton, C., Sybesma, W., Fitzgerald, G.F., and Ross, R.P. (2009). Reconstitution conditions for dried probiotic powders represent a critical step in determining cell viability. J. Appl. Microbiol. doi: 10.1111/j.1365-2672.2009.04533.x.

Ong, L., Henriksson, A., and Shah, N.P. (2006). Development of probiotic Cheddar cheese containing *Lactobacillus acidophilus, Lb. casei, Lb. paracasei* and *Bifidobacterium* spp. and the influence of these bacteria on proteolytic patterns and production of organic acid. Int. Dairy J. *16*, 446–456.

O'Riordan, K., and Fitzgerald, G.F. 1998. Evaluation of bifidobacteria for the production of antimicrobial compounds and assessment of performance in cottage cheese at refrigeration temperature. J. Appl. Microbiol. *85*, 103–114.

Özer, B., Kirmaci, H.A., Şenel, E., Atamer, M., and Hayaloğlu, A. (2009). Improving the viability of *Bifidobacterium bifidum* BB-12 and *Lactobacillus acidophilus* LA-5 in white-brined cheese by microencapsulation. Int. Dairy J. *19*, 22–29.

Palacios, M.C., Sanz, Y., Haros, M., and Rosell, C.M. (2006). Application of *Bifidobacterium* strains to the breadmaking process. Process Biochem. *41*, 2434–2440.

Petschow, B.W., and Talbott, R.D. 1990. Growth promotion of *Bifidobacterium* species by whey and casein fractions from human and bovine milk. J. Clin. Microbiol. *28*, 287–292.

Phillips, M., Kailasapathy, K., and Tran, L. (2006). Viability of commercial probiotic cultures (*L. acidophilus, Bifidobacterium* sp., *L. casei, L. paracasei* and *L. rhamnosus*) in cheddar cheese. Int. J. Food Microbiol. *108*, 276–280.

Picot, A., and Lacroix, C. (2004). Encapsulation of bifidobacteria in whey protein-based microcapsules and survival in simulated gastrointestinal conditions and in yoghurt. Int. Dairy J. *14*, 505–515.

Pichereau, V., Hartke, A., and Auffray, Y. (2000). Starvation and osmotic stress induced multiresistances: Influence of extracellular compounds. Int. J. Food Microbiol. *55*, 19–25.

Poste, L.M., Mackie, D.A., Butler, G., and Larmond, E. (1991). Laboratory Methods for Sensory Analysis of Food. Research Branch, Agriculture Canada, Ottawa, ON. Publication 1864/E.

Rodríguez-Huezo, M.E., Durán-Lugo, R., Prado-Barragán, L.A., Cruz-Sosa, F., Lobato-Calleros, C., Alvarez-Ramírez, J., and Vernon-Carter, E.J. (2007). Pre-selection of protective colloids for enhanced viability of *Bifidobacterium bifidum* following spray-drying and storage, and evaluation of aguamiel as thermoprotective prebiotic. Food Res. Int. *40*, 1299–1306.

Roy, D., Desjardins, M.L., and Mondou, F. (1995). Selection of bifidobacteria for use under cheesemaking condition. Milchwissenchaft *50*, 139.

Rozada-Sánchez, R., Sattur, A.P., Thomas, K., and Pandiella, S.S. (2008). Evaluation of *Bifidobacterium* spp. for the production of a potentially probiotic malt-based beverage. Process Biochem. *43*, 848–854.

Saarela, M., Virkajärvia, I., Alakomia, H.-L., Sigvart-Mattila, P., and Mättö, J. (2006). Stability and functionality of freeze-dried probiotic *Bifidobacterium* cells during storage in juice and milk. Int. Dairy J. *16*, 1477–1482.

Santivarangkna, C., Kulozik, U., and Foerst, P. (2007). Alternative drying processes for the Industrial preservation of lactic acid starter cultures. Biotechnol. Prog. *23*, 302–315.

Sheehan, V.M., Ross, P., and Fitzgerald, G.F. (2007). Assessing the acid tolerance and the technological robustness of probiotic cultures for fortification in fruit juices. Inn. Food Sci. Emerging Technol. *8*, 279–284.

Sheehan, V.M., Sleator, R.D., Hill, C., and Fitzgerald, G.F. (2007). Improving gastric transit, gastrointestinal persistence and therapeutic efficacy of the probiotic strain *Bifidobacterium breve* UCC2003. Microbiol. *153*, 3563–3571.

Shimakawa, Y., Matsubara, S., Yuki, N., Ikeda, M., and Ishikawa, F. (2003). Evaluation of *Bifidobacterium breve* strain Yakult-fermented soymilk as a probiotic food. Int. J. Food Microbiol. *81*, 131–136.

Sleator, R.D., and Hill, C. (2006). Patho-biotechnology: using bad bugs to do good things. Curr. Opinion Biotechnol. *17*, 211–216.

Schuller-Malyoth, R., Ruppert, A., and Müller, F. (1968). The microorganisms of the bifidus group (*Lactobacillus bifidus*). I. Historical review, nutritional, physiological and therapeutic aspects, morphology, culture procedures and taxonomy. Milchwissenschaft *23*, 356–360.

Shah, N., Lankaputhra, W.E.V., Britz, M., and Kyle, W.S.A. (1995). Survival of *Lactobacillus acidophilus* and *Bifidobacterium bifidum* in commercial yoghurt during refrigerated storage. Int. Dairy J. *5*, 515–521.

Stanton, C., Gardiner, G., Lynch, P.B., Collins, J.K., Fitzgerald, G., and Ross, R.P. (1998). Probiotic cheese. Int. Dairy J. *8*, 491–496.

Sultana, K., Godward, G., Reynolds, N., Arumugaswamy, R., Peiris, P., and Kailasapathy, K. (2000). Encapsulation of probiotic bacteria with alginate–starch and evaluation of survival in simulated gastrointestinal conditions and in yoghurt. Int. J. Food Microbiol. *62*, 47–55.

Sun, W., and Griffiths, M.W. (2000). Survival of bifidobacteria in yoghurt and simulated gastric juice following immobilization in gellan–xanthan beads. Int. J. Food Microbiol. *61*, 17–25.

Truelstrup Hansen, L., Allan-Wojtas, P.M., Jin, Y.-L., and Paulson, A.T. (2002). Survival of Ca-alginate microencapsulated *Bifidobacterium* spp. in milk and simulated gastrointestinal conditions. Food Microbiol. *19*, 35–45.

Vinderola, C.G., Bailo, N., and Reinheimer, J.A. (2000). Survival of probiotic microflora in Argentinian yoghurts during refrigerated storage. Food Res. Int. 33, 97–102.

Wang, Y.C., Yu, R.C., and Chou, C.C. (2006). Antioxidative activities of soymilk fermented with lactic acid bacteria and bifidobacteria. Food Microbiol. *23*, 128–135.

Wang, Y.C., Yu, R.C., and Chou, C.C. (2002). Growth and survival of bifidobacteria and lactic acid bacteria during the fermentation and storage of cultured soymilk drinks. Food Microbiol. *19*, 501–508.

Wang, Y.C., Yu, R.C., and Chou, C.C. (2004). Viability of lactic acid bacteria and bifidobacteria in fermented soymilk after drying, subsequent rehydration and storage. Int. J. Food Microbiol. *93*, 209–217.

Mobile Genetic Elements, Cloning Vectors and Genetic Manipulation of Bifidobacteria

13

Pablo Álvarez Martín, Simone Guglielmetti and Baltasar Mayo

Abstract

Growth difficulties, because of their fastidious nutritive nature and oxygen sensitivity, and a lack of efficient genetic tools have impeded until recently proper development of molecular studies in *Bifidobacterium*. These studies, however, are critical to uncover the cross-talk between bifidobacteria and their hosts' cells, and also to prove unequivocally the supposed beneficial activities supplied through the gastrointestinal tract of mammals either endogenously or after ingestion as probiotics. Analysis of gene sequences provided by whole genome sequencing projects has opened new avenues to decipher the genetic basis of bacteria–cell interactions and probiotic effects. However, the purposeful development of stable cloning and expression vectors based on robust replicons – either from temperate phages or resident plasmids – is additionally needed. This chapter addresses the current knowledge on the mobile genetic elements of bifidobacteria (phages, plasmids, and transposons) and reviews the different types of vectors already available for the *Bifidobacterium* species, together with the transformation procedures for introducing DNA into bifidobacterial cells. It also covers recent molecular studies performed with such vectors and incipient results on the genetic modification of these organisms, establishing the basis that would allow the use of bifidobacteria for many biotechnological applications. Pitfalls, drawbacks, and future needs in the field are also discussed.

Introduction

Bifidobacteria are among the dominant bacterial populations in the gastrointestinal tract (GIT) of humans and other mammals, interacting with epithelial cells in a close and lifelong relationship which starts soon after birth (Turroni *et al.*, 2009; Leahy *et al.*, 2005). Within the human and animal GIT, bifidobacteria are considered to exert a vast array of beneficial health effects, including the establishment of a healthy microbiota in infants, prevention and treatment of diarrhoea, alleviation of constipation and lactose intolerance symptoms, enhancement of immune functions, cholesterol reduction, and suppression of tumorigenesis (Leahy *et al.*, 2005; Turroni *et al.*, 2009). Thus, unsurprisingly bifidobacteria have become a common component of probiotic products designed for human or animal consumption (Tuohy *et al.*, 2003; Leahy *et al.*, 2005; Parvez *et al.*, 2006). Probiotic products represent a strong growth area within the functional foods market and are currently having a significant economic impact on the dairy sector. But, as stated by Stanton *et al.* (2001), 'long-term exploitation of probiotics as health promoters is dependent on several factors, including sound, scientifically-proven clinical evidence of health-promoting activity, accurate consumer information, effective marketing strategies, and, above all, a quality product that fulfils consumer expectations.' While clinical evidence for the purported beneficial effects is rapidly accumulating (Tuohy *et al.*, 2003; Leahy *et al.*, 2005), there still is a lack of fundamental knowledge on the molecular mechanisms by which bifidobacteria interact with other bacteria and host cells, while contributing to host health and well-being.

Genetics and mobilome of bifidobacteria

Compared with other microbes of industrial importance, the genetics of bifidobacteria is poorly understood (Ventura *et al.*, 2004). Bifidobacteria are fastidious microorganisms demanding rich media and mostly requiring strict anaerobic conditions for growth (Scardovi, 1986). Therefore, they are difficult to study in the laboratory. Genetic studies have further been hampered by a lack of appropriate bacterial replicons (of either plasmid or phage origin) with which to construct suitable genetic tools for molecular studies. Moreover, until recently, bifidobacteria were considered recalcitrant to transformation, and genetic engineering techniques were simply unknown.

Recently, whole genome sequencing has revolutionized the genetic, biochemical, and molecular biological research in bacteria and, indeed, in many higher organisms. Genome sequences constitute an essential step in generating primary information for downstream functional applications, such as comparative genomics, transcriptomics, and/or proteomics, which in turn can address fundamental and applied questions. Until now, the genome sequences of five bifidobacteria strains have been released and analysed, namely *Bifidobacterium longum* biotype *longum* NCC2705 (Schell *et al.*, 2002), *B. longum* biotype *infantis* ATCC15697 (Sela *et al.*, 2008), and three different strains of *Bifidobacterium animalis* subsp. *lactis*: DSM 10140 and B1–04 (Barrangou *et al.*, 2009), and AD011 (Kim *et al.*, 2009). In addition, other three genomes have recently been completed: *Bifidobacterium breve* UCC2003 (Leahy *et al.*, manuscript in preparation); *B. longum* biotype *longum* DJO10A (Lee *et al.*, 2008); and *Bifidobacterium adolescentis* ATCC15703 (NC_008618). In addition, several other ongoing whole bifidobacterial genome shotgun sequencing projects are listed in the NCBI database, namely *B. adolescentis* L2–32 (NZ_AAXD00000000), *Bifidobacterium angulatum* DSM20098 (NZ_ABYS00000000), *B. animalis* subsp. *lactis* HN019 (NZ_ABOT00000000), *Bifidobacterium bifidum* NCIMB41171 (NZ_ABQP00000000), *B. breve* DSM20213 (NZ_ACCG00000000), *Bifidobacterium catenulatum* DSM16992 (NZ_ABXY00000000), *Bifidobacterium dentium* ATCC 27678 (NZ_ABIX00000000), *Bifidobacterium gallicum* DSM20093 (NZ_ABXB00000000), *B. longum* biotype *infantis* ATCC55813 (NZ_ACHI00000000), *B. longum* biotype *infantis* CCUG52486 (NZ_ABQQ00000000), and *Bifidobacterium pseudocatenulatum* DSM20438 (NZ_ABXX00000000).

Genomes are the result of the adaptative evolution of microorganisms to their ecological niche (Ventura *et al.*, 2007). In this context, various genetic events, such as gene duplication, horizontal gene transfer (HGT), gene decay, and chromosomal rearrangements, have determined the shape of bacterial genomes. As suggested by Philippe and Douady (2003), plasmids, bacteriophages (phages), and transposons are considered major agents for shaping bacterial genomes through HGT processes. These latter mobile elements as well as others, such as group II introns and jumping genes, constitute what has recently been defined as the mobilome (Siefert, 2009). In spite of comparative genomics, sequencing data do not usually reveal clear-cut candidate genes involved in bacteria–bacteria, bacteria–host interactions, and those related to health-promoting functions. Operons and genes encoding several cell envelope-associated structures, such as exopolysaccharides (EPS) (Schell *et al.*, 2002; Barrangou *et al.*, 2009), fimbriae-like glycoproteins (Schell *et al.*, 2002), serpin-like protease inhibitors (Ivanov *et al.*, 2006), and proteins with collagen-binding motifs, have all been identified. In addition, genes coping with diverse stresses that bifidobacteria face in their environment (acid, bile), genes encoding adaptative functions to the intestinal niche, or others contributing to ecological fitness have also been identified (Schell *et al.*, 2002; Sela *et al.*, 2008; Barrangou *et al.*, 2009; Kim *et al.*, 2009). However, to manage this enormous wealth of genetic data and prove unequivocally the biological role of each particular operon or gene, cloning, expression, knock-out, and transfer of the determinants are needed, for which suitable vectors and other genetic tools based on the replication units and insertional machinery of phages, plasmids and/or transposons are essential.

Plasmids of bifidobacteria

Plasmids are genetic elements found in bacteria, archaea, and eukaryotic cells, which replicate autonomously of the bacterial chromosome. Despite this, plasmids make use of cellular enzymes to ensure replication and benefit from the cellular environment for many plasmid-associated functions (Hayes, 2003). Owing to their reduced size compared to that of the host chromosome and their genetic amenability, plasmids have been extremely useful to study the molecular biology of various organisms and unravelling the molecular mechanisms of many cellular metabolic pathways (Cohen, 1993). Some plasmids are selfish elements, while others provide their host with the ability to use particular carbohydrates and/or protein substances as a source of carbon and energy (Mills *et al.*, 2006). Plasmid-encoded properties further allow their harbouring host with the metabolic ability to degrade complex and recalcitrant (toxic) compounds found in particular ecosystems. All these properties have endowed plasmids of different genera with the title of primarily adaptative entities (Siezen *et al.*, 2005). Plasmids can also carry undesirable properties, such as antibiotic resistance (AR) genes, whose spread among pathogens is of great medical concern (Actis *et al.*, 1998). They can also encode determinants of virulence and pathogenesis. Apart from the traits they encode, important characteristics of plasmids include copy number (low, medium, high copy number), host range (narrow, broad), and capability to spread (conjugation, mobilization) (Hayes, 2003).

Analyses of bifidobacteria strains have indicated that extrachromosomal elements are scarcer than in other intestinal bacteria (Sgorbati *et al.*, 1982; Iwata and Morishita, 1989; Park *et al.*, 1997) and, where found, they are usually smaller in size. At present, around 25 cryptic bifidobacterial plasmid molecules have been sequenced and analysed (Table 13.1; http://www.ncbi.nlm.nih.gov/sites/entrez/). Plasmids from the genome-sequenced strains were retrieved during sequencing, while others were characterized by conventional cloning and sequencing strategies. A majority of these molecules (up to 16) were isolated from *B. longum* strains, which includes pMB1 (Rossi *et al.*, 1996), the first plasmid to be analysed at a molecular level from species of this genus. Three other plasmids have been characterized from strains of *B. breve*, two from *B. asteroides* and *B. bifidum*, and single plasmids have been analysed from strains of *B. pseudocatenulatum* and *B. catenulatum*. Strains from the species *B. longum*, *B. globosum*, *B. asteroides*, and *B. indicum* seem to harbour more plasmids than those from other species (Sgorbati *et al.*, 1982). The obtained sizes of the studied plasmids show a range between 1.8 kilobase pair (kbp) for pMB1 to 10.2 kbp for pNAC3. In most cases, strains harbour a single plasmid, except for three strains of *B. longum* in which two different plasmid molecules were identified (Table 13.1). In addition, *B. longum* biovar. *longum* NAL8 and *B. longum* FI10564 have been reported to contain three different plasmids each. The significance of these autonomously-replicating DNA elements in *Bifidobacterium* remains unclear, since no obvious phenotypic traits have been associated to plasmids, except for the production of the bacteriocin bifidocin B by *B. bifidum* NCFB 1454, which was associated to a 8 kbp plasmid following acriflavine curing (Yildirim *et al.*, 1999). More recently, a further plasmid of this species (pBIF10) has been found to contain a *tet*(Q), a gene encoding a ribosome protection protein providing tetracycline resistance (DQ093580).

The basic biology of bifidobacterial plasmids remains poorly understood; indeed, the mode of replication has been analysed for only a few of them (Moon *et al.*, 2009; Park *et al.*, 2008; Lee and O'Sullivan, 2006; Corneau *et al.*, 2004; O'Riordan and Fitzgerald, 1999; Park *et al.*, 1999). Furthermore, dissection of open reading frames (ORFs) and analysis of untranslated sequences and structures have been undertaken for only two plasmids: pBC1 (Álvarez-Martín *et al.*, 2007b) and pCIBA089 (Cronin *et al.*, 2007). Sequence comparison suggests that most bifidobacterial plasmids probably replicate by means of a rolling-circle mechanism, the exception represented by five or six plasmids that appear to use the theta-replicating mode (Moon *et al.*, 2009; Álvarez-Martín *et al.*, 2008; Cronin *et al.*, 2007; Klijn *et al.*, 2006; Lee and O'Sullivan, 2006; Rossi *et al.*, 1996). Phylogenetic analysis by comparing their replication (Rep) initiator proteins suggested that plasmids could be clustered into five different groups (Corneau *et al.*, 2004).

Table 13.1 Plasmids of the *Bifidobacterium* genus whose nucleotide sequence has been determined

Species, strain	Plasmid	Replication type	Size (kbp)	Relevant genotype or phenotype[a], and application	Reference or source[b]
B. asteroides	pCIBA089	Theta	2.1	Iteron (3DR), Derived vectors	Cronin *et al.* (2007)
B. asteroides	pAP1	RC	2.1	–	GenBank accession no. Y11549[b]
B. bifidum	pBIF10	RC	9.2	*tet*(Q), *tra*, *mobA, mobB*	GenBank accession no. DQ093580[b]
B. bifidum B80	pB80	RC	4.8	*mobA*, Derived vectors	Shkoporov *et al.* (2008a)
B. breve	pB21a	RC	5.2	FstK-like protein	Shkoporov *et al.* (2008a)
B. breve ATCC15698	pNBb1	RC	2.2	–	GenBank accession no. E17316[b]
B. breve NCFB 2258	pCIBb1	RC	5.7	Par, FstK-like protein	O'Riordan and Fitzgerald (1999)
B. catenulatum L48	pBC1	Theta	2.5	Iteron 31/2 DR, Derived vectors	Álvarez-Martín *et al.* (2007)
B. longum	pMG1	RC	3.6	*mobA*	Park *et al.* (2003)
B. longum A24	pBIFA24	RC	4.8	*membA*, *mobB*, Iteron 3DR, *oriT*	Park *et al.* (2007)
B. longum B2577	pMB1	Theta	1.8	Derived vectors	Rossi *et al.* (1996)
B. longum BK51	pTB6	RC	3.6	*mobA*, MembB, *oriT*, Derived vectors	Tanaka *et al.* (2005)
B. longum DJ10A	pDOJH10L	RC	10	*mobA, mobB*, *membA, membB*, Iteron 4DR	Lee and O'Sullivan (2006)
B. longum DJ10A	pDOJH10S	Theta	3.6	*mobB, oriT*, Derived vectors	Lee and O'Sullivan (2006)

Species, strain	Plasmid	Replication type	Size (kbp)	Relevant genotype or phenotype[a], and application	Reference or source[b]
B. longum FI10564	pFI2576	Theta	2.2	Iteron 6DR	Moon *et al.* (2009)
B. longum KJ	pKJ36	RC	3.6	*mobB*, *membB*	Park *et al.* (1997)
B. longum KJ	pKJ50	RC	4.9	*mobA, oriT*, Iteron 3DR, Derived vectors	Park *et al.* (1999)
B. longum NAL8	pNAL8M	RC	4.9	*membA*	Guglielmetti *et al.* (2007)
B. longum NAL8	pNAL8L	RC	3.4	Iteron 3DR, Derived vectors	Guglielmetti *et al.* (2007)
B. longum NC2705	pBLO1	RC	3.6	*mobA*	Schell *et al.* (2002)
B. longum RW041	pNAC2	RC	3.6	*mobA, oriT*, Iteron 4DR	Corneau *et al.* (2004)
B. longum RW041	pNAC3	RC	10.2	*traA*, Iteron 3DR	Corneau *et al.* (2004)
B. longum RW048	pNAC1	RC	3.5	Hypothetical R/M type II system, Iteron 3DR	Corneau *et al.*, 2004
B. longum VMKB44	pB44	RC	3.6	*mobA*, *membB*, Derived vectors	Shkoporov *et al.* (2008a)
B. longum subsp. *globosum*	pASV479	RC	4.8	FstK-like protein, Derived vectors	Sangrador-Vegas *et al.* (2007)
B. pseudocatenulatum VMKB4M	p4M	RC	4.4	FstK-like protein	GenBank accession no. NC_003527[b]

[a]Different of the replication functions: R/M, system or component of a restriction-modification system; *mob*, gene encoding a mobilization protein; *tra*, gene coding for a conjugation-associated protein; *mem*, gene encoding an integral transmembrane protein; *tet*(Q), tetracycline resistance gene encoding a ribosomal protection protein; *oriT*, origin of transference region; Par and FstK, related cell division proteins; DR, direct repeat. Plasmids used for the construction of derived vectors are indicated.

[b]Sequences submitted to GenBank under the accession number (no.) and available at the internet address: http://www.ncbi.nlm.gov/PubMed/

Homology of bifidobacterial Rep proteins has shown that, in some cases, their closest relatives are found in plasmids from other bacterial groups (Álvarez-Martín *et al.*, 2007a; Guglielmetti *et al.*, 2007). More surprisingly, the Rep protein encoded by plasmid p4M from *B. pseudocatenulatum* VMKB4M displays its highest level of similarity to the replication initiator protein of the eukaryotic circoviruses, a feature that is unprecedented in any known bacterial plasmid (Gibbs *et al.*, 2006).

Though differences in nucleotide sequences and gene organization have been encountered among bifidobacterial plasmids, Fig. 13.1 presents the physical and genetic map of pBC1 as an example of their general features. Analysis of the pBC1 sequence showed three predicted ORFs encoding polypeptides larger than 50 amino acids: *repB*, possibly encoding a replicase; a transcriptionally-coupled gene (*orfX*) bearing some homology to the transcriptionally-coupled *orfX* of theta-replicating plasmids of lactococci (Mills *et al.*, 2006); and a third ORF on the complementary strand, *copG*-like, which showed a conserved domain present in proteins of the CopG family. Comparison of the deduced RepB protein from pBC1 to other replication proteins available in public databases identified pMB1 from *B. longum* B2577 (Rossi *et al.*, 1996) as its closest relative. More exactly, the pBC1 Rep protein exhibits 81% amino acid identity to that of pMB1. Besides translated sequences, several recognized untranslated sequences were also identified in the pBC1 sequence (Álvarez-Martín *et al.*, 2007a). Upstream of *repB* A+T-rich stretches were identified, interspersed with C+G-rich boxes. Four inverted repeats (IRs) of variable length with small or no loops were observed immediately upstream of an imperfect direct repeat (DR) of 24 nt (ATGGACATCTCCATGTGGATGTCC), which was repeated three-and-a-half times. Upstream of the start codon of *repB*, a further IR was detected resembling that of the *oriV* origin of replication of the lambda phage. This kind of organization resembles the so-called DNA iteron structure observed in the *ori* region of some theta plasmids (del Solar *et al.*, 1998). The tandem repeat organization is similar in all plasmids but sequences of the repeats are variable, thus, conferring specific interaction between Rep proteins and DNA sequences.

Apart from Rep, many plasmids contains accessory ORFs encoding hypothetical proteins, some of which, such as the mobilization-like proteins (Table 13.1), may be involved, together

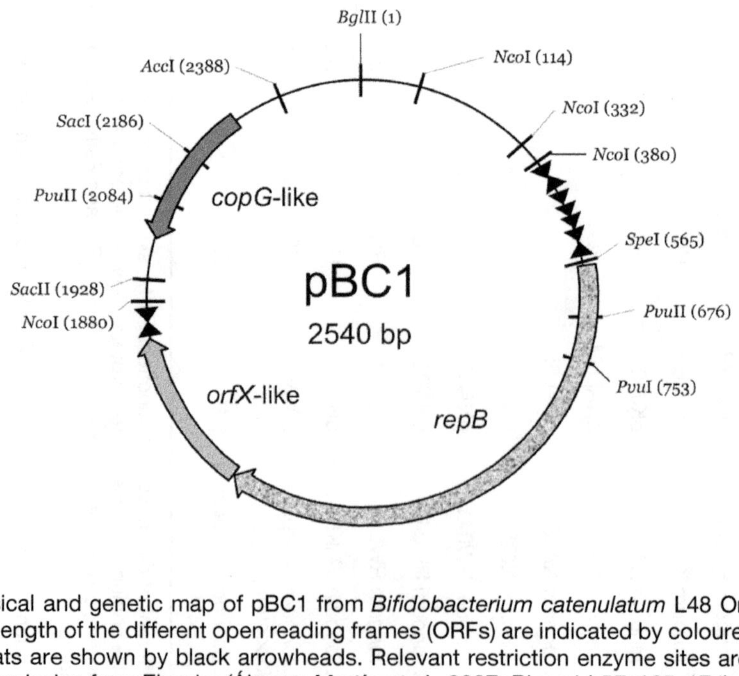

Figure 13.1 Physical and genetic map of pBC1 from *Bifidobacterium catenulatum* L48 Origin, orientation and approximate length of the different open reading frames (ORFs) are indicated by coloured arrows. Direct and reverse repeats are shown by black arrowheads. Relevant restriction enzyme sites are also depicted. Reprinted with permission from Elsevier (Álvarez-Martín *et al.*, 2007; Plasmid *57*, 165–174).

with their accompanying *oriT* sequences, in plasmid spread. Non-essential proteins, such as *OrfX* and *copG*, are frequently involved in the control of replication, copy number and/or plasmid stability (Álvarez-Martín *et al.*, 2007b; del Solar *et al.*, 1998). These variables, however, may also be influenced by the genetic background of the host cell by, for instance, interfering with resident plasmids or integrated plasmid remnants, such as those discovered in *B. longum* NCC2705 (Schell *et al.*, 2002).

Phages of bifidobacteria

Phages are widely distributed among eubacteria, where they are thought to influence the genomic evolution and adaptive capabilities of their hosts (Canchaya *et al.*, 2003). The first report of a bifidus-specific phage was that of Youseff *et al.* (1966), in which phages from rumen were detected. However, in this work the morphology and other characteristics of the phage particles were unreported. The work of Youseff *et al.* was followed by an electron microscopy observation of a lytic phage from the so-called *Bifidobacterium ruminale* RU 271 by Matteuzzi and Sozzi (1970). Further pioneering reports include that of Sgorbati *et al.* (1983), where inducible prophages from strains of *B. longum* were released by mitomycin C and further characterized by electron microscopy. High phage titres have been found in the rumen (Klieve and Bauchop, 1988), suggesting a similar high phage titre in the GIT content. However, no further studies on bifidobacterial phages appeared until the analysis of whole genomes.

Three highly related prophage-like elements have been reported to be present in the genome of *B. breve* UCC2003, *B. longum* biotype *longum* NCC2705, and *B. longum* biotype *longum* DJO10A (Ventura *et al.*, 2005a). These elements, designated Bbr-1, Bl-1, and Blj-1 respectively, share nucleotide and organization homology with double-stranded DNA bacteriophages infecting low G+C Gram-positive bacteria, arguing for a common evolution of phages within the GIT ecosystem (Ventura *et al.*, 2005a). The Blj-1 prophage is 36.9 kb long and is excised from the chromosome when *B. longum* DJO10A is exposed to mitomycin C or hydrogen peroxide (Ventura *et al.*, 2005a). Thus, Blj-1 appears to constitute the first inducible prophage whose sequence is entirely known. In contrast, Bbr-1 and Bl-1 elements are not inducible, suggesting they may represent non-functional prophages. Though defective, they may still constitute functional satellite phages, whose mobility depends on helper phages in a similar manner to that described for the cryptic mycophages Rv1 and Rv2 (Hendrix *et al.*, 1999). All three bifidobacterial prophages are integrated in a tRNAMet gene, which had not previously been shown to act as an *attB* site in Gram-positives (Campbell, 1992). Analysis of the distribution of this integration site in bifidobacterial species has revealed that *attB* sites are well conserved. In addition, in the genome of *B. longum* subsp. *infantis* strain ATCC15697 and those of *Bifidobacterium animalis* subsp. *lactis* DSM 10140 and Bl-04, prophage genes have also been encountered (Sela *et al.*, 2008; Barrangou *et al.*, 2009), although these remnant elements are not adjacent to tRNAMet sequences.

The use of conserved *attB* sequences and their associated *int* genes might allow the construction of efficient recombination modules analogous to the *Streptomyces* integrative plasmid pSE211 (Brown *et al.*, 1990). This module may represent an ideal source for integration systems, enabling future development of food-grade, single copy integration of foreign DNA at specific sites within the bifidobacterial chromosome without disturbing host functions. Similar systems have been developed for lactic acid bacteria, such as *Lactobacillus* (Martín *et al.*, 2000), and for high G+C bacteria, such as *Streptomyces* and *Mycobacterium* (Combes *et al.*, 2002).

Transposons and insertion sequence (IS) elements in bifidobacteria

Transposons and IS elements are mobile genetic units that move from one to another position by a process referred to as transposition. Transposition occurs via one of two mechanisms: cut-and-paste transposition or replicative transposition (Roberts *et al.*, 2008), leaving, respectively, one copy on the target DNA or two copies on both donor and target DNA. Both transposons and ISs are potent, broad spectrum mutators contributing to the shape of the function, structure and dynamics of genes and genomes (Philippe and Douady, 2003). If properly domesticated, transposons and

ISs can be converted into powerful genetic tools with which to explore the functionality of genes. In addition, transposons integrating at preferred, neutral sites can be used for genetic modification of bacteria. However, tools for bifidobacteria derived from such elements are not yet available.

Sixteen IS elements of five classes have been reported to be spread in the *B. longum* NCC2705 genome (Schell *et al.*, 2002). Although present at similar numbers in the genome of all other analysed strains, the IS was not considered in other works. Recently, an IS of 1047 bp long was identified in the upstream region of the tetracycline resistant gene *tet*(W) found on the chromosome of *B. longum* F8 (Kazimierczak *et al.*, 2006). A similar IS of 744 bp was also found to interrupt the structural *tet*(W) gene in the susceptible *B. longum* M21 strain (Ammor *et al.*, 2008). The spreading process of the ISs involves excision and integration into a new place, at which position a short nucleotide duplication is usually found. Interestingly, five out of the six bp sequences duplicated in F8 (CAATGC) seem to mirror the 5 bp duplication in M21 (GTTAC), suggesting the presence of active insertion sites in these bifidobacterial genomes.

A copy of the transposon Tn*5432*, which encodes resistance to erythromycin and clindamycin, has recently been identified in several *B. thermophilum* strains isolated from pig faeces (van Hoek *et al.*, 2008). Tn*5432* was first isolated from *Corynebacterium xerosis* and rescued copies on plasmids were shown to be able to transpose in *Corynebacterium glutamicum* causing several mutations (Tauch *et al.*, 1995). The transposition ability of Tn*5432* from *B. thermophilum* remains to be determined, but the development of a high-efficiency transposon mutagenesis system should facilitate the molecular analysis of bifidobacteria.

General and specialized vectors for bifidobacteria

Full exploitation of genomic data requires the use of general and specialized vectors for gene overexpression, integration, knock-out, and gene expression studies. Molecular tools are further required for the future modification of industrial *Bifidobacterium* strains by genetic engineering (Kullen and Klaenhammer, 2000), i.e. the construction of strains with enhanced probiotic characteristics and/or increased viability during storage. In addition, bifidobacteria have recently been appointed as biotechnological agents for *in situ* production and delivery of therapeutic compounds, such as antigens (for live vaccine development) and tumour-suppressing substances (Fujimori, 2006; Xu *et al.*, 2007), and as a means of increasing beneficial detoxifying activities into the GIT (Park *et al.*, 2007). Traditional and new applications will require utilization of robust genetic tools and improved genetic transformation techniques.

Some plasmids have provided the basis for the construction of *Escherichia coli*–*Bifidobacterium* shuttle vectors, mostly resulting from the direct cloning of whole plasmids into an *E. coli* vector containing selectable antibiotic resistance genes such as spectinomycin, erythromycin and chloramphenicol (Álvarez-Martín *et al.*, 2008; Guglielmetti *et al.*, 2007; Klijn *et al.*, 2006; Lee and O'Sullivan, 2006; Park *et al.*, 1999; Rossi *et al.*, 1996, 1998; Matsumura *et al.*, 1997; Missich *et al.*, 1994; Table 13.2). The plasmid pBC1 from *B. catenulatum* L48 has been used for the construction of a series of *E. coli*–*Bifidobacterium* shuttle vectors with innovative characteristics such as the presence of a tetracycline resistance gene of bifidobacterial origin (*tet*(W)) as the selection marker, or the cloning of the β-galactosidase complementing peptide gene for a convenient blue/white screening of recombinant clones in *E. coli* (Álvarez-Martín *et al.*, 2008). The functionality of the vectors was further checked by cloning and overexpression of an α-L-arabinofuranosidase gene from *B. longum* B667 in *E. coli* and *Bifidobacterium* strains. As plasmid maintenance constitutes a major problem for vector utilization, González Vara and co-workers studied the segregational and structural stability of pMB1-derived constructs in *B. animalis* by continuous culture (González Vara *et al.*, 2003). These authors reported a high correlation between instability and plasmid size, while no major deletions and rearrangements were observed. However, some constructs did not behave as expected (González Vara *et al.*, 2003), a result that agrees with observations by other authors (Álvarez-Martín *et al.*, 2007b), suggesting that beyond plasmid size, secondary structure of the constructs may further influence stability.

It is worth noting that in spite of a limited knowledge of plasmid biology, a number of vectors for heterologous expression of desirable foreign genes have already been developed. As an example, the reporter vector pMDY23 expresses the *gusA* gene of *E. coli* (Klijn *et al.*, 2006); vector pBES2 has been used to express the α-amylase gene of *B. adolescentis* in *B. longum* (Rhim *et al.*, 2006); pBLES100 (Matsumura *et al.*, 1997) has been used in tumour suppressor studies (Yazawa *et al.*, 2001) and for the expression of the *Salmonella* flagellin gene for mucosal immunization in a mouse model (Takata *et al.*, 2006); and pBV22210 has been used to express and deliver the anticancer protein endostatin, which may be useful in cancer therapy (Xu *et al.*, 2007).

Genetic modification of bifidobacteria

Since the mid-eighties, research efforts have focused on establishing effective protocols for the genetic modification of bifidobacteria. Currently, electrotransformation (electroporation) of bifidobacteria by plasmid DNA is commonly being reported, whereas little or nothing is known about other recombinant DNA technologies such as conjugation or homologous recombination. In fact, members of the genus *Bifidobacterium* have traditionally been considered refractory to efficient and reproducible genetic manipulation (Kullen and Klaenhammer, 2000). To various extents in different strains, several factors could contribute to bifidobacterial recalcitrance for acquiring exogenous DNA, such as the presence of a thick (multi-layered) and complex cell wall (Fischer *et al.*, 1997), intracellular restriction/modification barriers (O'Connell-Motherway *et al.*, 2009; Yasui *et al.*, 2009; Schell *et al.*, 2002; Hartke *et al.*, 1996), and sensitivity to environmental stresses (principally oxygen) during preparation of competent cells and transformation.

Genetic transformation by electroporation

Electroporation is the actual essential procedure to introduce exogenous DNA into bifidobacteria. The first scientific proof of genuine genetic transformation of *Bifidobacterium* dates back to 1994, when Missich *et al.* introduced pRM2 (Missich *et al.*, 1994), a derivative of the small *B. longum* cryptic plasmid pMB1 (Sgorbati *et al.* 1982), into a cured *B. longum* strain. The protocols available for preparing electro-competent cells and subsequent electroporation are based mainly on the comprehensive studies by Argnani *et al.* (1996) and Rossi *et al.* (1997), who considered and optimized several conditions such as growth medium, washing solutions, incubation temperatures, and voltage. The protocols used in subsequent research were simple adaptations of the above studies (Table 13.2).

Preparation of electro-competent cells

The preparation of electro-competent bacterial cells consists of weakening the cell wall and making the bacteria permeable to DNA during an electrical discharge, while preserving their viability. The general strategies for achieving this goal comprise using bacterial cells in the log growth phase, growth in the presence of high sugar concentrations, osmotic stabilizers in washing and electroporation buffers, or maintaining cells at low temperatures.

Growth phase

Since the composition of the cell wall plays a key role in DNA uptake, numerous studies have reported the importance of harvesting bacterial cells at a specific stage of their growth (Table 13.2). Some studies showed that bifidobacteria could be effectively transformed only when they are in the early exponential phase ($OD_{600\,nm}$ 0.2–0.4) (Argnani *et al.*, 1996; Rossi *et al.*, 1997; Missich *et al.*, 1994), whereas efficiencies dropped for older cells. In contrast, other researchers have observed maximal transformation efficiency with cells in the middle to late log phase. For instance, Matsumura *et al.* (1997) found that with the vector pBLESS100 the transformation efficiency of *B. longum* 105-A was about one order of magnitude higher when cells were approaching the stationary phase than when they were in the early-log phase. Similarly, in more recent studies mid-logarithmic phase cells were used (optical density at 600 nm 0.5 to 0.7) to effectively transform different *Bifidobacterium* species (Álvarez Martín *et al.*, 2008; Cronin *et al.*, 2007; Sangrador-Vegas *et al.*, 2007; MacConaill *et al.*, 2003).

Table 13.2 Summary of protocols developed for electrotransformation of bifidobacteria

Vector (size, kbp) (marker)	Replicon(s)	Growth medium[1]	Washing buffer	Electroporation buffer	Preincubation[2]	Voltage and resistance	Recovering medium	Transformation rate (transformants/μg DNA)	Reference
pDG7 (7.3 kbp) (Cm^R)	pMB1 pBR322	MRS, 0.5 M sucrose, 0.05% cysteine-HCl (OD_{600nm} 0.2)	0.5 M sucrose	1 mM ammonium citrate buffer pH6, 0.5 M sucrose	3.5 h at 4°C in electroporation buffer	12 kV/cm, 200 Ω	MRS, 0.5 M sucrose, 0.05% cysteine	*B. animalis* ATCC 27536 5.0×10^3 *B. breve* 4 1.3×10^4 *B. breve* AS 2.0×10^2 *B. bifidum* U3 3.0×10^2 *B. bifidum* ATCC 15696 7.4×10^3 *B. infantis* U1 2.5×10^2 *B. infantis* ATCC 27920 4.0×10^4 *B. longum* U2 2.6×10^3 *B. longum* Wiesby 2.7×10^4	Argnani *et al.* (1996)
pNC7 (4.9 kbp) (Cm^R)	pMB1 (non-replicative in *E. coli*)	Iwata Medium (IM; Iwata and Morishita, 1989), 16% Actilight®P (OD_{600nm} 0.2–0.3)	5 mM K-phosphate buffer pH7	KMR buffer (5 mM KH_2PO_4, 1 mM $MgCl_2$, 0.3 M raffinose, pH 4.8)	Overnight at 0°C in electroporation buffer	12.5 kV/cm, 200 Ω	IM, 6% Actilight®P	*B. animalis* ATCC 27536 3.0×10^4 *B. breve* MB226 6.6×10^4 *B. breve* MB252 2.3×10^4 *B. bifidum* MB254 7.2×10^4 *B. infantis* MB208 1.2×10^5 *B. infantis* MB263 9.3×10^3 *B. longum* MB231 2.8×10^2 *B. pseudocatenulatum* MB264 5.0×10^1 *B. ruminale* MB266 7.2×10^2 *B. dentium* MB269 3.6×10^1 *B. magnum* MB267 1.8×10^3	Rossi *et al.* (1997[3])
pRM2 (7.5 kbp) (Sp^R)	pMB1 pBR322	TPY + glucose (OD_{660nm} 0.6)	10% glycerol	10% glycerol	Freezing at −135°C and storage at −70°C	10 kV/cm, 200 Ω	TPY + glucose	*B. longum* B2577 3.8×10^2	Missich *et al.* (1994)

pBLES100 (9.1 kbp) (Sp^R)	pTB6 [4] pBR322	Briggs Medium (Briggs, 1953) supplemented with 2% lactose instead of glucose (cells in middle to late log phase)	10% glycerol	10% glycerol	Freezing at −135°C and storage at −70°C	10 kV/cm, 200 Ω	Briggs Medium	*B. longum* 105-A 2.2×10^4	Matsumura *et al.* (1997)
pBKJ50F (8.1 kbp) (Cm^R)	pKJ50 pBR322	According to Argnani *et al.* (1996). with the only modification of pulse at 10 kV/cm						*B. animalis* ATCC 27536 2.0×10^2	Park *et al.* (1999)
pBES2 (7.6 kbp) (Cm^R)	pMG1 [4] pUC (ColE1)	According to Argnani *et al.* (1996), with the addition of oxyrase® (Oxyrase inc. Ohio) in the recovering medium						*B. longum* MG1 7.3×10^3	Park *et al.* (2003)
pBRASTA101 (5.0 kbp) (Sp^R)	pTB6 [4] pUC (ColE1)	According to Missich *et al.* (1994) and Matsumura *et al.* (1997)						*B. longum* 105-A 2.5×10^6	Tanaka *et al.* (2005)
pFUN (8.1 kbp) (Ery^R)	pAMβ1 (*E. faecalis*) pBluescript (ColE1)	IM, (Mid-log phase cells, $OD_{600\,nm}$ 0.5 to 0.7)	0.5 M sucrose, 1 mM citrate buffer (pH 5.8)	0.5 M sucrose, 1 mM citrate buffer (pH 5.8)	-	10 kV/cm, 200 Ω	Iwata Medium	*B. breve* UCC2003 10^2–10^3	MacConaill *et al.* (2003)
pPKCm1 (6.2 kbp) (Cm^R)	pCIBA089 pBluescript (ColE1)	IM, (Mid-log phase cells, $OD_{600\,nm}$ 0.5 to 0.7)	0.5 M sucrose, 1 mM citrate buffer (pH 5.8)	0.5 M sucrose, 1 mM citrate buffer (pH 5.8)	-	10 kV/cm, 200 Ω	Iwata Medium	*B. breve* UCC2003 3.8×10^6 *B. animalis* subsp. *lactis* 10^1 *B. longum* NCIMB8809 10^2 *B. pseudolongum* NCIMB2244 10^2 *B. globosum* JCM5820 10^3 *B. pseudocatenulatum* LMG10505 10^3 *B. dentium* NCFB2843 10^4	Cronin *et al.* (2007)
pASV480 (9.0 kbp) (Cm^R)	pASV479 pBluescript (ColE1)	IM, (Mid-log phase cells, $OD_{600\,nm}$ 0.5 to 0.7)	0.5 M sucrose, 1 mM citrate buffer (pH 5.8)	0.5 M sucrose, 1 mM citrate buffer (pH 5.8)	–	10 kV/cm, 200 Ω	Reinforced Clostridial Medium (RCM)	*B. breve* NCIMB 8807 ~10^5 *B. breve* NCFB 2258 ~10^5	Sangrador-Vegas *et al.* (2007)

Table 13.2 Continued

Vector (size, kbp) (marker)	Replicon(s)	Growth medium[1]	Washing buffer	Electroporation buffer	Preincubation[2]	Voltage and resistance	Recovering medium	Transformation rate (transformants/μg DNA)	Reference
pAM4 (7.6 kbp) (Tet^R)	pBC1pUC	MRS, 0.05% cysteine (Mid-log phase cells, OD_{600nm} 0.5 to 0.7)	0.5 M sucrose, 1 mM citrate buffer (pH 5.8)	0.5 M sucrose, 1 mM citrate buffer (pH 5.8)	20 min in ice	10 kV/cm, 200 Ω	Reinforced Clostridial Medium (RCM)	*B. adolescentis* LMG10502 9.2×10^2 *B. animalis* LMG10508 4.0×10^1 *B. animalis* subsp. *lactis* Bb12 1.6×10^2 *B. breve* LMG13208 1.0×10^2 *B. breve* UCC2003 1.4×10^2 *B. dentium* F101 9.5×10^1 *B. longum* L25 6.6×10^1 *B. pseudolongum* LMG11571 6.3×10^1 *B. pseudocatenulatum* M115 1.0×10^5 *B. thermophilus* LMG11571 4.6×10^1 *C. glutamicum* LMG19741 3.0×10^0	Álvarez-Martín *et al.* (2008)

[1]Phase of growth at which bifidobacterial cells are collected before washing steps.

[2]Incubation step of competent cells before electroporation or storing at –80°C.

[3]Trasformation experiments were performed with vector DNA isolated from *B. animalis* MB209.

[4]pTB46 and pMG1 plasmids are isogenic to pB44, pNAC2, pBLO1, pDOJ10L, and pKJ36.

B. animalis ATCC 27536 is also known as *B. animalis* MB209; *B. infantis* stands for *Bifidobacterium longum* subsp. *infantis*.

Key of antibiotic resistance markers: Cm^R, chloramphenicol acetyl transferase (*cat*); Tet^R, tetracycline resistance (*tet*(W)); Sp^R, enterococcal spectinomycin resistance; Ery^R, erythromycin resistance.

Growth media

The addition of sugars in high concentrations to the growth medium is traditionally recognized as an effective strategy for the improvement of the transformation yield by affecting the composition of the cell wall. The best conditions for preparing electroporation-competent bifidobacteria were set by Argnani *et al.* (1997) by cultivating cells in MRS broth supplemented with 0.5 M sucrose and washing them in buffered sucrose solution at the same concentration. Rossi *et al.* (1996) showed a 100-fold increase in transformation efficiency when *Bifidobacterium* cells were grown in Iwata medium (IM) supplemented with raffinose 0.3 M or, especially, 16% Actilight®P, compared with cells grown in IM broth with or without a glucose supplement. Actilight®P is a commercial product comprising a mix of short-chain fructooligosaccharides (1-kestose, nystose, and fructosylnystose; FOS), which are metabolized by bifidobacteria and may protect them from stress (Guglielmetti *et al.*, 2008).

Electroporation buffers

Incubation of the cells in an electroporation buffer was found to be crucial for the competence of *Bifidobacterium*. Argnani *et al.* (1997) showed that a few-hour storage of bacterial cells before electroporation at 4°C in an electroporation buffer composed of 0.5 M sucrose and 1 mM ammonium citrate (pH 6.0) significantly improved their transformation efficiency (Table 13.2). They suggested that under these conditions the presence of low-molarity ammonium citrate (more than HEPES and phosphate buffers) as the osmotic stabilizer may support the right level of cell autolysis without affecting cell viability, which results in improved cell wall permeability for exogenous DNA. In addition, Rossi *et al.* (1996) showed the importance of the incubation step in the electroporation buffer at 0°C overnight. Their electroporation buffer, named KMR, was composed of KH_2PO_4 5 mM, $MgCl_2$ 1 mM, and raffinose 0.3 M (pH 4.8). The high salt concentration of KMR buffer may however favour *arcing* during the electrical discharge.

Efficiency of electro-transformation in bifidobacteria

In general, the rate of electroporation-mediated transformation in bifidobacteria is extremely low and constitutes the main limitation to successfully applying traditional genetic manipulation strategies. Wide variations in transformation efficiencies have been reported in the literature, ranging from about 10^0 (e.g. Rossi *et al.*, 1997) to more than 10^6 (e.g. Tanaka *et al.*, 2005) transformants per µg of recombinant DNA (Table 13.2). Besides the protocols adopted for preparing competent cells, differences may be because of strain-specific transformation rates (Table 13.2). Nonetheless, valuable recent progress has improved transformation rates thanks to studies on the restriction/modification systems of bifidobacteria, which were thought to be the main obstacle in the acquisition of foreign DNA by these bacteria (O'Connell-Motherway *et al.*, 2009).

Optimization strategies for transformation of Bifidobacterium

Preserving cell viability during electroporation is crucial for transformation efficiency. The main reason for cell mortality during these experiments is oxidative stress. To overcome this problem, Park *et al.* (2003) added Oxyrase® to the incubation buffer after the electric pulse. Oxyrase®, an enzyme system which removes oxygen from its environment, allowed *B. longum* MG1 to undergo plasmid transformation at a 100-fold improved efficiency.

In general, DNA introduced into bacteria by electroporation is more vulnerable to restriction nucleases than that transferred by conjugation or transfection. This is particularly important for bifidobacteria, whose immediate obstacle to acquiring exogenous DNA is the presence of restriction/modification (R–M) barriers. DNA R–M gene clusters coding for methyltransferases and restriction enzymes, belonging to either types I, II, or IV, have been recognized in all sequenced bifidobacterial genomes (O'Connell-Motherway *et al.*, 2009; Yasui *et al.*, 2009; Lee *et al.*, 2008; Schell *et al.*, 2002; Hartke *et al.*, 1996; Roberts, 1980; http://rebase.neb.com/rebase). Several studies have reported a significant increase in transformation efficiency when the transformed DNA was isolated from *Bifidobacterium* instead of *E. coli* (Rossi *et al.*, 1997, 1998; O'Connell-Motherway *et al.*, 2009; Yasui *et al.*, 2009). For instance, Rossi *et al.* (1998) found that certain strains of *B. longum*, *B. animalis*, or *B. Bifidum* could only be successfully transformed when the vector DNA was purified from bifidobacteria. Therefore, proper modification of plasmid DNA can help in bypassing the restriction barriers. This strategy was recently adopted in two independent studies by the research teams of Professor Suzuki (Yasui *et al.*, 2009) and Professor van Sinderen (O'Connell-Motherway *et al.*, 2009), seeking to boost the transformation efficiency of *B. adolescentis* ATCC15703 and *B. breve* UCC2003, respectively. Using this strategy, a shuttle vector was pre-methylated by *E. coli* cells, carrying genes assumed to encode the DNA modification enzymes of the target *Bifidobacterium* (Fig. 13.2). In fact, Yasui *et al.* (2009) developed a system called Plasmid Artificial Modification (PAM) and demonstrated its efficacy for *B. adolescentis* ATCC15703, a strain which can only be transformed at an extremely low level. In the genome of ATTC15703, two type II DNA methyltransferase genes were identified: BAD_1233 and BAD_1383. These were cloned in *E. coli* TOP10 (Invitrogen) by means of a low copy number vector, obtaining the so-called PAM host. The *E. coli* TOP10 laboratory strain was shown to be the most suitable host because it lacks the type IV restriction enzymes *mrr* and *mcrBC* (that degrade DNA methylated by R–M systems of other bacteria) and the methylases *dam*, *dcm*, and *hsdMS* (which can make the DNA sensitive to possible type IV restriction systems of bifidobacteria). Subsequently, an *E. coli*–*Bifidobacterium* shuttle vector, based on the pTB6 *B. longum* replicon (Tanaka *et al.*, 2005), was introduced into the PAM host. Transformation efficiency improved considerably when *B. adolescentis* was electroporated with the shuttle vector DNA isolated from the PAM host, jumping from $1–3 \times 10^0$ to 10^5 CFU/μg DNA compared with the unmodified vector.

The same approach was adopted by O'Connell-Motherway *et al.* (2009) with *B. breve* UCC2003 as the target host. In the annotated genome of this strain, they found three different R–M gene clusters, including the methylase genes *bbrIM*, *bbrIIM*, and *bbrIIIM*. The role of these modification genes in the acquisition of exogenous DNA by *B. breve* was studied in transformation experiments with the shuttle vector pAM5 based on the pBC1 replicon from *B. catenulatum* (Álvarez-Martín *et al.*, 2008). The authors observed a 1000-, 10-, and 5-fold higher transformation frequency for pAM5 DNA isolated from *E. coli* expressing M.BbrIII, M.BbrII, and M.BbrI, respectively, indicating that all three DNA methylation systems affected transformation efficiency.

The above studies demonstrated that artificial modification of DNA by target host methylases is an effective way to significantly increase the efficiency of electroporation. Genome analyses and experimental data, however, showed that *Bifidobacterium* strains harbour a very diverse range of R–M activities, even within the same species (O'Connell-Motherway *et al.*, 2009). Therefore, this strategy is at least partly limited to the strains whose whole genome sequence is known. To overcome this problem, O'Connell-Motherway *et al.* (2009) proposed the possibility of methylating exogenous DNA isolated from *E. coli* through incubation with crude cell extracts of the *Bifidobacterium* target host in the presence of S-adenosylmethionine. Experimental data of the practicability and effectiveness of this strategy are not yet available.

Conjugation in bifidobacteria

The R–M barriers of a bacterial strain can also be bypassed by introducing foreign DNA through conjugation. Conjugation may occur when *cis* (*oriT*) and *trans* (transfer proteins, Mob and Tra) elements found in mobilizing plasmids are recognized by the other mobilization components, which can be supplied by plasmids or the host cell. Putative Mob and Tra protein-coding genes and characteristic *cis* elements have been found in several bifidobacterial plasmids (Table 13.1), suggestive of their mobilization potential. Nevertheless, until now, no DNA transfer system based on conjugation is available for members of the genus *Bifidobacterium*, and no conjugation events have been irrefutably demonstrated. The only documented systematic attempt to achieve conjugation in bifidobacteria was made by Shkoporov *et al.* (2008a), who exploited the putative mobilization function of three different *Bifidobacterium* plasmids to develop genetic tools based on

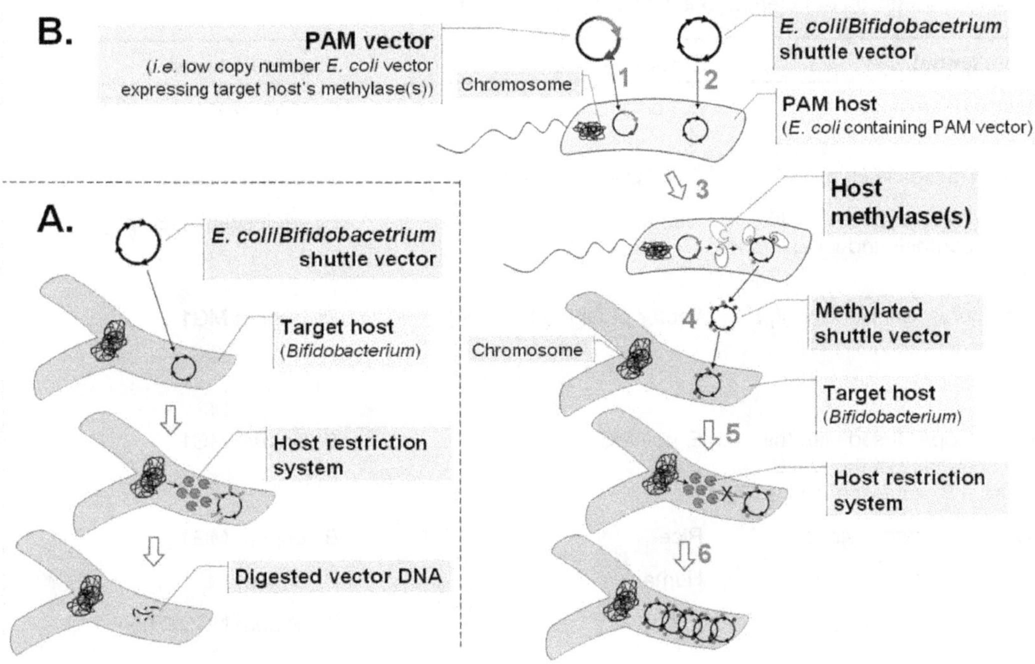

Figure 13.2 Conventional strategy of vector preparation for electrotransformation of bifidobacteria compared to the Plasmid Artificial Modification (PAM) strategy. (A) Conventional strategy, involving the extraction of shuttle vector DNA from *E. coli* and direct introduction in *Bifidobacterium* species by electroporation. (B) PAM strategy: 1 – preparation of the PAM host, consisting of an appropriate *E. coli* strain (e.g. *E. coli* TOP10) harbouring PAM vector (a low copy number vector coding for the methylase(s) of the target host); 2 – introduction of an *E. coli-Bifidobacterium* shuttle vector into the PAM host; 3 – modification of shuttle vector DNA by methylase(s) coded by PAM vector; 4 – extraction of the now methylated vector and introduction into the target host (*Bifidobacterium*) by electroporation; 5 – target host restriction system(s) cannot digest the methylated shuttle vector; 6 – shuttle vector replicates inside target host.

the well-characterized intergeneric conjugative element RP4 (IncPa) (Simon *et al.*, 1983) to transfer DNA into *B. pseudocatenulatum*. Their experiments produced antibiotic-resistant clones which, though PCR-positive, did not contain the expected plasmid DNA. Consequently, the development of effective conjugation systems for bifidobacteria, albeit potentially highly useful in the genetics and biotechnology of bifidobacteria, remains in its infancy.

Expression of heterologous genes among bifidobacteria

The development of systems for heterologous expression and secretion is strategically important to study the properties of *Bifidobacterium*, as well as for strain improvement and the construction of delivery systems used for the *in situ* production of peptide vaccines and anticarcinogenic proteins.

Heterologous genes from a diverse group of organisms have been expressed in bifidobacteria under the control of either heterologous or homologous promoters (Table 13.3). The current list includes genes and promoters mainly from Gram-negative and Gram-positive bacteria, but with several exceptions such as the luciferase gene from *Pyrophorus plagiophthalamus* (Guglielmetti *et al.*, 2008) and several human genes (Xu *et al.*, 2007). In addition to the genes reported in Table 13.3, several antibiotic resistance genes, used in preparing *Bifidobacterium* cloning vectors, were shown to be functional in *Bifidobacterium*, even under the control of their own regulatory elements. These include, for instance, the chloramphenicol acetyl transferase (*cat*) of plasmid pC194, the erythromycin resistance gene of pE194 from *Staphylococcus aureus*, and the spectinomycin resistance gene from *Enterococcus*

Table 13.3 Expression of heterologous genes in bifidobacteria

Protein (gene)	Origin	Expression host
Cytosine deaminase	*E. coli*	*B. longum* 105-A
Secreted nuclease (*nuc*)	*Staphylococcus aureus*	*B. breve* UCC2003
Endostatin (Liver cDNA); TNF-related apoptosis-inducing ligand (TRAIL)	Human	B. adolescentis; B. *longum*
Green fluorescent protein (*gfp*)	Vector pEGFP (Clontech)	*B. longum* MG1
Phytase (*appA*) fused with the signal sequence of the *amyB* gene from *B. adolescentis* Int-57	*E. coli* MC4100	*B. longum* MG1
Glutamate decarboxylase	Rice	*B. longum* MG1
PTEN tumour suppressor	Human	*B. longum* L17
β-glucuronidase (*gusA*)	*E. coli*	*B. longum* NCC2705
Flagellin (*fliC*)	*Salmonella typhimurium* ATCC14028	*B. animalis* ATCC27536
Interleukin 10 (*rhIL*-10)	Recombinant human protein	*B. longum* ATCC15707
β-glucuronidase (*gusA*)	*E. coli* (from promoter-probe vector pNZ272)	*B. breve* NCIMB8807
α-arabino furanosidase (*abfB*)	*B. longum* B667	*B. pseudocatenulatum* M115
Bacterial luciferase (*luxABCDE* operon)	*Photorhabdus luminescens*	*B. breve* UCC2003
Luciferase (*lucGR*)	*Pyrophorus plagiophthalamus*	*B. longum* NCC2705
Cholesterol oxidase (*choPA* operon)	*Streptomyces* spp.	*B. longum* MG1
Synthetic human basic fibroblast growth factor (FGF-2 gene) fused with signal peptide of Sec2 secreted protein from *B. breve* UCC2003	*pkFGFB* (Seeger *et al.*, 1995)	*B. breve* UCC2003
Listerial bile resistance mechanism BilE (*bilE* operon)	*L. monocytogenes* EGD-e	*B. breve* UCC2003

faecalis. In contrast, the thiostrepton resistance gene from *Streptomyces* is non-functional in bifidobacteria (Rossi *et al.*, 1998; Guglielmetti *et al.*, 2007). Similarly, the genes coding for a *Pseudomonas fluorescens* lipase, an α-amylase from *Bacillus licheniformis,* and a cholesterol oxidase from *Streptomyces* spp., although stably maintained in bifidobacteria, are not expressed under the control of their own promoters (Rossi *et al.*, 1998).

With respect to protein expression in bifidobacteria, excretion and secretion processes should be studied to develop export systems. In this context, MacConaill *et al.* (2003) investigated

Promoter	Reference
Promoter of *hup* gene, coding for the histone like protein of *B. longum*	Nakamura *et al.* (2002)
Seven different promoters from *B. breve* UCC2003	MacConaill *et al.* (2003)
Coliphage lambda P_RP_L promoter regions	Li *et al.* (2003); Fu *et al.* (2005); Xu *et al.* (2007)
Expressed from two promoters from *Bifidobacterium* spp. GE65 (Sequence analysis revealed identity with *L. johnsonii* genome; S. Guglielmetti, unpublished)	Ji *et al.* (2005)
Promoter of the *amyB* gene from *B. adolescentis* Int-57	Park M.S. *et al.* (2005)
Not known	Park K.B. *et al.* (2005)
Promoter region of hup gene from *B. longum*	Hou *et al.* (2006)
Putative promoters of genes BL1363, BL1613 and BL1518 from *B. longum* NCC2705	Klijn *et al.* (2006)
Promoter of *hup* gene from *B. longum*	Takata *et al.* (2006)
Promoter and terminator sequences from *hup* gene from *B. longum* NCC2705	Reyes-Escogido *et al.* (2007)
rRNA gene promoter from *B. breve* 8807	Sangrador-Vegas *et al.* (2007)
Native	Álvarez-Martín *et al.* (2008)
repC promoter from *B. catenulatum* plasmid pBC1 and promoter Phelp from *Listeria monocytogenes* (Riedel *et al.*, 2007)	Cronin *et al.* (2008)
Promoter from phage T5	Guglielmetti *et al.* (2008)
16S rRNA gene promoter from *B. longum* MG1	Park *et al.* (2008)
Promoter and terminator regions of *hup* gene from *B. longum* VMKB44 Promoter/TIR of *B. longum* VMKB44 gene *gap*	Shkoporov *et al.* (2008b)
Native	Watson *et al.* (2008)

protein export in *B. breve* UCC2003 by means of the export-specific vector pFUN (Poquet *et al.*, 1998), based on the use of the staphylococcal nuclease (Nuc) as a reporter enzyme in *Lactococcus lactis*. Owing to the removal of its native signal peptide, the Nuc reporter protein is translocation-competent but unable to direct its own secretion. In this study, translational fusions of chromosomal segments of *B. breve* UCC2003 were constructed to secrete a *nuc* gene without export signals. By this strategy, seven signal peptides were identified (MacConaill *et al.*, 2003).

A secretion system has more recently been developed for bifidobacteria based on the expression and secretion signals of the α-amylase gene (*amyB*) from *B. adolescentis* INT57 (Park *et al.*, 2005; Rhim *et al.*, 2006). The intracellular phytase gene from *E. coli* was introduced into this vector, transcriptionally fused to the signal sequence, and finally introduced by electroporation into *B. longum* MG1. Using this system, the phytase enzyme was successfully expressed and secreted by *B. longum* into the culture broth. Furthermore, the same research group employed this system for the expression and secretion of pediocin PA-1 from *Pediococcus acidilactici* K10 by *B. longum* MB1 (Moon *et al.*, 2005).

Information on regulated promoters, inducers, and repressors is extremely limited in bifidobacteria. Additional efforts are, therefore, needed to identify strong, weak, and regulated promoters for controlled gene expression in bifidobacteria under different environmental conditions. However, current knowledge on the expression of heterologous genes in *Bifidobacterium* has enabled the development of the reporter gene and drug delivery systems as promising tools for future research and pharmaceutical applications of bifidobacteria. A short summary of these systems is reported below.

Reporter gene systems

To study promoter strength and regulation analysis or identify genomic fragments containing active promoters, Klijn *et al.* (2006) developed the reporter vector pMDY23, based on the *E. coli* β-glucuronidase gene *gusA* and the small cryptic *B. longum* plasmid pNCC293. After introducing the vector in *B. longum* NCC2705, they then demonstrated the suitability of pMDY23 as a reporter plasmid for studying promoter activity (Klijn *et al.*, 2006; Table 13.3).

More recently, two studies showed the application of bioluminescence reporter genes in *Bifidobacterium* spp. (Guglielmetti *et al.*, 2008; Cronin *et al.*, 2008). Guglielmetti *et al.* transformed the human intestinal bacterium *B. longum* biovar *longum* NCC2705 by electroporation with a vector (pGBL8b) containing the insect luciferase gene *lucGR* from a click beetle (*Pyrophorus plagiophthalamus*). The same vector, however, was incapable of transforming *B. animalis* subsp. *lactis* Bb12 and *B. bifidum* MIMBb75 (Guglielmetti *et al.*, 2008 and unpublished results). The resulting bioluminescent *B. longum* was used to study variations in intracellular ATP concentration at acidic pH in the presence of different sugars, a technique proving a valuable tool for rapid and sensitive study of the physiological state of anaerobic bacterial cells under different environmental conditions. Nonetheless, the need to add exogenous D-luciferin substrate in this bacterial reporter system limits considerably its *in vivo* application.

Because of its lower light emission, the bacterial luciferase system (coded by the *luxABCDE* operon) is generally less sensitive than insect luciferases, yet bacterial luciferase requires as a substrate a long-chain fatty aldehyde, which is intracellularly synthesized by a fatty acid reductase complex encoded by *luxCDE*. Therefore, the intracellular expression of the *lux* operon circumvents the disadvantage of an exogenous addition of a luciferin substrate is, thus, more suitable for *in vivo* applications. Riedel *et al.* (2007) adopted the *lux* operon to develop the non-invasive luciferase reporter vector pLuxMC1, which was introduced into *B. breve* UCC2003. Once mice were orally inoculated with bioluminescent *B. breve*, the reporter system allowed tracking of the colonization potential and persistence of this probiotic strain in real-time *in vivo* (Cronin *et al.*, 2008), thus, contributing to the valuable identification of the caecum as a niche for *B. breve*.

Drug delivery systems

Owing to their safety and ability to colonize specific areas of the human GIT, bifidobacteria may turn out to be optimal vectors for *in situ* delivery of biologically-active substances. Of particular interest is the fact that certain species of anaerobic bacteria, including some of the genera *Clostridium* and *Bifidobacterium*, can selectively grow in the nearby hypoxic regions of solid tumours (Malmgren and Flanigan, 1955; Kimura *et al.*, 1980; Yazawa *et al.*, 2000, 2001). Many murine and human solid tumours, including most primary breast and uterine cervix tumours, are characterized by regions of hypoxia or necrosis. This fact was exploited by the Japanese research team of Professor Fujimori for the development of a strategy called Bifidobacterial Selective Targeting (BEST; Fujimori, 2006). The Fujimori team's BEST therapy involved the strain *B. longum* 105-A, which was genetically engineered via electrotransformation with pBLES100-S-eCD, a plasmid based on the shuttle vector pBLES100 (Matsumura *et al.*, 1997) and comprising the cytosine deaminase gene (CD) under the *hup* gene promoter of *B. longum* (Nakamura *et al.*, 2002; Table 13.3). The CD enzyme converts the non-toxic pro-drug 5-fluorocytosine (5-FC) to the chemotherapeutic 5-fluorouracil (5-FU), which is systemically administrated to treat solid tumours. However, in conventional therapy its clinical effectiveness has been limited by its high systemic toxicity, particularly towards the bone marrow. Fujimori's studies demonstrated that recombinant *B. longum* selectively produced CD in hypoxic autochthonous mammary tumour tissues in rats, and that CD could convert 5-FC into 5-FU *in vivo* both after intratumoral injection and also by systemic administration (Sasaki *et al.*, 2006). Furthermore, no adverse effects were observed in the animals during the use of *B. longum* as a gene delivery vector (Sasaki *et al.*, 2006), a finding supporting the potential effectiveness of this novel approach for cancer gene therapy in humans. More recently, the same research group developed an improved version of pBLES100-S-eCD able to display 10-fold increased CD activity in *B. longum* (Hamaji *et al.*, 2007) and demonstrated that the BEST system works well with different bifidobacterial species, represented by *B. breve* I-53–8w (Hidaka *et al.*, 2007). Bifidobacteria were also used as a delivery vehicle of CD by Chinese researchers, who expressed the CD gene in *B. longum* subsp. *infantis* by means of the pGEX-1λT vector. The recombinant bacterium was then used to inhibit melanoma in mice (Yi *et al.*, 2005). The functionality of the vector pGEX-1λT is surprising because it carries no bifidobacterial replicon but only the pBR322 *ori* region. Moreover, Yi *et al.* claimed that recombinant bifidobacteria were selected with ampicillin through the β-lactamase gene encoded by pGEX-1λT (Yi *et al.*, 2005), an antibiotic marker generally considered not to be active in Gram-positives.

The BEST strategy was also adopted by Professor Xu's research team, who employed *B. adolescentis* and *B. longum* as gene delivery vectors to transport the anti-angiogenic factor endostatin into a hypoxic solid liver tumour in mice (Li *et al.*, 2003; Fu *et al.*, 2005; Xu *et al.*, 2007). Originally, they claimed that expression of the human liver endostatin gene was achieved in *Bifidobacterium* spp. by means of the expression plasmid pBV220. The vector pBV220 (Zhang *et al.*, 1990) is a derivative of pBR322 and contains the P_RP_L promoters of the λ phage, the *cI857ts* gene encoding the temperature-sensitive λ repressor, and two strong transcriptional terminators (*rrnBT1T2*) of *E. coli*. This vector contains only the origin of replication of pBR322 and a unique antibiotic selection for ampicillin (β-lactamase gene). As for the vector pGEX-1λT, it is, therefore, surprising that pBV220 was found to be functional in bifidobacteria. The same authors reported, however, that this construct was highly unstable in *B. longum* (Xu *et al.*, 2007) and needed to be modified by introducing the pMB1 replicon and a chloramphenicol resistance gene. The resulting vector, called pBV22210, was more stable in *B. longum* than its parental pBV220 (Xu *et al.*, 2007). Very recently, the plasmid pBV22210 has also been used in *B. longum* as an expression vehicle of the extracellular domain of TNF-related apoptosis-inducing ligand (TRAIL). The resulting recombinant strain was shown to have a specific antitumour effect on mouse osteosarcoma (Hu *et al.*, 2009). Finally, the BEST strategy was adopted by Hou *et al.* (2006), who observed a significant inhibition of the growth of solid tumours in mice with a genetically engineered *B. longum* transformed with a pMB1-derived vector expressing the PTEN tumour suppressor gene (phosphatase and tensin homologue deleted on chromosome 10) under the *hup* gene promoter from *B. longum*.

Other gene delivery systems have recently been developed using bifidobacterial tools. The research group of Prof. Shirakawa developed a vaccine delivery system based on *B. animalis* ATCC27536, genetically modified by transformation with the vector pBLES100, in which they cloned the flagellin gene *fliC* of *Salmonella typhimurium* ATCC14028 under the *B. longum hup* gene promoter (Takata *et al.*, 2006). The group detected significantly higher levels of flagellin-specific IgA in the serum and stools of mice treated by oral administration with the recombinant *B. animalis* than in those treated with the original *B. animalis* strain (Takata *et al.*, 2006). Hence, their study supports the potential effectiveness of genetically modified bifidobacteria as oral vaccines. Bifidobacteria may also be used for the *in situ* delivery of human cytokines, as suggested by a recent study in which interleukin (IL)–10 was expressed in its mature form by *B. longum* ATCC15707 (Reyes Escogido *et al.*, 2007; Table 13.3).

Knock-out of bifidobacterial genes

The prime method for studying the function of a gene and its impact on overall cellular physiology and morphology consists of removing or disrupting the gene from its host chromosome (gene knock-out). This is generally accomplished by means of genetic systems based on homologous DNA recombination. For the members of the genus *Bifidobacterium*, a low occurrence of homologous recombination has been reported, which agrees well with the absence of the major prokaryotic DNA recombination pathway, encoded by *recBCD,* in the genome of some strains such as *B. longum* NCC2705 (Schell *et al.*, 2002). As previously discussed, bifidobacteria display relatively low transformation efficiency. Evidently, the recombination frequency in *Bifidobacterium* is, thus, generally lower than its transformation frequency, limiting effective chromosomal integration of recombinant DNA. For these reasons, the first successful homologous recombination in bifidobacteria has only recently been reported in the literature.

In 2008, the research group of Professor van Sinderen published successful knock-outs of *B. breve* UCC2003 genes by single-crossover recombination, employing two different strategies (O'Connell-Motherway *et al.*, 2008, 2009). First, O'Connell-Motherway *et al.* adopted the plasmid integration strategy described by Law *et al.* (1995) to disrupt the amylopullulanase gene *apuB* using two plasmids. The first contained the origin of replication but lacked the *rep* gene coding for its replication protein (Ori+/Rep– vector). The second vector, bearing a different antibiotic resistance marker, was a derivative of the lactococcal temperature-sensitive plasmid pVE6007 and harboured the *rep* gene of the former plasmid (Rep+ vector). An internal 1 kb DNA region of the *apuB* gene was cloned in the Ori+/Rep– plasmid, and both vectors (Ori+/Rep– and Rep+) were then introduced into *B. breve* UCC2003 under double antibiotic selection. In such recombinant cells, the Ori+/Rep– plasmid could replicate only in the presence of the other vector, which supplied the Rep protein *in trans*. Once the recombinant *B. breve* was cultivated at a high temperature (42°C) and with selection only for the Ori+/Rep– vector, the Rep+ plasmid was lost from *B. breve* cells because of its temperature sensitivity and segregational instability. Accordingly, under selective conditions the Ori+/Rep– plasmid has to integrate into the chromosome. Thanks to this approach, the authors found by replica plating some *B. breve apuB*-disruption isolates, which exhibited the expected phenotype and in which they verified plasmid integration by PCR and Southern hybridization. Although effective, O'Connell-Motherway *et al.* (2008) emphasized that the procedure is tedious, time consuming, and unreliable.

In a later publication, the same researchers described insertional mutagenesis of the *apuB* gene and the endogalactanase gene (*galA*) of *B. breve* UCC2003 by means of plasmid methylation (O'Connell-Motherway *et al.*, 2009). This strategy produced gene disruptions by single-crossover chromosomal integration of non-replicative plasmids containing DNA fragments of 476 and 744 bp, representing internal fragments of the *galA* gene, and a 939 bp internal fragment of the *apuB* gene. These three plasmids were first introduced into an *E. coli* host harbouring two *B. breve* UCC2003 methylase genes (*bbrIIM* and *bbrIIIM*). The resulting methylated plasmids were then introduced into *B. breve* UCC2003 by electroporation. Antibiotic-resistant transformants were isolated for all methylated vectors at a frequency of up to 50 per μg of transformed DNA when the longest homologous DNA insert was used. No transformants were obtained when unmethylated constructs were introduced into *B. breve* UCC2003. The expected integration of recombinant DNA in the chromosome was finally verified by genetic and phenotypic analyses. These results by O'Connell-Motherway *et al.* showed that methylation of the non-replicating plasmid by *B. breve* UCC2003 increased transformation to a level sufficiently high to allow site-specific homologous recombination, even with less than 500 bp of homologous DNA.

Arigoni and Delley (2008) also patented a method to obtain gene deletion in *Bifidobacterium* by two consecutive events of homologous recombination. The authors reported the deletion of the tetracycline resistance gene *tet*(W) from *B. animalis* subsp. *lactis* NCC2818 (commercially known as Bb12) and *B. longum* subsp. *longum* NCC2705. Their method comprised the following steps: i) DNA fragments of approximately 3 kb flanking the *tet*(W) gene were amplified and joined via the start and stop codon of *tet*(W); ii) the resultant 6 kb DNA fragment was cloned in pJL74 (LeDeaux and Grossman, 1995), a spectinomycin resistance vector unable to replicate in bifidobacteria; and iii) the resulting plasmid was introduced into *B. animalis* subsp. *lactis* NCC2818 according to the method of Argnani *et al.* (1996). Spectinomycin-resistant (spec^R) colonies were isolated, resulting in the plasmid being integrated into the *B. animalis* genome by homologous recombination via a single-crossover event. Spec^R transformants were then cultivated for about 100 generations without antibiotic selection to promote loss of the integrated plasmid. The spectinomycin-sensitive (spec^S) colonies were then selected by replica plating on MRS agar with or without added spectinomycin. Twenty-one per cent of the tested colonies were spec^S, indicating that a second crossover event had occurred and resulted in the excision of the vector from the chromosome. Finally, out of the 135 Spec^S colonies tested by PCR, two had resulted in the deletion of *tet*(W). The same strategy was used to delete *tet*(W) from *B. longum* NCC2705 with two positive recombinant colonies out of 12 colonies tested in a final PCR screening for *tet*(W) deletion. This study showed for the first time an *in frame* deletion of a specific entire gene in *Bifidobacterium*, achieved by targeted double-crossover recombination. The strategy, as that of O'Connell-Motherway *et al.*, involves extensive and time consuming work and depends greatly on the possibility of attaining high transformation efficiencies. The introduction of a PAM system to the strategy presented by Arigoni and Delley is likely to be a powerful tool for future gene deletion/replacement applications.

Conclusions

A significant number of bifidobacterial plasmids have already been sequenced and analysed from several *Bifidobacterium* species, which has allowed the development of some first generation genetic vectors. However, there is still a lack of fundamental knowledge about the basic biology of bifidobacterial plasmids for them to be used with confidence. Dissection of translated and untranslated sequences will help define the functionality of the different elements found, helping, at the same time, in the design of high-copy and low-copy number vectors for fine-tuning the expression of homologous and heterologous genes, while increasing the stability of the constructs. Polishing and refining current vectors and broadening their positive selection will be highly useful for molecular studies, as well as for the development of compatible systems allowing the introduction of two vectors in a single cell. The construction of food-grade vectors with no foreign DNA and free from antibiotic resistance markers would facilitate the future industrial use of genetically modified bifidobacteria strains. In spite of the improved transformation efficiency obtained by the laborious method reported by O'Connell-Motherway *et al.*, there is an urgent need for more reliable transformation protocols resulting in an enhanced transformation frequency. Strict anaerobic conditions during the preparation of competent cells and electroporation should increase the viability of the bacteria through the process, increasing the recovery of electrotransformants.

Acknowledgements

Work at the author's laboratory has been supported by projects from the Spanish Ministry of Science and Innovation (MICINN) (Ref. AGL2000–1474 and AGL2007–61869-ALI) and FICYT (Ref. IB08–005) to B.M., and to S.G.P.A-M. awarded a postdoctoral grant from MICINN (Ref. 2008-0082).

References

Actis, L.A., Tolmasky, M.E., and Crossa, J.H. (1998). Bacterial plasmids: replication of extrachromosomal genetic elements encoding resistance to antimicrobial compounds. Frontiers Biosci. 3, d43–d62.

Álvarez-Martín, P., Flórez, A.B., and Mayo, B. (2007a). Screening for plasmids among human bifidobacteria species: Sequencing and analysis of pBC1 from *Bifidobacterium catenulatum* L48. Plasmid *57*, 165–174.

Álvarez-Martín, P., O'Connell-Motherway, M., van Sinderen, D., and Mayo, B. (2007b). Functional analysis of the pBC1 replicon from *Bifidobacterium catenulatum* L48. Appl. Microbiol. Biotechnol. *76*, 1395–1402.

Álvarez-Martín, P., Belén Flórez, A., Margolles, A., del Solar, G., and Mayo, B. (2008). Improved cloning vectors for bifidobacteria, based on the *Bifidobacterium catenulatum* pBC1 replicon. Appl. Environ. Microbiol. *74*, 4656–4665.

Ammor, M.S., Flórez, A.B., Alvarez-Martín, P., Margolles, A., and Mayo, B. (2008). Analysis of tetracycline resistance *tet*(W) genes and their flanking sequences in intestinal *Bifidobacterium* species. J. Antimicrob. Chemother. *62*, 688–693.

Argnani, A., Leer, R.J., van Luijk, N., and Pouwels, P.H. (1996). A convenient and reproducible method to genetically transform bacteria of the genus *Bifidobacterium*. Microbiol. *142*, 109–114.

Arigoni, F., and Delley, M. (2008). Genetic remodelling in *Bifidobacterium*. International Patent WO 2008/019886 A1.

Barrangou, R., Briczinski, E.P., Traeger, L.L., Loquasto, J.R., Richards, M., Horvath, P., Coûté-Monvoisin, A.C., Leyer, G., Rendulic, S., Steele, J.L., *et al.* (2009). Comparison of the complete genome sequences of *Bifidobacterium animalis* subsp. *lactis* DSM 10140 and Bl-04. J. Bacteriol. *191*, 4144–4151.

Briggs, M. (1953). An improved medium for lactobacilli. J. Dairy Research *20*, 36–39.

Brown, D.P., Idler, K.B., and Katz, L. (1990). Characterization of the genetic elements required for site-specific integration of plasmid pSE211 in *Saccharopolyspora erythraea*. J. Bacteriol. *172*, 1877–1888.

Campbell, A.M. (1992). Chromosomal insertion sites for phages and plasmids. J. Bacteriol. *174*, 7495–7499.

Canchaya, C., Proux, C., Fournous, G., Bruttin, A., and Brussow, H. (2003). Prophage genomics. Microbiol. Mol. Biol. Rev. *67*, 238–276.

Cohen, S.N. (1993). Bacterial plasmids: their extraordinary contribution to molecular genetics. Gene *135*, 67–76.

Combes, P., Till, R., Bee, S., and Smith, M.C.M. (2002). The *Streptomyces* genome contains multiple pseudo-*attB* sites for the phage C31-encoded site-specific recombination system. J. Bacteriol. *184*, 5746–5752.

Corneau, N., Emond, E., and LaPointe, G. (2004). Molecular characterization of three plasmids from *Bifidobacterium longum*. Plasmid *51*, 87–100.

Cronin, M., Knobel, M., O'Connell-Motherway, M., Fitzgerald, G.F., and van Sinderen, D. (2007). Molecular dissection of a bifidobacterial replicon. Appl. Environ. Microbiol. 73, 7858–7866.

Cronin, M., Sleator, R.D., Hill, C., Fitzgerald, G.F., and van Sinderen, D. (2008). Development of a luciferase-based reporter system to monitor *Bifidobacterium breve* UCC2003 persistence in mice. BMC Microbiol. *8*, 161.

del Solar, G., Giraldo, R., Ruiz-Echevarría, M.J., Espinosa, M., and Díaz-Orejas, R. (1998). Replication and control of circular bacterial plasmids. Microbiol. Mol. Biol. Rev. *62*, 434–464.

F.A.O/W.H.O (Food and Agricultural Organization of the United Nations and World Health Organization). (2006). Probiotics in food. Health and nutritional properties and guidelines for evaluation. Roma. (ftp://ftp.fao.org/docrep/fao/009/a0512e/a0512e00.pdf).

Fischer, W. (1987). Analysis of the lipoteichoic acid-like macroamphiphile from *Bifidobacteriuim bifidum* subspecies *pennsylvanium* by one- and two-dimensional 1H- and 13C-NMR spectroscopy. Eur. J. Biochem. *165*, 647–652.

Fu, G.F., Li, X., Hou, Y.Y., Fan, Y.R., Liu, W.H., and Xu, G.X. (2005). *Bifidobacterium longum* as an oral delivery system of endostatin for gene therapy on solid liver cancer. Cancer Gene Ther. *12*, 133–140.

Fujimori, M. (2006). Genetically engineered *Bifidobacterium* as a drug delivery system for systemic therapy of metastatic breast cancer patients. Breast Cancer *13*, 27–31.

Gibbs, M.J., Smeianov, V.V., Steele, J.L., Upcroft, P., and Efimov, B.A. (2006). Two families of *rep*-like genes that probably originated by interspecies recombination are represented in viral, plasmid, bacterial, and parasitic protozoan genomes. Mol. Biol. Evol. *23*, 1097–1100.

González Vara, A., Rossi, M., Altomare, L., Eikmanns, B., and Matteuzzi, D. (2003). Stability of recombinant plasmids on the continuous culture of *Bifidobacterium animalis* ATCC 27536. Biotechnol. Bioengineering *84*, 145–150.

Guglielmetti, S., Karp, M., Mora, D., Tamagnini, I., and Parini, C. (2007). Molecular characterization of *Bifidobacterium longum* biovar *longum* NAL8 plasmids and construction of a novel replicon screening system. Appl. Microbiol. Biotechnol. *74*, 1053–1061.

Guglielmetti, S., Ciranna, A., Mora, D., Parini, C., and Karp, M. (2008). Construction, characterization and exemplificative application of bioluminescent *Bifidobacterium longum* biovar *longum*. Int. J. Food Microbiol. *124*, 285–290.

Hamaji, Y., Fujimori, M., Sasaki, T., Matsuhashi, H., Matsui-Seki, K., Shimatani-Shibata, Y., Kano, Y., Amano, J., and Taniguchi, S. (2007). Strong enhancement of recombinant cytosine deaminase activity in *Bifidobacterium longum* for tumor-targeting enzyme/prodrug therapy. Biosci. Biotechnol. Biochem. *71*, 874–883.

Hartke, A., Benachour, A., Bouibonnes, P., and Auffray, Y. (1996). Characterisation of a complex restriction/modification system detected in a *Bifidobacterium longum* strain. Appl. Microbiol. Biotechnol. *45*, 132–136.

Hayes, F. (2003). The function and organization of plasmids. Methods Mol Biol. *235*, 1–17.

Hendrix, R.W., Smith, M.C.M., Burns, R.N., Ford, M.E., and Hatfull, G.F. (1999). Evolutionary relationships among diverse bacteriophages and prophages: all the world's a phage. Proc. Natl. Acad. Sci. U.S.A *96*, 2192–2197.

Hidaka, A., Hamaji, Y., Sasaki, T., Taniguchi, S., and Fujimori, M. (2007). Exogenous cytosine deaminase gene expression in *Bifidobacterium breve* I-53-8w for tumor-targeting enzyme/prodrug therapy. Biosci. Biotechnol. Biochem. *71*, 2921–2926.

Hou, X., and Liu, J.E. (2006). Construction of *Escherichia coli-Bifidobacterium longum* shuttle vector and expression of tumor suppressor gene PTEN in *B. longum*. Wei Sheng Wu Xue Bao. *46*, 347–352.

Hu, B., Kou, L., Li, C., Zhu, L.P., Fan, Y.R., Wu, Z.W., Wang, J.J., and Xu, G.X. (2009). *Bifidobacterium longum* as a delivery system of TRAIL and endostatin cooperates with chemotherapeutic drugs to inhibit hypoxic tumor growth. Cancer Gene Ther. doi: 10.1038/cgt.2009.7.

Ivanov, D., Emonet, C., Foata, F., Affolter, M., Delley, M., Fisseha, M., Blum-Sperisen, S., Kochhar, S., and Arigoni, F. (2006). A serpin from the gut bacterium *Bifidobacterium longum* inhibits eukaryotic elastase-like serine proteases. J. Biol. Chem. *281*, 17246–17252.

Iwata, M., and Morishita, T. (1989). The presence of plasmids in *Bifidobacterium breve*. Lett. Appl. Microbiol. *9*, 165–168.

Ji, G.E., Park, M.S., Kang, Y.H., and Seo, J.M. (2005). Plasmid originated from *Bifidobacterium*, recombinant expression vector using the plasmid and transformation method. United States Patent 6887702.

Kazimierczak, K.A., Flint, H.J., and Scott, K.P. (2006). Comparative analysis of sequences flanking *tet*(W) resistance genes in multiple species of gut bacteria. Antimicrob. Agents Chemother. *50*, 2632–2639.

Kim, J.F., Jeong, H., Yu, D.S., Choi, S.H., Hur, C.G., Park, M.S., Yoon, S.H., Kim, D.W., Ji, G.E., Park, H.S., and Oh, T.K. (2009). Genome sequence of the probiotic bacterium *Bifidobacterium animalis* subsp. *lactis* AD011. J. Bacteriology. *191*, 678–679.

Kimura, N.T., Taniguchi, S., and Aoki, K. (1980). Selective localization and growth of *Bifidobacterium bifidum* in mouse tumors following intravenous administration. Cancer Res. *40*, 2061–2068.

Klieve, A.V., and Bauchop, T. (1988). Morphological diversity of rumial bacteriophages from sheep and cattle. Appl. Environ. Microbiol. *54*, 1637–1641.

Klijn, A., Moine, D., Delley, M., Mercenier, A., Arigoni, F., and Pridmore, R.D. (2006). Construction of a reporter vector for the analysis of *Bifidobacterium longum* promoters. Appl. Environ. Microbiol. *72*, 7401–7405.

Kullen, M.J., and Klaenhammer, T.R. (2000). Genetic modification of intestinal lactobacilli and bifidobacteria. Curr. Issues Mol. Biol. *2*, 41–50.

Law, J., Buist, G., Haandrikman, A., Kok, J., Venema, G., and Leenhouts, K. (1995). A system to generate chromosomal mutations in *Lactococcus lactis* which allows fast analysis of targeted genes. J. Bacteriol. *177*, 7011–7018.

Leahy, S.C., Higgins, D.G., Fitzgerald, G.F., and van Sinderen, D. (2005). Getting better with bifidobacteria. J. Appl. Microbiol. *98*, 1303–1315.

LeDeaux, J.R., and Grossman, A.D. (1995). Isolation and characterization of *kinC*, a gene that encodes a sensor kinase homologous to the sporulation sensor kinases KinA and KinB in *Bacillus subtilis*. J. Bacteriol. *177*, 166–175.

Lee, J.H., and O'Sullivan, D.J. (2006). Sequence analysis of two cryptic plasmids from *Bifidobacterium longum* DJO10A and construction of a shuttle cloning vector. Appl. Environ. Microbiol. *72*, 527–535.

Lee, J.H., Karamychev, V.N., Kozyavkin, S.A., Mills, D., Pavlov, A.R., Pavlova, N.V., Polouchine, N.N., Richardson, P.M., Shakhova, V.V., Slesarev, A.I., *et al.* (2008). Comparative genomic analysis of the gut bacterium *Bifidobacterium longum* reveals loci susceptible to deletion during pure culture growth. BMC Genomics. *9*, 247.

Li, X., Fu, G.F., Fan, Y.R., Liu, W.H., Liu, X.J., Wang, J.J., and Xu, G.X.(2003). *Bifidobacterium adolescentis* as a delivery system of endostatin for cancer gene therapy: selective inhibitor of angiogenesis and hypoxic tumor growth. Cancer Gene Ther. *10*, 105–111.

MacConaill, L.E., Fitzgerald, G.F., and van Sinderen, D. (2003). Investigation of protein export in *Bifidobacterium breve* UCC2003. Appl. Environ. Microbiol. *69*, 6994–7001.

Malmgren, R.A., and Flanigan, C.C. (1955). Localization of the vegetative from of *Clostridium tetani* in mouse tumors following intravenous spore administration. Cancer Res. *15*, 473–478.

Martín, M.C., Alonso, J.C., Suárez, J.E., and Álvarez, M.A. (2000). Generation of food-grade recombinant lactic acid bacterium strains by site-specific recombination. Appl. Environ. Microbiol. *66*:,2599–2604.

Matsumura, H., Takeuchi, A., and Kano, Y. (1997). Construction of *Escherichia coli-Bifidobacterium longum* shuttle vector transforming *B. longum* 105-A and 108-A. Biosci. Biotechnol. Biochem. *61*, 1211–1212.

Matteuzzi, D., and Sozzi, T. 1970. *Bifidobacterium* bacteriophage from calf rumen. Zeitschrift für Allg. Mikrobiologie *11*, 57–58.

Mills, S., McAuliffe, O.E., Coffey, A., Fizgerald, G.F., and Ross, R.P. (2006). Plasmids of lactococci. Genetic accessories or genetic necessities. FEMS Microbiol. Rev. *30*, 243–273.

Missich, R., Sgorbati, B., and Leblanc, D.J. (1994). Transformation of *Bifidobacterium longum* with pRM2, a constructed *Escherichia coli-B. longum* shuttle vector. Plasmid 32, 208–211.

Moon, G.S., Pyun, Y.R., Park, M.S., Ji, G.E., and Kim, W.J. (2005). Secretion of recombinant pediocin PA-1 by *Bifidobacterium longum*, using the signal sequence for bifidobacterial alpha-amylase. Appl. Environ. Microbiol. *71*, 5630–5632.

Moon, G.S., Wegmann, U., Gunning, A.P., Gasson, M.J., and Narbad, A. (2009). Isolation and characterization of a Theta-type cryptic plasmid from *Bifidobacterium longum* FI10564. J. Microbiol. Biotechnol. *19*, 403–408.

Nakamura, T., Sasaki, T., Fujimori, M., Yazawa, K., Kano, Y., Amano, J., and Taniguchi, S. (2002). Cloned cytosine deaminase gene expression of *Bifidobacterium longum* and application to enzyme/pro-drug therapy of hypoxic solid tumors. Biosci. Biotechnol. Biochem. *66*, 2362–2366.

O'Connell Motherway, M., Fitzgerald, G.F., Neirynck, S., Ryan, S., Steidler, L., and van Sinderen, D. (2008). Characterization of ApuB, an extracellular type II amylopullulanase from *Bifidobacterium breve* UCC2003. Appl. Environ. Microbiol. *74*, 6271–6279.

O'Connell Motherway, M., O'Driscoll, J., Fitzgerald, G.F., and van Sinderen, D. (2009). Overcoming the restriction barrier to plasmid transformation and targeted mutagenesis in *Bifidobacterium breve* UCC2003. Microbial Biotechnology. *2*, 321–332.

O'Riordan, K., and Fitzgerald, G.F. (1999). Molecular characterisation of a 5.75-kb cryptic plasmid from *Bifidobacterium breve* NCFB 2258 and determination of mode of replication. FEMS Microbiol. Lett. *174*, 285–294.

Park, K.B., Ji, G.E., Park, M.S., and Oh, S.H. (2005). Expression of rice glutamate decarboxylase in *Bifidobacterium longum* enhances gamma-aminobutyric acid production. Biotechnol. Lett. *27*, 1681–1684.

Park, M.S., Lee, K.H., and Ji, G.E. (1997). Isolation and characterization of two plasmids from *Bifidobacterium longum*. Lett. Appl. Microbiol. *25*, 5–7.

Park, M.S., Shin, D.W., Lee, K.H., and Ji, G.E. (1999). Sequence analysis of plasmid pKJ50 from *Bifidobacterium longum*. Microbiology *145*, 585–592.

Park, M.S., Moon, H.W., and Ji, G.E. (2003). Molecular characterization of plasmid from *Bifidobacterium longum*. J. Microbiol. Biotechnol. *13*, 457–462.

Park, M.S., Seo, J.M., Kim, J.Y., and Ji, G.E. (2005). Heterologous gene expression and secretion in *Bifidobacterium longum*. Lait *85*, 1–8.

Park, M.S., Kwon, B., Shim, J.J., Huh, C.S., and Ji, G.E. (2007). Heterologous expression of cholesterol oxidase in *Bifidobacterium longum* under the control of 16S rRNA gene promoter of bifidobacteria. Biotechnol. Lett. *30*, 165–172.

Park, Y.S., Kim, K.H., Park, J.H., Oh, I.K., and Yoon, S.S. (2008). Isolation and molecular characterization of a cryptic plasmid from *Bifidobacterium longum*. Biotechnol. Lett. *30*, 145–151.

Parvez, S., Malik, K.A., Ah Kang, S., and Kim, H.Y. (2006). Probiotics and their fermented food products are beneficial for health. J. Appl. Microbiol. *100*, 1171–1185.

Philippe, H., and Douady, C.J. (2003). Horizontal gene transfer and phylogenetics. Curr. Opin. Microbiol. *6*, 498–505.

Poquet, I., Ehrlich, S.D., and Gruss, A. (1998). An export-specific reporter designed for Gram-positive bacteria: application to *Lactococcus lactis*. J. Bacteriol. *180*, 1904–1912.

Reyes Escogido, M.L., De León Rodríguez, A., and Barba de la Rosa, A.P. (2007). A novel binary expression vector for production of human IL-10 in *Escherichia coli* and *Bifidobacterium longum*. Biotechnol. Lett. *29*, 1249–1253.

Rhim, S.L., Park, M.S., and Ji, G.E. (2006). Expression and secretion of *Bifidobacterium adolescentis* amylase by *Bifidobacterium longum*. Biotechnol. Lett. *23*, 163–168.

Riedel, C.U., Monk, I.R., Casey, P.G., Morrissey, D., O'Sullivan, G.C., Tangney, M., Hill, C., and Gahan, C.G. (2007). Improved luciferase tagging system for *Listeria monocytogenes* allows real-time monitoring *in vivo* and *in vitro*. Appl. Environ. Microbiol. *73*, 3091–3094.

Roberts, A.P., Chandler, M., Courvalin, P., Guédon, G., Mullany, P., Pembroke, T., Rood, J.I., Smith, C.J., Summers, A.O., Tsuda, M., and Berg, D.E. (2008). Revised nomenclature for transposable genetic elements. Plasmid *60*, 167–173.

Roberts, R.J. (1980). Restriction and modification enzymes and their recognition sequences. Nucleic. Acids Res. *8*, r63–r80.

Rossi, M., Brigidi, P., Gonzalez Vara y Rodriguez, A., and Matteuzzi, D. (1996). Characterization of the plasmid pMB1 from *Bifidobacterium longum* and its use for shuttle vector construction. Res. Microbiol. *147*, 133–143.

Rossi, M., Brigidi, P., and Matteuzzi, D. (1997). An efficient transformation system for *Bifidobacterium* spp. Lett. Appl. Microbiol. *24*, 33–36.

Rossi, M., Brigidi, P., and Matteuzzi, D. (1998). Improved cloning vectors for *Bifidobacterium* spp. Lett. Appl. Microbiol. *26*, 101–104.

Sangrador-Vegas, A., Stanton, C., van Sinderen, D., Fitzgerald, G.F., and Ross, R.P. (2007). Characterization of plasmid pASV479 from *Bifidobacterium pseudolongum* subsp. *globosum* and its use for expression vector construction. Plasmid *58*, 140–147.

Sasaki, T., Fujimori, M., Hamaji, Y., Hama, Y., Ito, K., Amano, J., and Taniguchi, S. (2006). Genetically engineered *Bifidobacterium longum* for tumor-targeting enzyme-prodrug therapy of autochthonous mammary tumors in rats. Cancer Sci. *97*, 649–657.

Scardovi, V. (1986). Genus *Bifidobacterium*. Orla-Jensen 1924, 472^{AL}. In: Bergey's Manual of Systematic Bacteriology, Vol. 2. Sneath, P.H.A., Mair, N.S., Sharpe, M.E., and Holt, J.G., eds. (Baltimore, USA: Williams and Wilkins), pp. 1418–1434.

Schell, M.A., Karmirantzou, M., Snel, B., Vilanova, D., Berger, B., Pessi, G., Zwahlen, M.C., Desiere, F., Bork, P., Delley, M., *et al.* (2002). The genome sequence of *Bifidobacterium longum* reflects its adaptation to the human gastrointestinal tract. Proc. Natl. Acad. Sci. U.S.A 99, 14422–14427.

Sela, D.A., Chapman, J., Adeuya, A., Kim, J.H., Chen, F., Whitehead, T.R., Lapidus, A., Rokhsar, D.S., Lebrilla, C.B., German, J.B., *et al.* (2008). The genome sequence of *Bifidobacterium longum* subsp. *infantis* reveals adaptations for milk utilization within the infant microbiome. Proc. Natl. Acad. Sci. U.S.A *105*, 18964–18969.

Sgorbati, B., Scardovi, V., and Leblanc, D.J. (1982). Plasmids in the genus *Bifidobacterium*. J. Gen. Microbiol. *128*, 2121–2131.

Sgorbati, B., Smiley, M.B., and Sozzi, T. (1983). Plasmids and phages in *Bifidobacterium longum*. Microbiologica. *6*, 169–173.

Shkoporov, A.N., Efimov, B.A., Khokhlova, E.V., Steele, J.L., Kafarskaia, L.I., and Smeianov, V.V. (2008a). Characterization of plasmids from human infant *Bifidobacterium* strains: sequence analysis and construction of *E. coli-Bifidobacterium* shuttle vectors. Plasmid *60*, 136–148.

Shkoporov, A.N., Efimov, B.A., Khokhlova, E.V., Kafarskaia, L.I., and Smeianov, V.V. (2008b). Production of human basic fibroblast growth factor (FGF-2) in *Bifidobacterium breve* using a series of novel expression/secretion vectors. Biotechnol. Lett. *30*, 1983–1988.

Siefert, J.L. (2009). Defining the mobilome. Methods Mol. Biol. *532*, 13–27.

Siezen, R.J., Renckens, B., van Swam, I., Peters, S., van Kranenburg, R., Klerebeezem, M., and de Vos, W.M. (2005). Complete sequences of four plasmids of *Lactococcus lactis* subsp. *cremoris* SK11 reveal extensive adaptation to the dairy environment. Appl. Environ. Microbiol. *71*, 8371–8382.

Simon, R., Priefer, U., and Puhler, A. (1983). A broad host range mobilization system for *in vivo* genetic engineering: transposon mutagenesis in Gram-negative bacteria. Bio/Technology. *1*, 784–791.

Stanton, C., Gardiner, G., Meehan, H., Collins, K., Fitzgerald, G., Lynch, P.B., and Ross, R.P. (2001). Market potential for probiotics. Am. J. Clin. Nutr. 73, 476S-483S.

Takata, T., Shirakawa, T., Kawasaki, Y., Kinoshita, S., Gotoh, A., Kano, Y., and Kawabata, M. (2006). Genetically engineered *Bifidobacterium animalis* expressing the *Salmonella* flagellin gene for the mucosal immunization in a mouse model. J. Gene Med. *8*, 1341–1346.

Tanaka, K., Samura, K., and Kano, Y. (2005). Structural and functional analysis of pTB6 from *Bifidobacterium longum*. Biosci. Biotechnol. Biochem. *69*, 422–425.

Tauch, A., Kassing, F., Kalinowski, J., and Pühler, A. (1995). The *Corynebacterium xerosis* composite transposon Tn*5432* consists of two identical insertion sequences, designated IS1249, flanking the erythromycin resistance gene *ermCX*. Plasmid *34*, 119–131.

Tuohy, K.M., Probert, H.M., Smejkal, C.W., and Gibson, G.R. (2003). Using probiotics and prebiotics to improve gut health. Drug Discov. Today. *8*, 692–700.

Turroni, F., van Sinderen, D., and Ventura, M. (2009). Bifidobacteria: from ecology to genomics. Front. Biosci. *14*, 4673–4684.

van Hoek, A.H.A.M., Mayrhofer, S., Domig, K.J., and Aarts, H.J.M. (2008). Resistance determinant *erm*(X) is borne by transposon Tn*5432* in *Bifidobacterium thermophilum* and *Bifidobacterium animalis* subsp. *lactis*. Int. J. Antimicrob. Agents *31*, 544–548.

Ventura, M., van Sinderen, D., Fitzgerald, G.F., and Zink, R. (2004). Insights into the taxonomy, genetics and physiology of bifidobacteria. Antonie van Leeuwenhoek. *86*, 205–223.

Ventura, M., Lee, J.H., Canchaya, C., Zink, R., Leahy, S., Moreno-Muñoz, J.A., O'Connell-Motherway, M., Higgins, D., Fitzgerald, G.F., O'Sullivan, D.J., and van Sinderen, D. (2005a). Prophage-like elements in bifidobacteria: insights from genomics, transcription, integration, distribution, and phylogenetic analysis. Appl. Environ. Microbiol. *71*, 8692–8705.

Ventura, M., Zhang, Z., Cronin, M., Canchaya, C., Kenny, J.G., Fitzgerald, G.F., and van Sinderen, D. (2005b). The ClgR protein regulates transcription of the *clpP* operon in *Bifidobacterium breve* UCC2003. J. Bacteriol. *187*, 8411–8416.

Ventura, M., Canchaya, C., Tauch, A., Chandra, G., Fitzgerald, G.F., Chater, K.F., and van Sinderen, D. (2007). Genomics of *Actinobacteria*: tracing the evolutionary history of an ancient phylum. Microbiol. Mol. Biol. Rev. *71*, 495–548.

Watson, D., Sleator, R.D., Hill, C., and Gahan, C.G. (2008). Enhancing bile tolerance improves survival and persistence of *Bifidobacterium* and *Lactococcus* in the murine gastrointestinal tract. BMC Microbiol. *8*, 176.

Xu, Y.F., Zhu, L.P., Hu, B., Fu, G.F., Zhang, H.Y., Wang, J.J., and Xu, G.X. (2007). A new expression plasmid in *Bifidobacterium longum* as a delivery system of endostatin for cancer gene therapy. Cancer Gene Ther. *14*, 151–157.

Yasui, K., Kano, Y., Tanaka, K., Watanabe, K., Shimizu-Kadota, M., Yoshikawa, H., and Suzuki, T. (2009). Improvement of bacterial transformation efficiency using plasmid artificial modification. Nucleic. Acids Res. *37*, e3.

Yazawa, K., Fujimori, M., Amano, J., Kano, Y., and Taniguchi, S. (2000). *Bifidobacterium longum* as a delivery system for cancer gene therapy: selective localization and growth in hypoxic tumors. Cancer Gene Ther. 7, 269–274.

Yazawa, K., Fujimori, M., Nakamura, T., Sasaki, T., Amano, J., Kano, Y., and Taniguchi, S. (2001). *Bifidobacterium longum* as a delivery system for gene therapy of chemically induced rat mammary tumors. Breast Cancer Res. Treat. *66*, 165–170.

Yi, C., Huang, Y., Guo, Z.Y., and Wang, S.R. (2005). Antitumor effect of cytosine deaminase/5-fluorocytosine suicide gene therapy system mediated by *Bifidobacterium infantis* on melanoma. Acta Pharmacol. Sin. *26*, 629–634.

Yildirim, Z., Winters, D.K., and Johnson, M.G. (1999). Purification, amino acid sequence and mode of action of bifidocin B produced by *Bifidobacterium bifidum* NCFB 1454. J. App. Microbiol. *86*, 45–54.

Youseff, M., Müller-Beuthow, W., and Haenel, H. (1966). Isolierung von Backeiophagen für anaerobe Laktobazillen aus dem Sthul des Menschen. Naturwissenchafen 53, 589.

Zhang, Z.Q., Yao, L.H., and Hou, Y.D. (1990). Construction and application of a high level expression vector containing P_RP_L promotor. Chin. J. Virol. *6*, 111–116.

www.ingramcontent.com/pod-product-compliance
Lightning Source LLC
Chambersburg PA
CBHW081415220926
51613CB00010B/224
* 9 7 8 1 9 0 4 4 5 5 6 8 4 *